Universitext

Series Editors
Nathanaël Berestycki, Universität Wien, Vienna, Austria
Carles Casacuberta, Universitat de Barcelona, Barcelona, Spain
John Greenlees, University of Warwick, Coventry, UK
Angus MacIntyre, Queen Mary University of London, London, UK
Claude Sabbah, École Polytechnique, CNRS, Université Paris-Saclay, Palaiseau, France
Endre Süli, University of Oxford, Oxford, UK

Universitext is a series of textbooks that presents material from a wide variety of mathematical disciplines at master's level and beyond. The books, often well classtested by their author, may have an informal, personal, or even experimental approach to their subject matter. Some of the most successful and established books in the series have evolved through several editions, always following the evolution of teaching curricula, into very polished texts.

Thus as research topics trickle down into graduate-level teaching, first textbooks written for new, cutting-edge courses may find their way into *Universitext*.

Max Koecher • Aloys Krieg

Elliptic Functions and Modular Forms

 Springer

Max Koecher (Deceased)

Aloys Krieg
RWTH Aachen University
Aachen, Germany

ISSN 0172-5939 ISSN 2191-6675 (electronic)
Universitext
ISBN 978-3-662-71223-8 ISBN 978-3-662-71224-5 (eBook)
https://doi.org/10.1007/978-3-662-71224-5

Mathematics Subject Classification (2020): 11F03, 11F11, 11F20, 11F25, 11F27, 30B50, 33E05

Translation from the German language edition: "Elliptische Funktionen und Modulformen" by Max Koecher and Aloys Krieg, © Springer-Verlag Berlin Heidelberg 2007. Published by Springer Berlin Heidelberg. All Rights Reserved.

© The Editor(s) (if applicable) and The Author(s), under exclusive license to Springer Nature Switzerland AG 2025

This work is subject to copyright. All rights are solely and exclusively licensed by the Publisher, whether the whole or part of the material is concerned, specifically the rights of translation, reprinting, reuse of illustrations, recitation, broadcasting, reproduction on microfilms or in any other physical way, and transmission or information storage and retrieval, electronic adaptation, computer software, or by similar or dissimilar methodology now known or hereafter developed.
The use of general descriptive names, registered names, trademarks, service marks, etc. in this publication does not imply, even in the absence of a specific statement, that such names are exempt from the relevant protective laws and regulations and therefore free for general use.
The publisher, the authors and the editors are safe to assume that the advice and information in this book are believed to be true and accurate at the date of publication. Neither the publisher nor the authors or the editors give a warranty, expressed or implied, with respect to the material contained herein or for any errors or omissions that may have been made. The publisher remains neutral with regard to jurisdictional claims in published maps and institutional affiliations.

This Springer imprint is published by the registered company Springer-Verlag GmbH, DE, part of Springer Nature.
The registered company address is: Heidelberger Platz 3, 14197 Berlin, Germany

If disposing of this product, please recycle the paper.

Preface

The present textbook is a revision and translation of the original German language edition. In particular, we

- change the notation and references within the book to a standard form,
- change and update the references to standard English language textbooks,
- add several new exercises,
- derive the NIELSEN-SCHREIER Theorem on generators of subgroups,
- determine the commutator subgroup of the modular group,
- introduce nearly holomorphic modular forms,
- describe modular functions for subgroups of the modular group,
- introduce three motivations for HECKE operators,
- fix a gap in the proof of HECKE's Converse Theorem,
- show that the modular group is maximal discrete,
- describe the most relevant results from complex analysis which are used in the text in Appendix A and B.

There is a website for this book available under

http://www.mathA.rwth-aachen.de/Elliptic_Modular_Forms

which leads to a Moodle space. It is intended to function as an interactive book. Here the reader can find, e.g. hints at the solutions of some exercises as well as additional explanations and a list of typos. Moreover the reader can pose questions to be answered by the author or in a forum and can find links to videos on some parts of the book.

A first version of the translation was done using DeepL. I would like to thank Barbara GIESE and Melanie THELLMANN for their careful preparation of the TEX-version and Kai TANGELDER for the preparation of the figures. My thanks are due to Nicola LÜSSEM and Brandon WILLIAMS for their help on the translation and Kilian RAUSCH for the proof-reading as well as Springer-Verlag for the kind support.

Aachen, March 2025   Aloys KRIEG

Preface to the first German language edition

The present text originated from the lecture manuscript I prepared in the winter semester 88/89 in Hemer. A new version of the manuscript (April 1989) was kindly reviewed by E. NEHER (Ottawa), H.P. PETERSSON (Hagen) and J. ELSTRODT (Münster), who made valuable comments. I thank especially J. ELSTRODT for many additional remarks.

Münster, June 1989   Max KOECHER

This book is intended for undergraduate students of mathematics. Our aim is to give an introduction into the classical theory of elliptic functions and elliptic modular forms following a one-semester course on complex analysis. It should be explicitly noted that Chapters II through V on modular forms can also be read independently of Chapter I on elliptic functions. This book originated from the elaboration of the above mentioned course of M. KOECHER. After his death, I took over the task to revise and complete the draft in KOECHER's style. In addition, on the basis of this manuscript, I gave courses at the University of Münster (winter semester 89/90 and 91/92) and at the RWTH Aachen (winter semester 93/94 and 95/96). I would like to thank K. HAVERKAMP and F. BÜHLER for their help with the corrections, the drawings and the exercises, G. WECKERMANN and A. SEVES for the preparation of the print-ready TEX-template and finally the publisher for his kindness.

Aachen, May 1997   Aloys KRIEG

Contents

I	**Elliptic functions**		1
	Introduction		1
	§ 1	Periods and lattices	11
	§ 2	The field of elliptic functions	24
	§ 3	The Weierstrass $\wp$-Function	34
	§ 4	The dependence on the lattice	47
	§ 5	Elliptic curves and the Addition Theorem of the $\wp$-Function	61
	§ 6	Product expansions	75
	§ 7*	$\wp$-partial values, algebraic dependence and complex multiplication	93
	§ 8*	Miscellaneous remarks	102
II	**Geometry in the upper half-plane and the action of the modular group**		109
	Introduction		109
	§ 1	The upper half-plane	111
	§ 2	The modular group	126
	§ 3	Subgroups of the modular group	136
	§ 4*	Discontinuous groups	147
III	**Modular forms**		153
	Introduction		153
	§ 1	The elementary theory	156
	§ 2	Examples	163
	§ 3	The weight formula	172
	§ 4	Modular forms	177
	§ 5	Modular functions	189
	§ 6	The Dedekind eta–function	194
	§ 7	Modular forms for congruence subgroups	203
IV	**The Hecke–Petersson theory**		215
	Introduction		215
	§ 1	Hecke operators	218

	§ 2	The algebra of Hecke operators 232
	§ 3	Petersson scalar product 242
	§ 4	Dirichlet series with functional equation 252
	§ 5*	Maximal discrete groups 270

V Theta series 275
    Introduction 275
    § 1 Integral and positive definite matrices 276
    § 2 Theta series as modular forms 287
    § 3 A special case of Siegel's Main Theorem 300
    § 4* Harmonic polynomials and quadratic forms of higher level 310
    § 5* The Epstein zeta function and applications 324

Appendix A 339

Appendix B 345

Bibliography 351

List of mathematicians 355

Table of symbols 357

Index 361

Chapter I
Elliptic functions

Introduction

1. Preliminary remark. If we think of the concrete functions that we learned as students in one or two semester courses on complex analysis, we would note down the

(i) rational functions,
(ii) trigonometric and hyperbolic functions together with the exponential function as well as their inverse functions, and possibly the
(iii) gamma function.

The same modest result can be obtained by looking through the standard textbooks on complex analysis for concrete functions. In all three example classes mentioned, we actually do not need any additional theory to deal with the functions, we can derive them directly without complex analysis, i.e. without using the theorems of that theory: we only need to extend the usual justifications of real analysis to complex arguments. Historically, (i) and (ii) as well as the essential parts of (iii) were known by the beginning of the 19th century. It was not until the 19th century that complex analysis began to triumph in analysis, when a theory had to be developed to deal with new classes of functions. Such a class of functions is examined in Chapter I.

2. Integration problems. In the late 18th and early 19th century, the problem of calculating integrals of the form

$$F(x) := \int_a^x \frac{dt}{\sqrt{p(t)}}, \qquad (0.1)$$

where $p(t)$ is a real polynomial of degree 3 or 4 in t that takes only positive values in the interval, was considered again and again. The analogous question for polynomials p of degree 1 or 2 can be solved by elementary means or with the functions (ii). In the case (0.1), however, elementary integration does not succeed in general! Integrals of the form (0.1) occur in the determination of the arc length of ellipses (cf. sect.

5), hyperbolas and the lemniscate ($p(t) = 1 - t^4$, cf. sect. 4). Therefore, they are (misleadingly) called *elliptic integrals*. Such elliptic integrals are first mentioned around 1655 by J. WALLIS (1616–1703) and then by Jakob BERNOULLI (1655–1705).

To avoid the problem of possible zeros of p, we assume that F (locally) has a strictly monotonically increasing inverse function G. An elementary transformation then leads to the (equivalent) problem of solving the differential equation

$$G'^2 = F'(G)^{-2} = p(G). \tag{0.2}$$

From this differential equation it is easy to see that the case of a polynomial p of degree 4 can always be reduced to the case of a polynomial of degree ≤ 3:

0.1 Proposition. *Let $p(t)$ be a complex polynomial of degree 4 and $r \in \mathbb{C}$ be a first order zero of $p(t)$, as well as*

$$q(t) := p'(r) \cdot t^3 + \frac{1}{2}p''(r) \cdot t^2 + \frac{1}{6}p'''(r) \cdot t + \frac{1}{24}p^{(iv)}(r) \in \mathbb{C}[t].$$

In this case, G is a solution of the differential equation $G'^2 = p(G)$ if and only if either $G \equiv r$ or

$$H := \frac{1}{G - r}$$

satisfies the differential equation

$$H'^2 = q(H). \tag{0.3}$$

Proof Observe that

$$G \not\equiv r \text{ and } H = 1/(G - r) \text{ is equivalent to } H \not\equiv 0 \text{ and } G = r + 1/H. \quad (*)$$

If $G \not\equiv r$ is a solution of $G'^2 = p(G)$, a simple calculation yields

$$H'^2 = (G - r)^{-4} \cdot G'^2 = (G - r)^{-4} \cdot p(G) = H^4 \cdot p\left(\frac{1}{H} + r\right) = q(H),$$

hence (0.3). Now let H be a solution of (0.3). As $q(0) = \frac{1}{24}p^{(iv)}(0) \neq 0$, we get $H \not\equiv 0$. Using (*), an analogous calculation yields

$$G'^2 = H^{-4} \cdot H'^2 = (G - r)^4 \cdot q\left(\frac{1}{G - r}\right) = p(G). \quad \square$$

If q has degree 3, we can still replace H by $\alpha H + \beta$, $\alpha, \beta \in \mathbb{C}$, $\alpha \neq 0$, if necessary. Then q can be assumed to be in the so-called WEIERSTRASS *normal form*

$$q(t) = 4t^3 - c_2 t - c_3 \tag{0.4}$$

I Elliptic functions

The way to go from a 4th degree polynomial to a 3rd degree polynomial as shown above was described by A. CAYLEY (1821–1895) ([9], vol. IV, 60-69) in 1856. This reduction was covered by K.T.W. WEIERSTRASS (1815–1897) in his lectures at the University of Berlin from about 1862 ([88], vol. V).
Another way of reducing to a simple standard form was mentioned already in 1825 by A.-M. LEGENDRE (1752–1833) (cf. [54], vol. I, 4–11, or H. WEBER [87], §4):
Let $p(t)$ have only simple zeros. Let $a, b \in \mathbb{C}$, $a \neq b$ with $p(a) = p(b) = 0$. Now we set:

$$H := \begin{cases} \sqrt{\gamma \cdot (G - a)}, & \text{if } \deg p(t) = 3, \\ \sqrt{\gamma \cdot (G - a)/(G - b)}, & \text{if } \deg p(t) = 4, \end{cases}$$

with some $0 \neq \gamma \in \mathbb{C}$. From (0.2) we obtain for H a differential equation of the form

$$H'^2 = AH^4 + BH^2 + C, \quad A, B, C \in \mathbb{C}, \quad AC \neq 0.$$

By a suitable choice of $0 \neq \gamma \in \mathbb{C}$ we obtain a reduction to the so-called LEGENDRE *normal form*

$$H'^2 = C \cdot q(H), \quad q(t) = (1 - t^2) \cdot (1 - k^2 t^2), \quad k^2 \in \mathbb{C}, \; k^2 \neq 0, 1. \tag{0.5}$$

The number k is called the *modulus* of the corresponding elliptic integral. If we start from the WEIERSTRASS normal form (0.4) with

$$p(t) = 4(t - a)(t - b)(t - c)$$

we get

$$k^2 = (b - a)/(c - a).$$

If $p(t)$ has degree 4, then k^2 is the *cross-ratio* of the zeros. Correspondingly, an integral of the form

$$\int_0^x \frac{dt}{\sqrt{(1 - t^2)(1 - k^2 t^2)}} \tag{0.6}$$

is called an integral (of first kind) in LEGENDRE *normal form*. With the substitution $t = \sin \psi$, we can reduce (0.6) to integrals of the form

$$\int_0^\varphi \frac{d\psi}{\sqrt{1 - k^2 \sin^2 \psi}}, \quad \varphi = \arcsin x, \tag{0.7}$$

which are also named after LEGENDRE.

3. FAGNANO's example. We will give an example that is over 250 years old to show that solutions of the differential equation (0.2) can have unexpected properties: let $p(t) := 1 - t^4$, i.e. $k = i$ in (0.6), such that we obtain the FAGNANO *integral*

$$F(x) := \int_0^x \frac{dt}{\sqrt{1-t^4}}, \quad 0 \le x \le 1. \tag{0.8}$$

Since to $1-t^4 = (1-t)(1+t+t^2+t^3) \ge 1-t$ for all $0 \le t \le 1$, the improper integral

$$\sigma := F(1) \le \int_0^1 \frac{1}{\sqrt{1-t}} dt = 2$$

exists. The function $F : [0,1] \to [0,\sigma]$ is strictly monotonically increasing and continuous and therefore has a strictly monotonically increasing inverse function $G : [0,\sigma] \to [0,1]$ with

$$G'^2 = 1 - G^4, \quad G(0) = 0, \quad G'(0) = 1. \tag{0.9}$$

Conte G.C. FAGNANO (1682–1766) [23] discovered an unexpected property of F around 1750:

0.2 FAGNANO's Theorem. *For all sufficiently small $x \ge 0$,*

$$2F(x) = F\left(2x \cdot \frac{\sqrt{1-x^4}}{1+x^4}\right).$$

We rewrite this in terms of the inverse function G and get the

0.3 Corollary. *For all sufficiently small $u \ge 0$,*

$$G(2u) = \frac{2G(u)G'(u)}{1+G^4(u)}.$$

Proof The approach $t^2 = 2s^2/(1+s^4)$ leads to

$$1 - t^4 = ((1-s^4)/(1+s^4))^2 \quad \text{and} \quad t \cdot dt = 2s(1-s^4)/(1+s^4)^2 \cdot ds,$$

hence to

$$\frac{dt}{\sqrt{1-t^4}} = \sqrt{2} \cdot \frac{ds}{\sqrt{1+s^4}}.$$

Now the substitution $s^2 = 2r^2/(1-r^4)$, $t = 2r \cdot \sqrt{1-r^4}/(1+r^4)$ lead to

$$\frac{ds}{\sqrt{1+s^4}} = \sqrt{2} \cdot \frac{dr}{\sqrt{1-r^4}}, \quad \text{thus} \quad \frac{dt}{\sqrt{1-t^4}} = 2 \cdot \frac{dr}{\sqrt{1-r^4}}.$$

But this is the assertion. □

With the function G in (0.9), we can illustrate the derivation of (0.4): we set

$$H := \frac{1}{G+1} - \frac{1}{2}, \quad \text{hence} \quad G = \frac{1-2H}{1+2H}$$

I Elliptic functions

and verify the WEIERSTRASS normal form $H'^2 = 4H^3 + H$. With the substitution $G = 1/\sqrt{H}$, on the other hand we obtain $H'^2 = 4H^3 - 4H$. Compare here with WEIERSTRASS' work on the *lemniscate function* ([88], vol. VI, 183–219).

FAGNANO's work *Produzioni matematiche* was submitted to L. EULER for review at the Berlin Academy on Dec. 23, 1751. JACOBI calls this day the *birthday* of elliptic functions. Indeed, this work gave EULER the stimulus for his most important investigations on elliptic integrals, in particular for the discovery of the addition theorems (P. STÄCKEL and W. AHRENS [80], 23, 31 and 34), which JACOBI mentioned in a letter to Fuss:

> ... Bei dieser Gelegenheit habe ich auch einen für die Geschichte der Mathematik ungemein wichtigen Tag gefunden, an welchem unsere Akademie EULER auffordert, das von FAGNANI ihr übersandte Werk zu prüfen, ehe man dem Verfasser antwortet. Aus dieser Prüfung ist die Theorie der elliptischen Functionen entstanden (Brief von JACOBI an FUSS vom 24. Okt., 1847).

> ... On this occasion, I have also found a day of immense importance for the history of mathematics, on which our academy asks EULER to examine the work sent to it by FAGNANI before we answer the author. From this examination the theory of elliptic functions arose (Letter from JACOBI to FUSS from Oct. 24, 1847).

In 1761 L. EULER (1707–1783) [22], vol. 20, 58–107) took up this problem and proved a general *Addition Theorem*:

0.4 Euler's Addition Theorem. *For all sufficiently small* $x, y \geq 0$,

$$F(x) + F(y) = F\left(\frac{x\sqrt{1-y^4} + y\sqrt{1-x^4}}{1 + x^2 y^2}\right).$$

Passing to the inverse function results in the

0.5 Corollary. *For all sufficiently small* $u, v \geq 0$,

$$G(u + v) = \frac{G(u)G'(v) + G(v)G'(u)}{1 + G^2(u)G^2(v)}.$$

Soon thereafter, EULER wrote that the elliptic integrals should be introduced into calculus as transcendentals of independent interest. LEGENDRE supported this view.

The example of FAGNANO is typical in the sense that the inverse functions of integrals of the form (0.1), i.e. the solutions of differential equations of the form (0.2), form a class of interesting functions. It is natural to go into the complex numbers and allow meromorphic functions in general. From the unique solvability of the initial value problem (0.9), we obtain

$$G(iu) = iG(u) \qquad (0.10)$$

Corollary 0.5 then shows that the function G will be *doubly periodic*: for all $v \in \mathbb{C}$ with $G'(v) = 1$, we have $G(v) = 0$ by (0.9) and thus $G(u + v) = G(u)$ for all $u \in \mathbb{C}$. Then v and iv are periods of G because of (0.10).

The discovery of the double periodicity of those inverse functions is the basis of the development of the theory of elliptic functions by N.H. ABEL (from 1826) and C.G.J. JACOBI (from 1827). As early as the end of the 18th century, C.F. GAUSS possessed essential results on elliptic functions. However, he did not publish anything, except for a short hint in *Disquisitiones arithmeticae* ([31], vol. I, art. 335 (1801)), an application to arithmetic ([31], vol. II, 9–45 (1808)) and an application to secular perturbation theory ([31], vol. III, 331–355 (1818)). The fact that many essential results were indeed known to GAUSS became clear only during the examination and evaluation of his legacy ([31],vol. III, 361–490, and vol. VIII, 35–117; F. KLEIN and M. BRENDEL [44]). The approach followed in this first chapter is due to J. LIOUVILLE (1809–1882) and H. BURKHARDT (1861–1914), which does not start from the elliptic integrals as K.T.W. WEIERSTRASS ([88], vol.V) did, but instead focuses on the double periodicity. In the literature, this construction probably appears for the first time in the book of H. BURKHARDT [8]. By the classical book of A. HURWITZ and R. COURANT [40] from 1922 this approach became worldwide spread.

4. The lemniscate and its arc length. Let two different points p, q be given in the Euclidean plane $\mathbb{R}^2$. By a *lemniscate*, we mean the set of all points $z \in \mathbb{R}^2$ for which

$$|z - p| \cdot |z - q| = \frac{1}{4}|p - q|^2 \tag{0.11}$$

holds. Obviously, the curve passes through the midpoint $\frac{1}{2}(p+q)$ of the line from p to q. Therefore, up to an affine transformation, we may take the midpoint to be the origin as well as

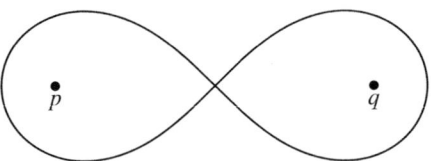

Figure 1: The lemniscate

$$p = \begin{pmatrix} -a \\ 0 \end{pmatrix}, q = \begin{pmatrix} a \\ 0 \end{pmatrix}, a > 0 \text{ and } z = \begin{pmatrix} x \\ y \end{pmatrix},$$

and therefore

$$((x + a)^2 + y^2) \cdot ((x - a)^2 + y^2) = a^4.$$

This is equivalent to $(x^2 + y^2)^2 = 2a^2(x^2 - y^2)$. Now it is helpful to normalize $2a^2 = 1$, i.e. the distance of both points p, q to be $\sqrt{2}$:

$$(x^2 + y^2)^2 = x^2 - y^2. \tag{0.12}$$

Upon introducing polar coordinates $x = r\cos\varphi$, $y = r\sin\varphi$, $r \geq 0$, $0 \leq \varphi < 2\pi$, (0.12) becomes equivalent to

$$r^2 = \cos 2\varphi, \ 0 \leq \varphi < 2\pi. \tag{0.13}$$

In calculus, we learned that the arc length $s = s(t)$ of a curve given in polar coordinates by $r = r(t)$ and $\varphi = \varphi(t)$ is determined by the differential equation

I Elliptic functions

$$s'^2 = x'^2 + y'^2 = r'^2 + r^2 \cdot \varphi'^2. \tag{0.14}$$

0.6 Proposition. *The arc length of the lemniscate is given by*

$$s'^2 = \frac{r'^2}{1 - r^4}.$$

Proof From (0.13) it follows that $rr' = -\sin 2\varphi \cdot \varphi'$ and therefore

$$r^2 r'^2 = \sin^2 2\varphi \cdot \varphi'^2 = (1 - \cos^2 2\varphi) \cdot \varphi'^2 = (1 - r^4) \cdot \varphi'^2.$$

By (0.14), we get

$$s'^2 = r'^2 \cdot \left(1 + \frac{r^4}{1 - r^4}\right) = \frac{r'^2}{1 - r^4}. \qquad \square$$

If we restrict ourselves to the first quadrant, we can choose r as a parameter. Then the arc length of the lemniscate is given by

$$F(R) := \int_0^R \frac{dr}{\sqrt{1 - r^4}},$$

hence by the FAGNANO *integral* (0.8), which will be calculated in §4.

FAGNANO's Theorem 0.2 states that twice the length of the arc between the origin and the point P equals the length of the arc between the origin and the point Q if the associated parameters $r = r_P$ and $R = r_Q$ satisfy the relation

$$R = 2r \cdot \sqrt{1 - r^4}/(1 + r^4).$$

The geometric meaning of this formula lies in the fact that the squares of R and r are obtained from one another by rational operations and taking square roots. Because

$$R^2 = 4r^2 \cdot \frac{1 - r^4}{(1 + r^4)^2} \quad \text{resp.} \quad r^2 = \frac{2t^2}{1 + t^4} \quad \text{with} \quad t^2 = \frac{2R^2}{1 - R^4}$$

we get the

0.7 Theorem. *The duplication of the arc of the lemniscate is possible by a construction with compass and ruler.*

FAGNANO appreciated this doubling so much that he expressed the wish (later carried out) to carve a lemniscate on his tombstone!

Reference: C.L. SIEGEL [78], vol. III, 249–251, and [77].

5. The arc length of the ellipse. Following the work on the FAGNANO integral (0.8), EULER examined an integral involving the arc length of the ellipse in 1761 ([22], vol. 20, 153–200). Considering the integral

$$I(x) := \int_0^x \frac{1 + At^2 + Bt^4}{\sqrt{1 + Ct^2 + Dt^4}}\, dt \tag{0.15}$$

EULER demonstrated the functional equation

$$I(x) + I(y) - I(z) = xyz \cdot \left(-A - \frac{x^2 + y^2 + z^2}{2}B + \frac{x^2 y^2 z^2}{6}BD\right) \tag{0.16}$$

with

$$z := \frac{x\sqrt{1 + Cy^2 + Dy^4} + y\sqrt{1 + Cx^2 + Dx^4}}{1 - Dx^2 y^2}. \tag{0.17}$$

As we can see, in the case $A = B = 0$ there is a perfect analogon to EULER's Addition Theorem 0.4. If an ellipse is given by

$$\frac{x^2}{a^2} + \frac{y^2}{b^2} = 1, \quad a > 0, \quad b > 0,$$

then in the first quadrant we have

$$y = \frac{b}{a}\sqrt{a^2 - x^2}, \quad y' = -\frac{b}{a}\frac{x}{\sqrt{a^2 - x^2}}, \quad 0 \le x < a.$$

This gives an integral for the arc length of the form

$$\frac{1}{a}\int_0^x \frac{\sqrt{a^4 + (b^2 - a^2)s^2}}{\sqrt{a^2 - s^2}}\, ds = a\int_0^{x/a} \frac{\sqrt{1 + kt^2}}{\sqrt{1 - t^2}}\, dt, \quad k := \frac{b^2 - a^2}{a^2},$$

when we have substituted $s = at$. We can also write this as

$$a\int_0^{x/a} \frac{1 + kt^2}{\sqrt{(1 - t^2)(1 + kt^2)}}\, dt,$$

(0.16) gives us a simple functional equation for the arc length.

6. ABEL and JACOBI: A theory emerges. The *Journal für die reine und angewandte Mathematik* or *Crelles Journal* was the leading international mathematical journal in the mid-19th century. It was founded in 1826 by the senior civil engineer A.L. CRELLE (1780–1855; from 1828 consultant in the Prussian Ministry of Culture). The first volume contains 7 papers by N.H. ABEL (1802–1829). Among them is the famous paper on the unsolvability of the general equation of 5th degree by radicals. In the same volume, we find a paper by C.G.J. JACOBI (1804–1851), while in the second volume, 9 papers by JACOBI and 4 by ABEL are printed.
Both in ABEL's fundamental work *Recherches sur les fonctions elliptiques* from 1827/28 ([1], vol. I, 263–388) and in JACOBI's famous book *Fundamenta nova*

I Elliptic functions

theoriae functionum ellipticarum from 1829 ([42], vol. I, 49–239) a theory of elliptic functions is developed from a theory of the elliptic integrals by using the inverse functions of the elliptic integrals in LEGENDRE normal form, as they were developed by JACOBI at about the same time. The theory is a natural generalization of the trigonometric functions. An essential part discusses the multiplication of elliptic functions in the sense of sect. 8 in §6. It appears today that ABEL was aware of the essential insights some time before JACOBI.

In his *Fundamenta nova*, JACOBI uses the name *amplitude* for the upper limit φ of the LEGENDRE integral (0.7)

$$u := u(\varphi) := \int_0^\varphi \frac{d\psi}{\sqrt{1 - k^2 \sin^2 \psi}} \tag{0.18}$$

([54], vol. I), he writes $\varphi = am\, u$ for the inverse function of (0.18) and names the trigonometric functions

$$\sin \varphi = \sin(am\, u), \quad \cos \varphi = \cos(am\, u), \quad \sqrt{1 - k^2 \sin^2 \varphi} = \Delta(am\, u)$$

elliptic functions. Following Chr. GUDERMANN (1798–1851; *Crelles Journal* **18**, **19**, **20**, **21**, **23** and **25**), we denote them by

$$sn\, u\,, \quad cn\, u\,, \quad dn\, u. \tag{0.19}$$

JACOBI showed that the functions (0.19) have two "independent" periods, namely $4K$ and $2iK$ with $K := u(\pi/2)$.

In the memorial speech to JACOBI, P.G.L. DIRICHLET compares the results of ABEL and JACOBI in a remarkable way (C.G.J. JACOBI [42], vol. I, 1–28). He writes on p. 10:

> Obgleich die Umgestaltung der Theorie der elliptischen Functionen, welche man Abel und Jacobi verdankt, aus dem Zusammenwirken mehrerer sich gegenseitig unterstützender Gedanken hervorgegangen ist, so scheint doch zweien dieser Gedanken die größte Wichtigkeit zugeschrieben werden zu müssen, weil sie alle Theile der neuen Theorie innig durchdringen. Während die früheren Bearbeiter dieses Gegenstandes das elliptische Integral der ersten Gattung als eine Function seiner Grenze ansahen, erkannten Abel und Jacobi unabhängig von einander, wenn auch der erstere einige Monate früher, die Nothwendigkeit die Betrachtungsweise umzukehren und die Grenze nebst zwei einfachen von ihr abhängigen Größen, die so unzertrennlich mit ihr verbunden sind wie der Sinus zum Cosinus gehört, als Functionen des Integrals zu behandeln, gerade wie man schon früher zur Erkenntniss der wichtigsten Eigenschaften der vom Kreise abhängigen Transcendenten gelangt war, indem man den Sinus und Cosinus als Functionen des Bogens und nicht diesen als eine Function von jenen betrachtete.

> Ein zweiter Abel und Jacobi gemeinsamer Gedanke, der Gedanke das Imaginäre in diese Theorie einzuführen, war von noch größerer Bedeutung und Jacobi hat es später oft wiederholt, dass die Einführung des Imaginären allein alle Räthsel der früheren Theorie gelöst habe.

The revolution of the theory of elliptic functions, which we owe to Abel and Jacobi, has arisen from the interaction of several ideas supporting each other. Nevertheless it seems that the greatest importance has to be ascribed to two of these ideas, which permeate all parts of the new theory intimately. Whereas the earlier workers in this subject regarded the elliptic integral of the first kind as a function of its endpoint, Abel and Jacobi independently of each other, although the former a few months earlier, recognized the necessity of looking at its inverse and viewing the boundary together with two simple quantities, which are as inseparably connected to it as the sine is to the cosine, as functions of the integral, just as one had earlier recognized the most important properties of the transcendentals dependent on the circle by considering the sine and cosine as functions of the arc and not the first one as a function of the latter.

A second idea common to Abel and Jacobi is the idea of introducing imaginary numbers into this theory. This was of even greater importance and Jacobi often repeated later that the introduction of imaginary numbers alone solved all the mysteries of the earlier theory.

Niels Henrik ABEL was born in 1802 as the son of a penniless priest in the Norwegian village of Find near Stavanger. From 1822 he attended the University of Christiania. He read classical mathematical literature as a schoolboy and noticed that not all of the published theorems were backed up by rigorous proofs. He is said to have decided even then to work on closing these gaps. His teacher HOLMBOE, who published ABEL's *Œuvres* as early as 1839, recognized his extraordinary talent. ABEL caused a sensation by a publication in which he falsely claimed to have a general method for solving the general equation of the fifth degree. The proof that such a method does not exist was published by himself in 1824 on a pamphlet ([1], vol. I, 28–33).
On the basis of this success, he received a scholarship for a trip abroad, which first took him to Berlin to A.L. CRELLE, the editor of the "Journal". He then traveled to Paris to visit CAUCHY, where he fell ill. He returned to Oslo in 1827 and died there of tuberculosis in 1829.

Carl Gustav Jacob JACOBI was born in Potsdam in 1804, as a son of a wealthy banker. From 1821 he studied at the University of Berlin, where he also began his academic career as a private lecturer at the age of only 20. From 1826 to 1844 he worked at the University of Königsberg, where he met the astronomer F. BESSEL. JACOBI wrote numerous papers on elliptic functions, analysis, number theory, geometry, and mechanics. His work on elliptic functions led him to the French school of LEGENDRE, FOURIER and POISSON, which he visited twice in Paris. In 1843, JACOBI became seriously ill with diabetes. From then on, he was rarely able to give his highly appreciated lectures. After a stay in a sanatorium in Italy in 1844, he returned to Berlin as a member of the Prussian Academy of Sciences, where he died in 1851.

0.8 Exercises.
1) Give an explicit solution of (0.1) if $p(t)$ is a polynomial of degree 1 or 2.
2) Derive the result of Proposition 0.1, if $p(t)$ has roots of higher multiplicity.
3) Show that $F(x)$ in (0.8) satisfies $1 < F(1) < \pi/2$.

§ 1 Periods and lattices

1. Meromorphic functions. A function f is called *meromorphic* on $\mathbb{C}$ if there exists a closed discrete subset D_f of $\mathbb{C}$ such that

(i) $f : \mathbb{C} \setminus D_f \to \mathbb{C}$ is holomorphic and
(ii) f has poles at the points of D_f.

Here, a closed subset D of $\mathbb{C}$ is called *discrete* if, for every $c \in \mathbb{C}$, there is a neighborhood U of c for which $D \cap U$ is finite. For U we can take any bounded neighborhood of c. Equivalently, we can say that the set

$$\{z \in D \,;\, |z| \leq \rho\} \text{ is finite for every } \rho > 0 \,. \tag{1.1}$$

Furthermore, f has a *pole* at $c \in D_f$ if and only if f is not holomorphically continuable to c, but if there is a positive integer m and a neighborhood U of c such that

$$(z - c)^m \cdot f(z) \text{ is bounded on } U \setminus \{c\} \,. \tag{1.2}$$

If we include the behavior of zeros at the holomorphic points of f, then $f \not\equiv 0$ is meromorphic on $\mathbb{C}$ if and only if for every $c \in \mathbb{C}$ there exists an $n \in \mathbb{Z}$, a neighborhood U of c and a holomorphic function $g : U \to \mathbb{C}$ with the property

$$f(z) = (z - c)^n \cdot g(z) \quad \text{for every } z \in U \setminus \{c\} \text{ and } g(c) \neq 0. \tag{1.3}$$

Then, of course, $n =: \mathrm{ord}_c f$ is the *order* of f at c. If $n < 0$, then we have a pole of *multiplicity* $-n$. If $n > 0$, we have a zero of *multiplicity* n. The set of meromorphic functions on $\mathbb{C}$ is denoted by $\mathcal{M}$. The reader is strongly encouraged to collect examples of meromorphic functions beyond the rational functions. Because of (1.2), αf, $\alpha \in \mathbb{C}$, $f + g$ and $f \cdot g$ are clearly meromorphic if f and g are, and

$$D_{\alpha f} = D_f \,,\, \alpha \neq 0,\, D_{f+g} \subseteq D_f \cup D_g \,,\, D_{fg} \subseteq D_f \cup D_g.$$

According to the Identity Theorem A.1, the set of roots of $0 \neq f \in \mathcal{M}$ is closed and discrete in $\mathbb{C}$. Thus $1/f$ is also meromorphic on $\mathbb{C}$, which implies the following

1.1 Theorem. *The meromorphic functions $\mathcal{M}$ on $\mathbb{C}$ form a field.* We add a general

1.2 Remark. a) Meromorphic functions on $\mathbb{C}$ are (contrary to common usage) not mappings defined on $\mathbb{C}$. But with the usual extension of the definition $f(c) := \infty$ for all $c \in D_f$, we can identify every meromorphic function f with a mapping $f : \mathbb{C} \to \mathbb{P}(\mathbb{C}) := \mathbb{C} \cup \{\infty\}$. The identification is compatible with the usual rules for calculating with infinity:

$$c + \infty = \infty, \quad \frac{\alpha}{0} = \infty, \quad \frac{\alpha}{\infty} = 0, \quad \alpha \cdot \infty = \infty \quad \text{for } \alpha \neq 0 \quad \text{etc..}$$

The notation $\mathbb{P}(\mathbb{C})$ is intended to remind us that $\mathbb{C} \cup \{\infty\}$ can be identified in a canonical way with the *projective space* $\mathbb{P}_1(\mathbb{C})$. As a warning, note that neither $0 \cdot \infty$

nor $\infty \pm \infty$ are defined: these expressions do not make any sense.

b) Algebraically, $\mathcal{M}$ is the quotient field of the ring of all entire functions: for every $f \in \mathcal{M}$ there exist functions p and q, $q \not\equiv 0$, that are holomorphic on $\mathbb{C}$ and satisfy

$$f(z) = p(z)/q(z) \quad \text{for all} \quad z \in \mathbb{C} \setminus D_f$$

(cf. Corollary A.11).

2. Periods of meromorphic functions. First we introduce an abbreviation: For $\omega \in \mathbb{C}$ and $D \subseteq \mathbb{C}$ the subset $D + \omega$ of $\mathbb{C}$ is defined to be

$$D + \omega := \{d + \omega \, ; \, d \in D\}.$$

Obviously,

$$(D + \omega) + \omega' = D + (\omega + \omega') \quad \text{for} \quad \omega, \omega' \in \mathbb{C}. \tag{1.4}$$

Let f be a meromorphic function on $\mathbb{C}$. An $\omega \in \mathbb{C}$ is called a *period* of f if

(P.1) $D_f + \omega = D_f$ and

(P.2) $f(z + \omega) = f(z)$ for all $z \in \mathbb{C} \setminus D_f$.

Trivially, 0 is a period of f for every $f \in \mathcal{M}$. Let Per f denote *period set* of f. Using (1.4) we see immediately *that* Per f *is always a subgroup of the additive group* $(\mathbb{C}, +)$. If we require (P.2) for all $z \in \mathbb{C}$ for which both sides of (P.2) are defined, then (P.1) is clearly a consequence of (P.2). Of course, Per $f = \mathbb{C}$ holds for any constant function f.

1.3 Lemma. *If $f \in \mathcal{M}$ is not constant, then* Per f *is a closed discrete subgroup of the additive group* $(\mathbb{C}, +)$.

Proof If Per f is not discrete or not closed, then there exist pairwise distinct periods $\omega_n \in $ Per f, $n \geq 1$, for which the limit $\omega := \lim_{n \to \infty} \omega_n$ exists. From (P.1), we conclude $D_f + \omega = D_f$, since D_f is closed in $\mathbb{C}$. If f is holomorphic at c, then f is also holomorphic at $c + \omega$ and $f(c) = f(c + \omega_n)$ holds for all n. But by the Identity Theorem A.1, f is constant and equal to $f(c) = f(c + \omega)$. □

A description of the possible period sets is contained in the

1.4 Fundamental Lemma. *If $f \in \mathcal{M}$ is not constant, then exactly one of the following three cases occurs:*

   (I) Per $f = \{0\}$.

  (II) *There is some $\omega_f \in \mathbb{C} \setminus \{0\}$ satisfying* Per $f = \mathbb{Z}\omega_f := \{m\omega_f \, ; \, m \in \mathbb{Z}\}$ *and ω_f is unique up to sign.*

 (III) *There are $\omega_1, \omega_2 \in \mathbb{C} \setminus \{0\}$ with the following properties:*

      (i) Per $f = \mathbb{Z}\omega_1 + \mathbb{Z}\omega_2 := \{m_1\omega_1 + m_2\omega_2 \, ; \, m_1, m_2 \in \mathbb{Z}\}$.

     (ii) ω_1, ω_2 *are linearly independent over* $\mathbb{R}$.

    (iii) $\tau := \omega_1/\omega_2$ *satisfies* Im $\tau > 0$, $|\text{Re } \tau| \leq 1/2$ *and* $|\tau| \geq 1$.

§ 1 Periods and lattices

Clearly, $\omega_1, \omega_2 \in \mathbb{C} \setminus \{0\}$ are linearly independent over $\mathbb{R}$ if and only if ω_1/ω_2 is not real.

Proof Let $\operatorname{Per} f \neq \{0\}$. Since $\operatorname{Per} f$ is closed and discrete by Lemma 1.3, there exists some $\omega_f \in \operatorname{Per} f$ with the property

$$0 < |\omega_f| = \inf\{|\omega| \, ; \, 0 \neq \omega \in \operatorname{Per} f\} \tag{$*$}$$

by (1.1). We first examine the periods on the line $\mathbb{R}\omega_f$.

Assertion. $\operatorname{Per} f \cap \mathbb{R}\omega_f = \mathbb{Z}\omega_f$.

Proof Obviously $\mathbb{Z}\omega_f \subseteq \operatorname{Per} f \cap \mathbb{R}\omega_f$ holds. For any $\omega \in \operatorname{Per} f \cap \mathbb{R}\omega_f$, there is an $\alpha \in \mathbb{R}$ with $\omega = \alpha \omega_f$. We choose $m \in \mathbb{Z}$ with $|\alpha - m| < 1$ and get

$$|\omega - m\omega_f| = |\alpha - m| \cdot |\omega_f| < |\omega_f|.$$

Along with ω and ω_f, $\omega - m\omega_f$ is an element of $\operatorname{Per} f \cap \mathbb{R}\omega_f$, and $\omega = m\omega_f$ follows from $(*)$. Therefore, $\operatorname{Per} f \cap \mathbb{R}\omega_f \subseteq \mathbb{Z}\omega_f$. □

If $\operatorname{Per} f$ lies on a straight line through 0, i.e. on $\mathbb{R}\omega_f$, then (II) follows from the assertion. As for uniqueness, note that $\mathbb{Z}\omega = \mathbb{Z}\omega'$ for $\omega, \omega' \in \mathbb{C}$ already implies $\omega' = \pm\omega$.

We may thus assume $\mathbb{Z}\omega_f \neq \operatorname{Per} f$. According to (1.1), there also exists an element $\omega_1 \in \operatorname{Per} f \setminus \mathbb{Z}\omega_f$ with

$$|\omega_1| = \inf\{|\omega| \, ; \, \omega \in \operatorname{Per} f \setminus \mathbb{Z}\omega_f\}. \tag{$**$}$$

Let $\omega_2 := \omega_f$. Then $\tau := \omega_1/\omega_2 \notin \mathbb{R}$ by the assertion, so ω_1, ω_2 are linearly independent over $\mathbb{R}$. By replacing ω_1 by $-\omega_1$, if necessary, we may assume $\operatorname{Im} \tau > 0$ without loss of generality. Then $(*)$ leads to

$$|\omega_1| \geq |\omega_2|, \quad \text{hence} \quad |\tau| \geq 1.$$

Using $(**)$, we obtain

$$|\omega_1 \pm \omega_2| \geq |\omega_1|, \quad \text{hence} \quad |\tau \pm 1| \geq |\tau|$$

and thus $|\operatorname{Re} \tau| \leq 1/2$. Obviously, $\mathbb{Z}\omega_1 + \mathbb{Z}\omega_2 \subseteq \operatorname{Per} f$. Now let $\omega \in \operatorname{Per} f$. Since ω_1, ω_2 is an $\mathbb{R}$-basis of $\mathbb{C}$, there are $\alpha_1, \alpha_2 \in \mathbb{R}$ with $\omega = \alpha_1\omega_1 + \alpha_2\omega_2$. Now choose $m_j \in \mathbb{Z}$ with $\beta_j = \alpha_j - m_j, |\beta_j| \leq 1/2, j = 1, 2$. Then

$$\omega' := \omega - m_1\omega_1 - m_2\omega_2 = \beta_1\omega_1 + \beta_2\omega_2 \in \operatorname{Per} f.$$

If $\beta_1 = 0$, then $\omega' = 0$ already follows from the assertion, so assume $\beta_1 \neq 0$. Then of course $\omega' \in \operatorname{Per} f \setminus \mathbb{Z}\omega_f$ and

$$|\omega'|^2 = |\beta_1\omega_1 + \beta_2\omega_2|^2 = (\beta_1^2 \cdot |\tau|^2 + 2\beta_1\beta_2 \cdot \operatorname{Re} \tau + \beta_2^2) \cdot |\omega_2|^2$$
$$\leq (\beta_1^2 + |\beta_1||\beta_2| + \beta_2^2) \cdot |\tau|^2 \cdot |\omega_2|^2 \leq \tfrac{3}{4}|\omega_1|^2,$$

if we take (iii) into account. This contradicts (∗∗). Therefore, $\omega' = 0$ follows and Per $f = \mathbb{Z}\omega_1 + \mathbb{Z}\omega_2$, i.e. (III). □

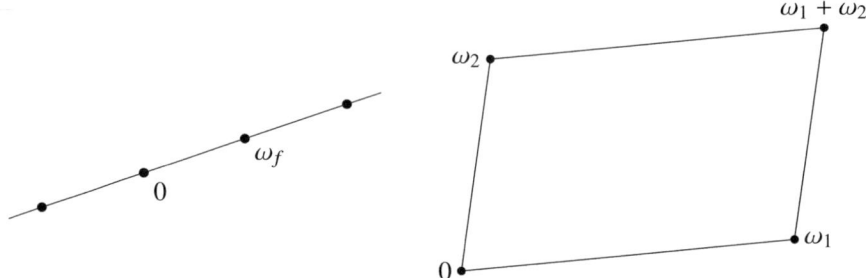

Figure 2: Period set Figure 3: Period lattice

Let V be a real vector space of finite dimension $n \geq 1$, e.g. $V = \mathbb{R}^n$. A subset Ω of V is called a *lattice* in V if there exists an $\mathbb{R}$-basis $(\omega_1, \omega_2, \ldots, \omega_n)$ of V with $\Omega = \mathbb{Z}\omega_1 + \mathbb{Z}\omega_2 + \ldots + \mathbb{Z}\omega_n$. We then call $(\omega_1, \omega_2, \ldots, \omega_n)$ a *basis* of Ω. A representation of $\omega \in \Omega$ in the form

$$\omega = m_1\omega_1 + m_2\omega_2 + \ldots + m_n\omega_n \quad \text{with} \quad m_1, m_2, \ldots, m_n \in \mathbb{Z}$$

is called a *linear combination* of ω by $\omega_1, \omega_2, \ldots, \omega_n$ over $\mathbb{Z}$. Obviously, if Ω is a lattice, then $\lambda\Omega := \{\lambda\omega \, ; \, \omega \in \Omega\}$ is also a lattice in V for any $0 \neq \lambda \in \mathbb{R}$. Thus, in case (III) Per f is a lattice in the $\mathbb{R}$-vector space $\mathbb{C}$, the so-called *period lattice* of f. In view of Lemma 1.3 and the Fundamental Lemma 1.4, we note the

1.5 Proposition. *Every lattice Ω in $\mathbb{C}$ is closed and discrete in $\mathbb{C}$.*

Proof For $\rho > 0$, let $S = \{\omega \in \Omega; |\omega| \leq \rho\}$. Replacing ρ by $\rho/|\omega_2|$ if necessary, we may assume without loss of generality $\Omega = \mathbb{Z}\tau + \mathbb{Z}$, $\tau = x + iy \in \mathbb{C}, y > 0$. If $\omega = m\tau + n \in S$ with $m, n \in \mathbb{Z}$, then on the one hand

$$\rho^2 \geq |m\tau + n|^2 = (mx + n)^2 + m^2y^2 \geq m^2y^2,$$

hence $|m| \leq \rho/y$. On the other hand

$$\rho \geq |mx + n| \geq |n| - |mx|, \quad \text{hence} \quad |n| \leq \rho(1 + |x|/y).$$

Therefore, S is finite. □

Another proof of the Fundamental Lemma 1.4 can be found in the book by A. HURWITZ and R. COURANT [40], II, 1, § 2, from 1922. Considering cases (II) and (III), we find the following general

1.6 Lemma. *For a subgroup $G \neq \{0\}$ of the additive group $(\mathbb{R}^n, +)$, the following assertions are equivalent:*

(i) *G is discrete.*
(ii) *There are linearly independent vectors $c_1, \ldots, c_r \in \mathbb{Z}^n$, $1 \leq r \leq n$, with $G = \mathbb{Z}c_1 + \ldots + \mathbb{Z}c_r$.*

§ 1 Periods and lattices

A *proof* can be found e.g. in M. NEWMAN [60].

The Fundamental Lemma 1.4 yields a constraint on the period set Per f of a non-constant meromorphic function f. Per $f = \{0\}$ can certainly be considered the 'general case' but it is of no interest in the present context. If $0 \neq \omega \in \mathbb{C}$, there is an entire function f with Per $f = \mathbb{Z}\omega$. We simply choose $f(z) = \exp(2\pi i z/\omega)$. Given an entire function f with Per $f = \mathbb{Z}\omega$, Theorem A.4 leads to a unique holomorphic function

$$F : \mathbb{C}\setminus\{0\} \to \mathbb{C}, \quad w \mapsto f\left(\frac{\omega}{2\pi i} \log w\right),$$

satisfying

$$f(z) = F(\exp(2\pi i z/\omega)) \quad \text{for all } z \in \mathbb{C}.$$

Before we will show in §3 that for every lattice Ω in $\mathbb{C}$ there are meromorphic functions f with $\Omega = \text{Per } f$, a discussion of the fundamental properties of lattices in $\mathbb{C}$ is useful. After the description of a different approach in sect. 3, this will be done in sections 4 to 6.

3*. JACOBI's approach. In 1835, C.G.J. JACOBI ([42], vol. II, 23–50) dealt with the question of how many independent periods a non-constant meromorphic function can have, and he showed that this number is at most two. We formulate his results in modern language as

1.7 JACOBI's Lemma. *If Ω is a discrete subgroup of $(\mathbb{C}, +)$ and $\omega_1, \omega_2, \omega_3 \in \Omega$ are given, then there are $m_1, m_2, m_3 \in \mathbb{Z}$, not all zero, such that*

$$m_1\omega_1 + m_2\omega_2 + m_3\omega_3 = 0.$$

In particular, any three elements of Ω are linearly dependent over $\mathbb{Q}$. For a proof we use the so-called

1.8 KRONECKER Approximation. *For every $N \in \mathbb{N}$ and all $\omega_1, \omega_2, \omega_3 \in \mathbb{C}$ there exist $m_1, m_2, m_3 \in \mathbb{Z}$, not all zero, with the properties*:

(i) $|m_1|, |m_2|, |m_3| \leq N$,

(ii) $|m_1\omega_1 + m_2\omega_2 + m_3\omega_3| < \dfrac{6\sqrt{2}}{\sqrt{N}} \cdot \max\{|\omega_1|, |\omega_2|, |\omega_3|\}$.

Proof We set $M := \max\{|\omega_1|, |\omega_2|, |\omega_3|\}$ and

$$\langle m, w \rangle := m_1\omega_1 + m_2\omega_2 + m_3\omega_3 \quad \text{for} \quad m = (m_1, m_2, m_3), \ w = (\omega_1, \omega_2, \omega_3).$$

The square in $\mathbb{C}$ with sides parallel to the coordinate axes, with center 0 and edge length $2K$ is denoted by

$$Q(K) := \{z \in \mathbb{C} \, ; \ |\text{Re } z| \leq K, \ |\text{Im } z| \leq K\}.$$

For all $m = (m_1, m_2, m_3)^{tr} \in \mathbb{Z}^3$ with

$$0 \leq m_1, m_2, m_3 \leq N \tag{*}$$

we have $\langle m, w \rangle \in Q(T)$ with $T := 3MN$. The edges of $Q(T)$ are now divided into t equal parts, such that we obtain a decomposition of $Q(T)$ into t^2 squares with the edge length $2T/t$. The number of $m \in \mathbb{Z}^3$ that satisfy $(*)$ is $(N+1)^3$. According to DIRICHLET's Pigeonhole Principle, if $(N+1)^3 > t^2$, then at least two points $\langle m', w \rangle$ and $\langle m'', w \rangle$ lie in the same square of edge length $2T/t$. By taking the difference we therefore obtain $0 \neq m \in \mathbb{Z}^3$ with (i) and

$$\langle m, w \rangle \in Q(2T/t), \quad \text{hence} \quad |\langle m, w \rangle| \leq \sqrt{2} \cdot 2T/t. \qquad (**)$$

We now choose some $t \in \mathbb{Z}$ with $(N+1)^{3/2} > t \geq (N+1)^{3/2} - 1$. Then $(N+1)^3 > t^2$ and $t \geq N^{3/2}$ hold, so (ii) follows from $(**)$. □

For the **proof** of JACOBI's Lemma 1.7 note that (1.1) leads to some $\rho > 0$ with $|\omega| \geq \rho$ for all $0 \neq \omega \in \Omega$. Therefore, for sufficiently large N the left-hand side of (ii) is equal to zero. □

From JACOBI's Lemma 1.7 we can now derive a new proof of the Fundamental Lemma 1.4. This proof uses an interesting argument: if Ω is a discrete subgroup of $\mathbb{C}$ and if $\omega_1, \omega_2 \in \Omega$ are linearly independent over $\mathbb{R}$, then JACOBI's Lemma 1.7 immediately yields

$$\Omega \subseteq \mathbb{Q}\omega_1 + \mathbb{Q}\omega_2 \qquad (1.5)$$

1.9 Proposition. *There exists an $N \in \mathbb{N}$ with the property*

$$\Omega \subseteq \frac{1}{N}(\mathbb{Z}\omega_1 + \mathbb{Z}\omega_2).$$

Proof Otherwise, (1.5) would yield $\omega \in \Omega$ with arbitrarily large denominators, i.e. we would have

$$0 \neq \omega_k' = \frac{1}{N_k}(r_k\omega_1 + s_k\omega_2) \in \Omega, \ k \in \mathbb{N},$$

with integers r_k, s_k, N_k satisfying $\gcd(r_k, s_k, N_k) = 1$ and $N_k \to \infty$ for $k \to \infty$. After the addition of suitable points from $\mathbb{Z}\omega_1 + \mathbb{Z}\omega_2$ suppose without loss of generality that $0 \leq r_k, s_k \leq N_k$. Thus we may assume $|\omega_k'| \leq |\omega_1| + |\omega_2|$ for all $k \in \mathbb{N}$. Then there would be a convergent subsequence of $(\omega_k')_{k \in \mathbb{N}}$ consisting of pairwise distinct elements. Since Ω is discrete, we obtain a contradiction. □

As is well known (cf. S. LANG [53], Theorem I.7.3), every subgroup of a finitely generated free abelian group is itself free. Thus according to Proposition 1.9, Ω is also free. Because $\mathbb{Z}\omega_1 + \mathbb{Z}\omega_2 \subseteq \Omega$, Ω is a lattice.

4. The group GL $(2; \mathbb{Z})$. The set

$$\text{Mat}(2; \mathbb{Z}) := \left\{ U = \begin{pmatrix} a & b \\ c & d \end{pmatrix} ; \ a, b, c, d \in \mathbb{Z} \right\}$$

is known to form a ring under matrix addition and multiplication with the identity element $I := \begin{pmatrix} 1 & 0 \\ 0 & 1 \end{pmatrix}$. The group of units of the ring Mat $(2; \mathbb{Z})$ is called the *general linear group of degree 2 over $\mathbb{Z}$* and is denoted by GL $(2; \mathbb{Z})$:

§ 1 Periods and lattices

$$\mathrm{GL}(2;\mathbb{Z}) := \{U \in \mathrm{Mat}(2;\mathbb{Z}) \,;\, \text{there is } V \in \mathrm{Mat}(2;\mathbb{Z}) \text{ with } UV = VU = I\} \quad (1.6)$$

1.10 Equivalence Theorem for GL(2; $\mathbb{Z}$). *For $U \in \mathrm{Mat}(2;\mathbb{Z})$ the following assertions are equivalent:*

(i) $U \in \mathrm{GL}(2;\mathbb{Z})$.
(ii) $\det U = \pm 1$.
(iii) U *is invertible over $\mathbb{Q}$ and* $U^{-1} \in \mathrm{Mat}(2;\mathbb{Z})$.
(iv) *The mapping* $U : \mathbb{Z}^2 \to \mathbb{Z}^2$, $x \mapsto Ux$, *is bijective.*
(v) *The mapping* $U : \mathbb{Z}^2 \to \mathbb{Z}^2$, $x \mapsto Ux$, *is surjective.*

Proof The implications (iii) $\Rightarrow$ (i) $\Rightarrow$ (iv) $\Rightarrow$ (v) are obvious.
(v) $\Longrightarrow$ (ii): There exist $u, v \in \mathbb{Z}^2$ with $Uu = \binom{1}{0}$ and $Uv = \binom{0}{1}$, i.e. $UV = I$ for $V := (u, v)$. Calculating determinants, $\det U \cdot \det V = 1$ follows, hence (ii).
(ii) $\Longrightarrow$ (iii): We use the well-known representation

$$U^{-1} = \frac{1}{\det U} \begin{pmatrix} d & -b \\ -c & a \end{pmatrix} \quad \text{for } U = \begin{pmatrix} a & b \\ c & d \end{pmatrix}. \qquad \square$$

Besides the group $\mathrm{GL}(2;\mathbb{Z})$, we consider the normal subgroup

$$\mathrm{SL}(2;\mathbb{Z}) := \{U \in \mathrm{GL}(2;\mathbb{Z}) \,;\, \det U = 1\} \quad (1.7)$$

of $\mathrm{GL}(2;\mathbb{Z})$, the so-called *special linear group of degree 2 over $\mathbb{Z}$*. Because of

$$\mathrm{GL}(2;\mathbb{Z}) = \mathrm{SL}(2;\mathbb{Z}) \cup \mathrm{SL}(2;\mathbb{Z}) \cdot \begin{pmatrix} 1 & 0 \\ 0 & -1 \end{pmatrix}$$

$\mathrm{SL}(2;\mathbb{Z})$ has index 2 in $\mathrm{GL}(2;\mathbb{Z})$. The group $\mathrm{SL}(2;\mathbb{Z})$ is larger than one might think: for $U = \begin{pmatrix} a & b \\ c & d \end{pmatrix} \in \mathrm{SL}(2;\mathbb{Z})$ we have $1 = \det U = ad - bc$, so that e.g. *c and d are coprime.* Conversely, we have the

1.11 Completion Lemma. *If $c, d \in \mathbb{Z}$ are coprime, then there is a matrix*

$$U = \begin{pmatrix} * & * \\ c & d \end{pmatrix} \in \mathrm{SL}(2;\mathbb{Z}).$$

Here U is unique up to multiplication on the left by matrices of the form $\begin{pmatrix} 1 & k \\ 0 & 1 \end{pmatrix}$ *with* $k \in \mathbb{Z}$.

Proof Since c, d are coprime, there exist $a, b \in \mathbb{Z}$ with $ad - bc = 1$, i.e., $U = \begin{pmatrix} a & b \\ c & d \end{pmatrix}$ belongs to $\mathrm{SL}(2;\mathbb{Z})$. If $V \in \mathrm{SL}(2;\mathbb{Z})$ is another matrix with $V = \begin{pmatrix} * & * \\ c & d \end{pmatrix}$, then

$$VU^{-1} = \begin{pmatrix} * & * \\ c & d \end{pmatrix} \begin{pmatrix} d & -b \\ -c & a \end{pmatrix} = \begin{pmatrix} * & * \\ 0 & 1 \end{pmatrix} \in \mathrm{SL}(2;\mathbb{Z}),$$

hence $VU^{-1} = \begin{pmatrix} 1 & k \\ 0 & 1 \end{pmatrix}$ for some $k \in \mathbb{Z}$. $\qquad \square$

The example $U = 2I$ shows that – in contrast to the analogous situation over a field – the condition "$U : \mathbb{Z}^2 \to \mathbb{Z}^2$ is injective" cannot be included in the Equivalence

Theorem for GL$(2;\mathbb{Z})$ 1.10.

The Completion Lemma 1.11 allows us to construct countless concrete examples: for example, the matrices

$$\begin{pmatrix} 3 & 1 \\ 2 & 1 \end{pmatrix}, \begin{pmatrix} 2 & 3 \\ 3 & 5 \end{pmatrix}, \begin{pmatrix} 3 & 10 \\ 2 & 7 \end{pmatrix}, \begin{pmatrix} 89 & 144 \\ 144 & 233 \end{pmatrix}, \begin{pmatrix} 514\,229 & 832\,040 \\ 832\,040 & 1\,346\,269 \end{pmatrix}$$

belong e.g. to SL$(2;\mathbb{Z})$.

The results of this section can easily be extended to $n \times n$ matrices. Compare V, §1 or M. NEWMAN [60], Chap. II.

1.12 Basis Lemma for Lattices. *Let Ω be a lattice in $\mathbb{C}$ and (ω_1, ω_2) a basis of Ω. For $\omega'_1, \omega'_2 \in \mathbb{C}$, the following holds.*
a) *If there exists $U \in \text{Mat}(2;\mathbb{Z})$ with*

$$\begin{pmatrix} \omega'_1 \\ \omega'_2 \end{pmatrix} = U \begin{pmatrix} \omega_1 \\ \omega_2 \end{pmatrix}, \tag{1.8}$$

then ω'_1 and ω'_2 belong to Ω.
b) *(ω'_1, ω'_2) is a basis of Ω if and only if the matrix U in (1.8) belongs to GL$(2;\mathbb{Z})$.*

Proof a) If ω'_1, ω'_2 are arbitrary points of Ω, then there exist $a, b, c, d \in \mathbb{Z}$ with

$$\omega'_1 = a\omega_1 + b\omega_2, \quad \omega'_2 = c\omega_1 + d\omega_2,$$

thus

$$\begin{pmatrix} \omega'_1 \\ \omega'_2 \end{pmatrix} = U \begin{pmatrix} \omega_1 \\ \omega_2 \end{pmatrix} \quad \text{with} \quad U = \begin{pmatrix} a & b \\ c & d \end{pmatrix} \in \text{Mat}(2;\mathbb{Z}). \tag{*}$$

Conversely, if $(*)$ is true, then ω'_1 and ω'_2 belong to Ω.
b) If (ω'_1, ω'_2) is a basis of Ω, then there is also a matrix $V \in \text{Mat}(2;\mathbb{Z})$ with the property $\begin{pmatrix} \omega_1 \\ \omega_2 \end{pmatrix} = V \begin{pmatrix} \omega'_1 \\ \omega'_2 \end{pmatrix}$. It follows that

$$\begin{pmatrix} \omega_1 \\ \omega_2 \end{pmatrix} = VU \begin{pmatrix} \omega_1 \\ \omega_2 \end{pmatrix} \quad \text{and} \quad \begin{pmatrix} \omega'_1 \\ \omega'_2 \end{pmatrix} = UV \begin{pmatrix} \omega'_1 \\ \omega'_2 \end{pmatrix}.$$

But since (ω_1, ω_2) and (ω'_1, ω'_2) are linearly independent over $\mathbb{R}$, we can conclude that $VU = UV = I$, so $U \in \text{GL}(2;\mathbb{Z})$ according to (1.6).
If $U \in \text{GL}(2;\mathbb{Z})$, then (ω'_1, ω'_2) in $(*)$ are linearly independent over $\mathbb{R}$. For any $\omega''_1, \omega''_2 \in \Omega$, by analogy, there is some $W \in \text{Mat}(2;\mathbb{Z})$ with

$$\begin{pmatrix} \omega''_1 \\ \omega''_2 \end{pmatrix} = W \begin{pmatrix} \omega_1 \\ \omega_2 \end{pmatrix}, \quad \text{so} \quad \begin{pmatrix} \omega''_1 \\ \omega''_2 \end{pmatrix} = WU^{-1} \begin{pmatrix} \omega'_1 \\ \omega'_2 \end{pmatrix}.$$

Therefore, ω''_1, ω''_2 are linear combinations of ω'_1, ω'_2 over $\mathbb{Z}$. Hence, ω'_1, ω'_2 is a basis of Ω. $\square$

§ 1 Periods and lattices

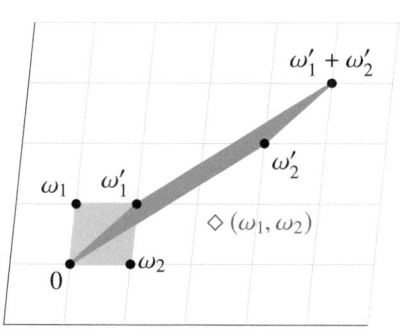
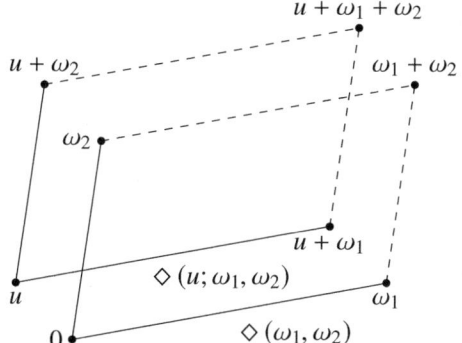

Figure 4: Fundamental parallelogram   Figure 5: Period parallelograms

Let Ω be a lattice in $\mathbb{C}$ and (ω_1, ω_2) a basis of Ω. For $u \in \mathbb{C}$, we define the *period parallelogram* (with respect to ω_1, ω_2 and with *base point* u) by

$$\Diamond(u; \omega_1, \omega_2) := \{u + \alpha_1\omega_1 + \alpha_2\omega_2 \; ; \; 0 \leq \alpha_1 < 1, 0 \leq \alpha_2 < 1\}. \quad (1.9)$$

In the case $u = 0$, we also write

$$\Diamond(\omega_1, \omega_2) := \Diamond(0; \omega_1, \omega_2) = \{\alpha_1\omega_1 + \alpha_2\omega_2 \; ; \; 0 \leq \alpha_1 < 1, 0 \leq \alpha_2 < 1\} \quad (1.10)$$

and call $\Diamond(\omega_1, \omega_2)$ a *fundamental parallelogram* of the lattice. Each period parallelogram $P := \Diamond(u; \omega_1, \omega_2)$ is a *fundamental domain* of $\mathbb{C}$ with respect to Ω in the following sense:

1.13 Proposition. *For every $z \in \mathbb{C}$, there is a unique $\omega \in \Omega$ satisfying $z + \omega \in P$. In particular, if z and $z + \omega$, $\omega \in \Omega$, belong to P, then $\omega = 0$.*

Proof Choose $\xi_1, \xi_2 \in \mathbb{R}$ such that $z - u = \xi_1\omega_1 + \xi_2\omega_2$ and reduce ξ_1 and ξ_2 modulo 1. □

There are many period parallelograms for each lattice, but the following holds.

1.14 Proposition. *The area of a period parallelogram $P = \Diamond(u; \omega_1, \omega_2)$ is equal to*

$$\operatorname{vol} \Omega := |\operatorname{Im}(\omega_1\overline{\omega_2})|.$$

It is independent of the choice of the basis (ω_1, ω_2) of Ω and the base point u.

Proof The area of the parallelogram P in Euclidean coordinates is given by

$$A := \left|\det \begin{pmatrix} \operatorname{Re} \omega_1 & \operatorname{Re} \omega_2 \\ \operatorname{Im} \omega_1 & \operatorname{Im} \omega_2 \end{pmatrix}\right|.$$

A computation yields $A = |\operatorname{Im}(\omega_1\overline{\omega_2})|$. Now if $\omega'_1 = a\omega_1 + b\omega_2$, $\omega'_2 = c\omega_1 + d\omega_2$ with $a, b, c, d \in \mathbb{Z}$, $ad - bc = \pm 1$, is another basis of Ω according to the Basis Lemma for Lattices 1.12, then

$$A' = |\text{Im}\,(\omega_1' \overline{\omega_2'})| = |\text{Im}\,(ad\omega_1\overline{\omega_2} + bc\omega_2\overline{\omega_1})| = |ad - bc| \cdot |\text{Im}\,(\omega_1\overline{\omega_2})|,$$

so $A' = A$. □

5. Factor group $\mathbb{C}/\Omega$. Let Ω be a subgroup of $(\mathbb{C}, +)$. We define an equivalence relation on $\mathbb{C}$ by

$$a \equiv b \pmod{\Omega} \iff a - b \in \Omega. \tag{1.11}$$

The equivalence classes can then be written in the form $a + \Omega$, $a \in \mathbb{C}$. As $(\mathbb{C}, +)$ is abelian, Ω is a normal subgroup of $\mathbb{C}$. The factor group is denoted by

$$\mathbb{C}/\Omega := \{a + \Omega\,;\, a \in \mathbb{C}\} \tag{1.12}$$

and the canonical projection is denoted by π:

$$\pi : \mathbb{C} \longrightarrow \mathbb{C}/\Omega\,,\ \pi(a) := a + \Omega. \tag{1.13}$$

If Ω is a lattice in $\mathbb{C}$, then according to Proposition 1.13, for any period parallelogram $P = \Diamond(u; \omega_1, \omega_2)$, the restriction

$$\pi|_P : P \longrightarrow \mathbb{C}/\Omega \quad \text{is bijective.} \tag{1.14}$$

As is well known, the addition in $\mathbb{C}/\Omega$ is given by

$$(a + \Omega) + (b + \Omega) := (a + b) + \Omega \tag{1.15}$$

Thus $\mathbb{C}/\Omega$ is an abelian group with zero element Ω. Inducing the topology of $\mathbb{C}$ via the canonical projection (1.13) to $\mathbb{C}/\Omega$, i.e., supplying $\mathbb{C}/\Omega$ with the quotient topology, $\mathbb{C}/\Omega$ becomes a compact topological space. By definition, a subset A of $\mathbb{C}/\Omega$ is open if $\pi^{-1}(A)$ is open in $\mathbb{C}$. Because of (1.14) we can think of $\mathbb{C}/\Omega$ as a *torus* in $\mathbb{R}^3$: all we have to do is to identify the opposite sides of a period parallelogram.

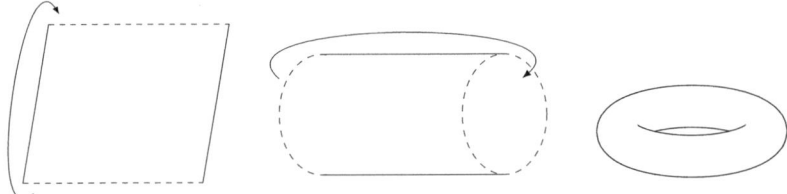

Figure 6: The torus $\mathbb{C}/\Omega$

We know that a discrete subgroup Ω of $(\mathbb{C}, +)$ is a lattice if and only if the factor group $\mathbb{C}/\Omega$ is compact. This can be realized by identifying the corresponding edges of a fundamental parallelogram.

6. Remarks on infinite series with multiple indices. Let $\mathbb{N} = \{n \in \mathbb{Z}; n > 0\}$ and $\mathbb{N}_0 = \{n \in \mathbb{Z}; n \geq 0\}$ stand for the *natural numbers*. Infinite series with multiple indices are usually written in the general form

§ 1 Periods and lattices

$$\sum_{g \in \mathbb{Z}^n} \alpha_g \quad \text{with} \quad \alpha_g \in \mathbb{C} \tag{1.16}$$

and for $n > 1$ are only considered when they are absolutely convergent. The series (1.16) is called *absolutely convergent* if, for every bijection $\varphi : \mathbb{N} \to \mathbb{Z}^n$, the series $\sum_{k \in \mathbb{N}} \alpha_{\varphi(k)}$ is absolutely convergent. The limit is written as

$$\sum_{g \in \mathbb{Z}^n} \alpha_g = \sum_{k=1}^{\infty} \alpha_{\varphi(k)}$$

and it does not depend on φ. Thus we also have the

1.15 Theorem. *An absolutely convergent series may be rearranged arbitrarily.*

We can define the absolute convergence of (1.16) without using φ. It is sufficient to find a constant $C > 0$ such that

$$\sum_{g \in S} |\alpha_g| < C \quad \text{for any finite subset } S \subseteq \mathbb{Z}^n.$$

7. Lattice invariants. Let Ω be a lattice in $\mathbb{C}$ with basis (ω_1, ω_2). Then

$$\delta := \delta(\omega_1, \omega_2) := \sup\{|z - w|\,;\ z, w \in \Diamond(\omega_1, \omega_2)\} \tag{1.17}$$

is the *diameter* of the corresponding fundamental parallelogram. Let $A_\rho(\Omega)$ denote the number of lattice points in the closed disk around 0 with radius $\rho > 0$, i.e.

$$A_\rho(\Omega) := \#\{\omega \in \Omega\,;\ |\omega| \leq \rho\}.$$

1.16 Proposition. *For any $\rho \geq \delta$,*

$$\frac{\pi}{\operatorname{vol} \Omega}(\rho - \delta)^2 \leq A_\rho(\Omega) \leq \frac{\pi}{\operatorname{vol} \Omega}(\rho + \delta)^2.$$

Proof We compare the areas of

$$K_\rho := \{z \in \mathbb{C}\,;\ |z| \leq \rho\} \quad \text{and} \quad M_\rho := \bigcup_{\omega \in \Omega,\, |\omega| \leq \rho} \Diamond(\omega; \omega_1, \omega_2).$$

By the definition of δ,

$$K_{\rho-\delta} \subseteq M_\rho \subseteq K_{\rho+\delta}.$$

Taking areas proves the assertion, because

$$\operatorname{area} K_\rho = \pi \rho^2 \quad \text{and} \quad \operatorname{area} M_\rho = \operatorname{vol} \Omega \cdot A_\rho(\Omega). \quad \square$$

Thus we obtain the

1.17 Convergence Lemma. *The series*

$$\sum_{0\neq\omega\in\Omega} |\omega|^{-\alpha}$$

converges if and only if $\alpha > 2$.

Proof Let $\alpha > 2$ and $\emptyset \neq S \subseteq \Omega \setminus \{0\}$ be finite and $M := \max\{|\omega|\,;\, \omega \in S\}$. From Proposition 1.16, we obtain a $c_2 > 0$ satisfying

$$A_{n+1}(\Omega) - A_n(\Omega) \leq \frac{\pi}{\mathrm{vol}\,\Omega} \cdot [(n+1+\delta)^2 - (n-\delta)^2] \leq c_2 n$$

for all $n \geq \delta$. Thus

$$c_1 := \sum_{0\neq\omega\in\Omega,\,|\omega|\leq\delta+1} |\omega|^{-\alpha}$$

yields

$$\sum_{\omega\in S} |\omega|^{-\alpha} \leq c_1 + \sum_{n\in\mathbb{N},\,\delta<n<M} (A_{n+1}(\Omega) - A_n(\Omega)) n^{-\alpha}$$

$$\leq c_1 + c_2 \sum_{n=1}^{\infty} n^{1-\alpha} =: C < \infty$$

(cf. Theorem B.1). The series trivially diverges for $\alpha \leq 0$. Let $0 < \alpha \leq 2$ and $N \in \mathbb{N}, N > 2\delta$. From Theorem 1.16 we get a $c_3 > 0$ satisfying

$$A_{kN}(\Omega) - A_{(k-1)N}(\Omega) \geq \frac{\pi}{\mathrm{vol}\,(\Omega)} \cdot [(kN-\delta)^2 - ((k-1)N+\delta)^2] \geq c_3 k$$

for all $k \in \mathbb{Z}, k \geq 2$. Let $S_n = \{\omega \in \Omega\,;\, 0 < |\omega| \leq nN\}$. Then we get

$$\sum_{\omega\in S_n} |\omega|^{-\alpha} \geq \sum_{k=2}^{n} (A_{kN}(\Omega) - A_{(k-1)N}(\Omega)) \cdot (kN)^{-\alpha} \geq c_3 N^{-\alpha} \cdot \sum_{k=2}^{n} k^{1-\alpha}.$$

Because the series $\sum_{k\geq 1} k^{1-\alpha}$ diverges for $\alpha \leq 2$ (cf. Theorem B.1), the series $\sum_{0\neq\omega\in\Omega} |\omega|^{-\alpha}$ also diverges for $\alpha \leq 2$. □

Another version of the Convergence Lemma 1.17 can be found in III, §2. The so-called EISENSTEIN *series*

$$E_k := E_k(\Omega) := \sum_{0\neq\omega\in\Omega} \omega^{-k} \quad \text{for } k \in \mathbb{Z},\ k > 2 \tag{1.18}$$

are therefore absolutely convergent. Since $-\omega$ belongs to $\Omega\setminus\{0\}$ whenever ω does, a rearrangement following Theorem 1.15 immediately yields $E_k = (-1)^k \cdot E_k$, i.e.

1.18 Proposition. *If $k \geq 3$ is odd, then $E_k(\Omega) = 0$.*

We cannot yet role out *all* E_k being equal to zero, although this would probably surprise the unbiased reader. We will see in Theorem 4.3 that all E_k, for $k \geq 4$ even, are in generally non-zero. It is one of the fascinating results of the soon-to-be-

§ 1 Periods and lattices 23

developed theory of elliptic functions that these EISENSTEIN series are polynomials over $\mathbb{Q}$ in E_4 and E_6 (cf. Corollary 3.12). For example,

$$7E_8 = 3E_4^2 \quad \text{and} \quad 11E_{10} = 5E_4E_6.$$

This is analogous to the known results about the RIEMANN *zeta function*

$$\zeta(s) := \sum_{n=1}^{\infty} n^{-s}, \quad s \in \mathbb{R}, \ s > 1.$$

If we consider a subgroup "of rank 1" instead of a lattice, e.g. $\Omega = \mathbb{Z}$, then instead of (1.18) we get the lattice invariants $F_k := 2\zeta(k)$, $k \geq 2$ even, which according to L. EULER, are known to be rational multiples of π^k as pointed out in Corollary B.5. So they can be written as monomials in F_2:

$$5F_4 = F_2^2, \quad 35F_6 = 2F_2^3.$$

1.19 Exercises.
1) Let $0 \neq \omega \in \mathbb{C}$.
a) There is a sequence $(f_n)_{n \in \mathbb{N}}$ of linearly independent entire functions with associated period sets $\text{Per} f_n \supseteq \mathbb{Z}\omega$.
b) The set of entire functions f with $\text{Per} f \supseteq \mathbb{Z}\omega$ is a ring isomorphic to the ring of holomorphic functions on $\mathbb{C} \setminus \{0\}$.
2) Let $\Omega = \mathbb{Z}\omega_1 + \mathbb{Z}\omega_2$ be a lattice in $\mathbb{C}$. For $\omega = m_1\omega_1 + m_2\omega_2 \in \Omega$ the following assertions are equivalent:

(i) There exists an $\omega' \in \Omega$ with $\Omega = \mathbb{Z}\omega + \mathbb{Z}\omega'$.
(ii) m_1 and m_2 are coprime.
(iii) If $n \in \mathbb{Z}$, $n > 1$, then $\frac{1}{n}\omega \notin \Omega$.

3) Determine $m_1, m_2, m_3 \in \mathbb{Z}$, not all zero, such that

$$|m_1\sqrt{2} + m_2(\sqrt{3} + i\sqrt{5}) + m_3 i\sqrt{7}| < 1 \quad (*)$$

and for which $m_1^2 + m_2^2 + m_3^2$ is minimal.
4) Let $K = \mathbb{Q}(\sqrt{d})$, $d \in \mathbb{Z}$, $d < 0$ square-free, be an imaginary-quadratic number field of discriminant D and Ω its ring of integers, i.e. $\Omega = \mathbb{Z} + \mathbb{Z}(1 + \sqrt{d})/2$ if $D = d$, $d \equiv 1 \pmod{4}$, and otherwise $\Omega = \mathbb{Z} + \mathbb{Z}\sqrt{d}$, $D = 4d$. Then Ω is a lattice in $\mathbb{C}$ and the area of a period parallelogram is $\sqrt{|D|}/2$.
5) $SL(2; \mathbb{Z})$ is generated by the matrices $\begin{pmatrix} 1 & 0 \\ 1 & 1 \end{pmatrix}$ and $\begin{pmatrix} 1 & 1 \\ 0 & 1 \end{pmatrix}$.
6) Let $\Omega = \mathbb{Z}\omega_1 + \mathbb{Z}\omega_2$ be a lattice in $\mathbb{C}$ and δ the diameter (1.17). Then

$$\delta(\omega_1, \omega_2) = \max\{|\omega_1|, |\omega_2|, |\omega_1 + \omega_2|, |\omega_1 - \omega_2|\}.$$

7) Let $\Omega = \mathbb{Z}\omega_1 + \mathbb{Z}\omega_2$ be a lattice in $\mathbb{C}$, $\tau := \omega_1/\omega_2 \in \mathbb{C} \setminus \mathbb{R}$, $|\text{Re}\,\tau| \leq 1/2$, $|\tau| \geq 1$, (cf. section 2). Then

$$\delta(\omega_1, \omega_2) \leq \delta(\omega_1', \omega_2') \quad \text{for every basis} \quad (\omega_1', \omega_2') \quad \text{of } \Omega.$$

8) Let Ω be a lattice in $\mathbb{C}$ and $C \in \mathbb{R}$. Then

$$\sup\{\delta(w_1, w_2);\ w_1, w_2 \text{ is a basis of } \Omega\} = \infty.$$

9) Let Ω be a lattice in $\mathbb{C}$ and $\mu = \inf\{|\omega|;\ 0 \neq \omega \in \Omega\}$. Then

$$\#\{\omega \in \Omega;\ |\omega| = \mu\} \in \{2, 4, 6\}.$$

Give examples for each of the three cases.

10) Let $a, b \in \mathbb{R}$ both be positive. For $\rho > 0$ let

$$e(\rho) := \#\left\{(x, y) \in \mathbb{Z} \times \mathbb{Z};\ \left(\frac{x}{a}\right)^2 + \left(\frac{y}{b}\right)^2 \leq \rho\right\}$$

be the number of integral points within the ellipse. Then there exists $c > 0$ such that

$$|e(\rho) - ab\pi\rho^2| \leq c\rho \quad \text{for all } \rho \geq 1.$$

11) Another proof of the Convergence Lemma 1.17 is the following:

  (i) It suffices to consider the lattice $\Omega = \mathbb{Z}i + \mathbb{Z}$.
  (ii) Let be $\alpha > 1$. Then there exist $0 < c' < c$ such that all $0 \neq m \in \mathbb{Z}$ satisfy

$$c' \cdot |m|^{1-\alpha} \leq \sum_{n \in \mathbb{Z}} (m^2 + n^2)^{-\alpha/2} \leq c \cdot |m|^{1-\alpha}.$$

  (iii) Derive the Convergence Lemma 1.17 from (i) and (ii).

12) Let $\Omega = \mathbb{Z}\omega_1 + \mathbb{Z}\omega_2$ be a lattice in $\mathbb{C}$. Show that the series

$$\sum_{0 \neq n \in \mathbb{Z}} \frac{1}{(n\omega_2)^2} + 2 \sum_{m=1}^{\infty} \sum_{n \in \mathbb{Z}} \frac{1}{(m\omega_1 + n\omega_2)^2}$$

is convergent by using Lemma B.3.

13) Let $\Omega = \mathbb{Z}\omega_1 + \mathbb{Z}\omega_2$ be a lattice in $\mathbb{C}$. Then, for $\alpha > 2$,

$$\sum_{0 \neq \omega \in \Omega} |\omega|^{-\alpha} = \sum_{0 \neq g \in \mathbb{Z}^2} (g^{tr} S g)^{-\alpha/2}, \quad S = \begin{pmatrix} |\omega_1|^2 & \operatorname{Re}(\omega_1 \overline{\omega_2}) \\ \operatorname{Re}(\omega_1 \overline{\omega_2}) & |\omega_2|^2 \end{pmatrix}.$$

14) Let $k \in \mathbb{Z}$, $k > 2$. Then $E_k(\Omega) = 0$ if $\Omega = \mathbb{Z}i + \mathbb{Z}$ and $k \not\equiv 0 \pmod 4$ or if $\Omega = \mathbb{Z}\frac{1}{2}(1 + i\sqrt{3}) + \mathbb{Z}$ and $k \not\equiv 0 \pmod 6$.

§2 The field of elliptic functions

It is not yet clear whether there exist meromorphic functions f with $\operatorname{Per} f = \Omega$ for a given lattice Ω in $\mathbb{C}$ (cf. sect. 2 in §1). Surprisingly, under the assumption of the

§2 The field of elliptic functions

existence of one suitable such function, we can already describe all such functions. The proof of existence will be given in §3. In this paragraph, let $\Omega = \mathbb{Z}\omega_1 + \mathbb{Z}\omega_2$ always denote a lattice in $\mathbb{C}$.

1. First properties of elliptic functions. A meromorphic function f on $\mathbb{C}$ is called *elliptic* or *doubly periodic* with respect to Ω if Ω is contained in the set $\operatorname{Per} f$ of the periods of f (cf. sect. 2 in §1): $\Omega \subseteq \operatorname{Per} f$. This means

$$D_f + \omega = D_f \quad \text{for all} \quad \omega \in \Omega, \tag{2.1}$$

$$f(z + \omega) = f(z) \quad \text{for all} \quad \omega \in \Omega \quad \text{and} \quad z \in \mathbb{C} \setminus D_f. \tag{2.2}$$

If we interpret (2.2) in such a way that the equation is valid for all z and ω for which both sides make sense, then (2.1) is a consequence of (2.2). Conditions (2.1) and (2.2) are certainly satisfied if they are true for a basis of Ω. Let $\mathcal{E}(\Omega)$ denote the set of elliptic functions with respect to Ω.

For $0 \neq f \in \mathcal{M}$ and $c \in \mathbb{C}$ there exists, as is well known (cf. A.3), a LAURENT *series development* of the form

$$f(z) = \sum_{n \geq m} a_n(z-c)^n, \quad a_m \neq 0, \quad m \in \mathbb{Z}.$$

This series converges normally, hence locally uniformly in a punctured neighborhood of c. Here the *order of f at c* is given by $\operatorname{ord}_c f := m$ as in (1.3) and the *residue of f at c* is $\operatorname{res}_c f := a_{-1}$.

For $f \in \mathcal{E}(\Omega)$, $\omega \in \Omega$ and z from a suitable neighborhood of $c + \omega$, we have

$$f(z) = f(z - \omega) = \sum_{n \geq m} a_n(z - [c + \omega])^n$$

and therefore

$$\operatorname{ord}_{c+\omega} f = \operatorname{ord}_c f \quad \text{as well as} \quad \operatorname{res}_{c+\omega} f = \operatorname{res}_c f. \tag{2.3}$$

In particular, $c + \omega$, $\omega \in \Omega$, is a pole (or zero) of $f \in \mathcal{E}(\Omega)$, whenever c is. Since poles cannot have limit points in compact sets, we obtain

2.1 Proposition. *The set of elliptic functions $\mathcal{E}(\Omega)$ with respect to Ω forms a subfield of the field $\mathcal{M}$ of all meromorphic functions on $\mathbb{C}$. It contains the constant functions. Each $f \in \mathcal{E}(\Omega)$ has only finitely many poles in each period parallelogram.*

(2.1) and (2.2) yield the simple, but important

2.2 Lemma. *Given $f \in \mathcal{E}(\Omega)$, then*

$$f'(z) \quad \text{and} \quad g(z) := f(nz + w) \quad \text{with} \quad n \in \mathbb{Z}, \ w \in \mathbb{C} \ \text{fixed},$$

also belong to $\mathcal{E}(\Omega)$.

2. The four Theorems of LIOUVILLE. In 1847 J. LIOUVILLE (1809–1882) noticed that elliptic functions are subject to considerable restrictions that are not immediately apparent (*J. Reine Angew. Math.* **88**, 277–310 (1880)):

2.3 Liouville's First Theorem. *If $f \in \mathcal{E}(\Omega)$ is holomorphic, then f is constant.*

Proof Let P be a period parallelogram. Since the closure $\overline{P}$ of P is compact, there exists some $C > 0$ with $|f(z)| \leq C$ for $z \in \overline{P}$. If $z \in \mathbb{C}$ is arbitrary, then by Proposition 1.13, there is some $\omega \in \Omega$ with $z + \omega \in \overline{P}$. Thus

$$|f(z)| = |f(z + \omega)| \leq C,$$

such that f is bounded on $\mathbb{C}$. According to Liouville's classical Theorem A.12 f is constant. □

2.4 Liouville's Second Theorem. *If $f \in \mathcal{E}(\Omega)$ and P is a period parallelogram, then*

$$\sum_{c \in P} \operatorname{res}_c f = 0. \qquad (2.4)$$

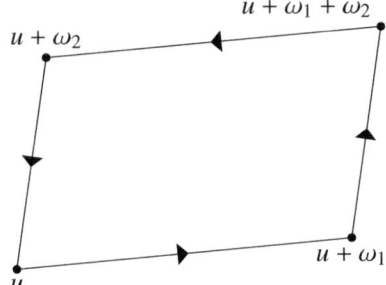

Figure 7: Path integration

Proof By (1.13) and Proposition 1.13 the sum (2.4) is finite, and it is independent of the choice of P due to (2.3). Let u be chosen such that there are no singularities on the boundary ∂P of P.
Now we integrate f over the boundary ∂P. The Residue Theorem A.5 leads to

$$\pm 2\pi i \sum_{c \in P} \operatorname{res}_c f$$

$$= \int_u^{u+\omega_1} f(z)\,dz + \int_{u+\omega_1}^{u+\omega_1+\omega_2} f(z)\,dz + \int_{u+\omega_1+\omega_2}^{u+\omega_2} f(z)\,dz + \int_{u+\omega_2}^{u} f(z)\,dz$$

$$= \int_u^{u+\omega_1} (f(z) - f(z+\omega_2))\,dz + \int_{u+\omega_2}^{u} (f(z) - f(z+\omega_1))\,dz \, .$$

But the right-hand side is zero because of $f \in \mathcal{E}(\Omega)$. □

2.5 Liouville's Third Theorem. *If $f \in \mathcal{E}(\Omega)$ is not constant, P is a period parallelogram and $w \in \mathbb{C}$, then*

$$\sum_{c \in P} \operatorname{ord}_c (f - w) = 0, \qquad (2.5)$$

i.e., counting with the appropriate multiplicities,

$$\text{the number of poles of } f = \text{number of roots of } f(z) = w$$
$$= \text{number of zeros of } f$$

§ 2 The field of elliptic functions

in P. In particular, each non-constant $f \in \mathcal{E}(\Omega)$ attains every complex value in P.

Proof By Lemma 2.2 and Proposition 2.1,

$$g(z) := \frac{f'(z)}{f(z) - w}$$

is again an elliptic function with respect to Ω and it satisfies $\mathrm{res}_c\, g = \mathrm{ord}_c\, (f - w)$. Therefore, the claim follows from LIUOVILLE's Second Theorem 2.4. Because f is not constant, $f - w$ has a root in P by (2.5), since $f - w$ has at least one pole in P by LIUOVILLE's First Theorem 2.3 and (2.3). □

2.6 LIOUVILLE's Fourth Theorem. If $0 \neq f \in \mathcal{E}(\Omega)$ and P is a period parallelogram, then

$$\sum_{c \in P} (\mathrm{ord}_c\, f) \cdot c \in \Omega. \qquad (2.6)$$

Proof We apply the method used in the proof of (2.4) to the integral

$$2\pi i \sum_{c \in P} (\mathrm{ord}_c\, f) \cdot c = \int_{\partial P} z \cdot \frac{f'(z)}{f(z)} dz$$

$$= \pm \left(\int_u^{u+\omega_1} z \cdot \frac{f'(z)}{f(z)} - (z + \omega_2) \cdot \frac{f'(z + \omega_2)}{f(z + \omega_2)} dz \right.$$

$$\left. + \int_{u+\omega_2}^u z \cdot \frac{f'(z)}{f(z)} - (z + \omega_1) \cdot \frac{f'(z + \omega_1)}{f(z + \omega_1)} dz \right)$$

$$= \pm \left(\omega_1 \cdot \int_u^{u+\omega_2} \frac{f'(z)}{f(z)} dz - \omega_2 \cdot \int_u^{u+\omega_1} \frac{f'(z)}{f(z)} dz \right).$$

Then $f(u) = f(u + \omega_j)$ leads to

$$\int_u^{u+\omega_j} \frac{f'(z)}{f(z)} dz \in 2\pi i \mathbb{Z} \quad \text{for } j = 1, 2,$$

hence (2.6). □

According to LIOUVILLE's Second Theorem 2.4, there are no elliptic functions with only one 1st order pole in P. So either we have to allow at least two 1st order poles or one pole of 2nd order with residue zero. Both approaches were carried out by K.T.W. WEIERSTRASS, but we first consider only the second case.

We reformulate LIOUVILLE's Fourth Theorem 2.6 using the notation of (1.11)

2.7 ABEL's Relation. Let $f \in \mathcal{E}(\Omega)$ be non-constant with zeros at $a_1, \ldots, a_r$ and poles at $b_1, \ldots, b_s$ in a period parallelogram P, where the multiplicity is indicated by the number of repetitions. Then $r = s$ and

$$a_1 + \ldots + a_r \equiv b_1 + \ldots + b_r (\mathrm{mod}\ \Omega). \tag{2.7}$$

This relation was already known to N.H. ABEL in 1826 ([1], vol. I, 145–211). Following a suggestion of JACOBI, (2.7) is therefore also called ABEL's *relation*. We call r the *order* of the elliptic function f. LIOUVILLE's First and Second Theorem 2.3 and 2.4 state that an elliptic function of order 0 is constant and that there is no elliptic function of order 1. We will see in §6 that $r \geq 2$ and (2.7) are also sufficient for the existence of an elliptic function with given roots and poles.

3. The Existence Theorem and first conclusions. As a first consequence of LIOUVILLE's Theorems, it is shown in this and the next section that we can obtain a description of all elliptic functions just from the hypothesis of the existence of *one suitable* elliptic function. There are no further analytic conclusions necessary. Only LIOUVILLE's Theorems are applied!

2.8 Existence Theorem. *There exists an elliptic function $\wp = \wp_\Omega$ that has exactly a pole of 2nd order in each point of Ω and is holomorphic otherwise. Its* LAURENT *development at 0 has the form*

$$\wp(z) = z^{-2} + a_1 z + \ldots. \tag{2.8}$$

The *proof* is postponed until the next paragraph (cf. sect. 1 in §3). Because of (2.3) it is clear that $\wp$ has residue zero at all poles. Because of LIOUVILLE's First Theorem 2.3, $\wp$ is uniquely determined as an elliptic function by (2.8). We call $\wp$ the WEIERSTRASS $\wp$-*function* (with respect to the lattice Ω).

2.9 Proposition. a) *$\wp$ is an even function, i.e. $\wp(-z) = \wp(z)$. Moreover, $a_1 = 0$ holds in (2.8).*
b) *$\wp'$ is an odd function that has poles of 3rd order exactly in the points of Ω and is holomorphic elsewhere.*

Proof a) The elliptic function $f(z) := \wp(-z) - \wp(z)$ is holomorphic at 0 with value 0 and therefore holomorphic everywhere. By LIOUVILLE's First Theorem 2.3, f is constant and therefore equal to 0 due to (2.8).
b) The claim follows from a) and the Existence Theorem 2.8. □

We can already describe all the roots of $\wp'$.

2.10 Lemma. *If $\omega \in \Omega$ but $\omega/2 \notin \Omega$, then $\omega/2$ is a simple root of $\wp'$, and every root of $\wp'$ is of this form.*

Proof Since $\wp'$ is an odd elliptic function, we have

$$\wp'(z + \omega) = \wp'(z) = -\wp'(-z).$$

If $\omega/2$ is not a pole of $\wp$ and $\wp'$, i.e. $\omega/2 \notin \Omega$, then we may set $z = -\omega/2$ and obtain

$$\wp'(\omega/2) = -\wp'(\omega/2), \quad \text{hence} \quad \wp'(\omega/2) = 0.$$

§ 2 The field of elliptic functions

Let ω_1, ω_2 be a basis of Ω. In the period parallelogram $P = \Diamond(\omega_1, \omega_2)$, then $\wp'$ has the roots

$$\omega_1/2, \ \omega_2/2, \ (\omega_1 + \omega_2)/2 . \tag{$*$}$$

By (2.8) $\wp'$ has only one 3rd order pole in P and it is at 0. According to LIOUVILLE's Third Theorem 2.5, the number of all roots of $\wp'$ in P (counted with their multiplicities) is equal to 3. Therefore, the roots $(*)$ are simple and they account for all the roots of $\wp'$ in P. If z is an arbitrary root of $\wp'$, then there exists some $\omega' \in \Omega$ with $z - \omega' \in P$. But then $z - \omega'$ is one of the points $(*)$ and thus z is of the form $\omega/2$ with $\omega \in \Omega$, but $\omega/2 \notin \Omega$. □

If f is an even function with $f(u) = 0$, then $f(-u) = 0$ also. In general we get the

2.11 Lemma. *Let P be a period parallelogram and $w \in \mathbb{C}$,*

$$w \neq \wp(\omega/2), \ \omega \in \Omega, \ \omega/2 \notin \Omega, \tag{2.9}$$

Then there are exactly two distinct points $u, v \in P$ with $\wp(u) = \wp(v) = w$. In this case $u + v \in \Omega$ holds. Conversely, if there are two distinct $u, v \in P$ with $\wp(u) = \wp(v) = w$, then (2.9) holds.

Proof We apply LIOUVILLE's Third Theorem 2.5 to $\wp$. According to (2.8), the number of roots of $\wp(z) = w$ in P (counted with their multiplicities) is equal to 2. We distinguish two cases:
a) There is only one $u \in P$ with $\wp(u) = w$. Then u is a double root of $\wp(z) = w$ and $\wp'(u) = 0$ follows. Because of (2.9) and Lemma 2.10 this case is excluded.
b) There are two distinct $u, v \in P$ with $\wp(u) = \wp(v) = w$. LIOUVILLE's Fourth Theorem 2.6 immediately yields $u + v \in \Omega$. □

Now we choose a basis ω_1, ω_2 of Ω, consider the attached period parallelogram $P := \Diamond(\omega_1, \omega_2)$ and use the standard notation for the *half-period values*

$$e_k := \wp(\omega_k/2), \ k = 1, 2, 3 \quad \text{with} \ \omega_3 := \omega_1 + \omega_2. \tag{2.10}$$

According to Lemma 2.10 and Lemma 2.11,

$$\wp(z) - e_k \text{ has exactly one double root in } P, \text{ namely at } z = \omega_k/2 \tag{2.11}$$

for $k = 1, 2, 3$ and

$$\wp(z) - w \quad \text{has two simple roots in } P \text{ for } w \neq e_1, e_2, e_3. \tag{2.12}$$

As $\omega_1/2, \omega_2/2, \omega_3/2$ are mutually distinct, (2.11) shows that

$$e_1, e_2, e_3 \quad \text{are also mutually distinct.} \tag{2.13}$$

These results already lead to the first *differential equation* for $\wp$:

2.12 Theorem. *For all $z \in \mathbb{C} \setminus \Omega$,*

$$\wp'^2(z) = 4 \cdot (\wp(z) - e_1) \cdot (\wp(z) - e_2) \cdot (\wp(z) - e_3).$$

Proof We consider $\wp'^2$ and the elliptic function

$$f(z) := 4 \cdot (\wp(z) - e_1) \cdot (\wp(z) - e_2) \cdot (\wp(z) - e_3).$$

According to (2.11), f has roots in P (in fact double roots) exactly at the points

$$\omega_1/2, \ \omega_2/2, \ \omega_3/2 = (\omega_1 + \omega_2)/2.$$

By Lemma 2.10, the same statement holds for $\wp'^2$. Since the poles in f cannot cancel, f has only one pole of order 6 at 0 in P. But by (2.8) the following holds

$$\wp(z) = z^{-2} + \ldots, \quad \wp'(z) = -2z^{-3} + \ldots, \quad \wp'^2(z) = 4z^{-6} + \ldots \quad (*)$$

and therefore $\wp'^2$ also has only one pole at $z = 0$ in P and it is of order 6. Thus $\wp'^2/f$ is an elliptic function without poles, i.e. constant by LIOUVILLE's First Theorem 2.3. Comparing the coefficient of z^{-6} with $(*)$ in the LAURENT series development around 0, we see that this constant is equal to 1. □

2.13 Remarks. a) The definition (2.10) depends, of course, on the choice of the basis of Ω. If the basis is transformed, the values e_1, e_2, e_3 are merely permuted.

b) It is also possible to construct a second form of the differential equation for $\wp$ described in sect. 3 in §3 if we are willing to consider the initially unkown coefficients of z^2 and z^4 in the LAURENT series development of $\wp$.

c) A simple function $f \in \mathcal{E}(\Omega)$ with two first order poles is e.g.

$$f(z) := \frac{\wp'(z) - \wp'(w)}{\wp(z) - \wp(w)}, \quad w \in \mathbb{C} \setminus \frac{1}{2}\Omega$$

(cf. Proposition 5.2). It is easy to see that f has only 1st order poles at the points of Ω and $-w + \Omega$. Another example of such an elliptic function is $1/\wp(z)$, whenever $e_1 e_2 e_3 \neq 0$. In this case we have two simple poles at the two different zeros of $\wp(z)$ in P.

4. The field $\mathcal{E}(\Omega)$. By (2.8) it is clear that the poles in a polynomial in $\wp$ do not cancel. Thus $\wp$ is not algebraic, i.e. it is transcendent over the field $\mathbb{C}$. Thus the field $\mathbb{C}(\wp)$ is isomorphic to the field of all rational functions over $\mathbb{C}$.

2.14 Theorem. a) *The even elliptic functions with respect to Ω are exactly the rational functions in $\wp$.*
b) $\mathcal{E}(\Omega) = \mathbb{C}(\wp)[\wp']$.
c) *The degree of field extension of $\mathcal{E}(\Omega)$ over $\mathbb{C}(\wp)$ is 2.*

Thus by Theorem 2.12, any $f \in \mathcal{E}(\Omega)$ can be uniquely written as

$$f = R(\wp) + Q(\wp) \cdot \wp', \tag{2.14}$$

§ 2 The field of elliptic functions

where R and Q are rational functions over $\mathbb{C}$. So we know the elliptic functions "as well as we know the function $\wp$".

Proof a) Let $f \in \mathcal{E}(\Omega)$ be even and non-constant. Let the order of f, i.e. the number of poles of f in the period parallelogram (counted with their multiplicities), be m. Let $P := \Diamond(\omega_1, \omega_2)$ be the corresponding fundamental parallelogram and $N := \{c \in P\,;\, f'(c) = 0\}$. Then N is finite.

(i) *The number m is even, $m = 2k$. For every complex number $u \notin f(N)$, there are mutually distinct points*

$$c_1, \ldots, c_k,\ c'_1, \ldots, c'_k \in P,\ c_j + c'_j \in \Omega \quad \text{for} \quad j = 1, \ldots, k, \qquad (*)$$

such that the roots of $f(z) = u$ are exactly the points $(*)$ with multiplicity 1.
By LIOUVILLE's Third Theorem 2.5 the number of roots of $f(z) = u$ is also equal to m. Let $c \in P$ be given such that $f(c) = u$. Since f is even, $f(-c) = u$ also holds and there is some $\omega \in \Omega$ with $c' = \omega - c \in P$ as well as $f(c') = u$. If c' were equal to c, we would obtain $f(c + z) = f(\omega - c + z) = f(-c + z) = f(c - z)$, hence $f'(c + z) = -f'(c - z)$ and thus $f'(c) = 0$, hence $u \in f(N)$ as a contradiction. Therefore, c and $c' = \omega - c$ are distinct and the roots of $f(z) = u$ occur in pairs. Each root of $f(z) = u$ has multiplicity 1 because of $u \notin f(N)$.

(ii) *f is a rational function in $\wp$.*
Now we choose $v \neq u$ not in $f(N)$ and use (i) to obtain

$$d_1, \ldots, d_k,\ d'_1, \ldots, d'_k,\ d_j + d'_j \in \Omega \quad \text{for} \quad j = 1, \ldots, k, \qquad (**)$$

such that the roots of $f(z) = v$ in P are exactly the points $(**)$ with multiplicity 1. The poles of f are canceled in the elliptic function

$$g(z) := \frac{f(z) - u}{f(z) - v}.$$

Thus $g(z)$ has

$$\begin{array}{l}\text{zeros in } P \text{ exactly at the points } (*), \text{ each of order 1,}\\ \text{and poles in } P \text{ exactly at the points } (**), \text{ each of order 1.}\end{array} \qquad (***)$$

As the points in $(*)$ and $(**)$ are mutually distinct,

$$c_j, c'_j, d_j, d'_j \notin \frac{1}{2}\Omega.$$

follows. Accordingly

$$h(z) := \frac{(\wp(z) - \wp(c_1)) \cdot \ldots \cdot (\wp(z) - \wp(c_k))}{(\wp(z) - \wp(d_1)) \cdot \ldots \cdot (\wp(z) - \wp(d_k))}$$

also fulfills the property $(***)$. The quotient g/h is therefore holomorphic, i.e. constant by LIOUVILLE's First Theorem 2.3. Thus $g \in \mathbb{C}(\wp)$ yields

$$f = \frac{vg - u}{g - 1} \in \mathbb{C}(\wp).$$

b) If $f \in \mathcal{E}(\Omega)$ is non-constant, we write

$$f = g + h\wp' \text{ with } g(z) := \frac{1}{2}(f(z) + f(-z)) \text{ and } h(z) = \frac{1}{2\wp'(z)}(f(z) - f(-z)).$$

Obviously, g and h belong to $\mathcal{E}(\Omega)$ and are even. Then, by a), g and h are rational functions in $\wp$.

c) The assertion follows from Theorem 2.12 because of $\wp' \notin \mathbb{C}(\wp)$. □

In algebraic formulation we obtain the

2.15 Corollary. *For independent indeterminates X, Y over $\mathbb{C}$*

$$\mathcal{E}(\Omega) \cong \mathbb{C}(X)[Y]/I(X,Y),$$

where $I(X, Y)$ denotes the principal ideal in $\mathbb{C}(X)[Y]$ generated by

$$Y^2 - 4(X - e_1)(X - e_2)(X - e_3).$$

Proof We define a homomorphism of rings

$$\Phi : \mathbb{C}(X)[Y] \to \mathcal{E}(\Omega) \text{ by } X \mapsto \wp, Y \mapsto \wp'.$$

According to Theorem 2.14, Φ is surjective. We now use division with remainder for $\varphi \in \mathbb{C}(X)[Y]$ in the form

$$\varphi(X, Y) = \left(Y^2 - 4(X - e_1)(X - e_2)(X - e_3)\right) \cdot q(X, Y) + r(X, Y)$$

with $q, r \in \mathbb{C}(X)[Y]$ and $r(X, Y) = r_1(X) + r_2(X) \cdot Y$ with $r_1(X), r_2(X) \in \mathbb{C}(X)$. Because of the differential equation in Theorem 2.12, φ is contained in the kernel of Φ if and only if $r(\wp, \wp') = 0$ holds. By (2.14), this means $r(X, Y) = 0$. This gives us ker $\Phi = I(X, Y)$ and the assumption follows from the homomorphism theorem for rings. □

According to Theorem 2.14, $\mathcal{E}(\Omega)$ has transcendence degree 1 over $\mathbb{C}$ in the algebraic sense (cf. J.S. MILNE [57]). Thus any two elements of $\mathcal{E}(\Omega)$ are algebraically dependent and we obtain the

2.16 Corollary. *Given two elliptic functions $f, g \in \mathcal{E}(\Omega)$, there exists a non-trivial polynomial $P(X, Y) \in \mathbb{C}[X, Y]$ satisfying*

$$P(f, g) = 0.$$

§ 2 The field of elliptic functions

As $\mathcal{E}(\Omega)$ is a field, $I(X,Y)$ is a maximal ideal in $\mathbb{C}(X)[Y]$ and hence the polynomial $Y^2 - 4(X - e_1)(X - e_2)(X - e_3)$ is irreducible in Y over $\mathbb{C}(X)$.

5*. Divisors. Given a mapping $\varphi : \mathbb{C}/\Omega \to \mathbb{Z}$ with finite support, the *degree* of φ is defined by

$$\deg \varphi := \sum_{c \in \mathbb{C}/\Omega} \varphi(c) \in \mathbb{Z}.$$

Now a *divisor* of $\mathbb{C}/\Omega$ is a mapping $\varphi : \mathbb{C}/\Omega \to \mathbb{Z}$ with finite support, that satisfies $\deg \varphi = 0$. The set $\mathrm{div}(\mathbb{C}/\Omega)$ of all divisors of $\mathbb{C}/\Omega$ forms an abelian group with respect to pointwise addition. Given $0 \neq f \in \mathcal{E}(\Omega)$ we can now assign the mapping

$$\varphi_f : \mathbb{C}/\Omega \longrightarrow \mathbb{Z}, \quad \varphi_f(c + \Omega) := \mathrm{ord}_c f.$$

By (2.3), this definition does not depend on the choice of the representative c. Then, because of LIOUVILLE's Third Theorem 2.5, φ_f is a divisor, the so-called *principal divisor* attached to $f \in \mathcal{E}(\Omega)$. We define a surjective homomorphism of groups

$$\Phi : \mathrm{div}(\mathbb{C}/\Omega) \longrightarrow \mathbb{C}/\Omega, \quad \Phi(\varphi) := \sum_{c \in \mathbb{C}/\Omega} \varphi(c) \cdot c,$$

and obtain

$$\mathrm{div}(\mathbb{C}/\Omega)/\ker \Phi \cong \mathbb{C}/\Omega.$$

For the proof that

$$\varphi_f \in \ker \Phi \quad \text{for all} \quad 0 \neq f \in \mathcal{E}(\Omega)$$

note that

$$\Phi(\varphi_f) = \sum_{c \in \mathbb{C}/\Omega} \varphi_f(c) \cdot c = \sum_{c \in P} (\mathrm{ord}_c f) \cdot (c + \Omega)$$

for any period parallelogram P of Ω. However, according to LIOUVILLE's Fourth Theorem 2.6, the right-hand side equals Ω. Conversely we will see in §6 that every $\varphi \in \ker \Phi$ is of the form $\varphi = \varphi_f$ for some $f \in \mathcal{E}(\Omega) \setminus \{0\}$.

2.17 Exercises.
1) An even $f \in \mathcal{E}(\Omega)$ already attains all its values in the triangle with vertices $0, \omega_1, \omega_2$.
2) Let f be an elliptic function that has exactly two simple poles in a and b in the period parallelogram P. Then $f(a + b - z) = f(z)$.
3) Describe all $f \in \mathcal{E}(\Omega)$ with simple poles exactly in the points $\omega/2$, $\omega \in \Omega \setminus 2\Omega$ and which are holomorphic elsewhere.
4) Let $\omega \in \Omega$ and $f \in \mathcal{E}(\Omega)$. If f is odd, then f has a pole or a zero at the point $z = \omega/2$ and it is of odd order. If f is even, then the order of f at the points $z = \omega/2$ is even.
5) Let $0 \neq f \in \mathcal{M}$ be such that for every $\omega \in \Omega$ there is a constant $c(\omega) \in \mathbb{C}$ with

$$f(z + \omega) = c(\omega) f(z) \quad \text{for all} \quad z \in \mathbb{C} \setminus D_f.$$

Then
(i) $\mathrm{ord}_{c+\omega} f = \mathrm{ord}_c f$ for all $c \in \mathbb{C}, \omega \in \Omega$.
(ii) $\sum_{c \in P} \mathrm{ord}_c f = 0$.
(iii) If f is an entire function, then there exist $a, b \in \mathbb{C}$ satisfying $f(z) = ae^{bz}$ for all $z \in \mathbb{C}$.
6) Let f be an entire function such that for every $\omega \in \Omega$ there is a constant $c(\omega) \in \mathbb{C}$ with $f(z+\omega) = f(z) + c(\omega)$. Then there exist $a, b \in \mathbb{C}$ with $f(z) = az + b$.
7) $\wp''(z) = 6\wp^2(z) + 2(e_1 e_2 + e_2 e_3 + e_3 e_1)$ holds for all $z \in \mathbb{C} \setminus \Omega$.
8) Determine a non-trivial polynomial $P(X, Y)$ with $P(\wp', \wp'') = 0$.
9) The degree of the field extension $\mathcal{E}(\Omega)$ over $\mathbb{C}(\wp')$ is 3.
10) We strengthen Remark 2.13 in the following way: for e_1, e_2, e_3 we define

$$\overline{e}_1 = (1, 0), \quad \overline{e}_2 = (0, 1), \quad \overline{e}_3 = (1, 1) \in (\mathbb{Z}/2\mathbb{Z})^2.$$

If $M = \begin{pmatrix} a & b \\ c & d \end{pmatrix} \in \mathrm{GL}(2; \mathbb{Z})$ and $\omega_1' = a\omega_1 + b\omega_2$, $\omega_2' = c\omega_1 + d\omega_2$ is another basis, consider $\overline{M} \in \mathrm{GL}(2; \mathbb{Z}/2\mathbb{Z})$, which is obtained from M by reducing all the components mod 2. $\overline{M}$ permutes the set $\{\overline{e}_1, \overline{e}_2, \overline{e}_3\}$. If we denote the corresponding elements of the new basis with a prime, then $e_j' = \overline{e}_j \cdot \overline{M}$, $j = 1, 2, 3$.
11) Consider $f(z)$ from Remark 2.13c) and calculate $f(w)$ and the residues of f at the points $z = 0$ and $z = -w$.
12) If $a_0, \ldots, a_n \in \mathbb{C}, a_1 = 0$, then there exists exactly one $f \in \mathcal{E}(\Omega)$ mit $D_f \subseteq \Omega$ and the LAURENT series development $a_n z^{-n} + \ldots + a_0 + \ldots$ at 0.
13) Let ω_1, ω_2 be a basis of the lattice Ω and $f \in \mathcal{E}(2\Omega)$. Then

$$g(z) := f(z) + f(z + \omega_1) + f(z + \omega_2) + f(z + \omega_1 + \omega_2)$$

belongs to $\mathcal{E}(\Omega)$. Which elliptic function is this, when f is the corresponding $\wp$-function?
14) Let $a, b \in \mathbb{C}$. There is an elliptic function $f \in \mathcal{E}(\Omega)$ that is holomorphic in a and b and satisfies $f(a) \neq f(b)$ if and only if $a \not\equiv b \pmod{\Omega}$.
15) There does not exist an $f \in \mathcal{E}(\Omega)$ with $\mathcal{E}(\Omega) = \mathbb{C}(f)$.
16) Give an invariant description (i.e. independent of (2.14)) of the GALOIS group of the field extension $\mathcal{E}(\Omega)/\mathbb{C}(\wp)$.

§ 3 The WEIERSTRASS $\wp$-Function

In this paragraph, let $\Omega = \mathbb{Z}\omega_1 + \mathbb{Z}\omega_2$ always be a lattice in $\mathbb{C}$.

1. Theorem on the construction of the $\wp$-function. For later purposes, we formulate the statement a bit more generally. For this we need the

3.1 Proposition. *Let $C \subseteq \{(\omega_1, \omega_2) \in \mathbb{C} \times \mathbb{C}; \ \omega_2 \neq 0, \ \omega_1/\omega_2 \notin \mathbb{R}\}$ be compact. Then there are positive constants α and β such that*

$$\alpha \cdot |m_1 i + m_2| \leq |m_1 \omega_1 + m_2 \omega_2| \leq \beta \cdot |m_1 i + m_2|$$

§ 3 The WEIERSTRASS $\wp$-Function

for all $m_1, m_2 \in \mathbb{R}$ and $(\omega_1, \omega_2) \in C$.

Proof By homogeneity we may assume $m_1^2 + m_2^2 = 1$, so $|m_1 i + m_2| = 1$. The continuous function $(\omega_1, \omega_2, m_1, m_2) \mapsto |m_1 \omega_1 + m_2 \omega_2|$ attains its minimum α and its maximum β on the compact set $C \times \{(m_1, m_2) \in \mathbb{R} \times \mathbb{R} \, ; \, m_1^2 + m_2^2 = 1\}$. As ω_1, ω_2 are linearly independent over $\mathbb{R}$, we get $m_1 \omega_1 + m_2 \omega_2 \neq 0$ for all $(m_1, m_2) \neq (0, 0)$. Thus α and β are positive. □

As an application, we obtain the

3.2 Convergence Theorem for the $\wp$-Function. *The series*

$$\wp(z; \omega_1, \omega_2) := z^{-2} + \sum_{0 \neq \omega \in \mathbb{Z}\omega_1 + \mathbb{Z}\omega_2} \left((z - \omega)^{-2} - \omega^{-2} \right) \tag{3.1}$$

converges absolutely uniformly on each compact subset of

$$\{(z, \omega_1, \omega_2) \in \mathbb{C} \times \mathbb{C} \times \mathbb{C} \, ; \, \omega_2 \neq 0, \, \omega_1/\omega_2 \notin \mathbb{R}, \, z \notin \mathbb{Z}\omega_1 + \mathbb{Z}\omega_2\}. \tag{3.2}$$

Proof Let C be a compact subset of (3.2). We choose $\rho > 0$ and a compact subset $C' \subseteq \{(\omega_1, \omega_2) \in \mathbb{C} \times \mathbb{C} \, ; \, \omega_2 \neq 0, \, \omega_1/\omega_2 \notin \mathbb{R}\}$ such that

$$C \subseteq C_\rho \times C', \quad C_\rho = \{z \in \mathbb{C} \, ; \, |z| \leq \rho\}.$$

For C' we choose α according to Proposition 3.1. Then all $(z, \omega_1, \omega_2) \in C$ and all $(m_1, m_2) \in \mathbb{Z} \times \mathbb{Z}$ with $|m_1 i + m_2| \geq (\rho + 1)/\alpha$ satisfy

$$|\omega| \geq \rho + 1 \quad \text{for} \quad \omega = m_1 \omega_1 + m_2 \omega_2$$

and

$$\left| \frac{1}{(z-\omega)^2} - \frac{1}{\omega^2} \right| = \left| \frac{2z\omega - z^2}{\omega^2(z-\omega)^2} \right| = \left| \frac{2 - z/\omega}{(1 - z/\omega)^2} \right| \cdot \frac{|z|}{|\omega|^3}$$

$$\leq \frac{3}{(1 - \rho/(\rho+1))^2} \cdot \frac{\rho}{|\omega|^3} \leq \frac{3\rho(\rho+1)^2}{\alpha^3 \cdot |m_1 i + m_2|^3}.$$

Since there are only finitely many $(m_1, m_2) \in \mathbb{Z} \times \mathbb{Z}$ with $|m_1 i + m_2| \leq (\rho + 1)/\alpha$, the assertion follows from the absolute convergence of the EISENSTEIN series $E_3(\mathbb{Z}i + \mathbb{Z})$ by the Convergence Lemma 1.17. □

For a fixed lattice Ω, we analogously obtain the

3.3 Lemma. *For $k \in \mathbb{N}, k \geq 3$, the series*

$$\sum_{\omega \in \Omega} (z - \omega)^{-k}$$

converges absolutely uniformly on every compact subset of $\mathbb{C} \setminus \Omega$.

Now we obtain the announced

3.4 Construction Theorem for the $\wp$-Function. *The series*

$$\wp(z) := \wp_\Omega(z) := z^{-2} + \sum_{0 \neq \omega \in \Omega} \left((z - \omega)^{-2} - \omega^{-2}\right), \quad z \in \mathbb{C} \setminus \Omega, \qquad (3.3)$$

converges absolutely uniformly by on every compact subset of $\mathbb{C}$ that does not contain a lattice point. The function $\wp$ is an even elliptic function with respect to Ω, it has poles of 2nd order in the lattice points of Ω with residue 0, and it is holomorphic in $\mathbb{C} \setminus \Omega$. The LAURENT *series development at 0 has the form*

$$\wp(z) = z^{-2} + a_2 z^2 + \ldots. \qquad (3.4)$$

This simultaneously proves the Existence Theorem 2.8.

We call $\wp(z)$ the WEIERSTRASS $\wp$-*function* (with respect to the lattice Ω). The idea of defining an elliptic function by a sum over all lattice points and thus making the double periodicity apparent goes back to F.G.M. EISENSTEIN (1823–1852) and K.T.W. WEIERSTRASS (1815–1897). See also the historical note in sect. 7. From postscripts we can learn that in WEIERSTRASS' lectures the "WEIERSTRASS $\wp$" gradually developed from an ordinary p.

Proof Use the abbreviations

$$f_\omega(z) := (z - \omega)^{-2} - \omega^{-2} \text{ for } 0 \neq \omega \in \Omega \text{ and } C_\rho := \{z \in \mathbb{C}; |z| \leq \rho\}, \rho > 0.$$

The absolute compact uniform convergence of the series (3.3) already follows from the Convergence Theorem 3.2.

Assertion 1. The series (3.3) represents a meromorphic function on $\mathbb{C}$ that has poles of 2nd order with residue 0 exactly in the points of Ω.

For the *proof*, let $\rho > 0$. It follows that

$$\wp(z) = z^{-2} + \sum_{|\omega| < \rho+1} f_\omega(z) + \sum_{|\omega| \geq \rho+1} f_\omega(z).$$

Here the first sum is finite and meromorphic on C_ρ, whereas the second is holomorphic on C_ρ according to the Convergence Theorem 3.2. So $\wp|_{C_\rho}$ has 2nd order poles with residue 0 exactly in the points of $\Omega \cap C_\rho$. □

Assertion 2. $\wp$ is an even function and (3.4) holds.

For the *proof* we replace ω by $-\omega$ in the sum (3.3). Because of the absolute convergence $\wp(-z) = \wp(z)$ follows. Now we note that $f_\omega(0) = 0$ for $\omega \neq 0$ and see that the LAURENT series development has the constant term 0. □

Assertion 3. $\wp(z + \omega) = \wp(z)$ holds for all $\omega \in \Omega$ and $z \in \mathbb{C}\setminus\Omega$.

For the *proof*, the Convergence Theorem 3.2 leads to

$$\wp'(z) = -2 \cdot \sum_{\omega \in \Omega} (z - \omega)^{-3}, \quad z \in \mathbb{C} \setminus \Omega,$$

§ 3 The WEIERSTRASS $\wp$-Function

and this series is absolutely convergent according to Lemma 3.3. Thus $\wp'$ is an elliptic function and $\wp'(z+\omega) = \wp'(z)$ follows for $\omega \in \Omega$. Therefore, if ω_1, ω_2 is a basis of Ω, then $\wp(z + \omega_j) = \wp(z) + C_j$ with constants C_j for $j = 1, 2$. Setting $z = -\omega_1/2$ or $z = -\omega_2/2$ we obtain $C_1 = C_2 = 0$, because $\wp$ is even. Thus $\wp(z + \omega_j) = \wp(z)$ holds for $j = 1, 2$ and $\wp$ has all $\omega \in \Omega$ as periods. □

3.5 Remarks. a) Knowing MITTAG–LEFFLER's Theorem (cf. J.B. CONWAY [13], VIII 3.2), we note that here we have repeated a proof of a special case of that theorem. Historically, MITTAG–LEFFLER's Theorem was modeled after the WEIERSTRASS construction of the $\wp$-function, however.
b) In the above assertion 3, the detour via $\wp'$ cannot be avoided without additional considerations. We owe to J. ELSTRODT the hint at a direct proof based on a careful consideration of the difference from (3.3) for $\wp(z + \omega) - \wp(z)$. For this, compare H. HANCOCK [35], Art. 270.

2. The LAURENT series development. As in (1.18), consider the EISENSTEIN series

$$E_k := E_k(\Omega) := \sum_{0 \neq \omega \in \Omega} \omega^{-k} \quad \text{for even } k \geq 4. \tag{3.5}$$

According to Proposition 1.18, these series are zero for odd $k \geq 3$. Finally we set

$$\mu := \mu(\Omega) := \min\{|\omega| \,;\, 0 \neq \omega \in \Omega\}$$

and obtain

3.6 Theorem. *For $z \in \mathbb{C}$ with $0 < |z| < \mu(\Omega)$,*

$$\wp(z) = z^{-2} + \sum_{n=2}^{\infty} (2n-1) E_{2n} \cdot z^{2n-2} = z^{-2} + 3E_4 \cdot z^2 + 5E_6 \cdot z^4 + \ldots . \tag{3.6}$$

Proof Due to

$$\frac{1}{(1-t)^2} = \frac{d}{dt}\left(\frac{1}{1-t}\right) = \sum_{m=1}^{\infty} m t^{m-1}, \quad |t| < 1,$$

we have

$$\frac{1}{(z-\omega)^2} - \frac{1}{\omega^2} = \frac{1}{\omega^2}\left(\frac{1}{(1-z/\omega)^2} - 1\right) = \sum_{m=2}^{\infty} m \cdot \frac{z^{m-1}}{\omega^{m+1}}, \quad |z| < \mu,$$

for $\omega \neq 0$ and therefore

$$\wp(z) = z^{-2} + \sum_{0 \neq \omega \in \Omega}\left(\sum_{m=2}^{\infty} m \cdot \frac{z^{m-1}}{\omega^{m+1}}\right), \quad 0 < |z| < \mu. \tag{*}$$

Because

$$\left| m \cdot \frac{z^{m-1}}{\omega^{m+1}} \right| \leq \gamma m \left(\frac{|z|}{\gamma} \right)^{m-1} \cdot |\omega|^{-3}$$

and due to the Convergence Lemma 1.17, the series $(*)$ is absolutely convergent in z and ω. By Theorem 1.15, we may rearrange the series and get

$$\wp(z) = z^{-2} + \sum_{m \geq 2} m E_{m+1} \cdot z^{m-1}, \quad 0 < |z| < \mu.$$

However, by Proposition 1.18 this is (3.6). □

3. The second differential equation. As already announced in §1, we obtain

3.7 Theorem. *The* WEIERSTRASS *$\wp$-function satisfies the differential equation*

$$\wp'^2 = 4\wp^3 - g_2 \wp - g_3, \qquad (3.7)$$

where g_2 und g_3 are defined by

$$g_2 := g_2(\Omega) := 60\, E_4(\Omega) \quad \text{and} \quad g_3 := g_3(\Omega) := 140\, E_6(\Omega). \qquad (3.8)$$

g_2 and g_3 are called the WEIERSTRASS *invariants* of the lattice Ω. We use the LANDAU symbol and write $O(z^k)$ for a function $f(z)$ that satisfies $|f(z)| \leq C \cdot |z|^k$ with a suitable C for z in a neighborhood of 0.

Proof Starting from

$$\wp(z) = z^{-2} + 3E_4 \cdot z^2 + 5E_6 \cdot z^4 + O(z^6)$$

in (3.6), we calculate

$$\begin{aligned}
\wp^2(z) &= z^{-4} + 6E_4 + 10E_6 \cdot z^2 + O(z^3), \\
\wp^3(z) &= z^{-6} + 9E_4 \cdot z^{-2} + 15E_6 + O(z), \\
\wp'(z) &= -2 \cdot z^{-3} + 6E_4 \cdot z + 20E_6 \cdot z^3 + O(z^4), \\
\wp'^2(z) &= 4 \cdot z^{-6} - 24E_4 \cdot z^{-2} - 80E_6 + O(z).
\end{aligned}$$

Using (3.8), we obtain

$$\wp'^2(z) - 4\wp^3(z) + g_2 \wp(z) + g_3 = O(z). \qquad (*)$$

Here, the left-hand side belongs to $\mathcal{E}(\Omega)$ and has poles at most where $\wp$ or $\wp'$ has a pole, hence in Ω. By $(*)$, however, the left-hand side is holomorphic at 0 and hence everywhere. LIOUVILLE's First Theorem 2.3 shows that this function is constant. By $(*)$ again, this constant is equal to 0. □

Differentiating (3.7), we have the following

3.8 Corollary. *It holds*

$$2\wp'' = 12\wp^2 - g_2.$$

3.9 Corollary. *If $k \in \mathbb{N}$, then*

§3 The Weierstrass $\wp$-Function

$$\wp^{(k)} \in \mathbb{Z}[E_4, E_6, \wp] + \mathbb{Z}[E_4, E_6, \wp]\wp'.$$

Proof The claim is true for $k = 0$ and 1. Because of (3.8), the statement follows for $k = 2$ from Corollary 3.8. Now an induction yields the assertion. □

3.10 Corollary. *For $f \in \mathcal{E}(\Omega)$, the following assertions are equivalent:*

(i) f is holomorphic in $\mathbb{C} \setminus \Omega$.
(ii) $f \in \mathbb{C}[\wp] + \mathbb{C}[\wp] \cdot \wp'$.

Proof (i) $\Longrightarrow$ (ii): Subtracting suitable $\alpha\wp^n$ resp. $\alpha\wp^n \cdot \wp'$, $\alpha \in \mathbb{C}$, $n \in \mathbb{N}$, from f, we can successively lower the order of the poles in the lattice points. Finally, note that $\mathrm{res}_0 f = 0$ according to Liouville's Second Theorem 2.4.
(ii) $\Longrightarrow$ (i): Clear. □

3.11 Corollary. *For $n \geq 4$ the following recursion formula holds:*

$$(n-3)(2n+1)(2n-1)E_{2n} = 3 \cdot \sum_{\substack{p \geq 2, q \geq 2 \\ p+q=n}} (2p-1)(2q-1)E_{2p}E_{2q}. \tag{3.9}$$

Proof Substitute the Laurent series development (3.6) in $\wp'' + 30E_4 = 6\wp^2$ according to Corollary 3.8:

$$30E_4 + \sum_{n \geq 2} (2n-1)(2n-2)(2n-3)E_{2n}z^{2n-4}$$
$$= 12 \sum_{n \geq 2} (2n-1)E_{2n}z^{2n-4} + 6 \sum_{p \geq 2} \sum_{q \geq 2} (2p-1)(2q-1)E_{2p}E_{2q}z^{2p+2q-4}.$$

A comparison of the coefficients immediately yields the assertion. □

$$7E_8 = 3E_4^2, \quad 11E_{10} = 5E_4E_6, \quad 143E_{12} = 42E_4E_8 + 25E_6^2 = 18E_4^3 + 25E_6^2$$

are particular examples of (3.9) and an induction yields

3.12 Corollary. *If $k \geq 8$, then*

$$E_k \in \mathbb{Q}[E_4, E_6].$$

3.13 Corollary. *Let Ω be a lattice in $\mathbb{C}$ with associated Weierstrass invariants g_2 and g_3. Let $D \subseteq \mathbb{C}$ be an open domain and f a non-constant meromorphic solution of the differential equation*

$$f'^2 = 4f^3 - g_2 f - g_3$$

on D. Then f is given by $f(z) = \wp(z + w)$, $z \in D$, with suitable $w \in \mathbb{C}$. If $f \in \mathcal{M}$ is such a solution, then Ω is the period lattice of f. The lattice Ω is uniquely determined by $g_2(\Omega)$ and $g_3(\Omega)$, hence also by $E_4(\Omega)$ and $E_6(\Omega)$.

Proof Let f be a non-constant solution of the given differential equation which is meromorphic in D. If f is holomorphic in a disk $U \subseteq D$ around u and f' is non-zero in U, then with a suitable choice of a root we have $f' = \sqrt{4f^3 - g_2 f - g_3}$. By Lemma 2.11, we can choose some $w \in \mathbb{C}$ with $\wp(w+u) = f(u)$ and may furthermore assume $\wp'(w+u) = f'(u)$ by replacing w by $-w - 2u$ if necessary. The functions $f(z)$ and $g(z) := \wp(z+w)$ satisfy the same first order differential equation and have the same value at the point u. Then $f(z) = g(z)$ follows for all $z \in U$ from the existence and uniqueness theorem (cf. W. WALTER [85]). The Identity Theorem A.1 implies $f(z) = g(z)$ for all $z \in D$. The final assertion follows from the fact that for the $\wp$-function the period lattice is equal to the set of poles, according to the Construction Theorem 3.4. □

Instead of (3.7), we can more generally look for solutions $w = f(z)$ of the so-called *binomial differential equation*

$$w'^n = R(z, w) \tag{3.10}$$

for given $n \in \mathbb{N}$. Here we have

3.14 Theorem of MALMQUIST and YOSIDA. *If* (3.10) *possesses a meromorphic and transcendental solution on* $\mathbb{C}$, *then* $R(z, w)$ *is a polynomial in* w *of degree* $\leq 2n$. A proof can be found in E. HILLE [38], Theorem 4.6.4. A classification of binomial differential equations is described by N. STEINMETZ, *Math. Ann.* **244**, 263–274 (1979).

4. A comparison of the differential equations. Besides

$$\wp'^2 = 4\wp^3 - g_2 \wp - g_3, \tag{3.11}$$

the differential equation

$$\wp'^2 = 4(\wp - e_1)(\wp - e_2)(\wp - e_3) \tag{3.12}$$

was derived in Theorem 2.12. Here, e_1, e_2, e_3 are given by

$$e_k = \wp(\omega_k/2), \quad k = 1, 2, 3, \quad \omega_3 := \omega_1 + \omega_2, \tag{3.13}$$

when ω_1, ω_2 is a basis of Ω due to (2.10). Since $\wp$ attains more than three different values, a comparison of the polynomials in $\mathbb{C}[X]$ yields the

3.15 Theorem. *It holds*

$$4X^3 - g_2 X - g_3 = 4(X - e_1)(X - e_2)(X - e_3).$$

Then Corollary 2.15 implies

3.16 Corollary. *If* X, Y *are independent indeterminates over* $\mathbb{C}$,

$$\mathcal{E}(\Omega) \cong \mathbb{C}(X)[Y]/I(X, Y),$$

where $I(X, Y)$ *denotes the principal ideal in* $\mathbb{C}(X)[Y]$ *generated by the polynomial* $Y^2 - 4X^3 + g_2 X + g_3$.

A comparison of the coefficients in the Theorem 3.15 yields

§ 3 The WEIERSTRASS $\wp$-Function

3.17 Corollary. *It holds*
$$0 = e_1 + e_2 + e_3, \tag{3.14}$$
$$g_2 = -4(e_1e_2 + e_2e_3 + e_3e_1), \tag{3.15}$$
$$g_3 = 4e_1e_2e_3. \tag{3.16}$$

3.18 Corollary. *It holds*
$$g_2^3 - 27g_3^2 = 16(e_1 - e_2)^2(e_2 - e_3)^2(e_3 - e_1)^2 \neq 0.$$

Proof Using (3.14) and (3.15), we first get
$$g_2 = 2(e_1^2 + e_2^2 + e_3^2) \quad \text{and} \quad g_2^2 = 16(e_1^2 e_2^2 + e_2^2 e_3^2 + e_3^2 e_1^2). \tag{$*$}$$
Then (3.14) and (3.15) yield
$$2(e_1 - e_2)^2 = 2(e_1^2 + e_2^2) - 4e_1e_2 = 2g_2 - 2e_3^2 + 4e_3(e_1 + e_2) = 2g_2 - 6e_3^2,$$
hence
$$(e_1 - e_2)^2 = g_2 - 3e_3^2.$$
Since the relations resulting from cyclic permutations of e_1, e_2, e_3 are also valid, we have
$$16(e_1 - e_2)^2(e_2 - e_3)^2(e_3 - e_1)^2 = 16(g_2 - 3e_1^2)(g_2 - 3e_2^2)(g_2 - 3e_3^2)$$
$$= 16g_2^3 - 3 \cdot 16g_2^2(e_1^2 + e_2^2 + e_3^2) + 9 \cdot 16g_2(e_1^2 e_2^2 + e_2^2 e_3^2 + e_3^2 e_1^2) - 27 \cdot 16e_1^2 e_2^2 e_3^2.$$
However, because of $(*)$ and (3.16), the right-hand side is equal to $g_2^3 - 27g_3^2$. According to (2.13), e_1, e_2, e_3 are mutually distinct. □

$$\Delta := \Delta(\Omega) := g_2^3 - 27g_3^2 \tag{3.17}$$

is called the *discriminant* and

$$j := j(\Omega) := (12g_2)^3/\Delta \tag{3.18}$$

the *absolute invariant* of the lattice Ω. An application of Corollary 3.17 and Corollary 3.18 yields

3.19 Corollary. *It holds*
$$j = -4 \cdot 12^3 \cdot \frac{(e_1e_2 + e_2e_3 + e_3e_1)^3}{(e_1 - e_2)^2(e_2 - e_3)^2(e_3 - e_1)^2}.$$

3.20 Corollary. *If* $\lambda := \dfrac{e_2 - e_3}{e_1 - e_3}$, *then*

$$j = 256 \cdot \frac{(1 - \lambda + \lambda^2)^3}{\lambda^2(1-\lambda)^2}.$$

3.21 Remarks. a) The discriminant Δ is (up to a factor) also also the discriminant of the polynomial $f(X) := 4X^3 - g_2 X - g_3$ in the algebraic sense (cf. S. LANG [53], V, § 10): There the discriminant of the polynomial f (except for a factor) is defined as the *resultant* of f and f', i.e. by

$$\det \begin{pmatrix} 4 & 0 & -g_2 & -g_3 & 0 \\ 0 & 4 & 0 & -g_2 & -g_3 \\ 12 & 0 & -g_2 & 0 & 0 \\ 0 & 12 & 0 & -g_2 & 0 \\ 0 & 0 & 12 & 0 & -g_2 \end{pmatrix} = -64\Delta.$$

b) The notation "absolute invariant" will be justified in §4.
c) λ in Corollary 3.20 already occurs in the Introduction in the transition from the WEIERSTRASS form to the LEGENDRE normal form of elliptic integrals

5. Conjugation stable lattices. A lattice Ω is called *conjugation stable* if $\overline{w} \in \Omega$ for all $w \in \Omega$, i.e. $\Omega = \overline{\Omega}$ holds. The most important examples of conjugation stable lattices are the following
a) the *rectangular lattice* $\Omega = \mathbb{Z}\omega_1 + \mathbb{Z}\omega_2$ with $\frac{1}{i}\omega_1, \omega_2 \in \mathbb{R}$,
b) the *hexagonal lattice* $\Omega = \mathbb{Z}\rho + \mathbb{Z}$ with $\rho = \frac{1}{2}(1 + i\sqrt{3})$.

We directly obtain the

3.22 Proposition. *If Ω is a conjugation stable lattice, then*

$$\overline{\wp_\Omega(z)} = \wp_\Omega(\overline{z}) \quad \text{and} \quad \overline{\wp'_\Omega(z)} = \wp'_\Omega(\overline{z}) \text{ for all } z \in \mathbb{C} \setminus \Omega.$$

Especially, we have for $z \in \mathbb{C} \setminus \Omega$:
a) $\wp_\Omega(z)$ *is real for* $z \in \mathbb{R}$ *and* $z \in i\mathbb{R}$,
b) $\wp'_\Omega(z)$ *is real for* $z \in \mathbb{R}$ *and purely imaginary for* $z \in i\mathbb{R}$.

We characterize conjugation stable lattices in the following

3.23 Theorem. *For a lattice $\Omega = \mathbb{Z}\omega_1 + \mathbb{Z}\omega_2$ the following assertions are equivalent:*

(i) $g_2(\Omega)$ *and* $g_3(\Omega)$ *are both real.*
(ii) *All* $E_k(\Omega)$, $k \geq 4$ *even, are real.*
(iii) *Among the quantities* e_1, e_2, e_3, *either two are complex conjugate and the third is real or all three are real.*
(iv) Ω *is conjugation stable.*

Proof (i) $\iff$ (ii): Use (3.8) and Corollary 3.12.
(i) $\iff$ (iv): The assertion follows from $\overline{g_2(\Omega)} = g_2(\overline{\Omega})$, $\overline{g_3(\Omega)} = g_3(\overline{\Omega})$ and the fact that Ω is uniquely determined by g_2 and g_3 according to Corollary 3.13.
(i) $\implies$ (iii): By Proposition 3.15, the values e_1, e_2, e_3 are exactly the roots of the real polynomial $4X^3 - g_2 X - g_3$. Therefore, at least one root is real.

§ 3 The WEIERSTRASS ℘-Function

(iii) ⟹ (i): Apply (3.15) and (3.16). □

Using conjugation stable lattices, the elliptic integrals in WEIERSTRASS normal form

$$\int \frac{dt}{\sqrt{q(t)}}, \quad q(t) = 4t^3 - c_2 t - c_3,$$

can be calculated for real c_2, c_3 (cf. §4).

6. The mapping defined by ℘ for a rectangular lattice. In this section, let Ω always be a rectangular lattice and

$$(\omega_1, \omega_2) \quad \text{a basis of } \Omega \text{ with } \tfrac{1}{i}\omega_1 > 0 \text{ and } \omega_2 > 0. \tag{3.19}$$

It is easy to see that ω_1 and ω_2 are uniquely determined.

3.24 Proposition. *Let $z \in \mathbb{C} \setminus \Omega$.*
a) *$\wp(z)$ is real if and only if there exists some $\omega \in \Omega$ satisfying*

$$z \in \frac{\omega}{2} + \mathbb{R} \quad or \quad z \in \frac{\omega}{2} + i\mathbb{R}. \tag{3.20}$$

b) *If $z \in \frac{\omega}{2} + \mathbb{R}$, $\omega \in \Omega$, then $\wp'(z)$ is real. If $z \in \frac{\omega}{2} + i\mathbb{R}$, $\omega \in \Omega$, then $\wp'(z)$ is purely imaginary.*

Proof $\tfrac{1}{2}(\omega - \overline{\omega}) \in \Omega$ holds due to (3.19). Then

$$\overline{\wp\left(\frac{\omega}{2} + z\right)} = \wp\left(\frac{\omega}{2} + \overline{z}\right), \quad \overline{\wp'\left(\frac{\omega}{2} + z\right)} = \wp'\left(\frac{\omega}{2} + \overline{z}\right) \tag{3.21}$$

follows from Proposition 3.22. As ℘ is even and ℘′ is odd, ℘ and ℘′ attain only real resp. purely imaginary values at the points (3.20).

Now let $z \in \Diamond(\omega_1, \omega_2)$ be such that z is not of the form (3.20). If $\wp(z)$ were real, then ℘ would take the same value at the four mutually distinct points

$$z, \ \omega_1 + \overline{z}, \ \omega_1 + \omega_2 - z, \ \omega_2 - \overline{z}$$

in $\Diamond(\omega_1, \omega_2)$ because of (3.21). This contradicts Lemma 2.11. □

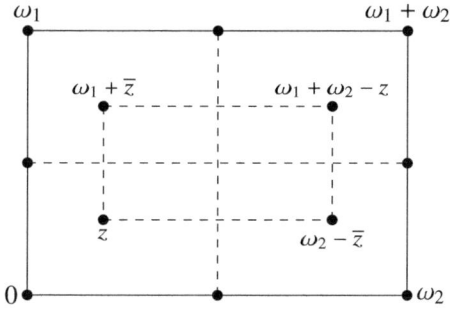

Figure 8: Rectangular lattice

3.25 Theorem. *The interior of the rectangle $0, \omega_1/2, (\omega_1+\omega_2)/2, \omega_2/2$ of the z-plane in $\mathbb{C}$ is conformally mapped by $z \mapsto w = \wp(z)$ onto the lower w-half-plane such that the contour of the rectangle is bijectively mapped onto the real axis (from $-\infty$ to $+\infty$).*

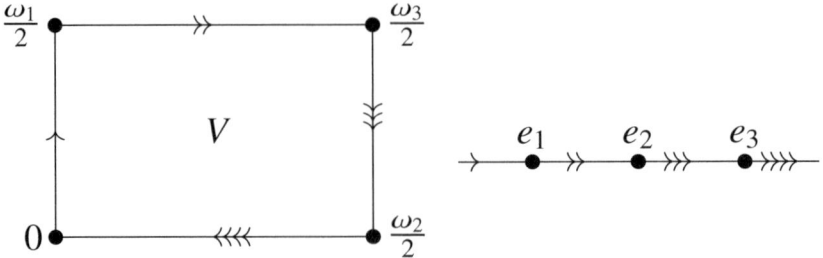

Figure 9: Mapping behavior of the quarter rectangle

Proof Let Q be the interior of the above quarter rectangle. For $z = x + iy$ with $0 < x \leq \varepsilon$, $0 < y \leq \varepsilon$, with a sufficiently small $\varepsilon > 0$,

$$\wp(z) = z^{-2} + O(\varepsilon^2) = \frac{x^2 - y^2 - 2ixy}{(x^2 + y^2)^2} + O(\varepsilon^2). \tag{$*$}$$

Because $\wp(Q)$ is simply connected, Proposition 3.24 implies

$$\wp(Q) \subseteq H \quad \text{with} \quad H := \{w \in \mathbb{C} \,;\, \operatorname{Im} w < 0\}, \tag{$**$}$$

Because $\wp(z + \omega/2) = \wp(-z - \omega/2) = \wp(-z + \omega/2)$,

$$\wp\left(\tfrac{\omega_3}{2} + Q\right) = \wp(Q) \subseteq H,$$
$$\wp\left(\tfrac{\omega_1}{2} + Q\right) = \wp\left(\tfrac{\omega_2}{2} + Q\right) \subseteq \overline{H} = \{w \in \mathbb{C} \,;\, \operatorname{Im} w > 0\} = \mathbb{H},$$

follows, so that the equality in $(**)$ holds due to Proposition 3.24 and LIOUVILLE's Third Theorem 2.5. According to Lemma 2.10, we have $\wp'(z) \neq 0$ for all $z \in Q$. Hence $\wp : Q \to H$ is biholomorphic. For sufficiently small $\varepsilon > 0$, therefore, the function $f(y) := \wp(iy)$, $0 < y \leq \varepsilon$, satisfies

$$f'(y) = i\wp'(iy) = 2y^{-3} + O(\varepsilon), \quad \text{hence} \quad f'(y) > 0,$$

and is monotonically increasing. From the differential equation

$$\wp'^2 = 4(\wp - e_1)(\wp - e_2)(\wp - e_3) \quad \text{and} \quad \wp'(iy) \neq 0$$

we get $f'(y) > 0$ for $0 < y < \omega_1/2i$. So f is strictly monotonically increasing on the whole interval $]0, \omega_1/2i]$ and maps it bijectively to $]-\infty, e_1]$.
On the other sides of Q we conclude in a similar manner. □

The images of the other quarter rectangles of $\Diamond(\omega_1, \omega_2)$ are mapped to the lower resp. upper w-half-planes according to the scheme

In particular we have

$$e_1 < e_3 < e_2.$$

§ 3 The WEIERSTRASS $\wp$-Function

Using Theorem 3.25, it is now possible to "solve" integrals of the form

$$\int \frac{dt}{\sqrt{4t^3 - gt - h}} \quad \text{with } g, h \in \mathbb{R}.$$

If we have found a rectangular lattice Ω with $g = g_2(\Omega)$ and $h = g_3(\Omega)$, then

$$\int_{\wp(s)}^{\wp(r)} \frac{dt}{\sqrt{4t^3 - gt - h}} = s - r$$

holds for $0 < r < s < \frac{\omega_2}{2}$. For a *proof* we substitute $t = \wp(z)$ and note that $\wp'(z) < 0$. The problem is to find a lattice Ω of the given type. Compare Corollary 4.11.

3.26 Remarks. a) Of course, we can also study the mapping of $\wp$ on a period parallelogram for an arbitrary lattice Ω. See E. GRAESER [32], 81–83.

b) By RIEMANN's Mapping Theorem, every rectangle in $\mathbb{C}$ is biholomorphically equivalent to the unit disk. Such a biholomorphic mapping can be constructed explicitly by Theorem 3.25 and the CAYLEY transformation (cf. Theorem II.1.4).

7. Ferdinand Gotthold Max EISENSTEIN (1823–1852) published six papers in *Crelles Journal* (cf. [20], vol. I, 299–478) in 1846/47. In these works he developed completely new ideas, which go far beyond the papers of his predecessors A.-M. LEGENDRE (1752–1833), N.H. ABEL (1802–1829) and C.G.J. JACOBI (1804–1851). His work, however, remained widely unnoticed.

L. KRONECKER (1823–1891) had intended to give a lecture at the first meeting of the Deutsche Mathematiker Vereinigung (En: German Mathematicians Society) (DMV) "about Eisenstein" in 1891. When he had to cancel for personal reasons, he wrote ([50], vol. V, 499) to the president of the DMV:

> "... über seine Arbeiten sprechen. Dabei müssten dann ausser den rein arithmetischen und analytisch-arithmetischen noch ganz besonders seine rein analytischen Untersuchungen über elliptische Functionen hervorgehoben werden, welche dem Bewusstsein der Jetztzeit ganz abhanden gekommen sind, ..."

> "... talk about his work. Thereby, besides his purely arithmetic and analytic-arithmetic, his purely analytic investigations on elliptic functions should be emphasized, which have been completely lost to the consciousness of the present time, ..."

He points out that, besides his arithmetic and analytic-arithmetic investigations the purely analytic investigations of elliptic functions are especially important, since they had not really been considered before. Indeed, these works do not seem to be mentioned in the 19th century except by L. KRONECKER and by A. HURWITZ (1859–1919) in a footnote ([39], vol. I, 31), by H. BURKHARDT and by F. KLEIN and R. FRICKE ([45], 24 und 150). In the preface to his book [8], BURKHARDT talks about the "EISENSTEIN–WEIERSTRASS partial fraction series". Only R. FRICKE (1861–1930) calls EISENSTEIN in his Encyclopedia Report [28] *a precursor of* WEIERSTRASS. Finally, H. HANCOCK [35], Art. 273, 280, 287, 291, mentions and honors EISENSTEIN's results in detail.

In 1976, the book *Elliptic functions according to Eisenstein and Kronecker* by A. WEIL [89] was published, in which WEIL paid convincing tribute to EISENSTEIN's input to the elliptic functions: the essential results about the $\wp$-function, its differential equation and the addition theorem and about the σ-function and the ζ-function (cf. §6) were already anticipated by EISENSTEIN in 1862. K.T.W. WEIERSTRASS mentioned corresponding results in his lectures at the Berlin University only in 1862 ([88], vol. V; the underlying manuscript was dictated by WEIERSTRASS to F. MERTENS in 1863). The $\wp$-function is still introduced here as the solution of the differential equation $\wp'^2 = 4\wp^3 - g_2\wp - g_3$. The fact that the convergence generating summands are missing in EISENSTEIN's definition of the analogon of the $\wp$-function (as well as σ and ζ) is no cause for doubt, because EISENSTEIN replaces this by a special summation rule of the corresponding conditionally convergent series.

We cannot understand today that WEIERSTRASS at no time referred to EISENSTEIN's preliminary work. If one wanted to be historically correct, one would have to rename all notations referring to WEIERSTRASS in the field of elliptic functions to EISENSTEIN–WEIERSTRASS. However, this has not become consensus, mainly because this fact has not (yet) found a place in the textbook literature.

EISENSTEIN's considerations are based on series over all lattice points analogously to the $\wp$-function and their elementary manipulation. As he first states, his method can excellently be applied to the foundation of trigonometric functions. This is demonstrated in the book by R. REMMERT and G. SCHUMACHER [67], chap. 11, §4.

A tale worth reading and an appreciation of EISENSTEIN can be found in the *Bulletin of the AMS* **82**, 658–663 (1976), by A. WEIL.

3.27 Exercises.
1) Express $e_1^2 + e_2^2 + e_3^2$ and $e_1^3 + e_2^3 + e_3^3$ in terms of g_2 and g_3.
2) Given $\omega \in \Omega$ with $\omega \notin 2\Omega$, define T by

$$T(z) = T_\omega(z) := \frac{\wp(\omega/4)) - \wp(\omega/2)}{\wp(z) - \wp(\omega/2)}, \quad z \in \Omega.$$

a) T is an even elliptic function with respect to Ω with roots of 2nd order at the points of Ω and poles of 2nd order at the points $\omega/2 + \Omega$.
b) Any even elliptic function can be rationally represented by T.
c) $T(z) \cdot T(z + \omega/2) = 1$.
3) Let ω_1, ω_2 be a basis of Ω. Then

$$\wp''(\omega_1/2) = 6e_1^2 - \frac{1}{2}g_2 = 2(e_1 - e_2)(e_1 - e_3),$$
$$\wp^{(iv)}(\omega_1/2) = 72e_1^3 - 6e_1g_2 = 24e_1(e_1 - e_2)(e_1 - e_3).$$

Derive analogous formulas for the values at the points $\omega_2/2$ and $\omega_3/2$.
4) Determine e_1, e_2, e_3 in the case that $g_3 = 0$ resp. $g_2 = 0$.
5) Let $\Omega = \mathbb{Z}\rho + \mathbb{Z}$, $\rho = \frac{1}{2}(1 + i\sqrt{3})$. Then

$$e_2 = \wp_\Omega(1/2) > 0, \quad e_1 = \wp_\Omega(\rho/2) = -\rho e_2, \quad e_3 = \wp_\Omega(-\bar\rho/2) = -\bar\rho e_2.$$

6) If $g_3 \neq 0$, consider the normalized homogeneous WEIERSTRASS equation

$$p(X, Y, Z) = \frac{4}{g_3}X^3 - \frac{g_2}{g_3}XZ^2 - Z^3 - \frac{1}{g_3}Y^2Z = 0.$$

We will show that this polynomial is represented by a determinant surface

$$p(X, Y, Z) = \det(XA + YB - ZI) = 0, \qquad (*)$$

where $A, B \in \mathrm{Mat}(3; \mathbb{C})$.

a) If $(*)$, holds, then A has the eigenvalues $\frac{1}{e_1}, \frac{1}{e_2}, \frac{1}{e_3}$ and B has eigenvalues $0, \pm 1/\sqrt{-g_3}$. The eigenvalues in each case are mutually distinct.

b) There is a solution of $(*)$ in which A is a diagonal matrix and B is skew symmetric.

7) Describe a biholomorphic mapping between the unit disk $\{z \in \mathbb{C};\ |z| < 1\}$ and the unit square $\{z \in \mathbb{C}\ ;\ 0 < x, y < 1\}$ in $\mathbb{C}$ explicitly.

8) Let $\Omega = \mathbb{Z}i\lambda + \mathbb{Z}\lambda, 0 \neq \lambda \in \mathbb{C}$ be a square lattice. Then the roots of the $\wp$-function are exactly the points of the set $\frac{1+i}{2}\lambda + \Omega$. Each root has order 2.

9) If Ω is a lattice such that the WEIERSTRASS $\wp$-function has only double roots, then there exists some $0 \neq \lambda \in \mathbb{C}$ with $\Omega = \mathbb{Z}i\lambda + \mathbb{Z}\lambda$.

10) Describe all conjugation stable lattices in $\mathbb{C}$.

11) If $E_j := \begin{pmatrix} e_j & e_j^2 + e_k e_\ell \\ 1 & -e_j \end{pmatrix}$, $\{j, k, \ell\} = \{1, 2, 3\}$, show that

a) $E_j^2 = (e_j - e_k)(e_j - e_\ell)I$ and $E_j E_k = (e_k - e_j)E_\ell$.

b) The set $\{\lambda E_\nu\ ;\ 0 \neq \lambda \in \mathbb{C},\ \nu = 1, 2, 3\} \cup \{\lambda I; 0 \neq \lambda \in \mathbb{C}\}$ is a subgroup of $\mathrm{GL}(2; \mathbb{C})$.

12) Show directly that a holomorphic solution

$$f : \{z \in \mathbb{C};\ |z - z_0| < r\} \to \mathbb{C}$$

of the differential equation $f'^2 = 4f^3 - g_2 f - g_3$ is uniquely determined by

$$f(z_0) = a,\ f'(z_0) = b \neq 0,\ b^2 = 4a^3 - g_2 a - g_3.$$

What about the case $b = 0$?

13) What can you say about the convergence of the series (3.1) if $\omega_2/\omega_1 \in \mathbb{R}\setminus\mathbb{Q}$?

§ 4 The dependence on the lattice

In this paragraph we want to study the dependence of the WEIERSTRASS $\wp$-function and the EISENSTEIN series E_k on the lattice Ω. For this purpose, let $\Omega = \mathbb{Z}\omega_1 + \mathbb{Z}\omega_2$ always be a lattice in $\mathbb{C}$.

1. Homogeneity and basis transformation. For every $0 \neq \lambda \in \mathbb{C}$, $\lambda\Omega$ is also a lattice in $\mathbb{C}$. (3.3) and (3.5) immediately imply

$$\wp_{\lambda\Omega}(\lambda z) = \lambda^{-2} \cdot \wp_{\Omega}(z) \quad \text{and} \quad E_k(\lambda\Omega) = \lambda^{-k} \cdot E_k(\Omega), \quad k \geq 3. \tag{4.1}$$

Using (3.8), (3.17) and (3.18), we also get

$$g_2(\lambda\Omega) = \lambda^{-4} \cdot g_2(\Omega), \quad g_3(\lambda\Omega) = \lambda^{-6} \cdot g_3(\Omega),$$

$$\Delta(\lambda\Omega) = \lambda^{-12} \cdot \Delta(\Omega), \quad j(\lambda\Omega) = j(\Omega). \tag{4.2}$$

4.1 Theorem. *For two given lattices Ω and Ω' in $\mathbb{C}$ the following assertions are equivalent:*

(i) *There exists some $0 \neq \lambda \in \mathbb{C}$ such that $\Omega' = \lambda\Omega$.*
(ii) $j(\Omega') = j(\Omega)$.

Proof (i) $\Longrightarrow$ (ii): Apply (4.2).
(ii) $\Longrightarrow$ (i): First, suppose $j(\Omega') = j(\Omega) \neq 0$. Then $g_2(\Omega) \neq 0$ and $g_2(\Omega') \neq 0$ hold because of (3.18). Thus there exists some $0 \neq \lambda \in \mathbb{C}$ with

$$g_2(\Omega') = \lambda^{-4} \cdot g_2(\Omega) = g_2(\lambda\Omega).$$

Using (4.2) and (3.17), we obtain

$$g_3(\Omega') = \pm \lambda^{-6} \cdot g_3(\Omega) = \pm g_3(\lambda\Omega).$$

Replacing λ by $i\lambda$, if necessary, we get $g_2(\Omega') = g_2(\lambda\Omega)$ and $g_3(\Omega') = g_3(\lambda\Omega)$. Then $\Omega' = \lambda\Omega$ is obtained from Corollary 3.13.
If $j(\Omega) = j(\Omega') = 0$, then $g_2(\Omega) = g_2(\Omega') = 0$ and $g_3(\Omega) \neq 0$ as well as $g_3(\Omega') \neq 0$ follow from Corollary 3.18. Then the assertion is obtained analogously. □

If (ω_1, ω_2) is a basis of Ω, we also write (cf. (3.1))

$$\wp(z; \omega_1, \omega_2) := \wp_{\Omega}(z) \quad \text{and} \quad E_k(\omega_1, \omega_2) := E_k(\Omega) \quad \text{for} \quad k \geq 3. \tag{4.3}$$

But since $\wp$ and E_k depend only on the lattice Ω and do not depend on the choice of a basis, the Basis Lemma for Lattices 1.12 immediately yields

$$\wp(z; \omega_1', \omega_2') = \wp(z; \omega_1, \omega_2) \text{ and } E_k(\omega_1', \omega_2') = E_k(\omega_1, \omega_2) \quad \text{for } k \geq 3, \tag{4.4}$$

whenever

$$\begin{pmatrix} \omega_1' \\ \omega_2' \end{pmatrix} = U \begin{pmatrix} \omega_1 \\ \omega_2 \end{pmatrix} \quad \text{with} \quad U = \begin{pmatrix} a & b \\ c & d \end{pmatrix} \in \text{GL}(2; \mathbb{Z}), \tag{4.5}$$

hence for

$$\omega_1' = a\omega_1 + b\omega_2, \quad \omega_2' = c\omega_1 + d\omega_2 \text{ and } a,b,c,d \in \mathbb{Z}, \quad ad - bc = \pm 1. \tag{4.6}$$

Given $0 \neq \lambda \in \mathbb{C}$ and $k \geq 3$, (4.1) can be reformulated as

$$\wp(\lambda z; \lambda\omega_1, \lambda\omega_2) = \lambda^{-2} \cdot \wp(z; \omega_1, \omega_2), \quad E_k(\lambda\omega_1, \lambda\omega_2) = \lambda^{-k} \cdot E_k(\omega_1, \omega_2). \tag{4.1'}$$

§ 4 The dependence on the lattice

As a basis of Ω, ω_1, ω_2 are linearly independent over $\mathbb{R}$, i.e. $\tau := \omega_1/\omega_2 \notin \mathbb{R}$. Since $(-\omega_1, \omega_2)$ is also a basis of Ω, we may assume without loss of generality $\operatorname{Im} \tau > 0$. Note that this is exactly the case if the triangle with vertices $(0, \omega_2, \omega_1)$ is positively oriented. Thus, we conclude from (4.4) and (4.1')

$$\wp(z; \omega_1, \omega_2) = \omega_2^{-2} \cdot \wp(z/\omega_2; \tau, 1) \tag{4.7}$$

$$E_k(\omega_1, \omega_2) = \omega_2^{-k} \cdot E_k(\tau, 1) \text{ for } k \geq 3.$$

Therefore, in order to study elliptic functions with respect to Ω, we may assume without significant restriction $\omega_2 = 1$, i.e.

$$\Omega = \mathbb{Z}\tau + \mathbb{Z} \quad \text{with} \quad \tau \in \mathbb{H}.$$

Here the *upper half-plane* $\mathbb{H}$ is defined by

$$\mathbb{H} = \{\tau \in \mathbb{C} \, ; \, \operatorname{Im} \tau > 0\}.$$

Because

$$\tau' := \frac{\omega_1'}{\omega_2'} = \frac{a\omega_1 + b\omega_2}{c\omega_1 + d\omega_2} = \frac{a\tau + b}{c\tau + d} \quad \text{and} \quad \operatorname{Im} \tau' = \frac{ad - bc}{|c\tau + d|^2} \cdot \operatorname{Im} \tau, \tag{4.8}$$

we then may use for the transition from the basis $(\tau, 1)$ of Ω to the basis $(\tau', 1)$ of the lattice $\frac{1}{c\tau+d}\Omega$ with $\tau' \in \mathbb{H}$ only matrices $U = \begin{pmatrix} a & b \\ c & d \end{pmatrix}$ from the special linear group over $\mathbb{Z}$, i.e. $\mathrm{SL}(2;\mathbb{Z}) := \{U \in \mathrm{GL}(2;\mathbb{Z}) \, ; \, \det U = 1\}$ (cf. (1.7)). Because of (4.6) and (4.1') we can write (4.4) in the form

$$\wp\left(\frac{z}{c\tau+d}; \frac{a\tau+b}{c\tau+d}, 1\right) = (c\tau + d)^2 \cdot \wp(z; \tau, 1) \tag{4.9}$$

resp.

$$E_k\left(\frac{a\tau+b}{c\tau+d}, 1\right) = (c\tau + d)^k \cdot E_k(\tau, 1) \quad \text{for } k \geq 4. \tag{4.10}$$

Here $a, b, c, d \in \mathbb{Z}$ satisfy $ad - bc = 1$.

2. A series expansion for E_k. To derive such an expansion, we start from a lattice of the form

$$\Omega = \mathbb{Z}\tau + \mathbb{Z} \quad \text{with} \quad \operatorname{Im} \tau > 0, \text{ i.e. } \tau \in \mathbb{H}, \tag{4.11}$$

and consider τ here as arbitrary but fixed. In the notation of sect. 1 we have

$$E_k(\tau) := E_k(\tau, 1) = \sum_{m,n}{}' (m\tau + n)^{-k}, \quad k \geq 4 \text{ even}, \tag{4.12}$$

where the prime in the sum shall mean that we sum up over

$$(0, 0) \neq (m, n) \in \mathbb{Z} \times \mathbb{Z}.$$

Of course, E_k can be regarded as a mapping $E_k : \mathbb{H} \to \mathbb{C}$. The essential tool now arises in the following generalization of the sine partial fraction expansion:

4.2 Proposition. *Given $\tau \in \mathbb{H}$ and an integer $k \geq 2$, then*

$$\sum_{n \in \mathbb{Z}} (\tau + n)^{-k} = \frac{(-2\pi i)^k}{(k-1)!} \cdot \sum_{r=1}^{\infty} r^{k-1} e^{2\pi i r \tau} \tag{4.13}$$

holds.

Proof By differentiating the cotangent partial fractional expansion (cf. Lemma B.3), we obtain the expansion

$$\left(\frac{\pi}{\sin \pi \tau}\right)^2 = \sum_{n \in \mathbb{Z}} (\tau + n)^{-2}, \quad \tau \in \mathbb{C}, \quad \tau \notin \mathbb{Z}.$$

The right hand side converges uniformly in every compact subset of $\mathbb{C}$ that does not contain a point of $\mathbb{Z}$. If we choose $\tau \in \mathbb{H}$, then $|e^{2\pi i \tau}| = e^{-2\pi \operatorname{Im} \tau} < 1$ leads to

$$\left(\frac{\pi}{\sin \pi \tau}\right)^2 = \left(\frac{2\pi i}{e^{\pi i \tau} - e^{-\pi i \tau}}\right)^2$$

$$= (-2\pi i)^2 e^{2\pi i \tau} \frac{1}{(1 - e^{2\pi i \tau})^2} = (-2\pi i)^2 \cdot \sum_{r=1}^{\infty} r e^{2\pi i r \tau}.$$

Hence (4.13) has been proved for $k = 2$. Since both sides converge locally uniformly in τ, we obtain the general case by repeated differentiation with respect to τ. □

The left-hand side of (4.13) is obviously periodic in τ with period 1; the right-hand side of (4.13) expresses this fact in the form of a FOURIER series. Thus we immediately obtain the FOURIER *series expansion* of the EISENSTEIN series.

4.3 Theorem. *For all $\tau \in \mathbb{H}$ and all even $k \geq 4$,*

$$E_k(\tau) = 2\zeta(k) + 2\frac{(2\pi i)^k}{(k-1)!} \cdot \sum_{m=1}^{\infty} \sigma_{k-1}(m) \cdot e^{2\pi i m \tau}, \quad \tau \in \mathbb{H}, \tag{4.14}$$

holds with

$$\zeta(s) := \sum_{m=1}^{\infty} m^{-s}, \quad s > 1, \quad \text{and} \quad \sigma_s(m) := \sum_{d \in \mathbb{N}, d | m} d^s, \quad s \in \mathbb{R}.$$

The series (4.14) converges absolutely uniformly for $\varepsilon > 0$ on any upper half-plane $\{\tau \in \mathbb{H}\,;\, \operatorname{Im} \tau \geq \varepsilon\}$. The functions E_k are holomorphic on $\mathbb{H}$ and satisfy

$$E_k\left(\frac{a\tau + b}{c\tau + d}\right) = (c\tau + d)^k \cdot E_k(\tau) \quad \text{for all} \quad \begin{pmatrix} a & b \\ c & d \end{pmatrix} \in \operatorname{SL}(2; \mathbb{Z}). \tag{4.15}$$

§ 4 The dependence on the lattice

Because of (4.14) it is now clear that the E_k for even $k \geq 4$ do not vanish identically.

Proof Because of the absolute convergence according to the Convergence Lemma 1.17 for $\Omega = \mathbb{Z}\tau + \mathbb{Z}$, we can transform (4.12) into

$$E_k(\tau) = \sum_{n \neq 0} n^{-k} + \sum_{m \neq 0} \sum_{n \in \mathbb{Z}} (m\tau + n)^{-k} = 2\zeta(k) + 2 \sum_{m=1}^{\infty} \sum_{n \in \mathbb{Z}} (m\tau + n)^{-k}.$$

Now we insert Proposition 4.2 and get

$$E_k(\tau) = 2\zeta(k) + 2 \frac{(2\pi i)^k}{(k-1)!} \cdot \sum_{s=1}^{\infty} \sum_{r=1}^{\infty} r^{k-1} e^{2\pi i r s \tau}.$$

Finally, we combine the terms with $rs = m$ and get (4.14). The holomorphy follows from the locally uniform convergence of the series (4.14), and (4.15) is a reformulation of (4.10). □

Using the well-known formulas (cf. Corollary B.5)

$$\zeta(4) = \frac{\pi^4}{90}, \quad \zeta(6) = \frac{\pi^6}{945}, \tag{4.16}$$

we get

$$E_4(\tau) = \frac{\pi^4}{45} \left(1 + 240 \cdot \sum_{m=1}^{\infty} \sigma_3(m) \cdot e^{2\pi i m \tau} \right), \tag{4.17}$$

$$E_6(\tau) = \frac{2\pi^6}{945} \left(1 - 504 \cdot \sum_{m=1}^{\infty} \sigma_5(m) \cdot e^{2\pi i m \tau} \right). \tag{4.18}$$

As early as 1881, A. HURWITZ (1859–1919) showed in his dissertation ([39], vol. I, 1–66) that the algebraic equations that the series E_k satisfy according to Corollary 3.11 give rise to number theoretic statements. We note the simplest case as

4.4 Corollary. (HURWITZ *Identity*) *For all $m \in \mathbb{N}$, one has*

$$\sigma_7(m) = \sigma_3(m) + 120 \sum_{\substack{r,s \in \mathbb{N} \\ r+s=m}} \sigma_3(r)\sigma_3(s).$$

Proof We use the identity $7E_8 = 3E_4^2$ according to (3.9). If we insert here (4.17) and use

$$E_8(\tau) = 2\zeta(8) + 2 \frac{(2\pi)^8}{7!} \cdot \sum_{m=1}^{\infty} \sigma_7(m) \cdot e^{2\pi i m \tau}$$

then, after multiplying out, we obtain a power series identity in $q = e^{2\pi i \tau}$. A comparison of the coefficients yields $7\zeta(8) = 6\zeta^2(4)$ and

$$7\frac{2(2\pi)^8}{7!}\sigma_7(m) = 3\frac{\pi^8}{45\cdot 45}\left(480\sigma_3(m) + 240\cdot 240 \sum_{r+s=m}\sigma_3(r)\sigma_3(s)\right).$$

But this is the assertion. □

4.5 Remarks. a) The statement of Proposition 4.2 remains true with $\Gamma(k)$ instead of $(k-1)!$ for any real $k > 1$ and is then sometimes named after R. LIPSCHITZ (*J. Reine Angew. Math.* **105**, 127–156 (1889)).

b) The HURWITZ Identity is a statement about natural numbers and is as such a subject of elementary number theory. However, up to now no proof within elementary number theory is known. A proof, which works with formal (or convergent) power series with coefficients from $\mathbb{Z}$, goes back to D. ZAGIER and N. SKORUPPA (1978). We also compare N. SKORUPPA, *J. Number Theory* **43**, 68-73 (1993):
For an indeterminate (or real variable x with $|x| < 1$), set

$$F_n := F_n(x) := \frac{x^n}{1 - x^n}$$

and first notice the identity

$$\sum_n \sigma_r(n)x^n = \sum_m m^r F_m. \qquad (*)$$

Here and later, the sums must be carried out over all positive integers. Furthermore we verify

$$F_m F_n = F_{m+n}(F_m + F_n + 1).$$

Using the abbreviations

$$A_k := \sum_{m+n=k} mn F_m F_n, \quad B_k := \sum_{n-m=k} mn F_m F_n, \quad C_k := kF_k \cdot \sum_m mF_m,$$

we then prove that

$$A_k = 2F_k \cdot \sum_{m<k} m(k-m)F_m + \frac{k^3 - k}{6}F_k,$$

$$B_k = 2C_k + F_k \cdot \sum_{m<k} m(m-k)F_m - \sum_{m>k} m(m-k)F_m,$$

hence

$$A_k + 2B_k - 4C_k = \frac{k^3 - k}{6}F_k - 2\sum_{m>k} m(m-k)F_m.$$

Thus we obtain

§ 4 The dependence on the lattice

$$\left(\sum_n n^3 F_n\right)^2 = \sum_{m,n} \frac{mn}{12}((m+n)^4 + (m-n)^4 - 2m^4 - 2n^4)F_m F_n$$

$$= \sum_k \frac{k^4}{12}(A_k + 2B_k - 4C_k) = \sum_k \frac{k^4}{12}\frac{k^3 - k}{6}F_k - \sum_m \frac{m}{6}F_m \cdot \sum_{k<m} k^4(m-k).$$

If we now use the sum formula for the sums of the 4th and 5th powers, we get

$$120\left(\sum_n n^3 F_n\right)^2 = \sum_k (k^7 - k^3)F_k.$$

Because of (∗), this is the HURWITZ Identity 4.4.

3. The discriminant. As in (3.8) or (3.17), we introduce
$$g_2(\tau) := 60\, E_4(\tau), \quad g_3(\tau) := 140\, E_6(\tau) \quad \text{and} \quad \Delta(\tau) := g_2^3(\tau) - 27 g_3^2(\tau) \quad (4.19)$$

From (4.17) and (4.18) we then get

$$g_2(\tau) = \frac{(2\pi)^4}{12}\left(1 + 240 \cdot \sum_{m=1}^\infty \sigma_3(m) \cdot e^{2\pi i m \tau}\right), \quad (4.20)$$

$$g_3(\tau) = \frac{(2\pi)^6}{216}\left(1 - 504 \cdot \sum_{m=1}^\infty \sigma_5(m) \cdot e^{2\pi i m \tau}\right). \quad (4.21)$$

4.6 Theorem. *The discriminant $\Delta(\tau)$ has a FOURIER series expansion of the form*

$$\Delta(\tau) = (2\pi)^{12} \cdot \sum_{m=1}^\infty \tau(m) \cdot e^{2\pi i m \tau}, \quad \tau \in \mathbb{H}, \quad (4.22)$$

with coefficients $\tau(m) \in \mathbb{Z}$ and $\tau(1) = 1$. The discriminant $\Delta : \mathbb{H} \to \mathbb{C}$ is a holomorphic function satisfying $\Delta(\tau) \neq 0$ for all $\tau \in \mathbb{H}$ and

$$\Delta\left(\frac{a\tau + b}{c\tau + d}\right) = (c\tau + d)^{12} \cdot \Delta(\tau) \quad \text{for all} \quad \begin{pmatrix} a & b \\ c & d \end{pmatrix} \in \mathrm{SL}(2; \mathbb{Z}). \quad (4.23)$$

The notation of the coefficients in (4.22) with $\tau(m)$ is traditional, the two τ's should not confuse the reader. $\tau(m)$ is called the RAMANUJAN *tau function*.

Proof With the abbreviations

$$A := \sum_{m=1}^\infty \sigma_3(m) \cdot e^{2\pi i m \tau}, \quad B := \sum_{m=1}^\infty \sigma_5(m) \cdot e^{2\pi i m \tau},$$

(4.19) can be reformulated via (4.20) and (4.21) as

$$\Delta(\tau) = \frac{(2\pi)^{12}}{1728} \cdot ((1+240A)^3 - (1-504B)^2) = (2\pi)^{12} \cdot (e^{2\pi i \tau} + \ldots) \quad (*)$$

and on the right-hand side we have a power series in $q = e^{2\pi i \tau}$. In order to prove that the coefficients are integers, we first note that $d^3 \equiv d^5 \pmod{12}$ for $d \in \mathbb{Z}$ and consequently $\sigma_3(m) \equiv \sigma_5(m) \pmod{12}$ for $m \in \mathbb{N}$. Thus if we relate the congruence to the coefficients of a FOURIER series, we have $A \equiv B \pmod{12}$. Now we calculate modulo $1728 = 12^3$ and get

$$(1+240A)^3 - (1-504B)^2 \equiv 12^2(5A+7B) \equiv 0 \pmod{12^3}.$$

So the denominator in $(*)$ cancels in all coefficients.

As g_2 and g_3 are holomorphic (cf. Theorem 4.3) Δ is also holomorphic. We obtain $\Delta(\tau) \neq 0$ from Corollary 3.18. The identity (4.23) follows directly from (4.15) as well as (4.19). □

m	$\tau(m)$	prime factorization				
1	1	1				
2	-24	-2^3	3			
3	252	2^2	3^2	7		
4	$-1\,472$	-2^6			23	
5	4 830	2	3	5	7	23
6	$-6\,048$	-2^5	3^3	7		
7	$-16\,744$	-2^3		7	13	23
8	84 480	2^9	3	5	11	
9	$-113\,643$	$-$	3^4		23	61
10	$-115\,920$	-2^4	3^2	5	7	23

(4.24)

4. The absolute invariant. Besides the discriminant Δ, the absolute invariant j plays an important role according to (3.18):

$$j(\tau) := (12 g_2(\tau))^3 / \Delta(\tau), \quad \tau \in \mathbb{H}. \tag{4.25}$$

Note that $j(\tau)$ is defined for all $\tau \in \mathbb{C}$ with $\operatorname{Im} \tau > 0$ because of Theorem 4.6. g_2 and Δ are representable by FOURIER series in τ according to (4.20) and (4.22), i.e. as power series in $q := e^{2\pi i \tau}$. In order to derive such a series for j, we need the

4.7 Proposition. *If f and g are convergent power series in q, $|q| < 1$,*

$$f(q) = \sum_{n \geq 0} a_n q^n, \quad g(q) = \sum_{n \geq 0} b_n q^n, \quad a_n, b_n \in \mathbb{Z},$$

with $b_0 = 1$ and $g(q) \neq 0$ for $|q| < 1$, then f/g also is a convergent power series with coefficients in $\mathbb{Z}$ for $|q| < 1$.

Proof First f and g, hence also f/g are holomorphic for $|q| < 1$. Hence f/g can be expressed as a power series, whose coefficients are denoted by c_n. Using

§ 4 The dependence on the lattice

$$\left(\sum_{n\geq 0} c_n q^n\right) \cdot \left(\sum_{n\geq 0} b_n q^n\right) = \sum_{n\geq 0} a_n q^n$$

and $b_0 = 1$, the recursion formula

$$c_0 = a_0, \quad c_m = a_m - \sum_{n=0}^{m-1} c_n b_{m-n}, \quad m \geq 1.$$

follows. Hence all the c_m are integers. □

This yields

4.8 Theorem. *The absolute invariant* $j : \mathbb{H} \to \mathbb{C}$ *is holomorphic and has a* FOURIER *series expansion of the form*

$$j(\tau) = e^{-2\pi i \tau} + \sum_{m \geq 0} j_m \cdot e^{2\pi i m \tau} \tag{4.26}$$

$$= e^{-2\pi i \tau} + 744 + 196\,884 \cdot e^{2\pi i \tau} + \ldots$$

with $j_m \in \mathbb{Z}$ *for all* $m \geq 0$. *We have*

$$j\left(\frac{a\tau + b}{c\tau + d}\right) = j(\tau) \quad \text{for all} \quad \begin{pmatrix} a & b \\ c & d \end{pmatrix} \in \mathrm{SL}(2;\mathbb{Z}). \tag{4.27}$$

Proof The holomorphy follows from Theorem 4.3 and Theorem 4.6 due to (4.25). Splitting off a factor q from Δ, we can apply Proposition 4.7 and get (4.26) from (4.20) and (4.22). Finally, (4.27) is a consequence of (4.15) and (4.23). □

We will see later (Theorem 6.18) that the coefficients j_m are even all positive. The name "absolute invariant" is motivated by (4.27). However, the converse of (4.27) also holds:

4.9 Theorem. *For* $\tau, \tau' \in \mathbb{H}$ *the following assertions are equivalent:*

(i) $j(\tau) = j(\tau')$.
(ii) *There exists a matrix* $\begin{pmatrix} a & b \\ c & d \end{pmatrix} \in \mathrm{SL}(2;\mathbb{Z})$ *such that*

$$\tau' = \frac{a\tau + b}{c\tau + d}.$$

Proof (ii) $\Rightarrow$ (i): Apply (4.27).
(i) $\Rightarrow$ (ii): By assumption $j(\mathbb{Z}\tau' + \mathbb{Z}) = j(\mathbb{Z}\tau + \mathbb{Z})$ holds. Theorem 4.1 shows that $\mathbb{Z}\tau' + \mathbb{Z} = \mathbb{Z}\lambda\tau + \mathbb{Z}\lambda$ for some $0 \neq \lambda \in \mathbb{C}$, hence $(\tau', 1)$ and $(\lambda\tau, \lambda)$ are two bases of the same lattice. Therefore, by the Basis Lemma for Lattices 1.12, there exists some $M = \begin{pmatrix} a & b \\ c & d \end{pmatrix} \in \mathrm{GL}(2;\mathbb{Z})$ such that $\tau' = a\lambda\tau + b\lambda$, $1 = c\lambda\tau + d\lambda$, hence $\tau' = \frac{a\tau+b}{c\tau+d}$. But since τ and τ' are in $\mathbb{H}$, we conclude $\det M = 1$ from (4.8). □

A statement to be proved later in a stronger form in Corollary III.5.4 is already mentioned now as

4.10 Theorem. *For every* $c \in \mathbb{C}$ *there exists some* $\tau \in \mathbb{H}$ *with* $j(\tau) = c$.

Proof We assume that $j(\tau) \neq c$ holds for all $\tau \in \mathbb{H}$. Then
$$F(\tau) = \frac{j'(\tau)}{j(\tau) - c}$$
is holomorphic on $\mathbb{H}$. We consider the integral
$$\int_\gamma F(\tau)d\tau, \quad \gamma = \gamma_1 + \gamma_2 + \gamma_3 + \gamma_4 + \gamma_5,$$
with the path $\gamma = \partial G$ in the adjacent figure. (4.27) yields
$$F(\tau + 1) = F(\tau), \quad F(-1/\tau) = \tau^2 \cdot F(\tau).$$

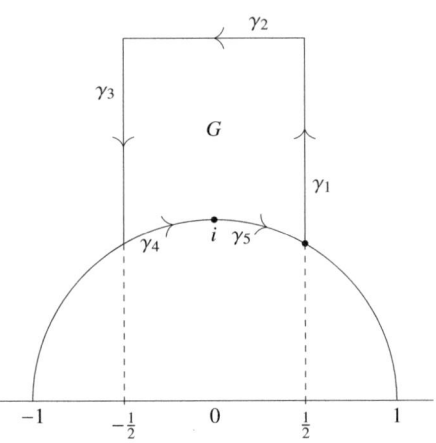

Figure 10: Integration path

This results in
$$\int_{\gamma_1} F(\tau)d\tau + \int_{\gamma_3} F(\tau)d\tau = \int_{\gamma_4} F(\tau)d\tau + \int_{\gamma_5} F(\tau)d\tau = 0.$$

According to Proposition 4.7 and (4.26), $F(\tau)$ has a FOURIER series expansion of the form
$$F(\tau) = \sum_{m \geq 0} a_m e^{2\pi i m \tau}, \quad a_0 = -2\pi i.$$

Thus we get $\int_{\gamma_2} F(\tau)d\tau = 2\pi i$. The Residue Theorem A.5 now implies
$$2\pi i \cdot \sum_{\tau \in G} \mathrm{ord}_\tau (j - c) = \int_\gamma F(\tau)d\tau = 2\pi i.$$

This is a contradiction. So there exists some $\tau \in \mathbb{H}$ with $j(\tau) = c$. □

Considering an arbitrary lattice Ω in $\mathbb{C}$ instead of a lattice $\mathbb{Z}\tau + \mathbb{Z}$, $\tau \in \mathbb{H}$, the absolute invariant $j(\Omega)$ according to (3.18) and Theorem 4.1 satisfies
$$j(\Omega) = j(\omega_1/\omega_2) \quad \text{if } \mathrm{Im}(\omega_1/\omega_2) > 0. \tag{4.28}$$

The meaning of Theorem 4.8 for elliptic functions is now described in

4.11 Corollary *If $c_2, c_3 \in \mathbb{C}$ with $c_2^3 - 27c_3^2 \neq 0$, then there is a uniquely determined lattice Ω in $\mathbb{C}$ such that*
$$c_2 = g_2(\Omega) \quad \text{and} \quad c_3 = g_3(\Omega).$$

Proof By Theorem 4.10, there is a lattice Ω satisfying

§ 4 The dependence on the lattice

$$j(\Omega) = \frac{(12c_2)^3}{c_2^3 - 27c_3^2}.$$

a) $c_2 = 0$: then $j(\Omega) = 0$, hence $g_2(\Omega) = 0$, $g_3(\Omega) \neq 0$. Now we choose $0 \neq \lambda \in \mathbb{C}$ such that $g_3(\Omega) = \lambda^6 c_3$. Because of (4.2) we conclude

$$g_3(\lambda\Omega) = \lambda^{-6} \cdot g_3(\Omega) = c_3 \quad \text{and} \quad g_2(\lambda\Omega) = \lambda^{-4} \cdot g_2(\Omega) = 0 = c_2.$$

b) $c_2 \neq 0$: then $j(\Omega) \neq 0$, hence $g_2(\Omega) \neq 0$. Now we choose a $0 \neq \lambda \in \mathbb{C}$ with $g_2(\Omega) = \lambda^4 c_2$. It follows that $g_2(\lambda\Omega) = c_2$ and $j(\lambda\Omega) = j(\Omega)$ leads to $c_3^2 = g_3^2(\lambda\Omega)$. The existence is derived by replacing λ by $i\lambda$ if necessary.

The uniqueness in both cases follows from Corollary 3.13. □

Due to Corollary 4.11, which is also called the *Inversion Theorem*, we can use the associated WEIERSTRASS $\wp$–function to solve the differential equations and integration problems mentioned in the Introduction.

5. Calculation of the FAGNANO integral. We now apply the results to calculate the FAGNANO integral described in (0.8) For this purpose let

$$\Gamma(s) := \int_0^\infty t^{s-1} e^{-t} dt, \quad s \in \mathbb{C}, \ \operatorname{Re}(s) > 0, \tag{4.29}$$

be the *gamma function* (cf. A.13).

4.12 Proposition. *It holds*

$$\int_1^\infty \frac{1}{\sqrt{4x^3 - 4x}} dx = \int_0^1 \frac{1}{\sqrt{1-t^4}} dt = \frac{\Gamma(1/4)^2}{4\sqrt{2\pi}}.$$

Proof The substitution $x = 1/t^2$ transforms the first integral into the second one. (4.29) leads to

$$\Gamma(1/4)^2 = \int_0^\infty \int_0^\infty (xy)^{-3/4} e^{-(x+y)} dx \, dy.$$

The substitution $x = r^2 \cos^2 \varphi$, $y = r^2 \sin^2 \varphi$ yields

$$\Gamma(1/4)^2 = 4\sqrt{2} \cdot \int_0^\infty e^{-r^2} dr \cdot \int_0^{\pi/2} \frac{1}{\sqrt{\sin(2\varphi)}} d\varphi$$

$$= 4\sqrt{2\pi} \cdot \int_0^{\pi/4} \frac{1}{\sqrt{\sin(2\varphi)}} d\varphi = 4\sqrt{2\pi} \cdot \int_0^1 \frac{1}{\sqrt{1-t^4}} dt,$$

if we apply A.14 and the substitution $t = \sqrt{\sin(2\varphi)}$. □

We first calculate the EISENSTEIN series of weight 4 for the lattice $\mathbb{Z}i + \mathbb{Z}$.

4.13 Theorem. *It holds*

$$g_2(\mathbb{Z}i + \mathbb{Z}) = \frac{\Gamma(1/4)^8}{16\pi^2} \quad \text{and} \quad g_3(\mathbb{Z}i + \mathbb{Z}) = 0.$$

Proof Because of $\mathbb{Z}i + \mathbb{Z} = i(\mathbb{Z}i + \mathbb{Z})$, we conclude from (4.2) that

$$g_3(\mathbb{Z}i + \mathbb{Z}) = g_3(i(\mathbb{Z}i + \mathbb{Z})) = -g_3(\mathbb{Z}i + \mathbb{Z}), \quad \text{hence} \quad g_3(\mathbb{Z}i + \mathbb{Z}) = 0.$$

On the other hand, (4.20) yields

$$g_2(\mathbb{Z}i + \mathbb{Z}) = g_2(i) = \frac{(2\pi)^4}{12}(1 + 240 \cdot \sum_{m=1}^{\infty} \sigma_3(m) \cdot e^{-2\pi m}) > 0.$$

Then according to (4.2) there exists some $\lambda \in \mathbb{R}$, $\lambda > 0$, such that $\Omega := \lambda(\mathbb{Z}i + \mathbb{Z})$ satisfies $g_2(\Omega) = 4$ and $g_3(\Omega) = 0$. Because of

$$4x^3 - 4x = 4x(x-1)(x+1)$$

it follows that $e_1 = -1$, $e_2 = 1$, $e_3 = 0$ from Theorem 3.15. Then Theorem 3.25 as well as $\wp = \wp_\Omega$, lead to

$$\int_1^\infty \frac{1}{\sqrt{4x^3 - 4x}} dx = \int_{\lambda/2}^0 \frac{\wp'(t)}{\sqrt{4\wp(t)^3 - 4\wp(t)}} dt = \int_0^{\lambda/2} dt = \frac{\lambda}{2},$$

when we use the differential equation in Theorem 2.12 and $\wp'(t) < 0$ on the interval $]0, \lambda/2[$. Proposition 4.12 leads to $\lambda = \Gamma(1/4)^2/(2\sqrt{2\pi})$. From (4.2), we conclude

$$g_2(\mathbb{Z}i + \mathbb{Z}) = \lambda^4 \cdot g_2(\Omega) = \frac{\Gamma(1/4)^8}{16\pi^2}.$$ □

4.14 Corollary *Let $\Omega = \mathbb{Z}i + \mathbb{Z}$ and $\wp = \wp_\Omega$ be the corresponding* WEIERSTRASS *$\wp$–function. If $0 < R \le 1$, then the* FAGNANO *integral (0.8) equals*

$$\int_0^R \frac{1}{\sqrt{1-t^4}} dt = \frac{\Gamma(1/4)^2}{2\sqrt{2\pi}} \xi,$$

where $\xi \in]0, 1/2]$ is determined by

$$\wp_\Omega(\xi) = \frac{\Gamma(1/4)^4}{8\pi R^2}.$$

Proof Setting $\lambda = \Gamma(1/4)^2/(2\sqrt{2\pi})$, it follows from Theorem 4.13 that

§ 4 The dependence on the lattice 59

$$\lambda\xi = \lambda \int_{\xi}^{0} \frac{\wp'(s)}{\sqrt{4\wp^3(s) - 4\lambda^4\wp(s)}} ds = \lambda \int_{\lambda^2/R^2}^{\infty} \frac{1}{\sqrt{4x^3 - 4\lambda^4 x}} dx = \int_{0}^{R} \frac{1}{\sqrt{1-t^4}} dt,$$

if we use the substitution $x = \lambda^2/t^2$. □

A corresponding result shall now be derived for the hexagonal lattice. In analogy with Proposition 4.12 we obtain the

4.15 Lemma. *It holds*

$$\int_{1}^{\infty} \frac{1}{\sqrt{4x^3 - 4}} dx = \int_{0}^{1} \frac{1}{\sqrt{1-t^6}} dt = \frac{\sqrt{\pi} \cdot \Gamma(1/6)}{6 \cdot \Gamma(2/3)}.$$

Proof The first equation follows by the substitution $x = t^{-2}$. Then (4.29) leads to

$$\Gamma(1/2) \cdot \Gamma(1/6) = \int_{0}^{\infty}\int_{0}^{\infty} x^{-1/2} y^{-5/6} e^{-(x+y)} dx\, dy$$

The substitution $x = r\cos^2\varphi$, $y = r\sin^2\varphi$ implies

$$\Gamma(1/2) \cdot \Gamma(1/6) = 2\int_{0}^{\infty}\int_{0}^{\pi/2} r^{-1/3}(\sin\varphi)^{-2/3} e^{-r} d\varphi\, dr = 6\Gamma(2/3) \cdot \int_{0}^{1} \frac{1}{\sqrt{1-t^6}} dt,$$

if we substitute $t = (\sin\varphi)^{1/3}$ in the last step. Because of $\Gamma(1/2) = \sqrt{\pi}$ (cf. Theorem A.13) the assertion follows. □

Now we calculate the EISENSTEIN series of weight 6 for the lattice $\mathbb{Z}(1 + i\sqrt{3})/2 + \mathbb{Z}$.

4.16 Theorem. *Let $\rho = \frac{1}{2}(1 + i\sqrt{3})$. Then*

$$g_2(\mathbb{Z}\rho + \mathbb{Z}) = 0 \quad and \quad g_3(\mathbb{Z}\rho + \mathbb{Z}) = \frac{4\pi^3 \cdot \Gamma(1/6)^6}{3^6 \cdot \Gamma(2/3)^6}.$$

Proof Using $\rho^2 = \rho - 1$ we calculate $\rho(\mathbb{Z}\rho + \mathbb{Z}) = \mathbb{Z}\rho + \mathbb{Z}$ and by (4.15)

$$g_2(\mathbb{Z}\rho + \mathbb{Z}) = g_2(\rho(\mathbb{Z}\rho + \mathbb{Z})) = \rho^{-4} \cdot g_2(\mathbb{Z}\rho + \mathbb{Z}), \quad \text{thus} \quad g_2(\mathbb{Z}\rho + \mathbb{Z}) = 0$$

by (4.2) because of $\rho^{-4} \neq 1$. On the other hand (4.21) yields

$$g_3(\mathbb{Z}\rho + \mathbb{Z}) = \frac{(2\pi)^6}{216}(1 - 504 \cdot \sum_{m=1}^{\infty}(-1)^m \sigma_5(m) \cdot q^m), \quad q = e^{-\pi\sqrt{3}}.$$

The estimate

$$\frac{\sigma_5(m+1)q^{m+1}}{\sigma_5(m)q^m} \leq \left(\frac{m+1}{m}\right)^5 \zeta(5) \cdot q \leq 2^5 \cdot \frac{\pi^4}{90} \cdot q < 1$$

shows that $(\sigma_5(m)q^m)_{m\geq 1}$ is a monotonically decreasing sequence with limit 0. Therefore, $g_3(\mathbb{Z}\rho + \mathbb{Z})$ is positive by to the LEIBNIZ criterion. Then (4.2) leads to the existence of some $\lambda \in \mathbb{R}$, $\lambda > 0$, such that the lattice $\Omega = \lambda(\mathbb{Z}\rho + \mathbb{Z}) = \mathbb{Z}\lambda\rho + \mathbb{Z}\lambda$ satisfies $g_2(\Omega) = 0$ and $g_3(\Omega) = 4$. As 1 is the only real root of $4x^3 - 4$ and $\wp(x) = \wp_\Omega(x)$ is real for $x \in \mathbb{R}\setminus\mathbb{Z}$ by Proposition 3.22, it follows that

$$e_2 = 1 \quad \text{and} \quad \lim_{x \downarrow 0} \wp_\Omega(x) = \infty.$$

Since the $\wp$-function is strictly monotonic on $]0, \lambda/2]$, we get

$$\int_1^\infty \frac{1}{\sqrt{4x^3 - 4}} dx = \int_{\lambda/2}^0 \frac{\wp'(t)}{\sqrt{4\wp(t)^3 - 4}} dt = \int_0^{\lambda/2} dt = \lambda/2,$$

using the differential equation in Proposition 3.22. Lemma 4.15 leads to the value $\lambda = \sqrt{\pi} \cdot \Gamma(1/6)/(3 \cdot \Gamma(2/3))$. Then (4.2) shows that

$$g_3(\mathbb{Z}\rho + \mathbb{Z}) = \lambda^6 \cdot g_3(\Omega) = \frac{4\pi^3 \cdot \Gamma(1/6)^6}{3^6 \cdot \Gamma(2/3)^6}. \qquad \square$$

4.17 Remark. From the properties of the gamma function (cf. Theorem A.13), we easily deduce

$$\int_0^1 \frac{1}{\sqrt{1 - t^6}} dt = \frac{\Gamma(1/3)^3}{4\pi\sqrt[3]{2}}, \quad g_3(\mathbb{Z}\rho + \mathbb{Z}) = \left(\frac{\Gamma(1/3)^3}{2\pi}\right)^6.$$

4.18 Exercises.
1) Given $z \in \mathbb{C} \setminus \mathbb{Z}$, the function $\tau \mapsto \wp(z; \tau, 1)$ is meromorphic on $\mathbb{H}$. Determine the poles and the principal parts of the LAURENT series expansion. Develop $\wp(z; \tau, 1)$ into a FOURIER series with respect to τ.
2) For pairwise distinct $\alpha, \beta, \gamma \in \mathbb{C}$ with $\alpha + \beta + \gamma = 0$, there is exactly one lattice Ω with $e_1 = \alpha$, $e_2 = \beta$ and $e_3 = \gamma$.
3) A lattice Ω satisfies $g_3(\Omega) = 0$ or $j(\Omega) = 1728$ if and only if there is some $0 \neq \lambda \in \mathbb{C}$ with $\Omega = \mathbb{Z}i\lambda + \mathbb{Z}\lambda$. The lattice Ω satisfies $g_2(\Omega) = 0$ or $j(\Omega) = 0$ if and only if there is some $0 \neq \lambda \in \mathbb{C}$ such that

$$\Omega = \mathbb{Z}\tfrac{1}{2}(1 + i\sqrt{3})\lambda + \mathbb{Z}\lambda.$$

4) $j(\Omega) \in \mathbb{R}$ holds if and only if there is some $0 \neq \lambda \in \mathbb{C}$ such that $\lambda\Omega$ is conjugation stable. $j(\tau) \in \mathbb{R}$ holds if and only if $2\text{Re}(\tau) \in \mathbb{Z}$.
5) e_1, e_2, e_3 are all real if and only if $g_2(\Omega)$ and $g_3(\Omega)$ are real and $\Delta(\Omega) > 0$.
6) Ω is a lattice such that the roots of $\wp_\Omega(z)$ are contained in $\tfrac{1}{2}\Omega$ if and only if

$\Omega = \mathbb{Z}i\lambda + \mathbb{Z}\lambda$ holds for some $0 \neq \lambda \in \mathbb{C}$.

7) As in Corollary 3.20, consider the function

$$\lambda : \mathbb{H} \to \mathbb{C}, \quad \lambda(\tau) := \frac{\wp(1/2; \mathbb{Z}\tau + \mathbb{Z}) - \wp((\tau+1)/2; \mathbb{Z}\tau + \mathbb{Z})}{\wp(\tau/2; \mathbb{Z}\tau + \mathbb{Z}) - \wp((\tau+1)/2; \mathbb{Z}\tau + \mathbb{Z})}.$$

Then λ is holomorphic with $\lambda(\tau) \neq 0, 1$ and satisfies

$$\lambda\left(\frac{a\tau+b}{c\tau+d}\right) = \lambda(\tau) \quad \text{for all} \quad \begin{pmatrix} a & b \\ c & d \end{pmatrix} \in \mathrm{SL}(2; \mathbb{Z}) \text{ with } b \equiv c \equiv 0 \pmod{2},$$

$$\lambda(\tau+1) = 1 - \lambda(\tau), \quad \lambda(-1/\tau) = 1/\lambda(\tau).$$

8) Let $\lambda : \mathbb{H} \to \mathbb{C}$ be defined as in Exercise 7. Then for $\gamma \in \mathbb{C}$ with $\gamma \neq 0, 1$ there exists some $\tau \in \mathbb{H}$ with $\lambda(\tau) = \gamma$.

9) Calculate the FOURIER expansion of e_1, e_2, e_3 for a lattice $\Omega = \mathbb{Z}\tau + \mathbb{Z}$, $\tau \in \mathbb{H}$.

10) Calculate an analogon of the HURWITZ identity for σ_9 and σ_{13}, in which only σ_3 and σ_5 occur.

11) Determine the lattice $\Omega = \mathbb{Z}\omega_1 + \mathbb{Z}\omega_2$ satisfying $e_1 = -1$, $e_2 = 1$, $e_3 = 0$.

12) Apply the results of sect. 3 in §3 in order to derive (4.16) as well as

$$\zeta(2n) = \sum_{p,q \geq 2, p+q=n} \frac{6(2p-1)(2q-1)}{(n-3)(2n-1)(2n+1)} \zeta(2p)\zeta(2q) \in \mathbb{Q}\pi^{2n} \quad \text{for } n \geq 4.$$

§ 5 Elliptic curves and the Addition Theorem of the $\wp$-Function

Unless explicitly stated otherwise, $\Omega = \mathbb{Z}\omega_1 + \mathbb{Z}\omega_2$ is also an arbitrary lattice in $\mathbb{C}$ in this paragraph, and $\wp(z) = \wp_\Omega(z) = \wp(z; \omega_1, \omega_2)$ is the corresponding WEIERSTRASS $\wp$-function.

1. The addition theorem. In this section we derive a formula for $\wp(z + w)$.

5.1 Addition Theorem of the $\wp$-Function. *Given $z, w \in \mathbb{C}$ with $z, w, z \pm w \notin \Omega$,*

$$\wp(z+w) + \wp(z) + \wp(w) = \frac{1}{4}\left(\frac{\wp'(z) - \wp'(w)}{\wp(z) - \wp(w)}\right)^2. \tag{5.1}$$

For this purpose we first of all derive the

5.2 Proposition. *Given $w \in \mathbb{C} \setminus \frac{1}{2}\Omega$,*

$$f(z) := f(z; w) := \frac{1}{2}\frac{\wp'(z) - \wp'(w)}{\wp(z) - \wp(w)} \tag{5.2}$$

is an elliptic function with respect to the lattice Ω with first order poles exactly at the points

$$z \in \Omega \quad \text{and} \quad z \in -w + \Omega \tag{5.3}$$

and principal parts

$$f(z; w) = -\frac{1}{z} - \wp(w) \cdot z + O(z^2) \quad \text{at} \quad z = 0, \tag{5.4}$$

$$f(z; w) = \frac{1}{z+w} + c(w) + O(z+w) \quad \text{at} \quad z = -w. \tag{5.5}$$

The coefficient $c(w)$ will be determined shortly.

Proof In addition to the points (5.3), f is not defined at $z \in w + \Omega$. Because of

$$\lim_{z \to w} f(z; w) = \frac{1}{2} \lim_{z \to w} \frac{(\wp'(z) - \wp'(w))/(z-w)}{(\wp(z) - \wp(w))/(z-w)} = \frac{1}{2} \frac{\wp''(w)}{\wp'(w)}$$

and Lemma 2.10, however, the singularities are removable here. For $z = 0$ we use $\wp(z) = z^{-2} + O(z^2)$, hence $\wp'(z) = -2z^{-3} + O(z)$ for the proof of (5.4). For $z = -w$, there is a pole of 1st order, whose residue turns out to be 1, by (2.12) and LIOUVILLE'S Second Theorem 2.4. □

1. **Proof** of the addition theorem and the proof of

$$c(w) = 0. \tag{5.5'}$$

(A second proof will follow in sect. 5.) Consider the elliptic function

$$g(z) := (f(z; w))^2 - \wp(z + w) - \wp(z) - \wp(w), \quad w \in \mathbb{C} \setminus \frac{1}{2}\Omega.$$

According to Proposition 5.2, g has poles at most at the points (5.3). Near $z = 0$,

$$g(z) = \left(z^{-2} + 2\wp(w)\right) - \wp(w) - z^{-2} - \wp(w) + O(z) = O(z). \tag{*}$$

Near $z = -w$,

$$g(z) = \frac{1}{(z+w)^2} + \frac{2c(w)}{z+w} - \frac{1}{(z+w)^2} + O(1) = \frac{2c(w)}{z+w} + O(1).$$

Therefore, g has at most 1st order poles at the points $-w + \Omega$. However, by LIOUVILLE'S Second Theorem 2.4, such poles cannot occur. Thus $c(w) = 0$ follows and g is constant by LIOUVILLE'S First Theorem 2.3. (*) leads to $g = 0$. The previously excluded points $\omega \in \Omega$, $w = \omega/2 \notin \Omega$ now follow from a continuity argument. □

5.3 Corollary. *For $z \in \mathbb{C} \setminus \frac{1}{2}\Omega$,*

$$\wp(2z) = -2\wp(z) + \frac{1}{4}\left(\frac{\wp''(z)}{\wp'(z)}\right)^2. \tag{5.6}$$

Proof Letting $w \to z$, we obtain (5.6) from (5.1). □

§ 5 Elliptic curves and the Addition Theorem of the $\wp$-Function

Analogously, we can proceed with $\wp(nz)$ for $n = 3, 4, \ldots$ by using the addition theorem. Note that all $\wp(nz)$ can be rationally expressed as even elliptic functions by $\wp(z)$ (cf. §2). This question will be discussed in §6.

5.4 Corollary. *Given $z, w \in \mathbb{C}$ with $z, w, z \pm w \notin \Omega$,*

$$\wp(z + w) - \wp(z - w) = -\frac{\wp'(z) \cdot \wp'(w)}{(\wp(z) - \wp(w))^2}. \tag{5.7}$$

Proof We replace w by $-w$ in (5.1) and subtract both equations. □

5.5 Corollary. *Given $z \in \mathbb{C}$ with $z, z + \omega_1/2 \notin \Omega$, then*

$$\wp(z + \omega_1/2) = \frac{e_1 \wp(z) + e_1^2 + e_2 e_3}{\wp(z) - e_1}.$$

Proof For $w := \omega_1/2$, i.e. $\wp'(w) = 0$, we insert the differential equation from Theorem 2.12 into (5.1) and use Corollary 3.17. □

We obtain analogous formulas by cyclic permutation of e_1, e_2, e_3.

5.6 Remarks. a) Proposition 5.2 leads directly to the formula

$$\frac{1}{2}\frac{d}{dz}\frac{\wp'(z) - \wp'(w)}{\wp(z) - \wp(w)} = \wp(z) - \wp(z + w) \quad \text{for } z, w \in \mathbb{C} \text{ with } z, w, z \pm w \notin \Omega,$$

because the difference is an elliptic function without poles and therefore 0 according to (5.4).

b) With the help of Proposition 5.2 we can construct elliptic functions that have 1st order poles at any two different points of a period parallelogram.

c) The Addition Theorem 5.1 for the inverse function of $\wp$ appears in the works of C.F. GAUSS in a special case ([31], vol. VIII, 93–95).

d) According to the Addition Theorem 5.1 and the differential equation, the product $\wp'(z) \cdot \wp'(w)$ is a polynomial in $\wp(z)$, $\wp(w)$ and $\wp(z + w)$. After squaring, by considering the differential equation (3.7), we obtain an *algebraic addition theorem* for $\wp$, i.e., there exists a polynomial $0 \neq P \in \mathbb{C}[X, Y, Z]$ satisfying

$$P(\wp(z), \wp(w), \wp(z + w)) = 0 \quad \text{for all} \quad z, w \in \mathbb{C} \text{ with } z, w, z + w \notin \Omega.$$

Explicitly we get

$$P(X, Y, Z) = \left[4(X + Y + Z)(X - Y)^2 - 4X^3 + g_2 X + g_3 - 4Y^3 + g_2 Y + g_3\right]^2 \\ - 4(4X^3 - g_2 X - g_3) \cdot (4Y^3 - g_2 Y - g_3).$$

According to H.A. SCHWARZ [72], WEIERSTRASS demonstrated the following converse in his lectures (though not in the version printed in [88], vol. V).

5.7 Theorem. *If a meromorphic function f on $\mathbb{C}$ satisfies an algebraic addition theorem, then f is either*

a) *a rational function or*
b) *a rational function in* $e^{2\pi i \alpha z}$ *with a suitable* $0 \neq \alpha \in \mathbb{C}$ *or*
c) *an elliptic function with respect to a suitable lattice.*

For a *proof*, see in W.F. OSGOOD [61], 515–516, or H. HANCOCK [35], chap. II.

2. Elliptic curves. In §1, we already considered the factor group

$$\mathbb{C}/\Omega = \{z + \Omega \; ; \; z \in \mathbb{C}\}. \tag{5.8}$$

The subset

$$\mathbb{E} := \mathbb{E}(\Omega) := \{(X, Y) \in \mathbb{C} \times \mathbb{C} \; ; \; Y^2 = 4X^3 - g_2 X - g_3\} \subseteq \mathbb{C} \times \mathbb{C} \tag{5.9}$$

is called the (affine) *elliptic curve* with respect to Ω. The WEIERSTRASS $\wp$-function allows a parametrization of the elliptic curve (5.9) by the factor group (5.8):

5.8 Lemma. *The mapping*

$$\Phi : (\mathbb{C}/\Omega) \setminus \{\Omega\} \longrightarrow \mathbb{E}(\Omega) \, , \quad \Phi(z + \Omega) := (\wp(z), \wp'(z)), \tag{5.10}$$

is a bijection.

Proof The differential equation (3.7) shows that the image of Φ is contained in $\mathbb{E}$. Given $(X, Y) \in \mathbb{E}$, we choose $z \in \mathbb{C}$ with $\wp(z) = X$ according to Lemma 2.11. Then

$$Y^2 = 4X^3 - g_2 X - g_3 = \wp'(z)^2$$

follows. If necessary, replace z by $-z$, such that $\wp'(z) = Y$ holds in addition to $\wp(z) = X$. Thus (X, Y) is contained in the image of Φ and Φ is surjective.

Now, given $z_1, z_2 \in \mathbb{C} \setminus \Omega$ with $(\wp(z_1), \wp'(z_1)) = (\wp(z_2), \wp'(z_2))$, we obtain

$$z_1 \equiv z_2 \, (\text{mod } \Omega), \quad \text{if} \quad \wp'(z_1) \neq 0,$$

and

$$z_1, z_2 \equiv \omega_1/2, \omega_2/2, \omega_3/2 \, (\text{mod } \Omega), \quad \text{if} \quad \wp'(z_1) = 0,$$

from (2.12) and (2.11) respectively. But since $\wp(\omega_k/2) = e_k$, $k = 1, 2, 3$, are pairwise distinct (cf. (2.13)), it also follows in this case that $z_1 \equiv z_2 \, (\text{mod } \Omega)$. Thus Φ is injective, too. $\square$

Besides $\mathbb{E} = \mathbb{E}(\Omega)$, we consider the "closure"

$$\overline{\mathbb{E}} := \overline{\mathbb{E}}(\Omega) := \mathbb{E} \cup \{O\} \quad \text{with} \quad O := (\infty, \infty) \tag{5.11}$$

and extend the mapping Φ to a bijection

$$\Phi : \mathbb{C}/\Omega \longrightarrow \overline{\mathbb{E}}(\Omega), \quad \Phi(z + \Omega) := \begin{cases} (\wp(z), \wp'(z)) & \text{for } z \notin \Omega, \\ O & \text{for } z \in \Omega. \end{cases} \tag{5.12}$$

§ 5 Elliptic curves and the Addition Theorem of the ℘-Function 65

By means of the bijective mapping Φ we can now transfer the group structure of $\mathbb{C}/\Omega$ (cf. §1) onto the set $\overline{\mathbb{E}}$: given $P, Q \in \overline{\mathbb{E}}$, an addition is defined by

$$P + Q := \Phi(\Phi^{-1}(P) + \Phi^{-1}(Q)), \tag{5.13}$$

where the addition in $\mathbb{C}/\Omega$ is given by $(u + \Omega) + (v + \Omega) := (u + v) + \Omega$. This directly yields the

5.9 Theorem. $\overline{\mathbb{E}}(\Omega)$ *becomes an abelian group by means of the addition* (5.13) *with O as identity element.*

$$\Phi : \mathbb{C}/\Omega \to \overline{\mathbb{E}}(\Omega)$$

is an isomorphism of groups. Given $z \in \mathbb{C}\setminus\Omega$,

$$-(\wp(z), \wp'(z)) = (\wp(-z), \wp'(-z)) = (\wp(z), -\wp'(z)).$$

Given $u, v \in \mathbb{C}$ with $u, v, u + v \notin \Omega$, we have

$$(\wp(u), \wp'(u)) + (\wp(v), \wp'(v)) = (\wp(u + v), \wp'(u + v)). \tag{5.14}$$

Of course, this still does not clarify how to calculate the sum (5.13) for $P, Q \in \overline{\mathbb{E}}$. According to the Addition Theorem 5.1, we can at least calculate the first component of the point $P + Q$ from the components of P and Q with (5.14). In sections 4 and 5 explicit formulas for the components of $P + Q$ are derived in a geometric way. We thus, among other things, obtain a new proof of the Addition Theorem 5.1.

5.10 Remarks. a) The exceptional role of the point $O = (\infty, \infty)$ disappears if the elliptic curve is described "projectively": let $\mathbb{P}(\mathbb{C})$ denote the 2-dimensional *projective space* over $\mathbb{C}$. Then there is a canonical surjective mapping

$$\pi : (\mathbb{C} \times \mathbb{C} \times \mathbb{C}) \setminus \{0\} \to \mathbb{P}(\mathbb{C})$$

satisfying $\pi(X, Y, Z) = \pi(X', Y', Z')$ if and only if $\mathbb{C}(X, Y, Z) = \mathbb{C}(X', Y', Z')$. The *projective elliptic curve* with respect to Ω is defined by

$$\mathbb{PE} := \mathbb{PE}(\Omega) := \{\pi(X, Y, Z) \; ; \; Y^2 Z = 4X^3 - g_2 X Z^2 - g_3 Z^3\}.$$

Obviously, $(X, Y) \mapsto \pi(X, Y, 1)$ defines an injection $\mathbb{E} \to \mathbb{PE}$ and the point O appears as $\pi(0, 1, 0)$.
b) If we consider $\mathbb{C}/\Omega$ as a RIEMANN surface and $\overline{\mathbb{E}}$ as a curve in $\mathbb{P}(\mathbb{C})$, then (5.12) defines a biholomorphic mapping.

3. Examples. a) *The real part of* $\mathbb{E}$. If Ω is conjugation stable (cf. §3), then g_2 and g_3 are real by Theorem 3.23 and we can draw the "real part" of $\mathbb{E}$. Depending on whether $4X^3 - g_2 X - g_3$ has three or one real roots, we get the two images:

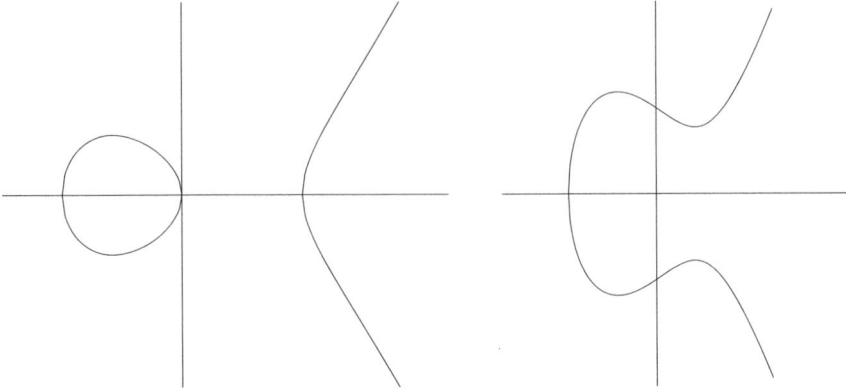

Figure 11: Elliptic curves

b) *A* FERMAT *curve.* Choose Ω according to Corollary 4.11 with $g_2 = 0$, $g_3 = 12^3$, hence $\wp'^2 = 4\wp^3 - 1728$. In fact,

$$\Omega = \frac{\Gamma(1/3)^3}{4\pi\sqrt{3}}(\mathbb{Z}\rho + \mathbb{Z})$$

follows from Remark 4.17. A calculation shows that in this case the mapping

$$(X, Y) \mapsto (U, V), \quad U := \frac{72 + Y}{12X}, \quad V := \frac{72 - Y}{12X},$$

is a bijection from

$$\mathbb{E}' := \{(X, Y) \in \mathbb{C} \times \mathbb{C} \,;\, Y^2 = 4X^3 - 1728, \, X \neq 0\}$$

onto the FERMAT curve

$$\mathbb{F}_3 := \{(U, V) \in \mathbb{C} \times \mathbb{C} \,;\, U^3 + V^3 = 1\}.$$

The inverse is given by

$$(U, V) \mapsto (X, Y) \quad \text{with} \quad X := \frac{12}{U + V} \quad \text{and} \quad Y := 72\frac{U - V}{U + V}.$$

4. Intersection formulas. Points P of $\mathbb{C} \times \mathbb{C}$ shall always be written in the form $P = (X_P, Y_P)$. Given $P, Q \in \mathbb{E}$ with

$$X_P \neq X_Q, \tag{5.15}$$

§ 5 Elliptic curves and the Addition Theorem of the ℘-Function

we consider the complex line $\Gamma = \Gamma_{P,Q}$ through P and Q:

$$Y = a_{P,Q} X + b_{P,Q}. \qquad (5.16)$$

Here, $a_{P,Q}$ and $b_{P,Q}$ are of course given by

$$a_{P,Q} := \frac{Y_P - Y_Q}{X_P - X_Q}, \qquad (5.17)$$

$$b_{P,Q} := \frac{X_P Y_Q - X_Q Y_P}{X_P - X_Q}.$$

The line Γ will intersect the curve $\mathbb{E}$ in another point. In the case that $\mathbb{E}$ has a real part, the situation is approximately as in the figure on the right-hand side. To make this more precise, we define a point $P \bullet Q$ in $\mathbb{C} \times \mathbb{C}$ with the coordinates

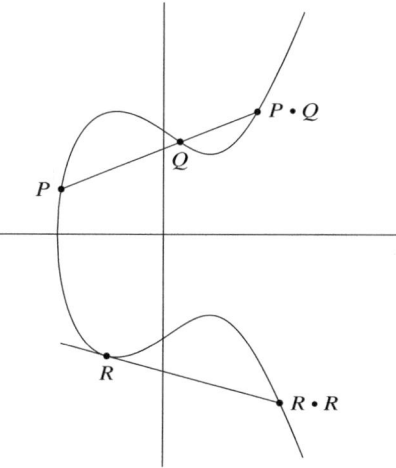

Figure 12: Lines and elliptic curves

$$X_{P \bullet Q} := \frac{1}{4} a_{P,Q}^2 - X_P - X_Q, \quad Y_{P \bullet Q} := a_{P,Q} X_{P \bullet Q} + b_{P,Q}. \qquad (5.18)$$

Obviously, $P \bullet Q$ lies on Γ.

5.11 Lemma. *For all $X \in \mathbb{C}$*

$$4X^3 - g_2 X - g_3 = 4(X - X_P)(X - X_Q)(X - X_{P \bullet Q}) + (a_{P,Q} X + b_{P,Q})^2 \qquad (5.19)$$

holds.

Proof First, the coefficients of X^3 in (5.19) coincide and because of (5.18), the coefficients of X^2 as well. Thus (5.19) reduces to an equation of the form

$$g_2 X + g_3 = AX + B \quad \text{with} \quad A, B \in \mathbb{C}.$$

However, since (5.19) is correct for $X = X_P$ and $X = X_Q$ due to the choice of Γ, it follows that $A = g_2$ and $B = g_3$, hence the validity of (5.19). □

This yields

5.12 Corollary. *Given $P, Q \in \mathbb{E}$ with $X_P \neq X_Q$, then $P \bullet Q \in \mathbb{E}$.*

Thus the point $P \bullet Q$ is the third intersection of the line Γ with $\mathbb{E}$. The formulas (5.18) are therefore called *intersection formulas*. If $P \neq Q$ holds with $X_P = X_Q$, then we have $P \bullet Q = O$. In the case $P = Q$ we now proceed analogously: instead of the connecting line we have to take the tangent line through a point $P \in \mathbb{E}$ with $Y_P \neq 0$. Setting

$$a_P := \frac{12X_P^2 - g_2}{2Y_P}, \quad b_P := Y_P - a_P X_P \qquad (5.20)$$

then $Y = a_P X + b_P$ is the tangent line through P to $\mathbb{E}$. Let $P \bullet P \in \mathbb{C} \times \mathbb{C}$ be given by

$$X_{P \bullet P} := \frac{1}{4} a_P^2 - 2X_P, \quad Y_{P \bullet P} := a_P X_{P \bullet P} + b_P. \qquad (5.21)$$

5.13 Lemma. *For all $X \in \mathbb{C}$ one has*

$$4X^3 - g_2 X - g_3 = 4(X - X_P)^2 (X - X_{P \bullet P}) + (a_P X + b_P)^2. \qquad (5.22)$$

Proof Again the coefficients of X^3 and, by (5.21), the coefficients of X^2 coincide. But (5.22) holds for $X = X_P$ and, because of (5.20), the differentiated equation of (5.22) holds. □

Thus we obtain the

5.14 Corollary. *For $P \in \mathbb{E}$ with $Y_P \neq 0$ one has $P \bullet P \in \mathbb{E}$.*

Thus, the point $P \bullet P$ is the second point of intersection of the tangent to $\mathbb{E}$ through P and (5.21) are the corresponding *intersection formulas*. If $Y_P = 0$, then we have $P \bullet P = O$.

5. Application to the $\wp$-function. Using the bijection (5.12),

$$\Phi : \mathbb{C}/\Omega \longrightarrow \overline{\mathbb{E}}(\Omega), \quad \Phi(z + \Omega) := \begin{cases} (\wp(z), \wp'(z)) & \text{for } z \notin \Omega, \\ O & \text{for } z \in \Omega, \end{cases} \qquad (5.23)$$

we define for $u, v \in \mathbb{C}$ points P, Q of $\overline{\mathbb{E}}$ by

$$P := \Phi(u + \Omega), \quad Q := \Phi(v + \Omega). \qquad (5.24)$$

5.15 Lemma. *If $u, v, w \in \mathbb{C} \setminus \Omega$ satisfy $u + v + w \in \Omega$, and $u + \Omega, v + \Omega, w + \Omega$ are pairwise distinct, then*

$$P \bullet Q = \Phi(w + \Omega).$$

Proof The elliptic function $f(z) := \wp'(z) - (a_{P,Q}\wp(z) + b_{P,Q})$ has a 3rd order pole at 0, hence also 3 zeros in $\mathbb{C}/\Omega$. By construction $f(u) = f(v) = 0$ and from LIOUVILLE's Fourth Thoerem 2.6 it follows that $f(w) = 0$. Then P, Q and $R := \Phi(w + \Omega)$ are points of intersection of a line with $\mathbb{E}$. For example, if $P \bullet Q$ were equal to P, we get $a_{P,Q} = a_P$ and $b_{P,Q} = b_P$ from (5.17), (5.18), (5.20) and (5.21). Then the line would be tangent to $\mathbb{E}$ and contain only the two points P and Q as points of intersection with $\mathbb{E}$. Because P, Q, R are pairwise distinct, this is not possible and we get $R = P \bullet Q$. □

2. Proof of the Addition Theorem 5.1. If $u, v, w \in \mathbb{C} \setminus \Omega$ are given such that the cosets $u + \Omega, v + \Omega, w + \Omega$ are pairwise distinct, then Lemma 5.15, (5.17) and (5.18) imply

§ 5 Elliptic curves and the Addition Theorem of the ℘-Function

$$\wp(u+v) = \wp(-w) = \wp(w) = \frac{1}{4}a_{P,Q}^2 - X_P - X_Q$$
$$= \frac{1}{4}\left(\frac{\wp'(u) - \wp'(v)}{\wp(u) - \wp(v)}\right)^2 - \wp(u) - \wp(v).$$

The remaining cases follow from a continuity argument. □

Finally, for $P \in \mathbb{C} \times \mathbb{C}$ we set

$$P^* := (X_P, -Y_P), \quad \text{if } P = (X_P, Y_P). \tag{5.25}$$

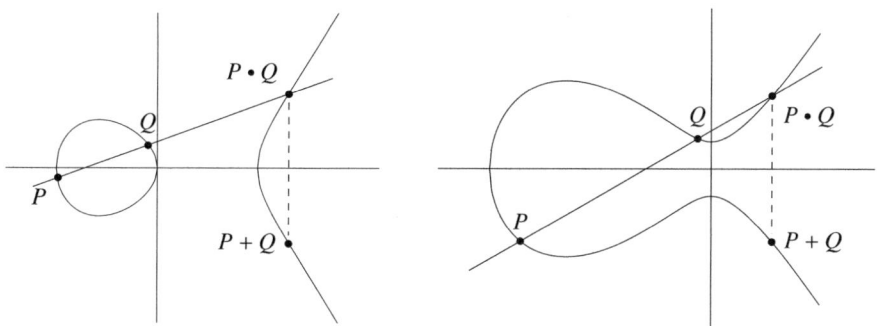

Figure 13: Addition on elliptic curves

The final result is the

5.16 Theorem. *The addition* $(P, Q) \mapsto P + Q$ *on* $\mathbb{E}$ *is described by*

$$P + Q = (P \bullet Q)^*, \quad \text{if } X_P \neq X_Q, \tag{5.26}$$

and by

$$2P = (P \bullet P)^*, \quad \text{if } Y_P \neq 0, \tag{5.27}$$

i.e., by

$$X_{P+Q} := \frac{1}{4}a_{P,Q}^2 - X_P - X_Q, \quad Y_{P+Q} := -a_{P,Q}X_{P+Q} - b_{P,Q}, \tag{5.28}$$

if $X_P \neq X_Q$, *and by*

$$X_{2P} := \frac{1}{4}a_P^2 - 2X_P, \quad Y_{2P} := -a_P X_{2P} - b_P, \tag{5.29}$$

if $Y_P \neq 0$. *Furthermore,*

$$-P = P^* = (X_P, -Y_P). \tag{5.30}$$

Proof If $\wp(u) \neq \wp(v)$ then (5.13) and Lemma 5.15 lead to

$$P + Q = \Phi(\Phi^{-1}(P) + \Phi^{-1}(Q)) = \Phi(u + v + \Omega) = \Phi(-w + \Omega)$$
$$= (\wp(w), -\wp'(w)) = (\Phi(w + \Omega))^* = (P \bullet Q)^*.$$

The remaining assertions now follow from (5.18) or (5.21) and Theorem 5.9. □

5.17 Remarks. a) Formula (5.28) contains the addition theorem for the derivative of the $\wp$-function. The addition is fully described by Theorem 5.16 if we note that $X_P = X_Q$ already implies $Y_P = \pm Y_Q$ and $(X_P, Y_P) + (X_P, -Y_P) = O$.

b) In the vector space $\mathbb{C} \times \mathbb{C}$, three points $(a_1, a_2), (b_1, b_2)$ and (c_1, c_2) lie on a line if and only if

$$\det \begin{pmatrix} 1 & 1 & 1 \\ a_1 & b_1 & c_1 \\ a_2 & b_2 & c_2 \end{pmatrix} = 0$$

holds. For $P, Q \in \mathbb{E}$, the three points P, Q and $P \bullet Q$ lie on a line by definition. According to Lemma 5.15, for $u, v, w \in \mathbb{C} \setminus \Omega$ with $u + v + w \in \Omega$, it follows that

$$\det \begin{pmatrix} 1 & 1 & 1 \\ \wp(u) & \wp(v) & \wp(w) \\ \wp'(u) & \wp'(v) & \wp'(w) \end{pmatrix} = 0.$$

6*. Rational points. Let Ω be a lattice such that g_2 and g_3 *are rational numbers*. Then consider the subset

$$\mathbb{E}_\mathbb{Q} := \mathbb{E}_\mathbb{Q}(\Omega) := \{(X, Y) \in \mathbb{E} \; ; \; X, Y \in \mathbb{Q}\} \tag{5.31}$$

of *rational points* of $\mathbb{E}$. The formulas (5.28), (5.29) and (5.30) show that

$$\overline{\mathbb{E}}_\mathbb{Q} := \mathbb{E}_\mathbb{Q}(\Omega) \cup \{O\}, \quad O := (\infty, \infty), \tag{5.32}$$

is a subgroup of $\overline{\mathbb{E}}$.

5.18 Examples. a) *A* FERMAT *curve.* According to the example in sect.3, the rational points

$$\mathbb{E}_\mathbb{Q} := \{(X, Y) \in \mathbb{Q} \times \mathbb{Q} \; ; \; Y^2 = 4X^3 - 1728\}$$

and

$$\mathbb{F}_{3,\mathbb{Q}} := \{(U, V) \in \mathbb{Q} \times \mathbb{Q} \; ; \; U^3 + V^3 = 1\}$$

are birationally related to each other by $X = \frac{12}{U+V}$, $Y = 72\frac{U-V}{U+V}$. $\mathbb{F}_{3,\mathbb{Q}}$ is known to have only the solutions $(1, 0)$ and $(0, 1)$, since the FERMAT problem is unsolvable by the Theorem of WILES (*Ann. Math.* **141**, 443–551 (1995)). However, the case $n = 3$ can be proved in an elementary way (cf. H.M. EDWARDS [18]). Hence $(12, 72)$ and $(12, -72)$ are the only rational points contained in $\mathbb{E}$. Thus $\overline{\mathbb{E}}_\mathbb{Q}$ is a group of order 3.

b) Consider

$$\mathbb{E} := \{(X, Y) \in \mathbb{C} \times \mathbb{C} \; ; \; Y^2 = 4X^3 - 4X + 4\}.$$

Obviously $P := (1, 2) \in \mathbb{E}_\mathbb{Q}$ gives $2P = (-1, 2) \in \mathbb{E}_\mathbb{Q}$. Using (5.28), we calculate successively

§ 5 Elliptic curves and the Addition Theorem of the ℘-Function

$$3P = (0, -2), \quad 4P = (3, -10), \quad 5P = (5, 22),$$
$$6P = \left(\frac{1}{4}, \frac{7}{4}\right), \quad 7P = \left(-\frac{11}{9}, -\frac{34}{27}\right), \quad 8P = \left(\frac{19}{25}, -\frac{206}{125}\right). \tag{5.33}$$

We can show that $\overline{\mathbb{E}}$ is an infinite cyclic group, which is generated by P.
c) The group belonging to $Y^2 = 4X^3 - 28X + 25$ has rank 3.

5.19 Remarks. a) According to a Theorem of L.J. MORDELL (*Proc. Cambridge Phil. Soc.* **21**, 179–192 (1922)), the groups $\overline{\mathbb{E}}_\mathbb{Q}$ are always finitely generated. This result was already conjectured by H. POINCARÉ.
b) Besides the rational points of $\mathbb{E}$, we can also search for the *integer points* of $\mathbb{E} \cap (\mathbb{Z} \times \mathbb{Z})$. L.J. MORDELL showed in 1923 (*Proc. London Math. Soc.* (2) **21**, 415–419) that this set is always finite. Compare C.L. SIEGEL [78], vol. I, 207–208. Because of (5.33) it is clear that the set of integer points of $\mathbb{E}$ in general is not a subgroup of $\overline{\mathbb{E}}_\mathbb{Q}$. Integer points can easily be calculated using SAGE.
c) H.W. LENSTRA (*Proc. Int. Congr. Berkeley* 1986, 99–120) has developed a method for the factorization of large natural numbers using elliptic curves. This method has shown to be very effective. We start from a natural number n, choose natural numbers x, y, g_2 with a random number generator and define $g_3 := 4x^3 - g_2 x - y^2$. In the case $g_2^3 - 27g_3^2 \neq 0$, the equation

$$Y^2 = 4X^3 - g_2 X - g_3$$

defines an elliptic curve containing the point $P := (x, y)$. With the formulas (5.18) and (5.21), we can now successively calculate the multiples kP on the elliptic curve (mod n) as long as the denominators in (5.17) and (5.20) are coprime to n. Thus, the method yields either the points kP on the elliptic curve (mod n) or a divisor of n.

References: H. COHEN [11]. D.H. HUSEMÖLLER [41]. J.H. SILVERMAN [79]. D. ZAGIER, Lösungen von Gleichungen in ganzen Zahlen, in *Miscellanea mathematica*, Springer–Verlag, Berlin–Heidelberg–New York 1991, 311–326.

7*. Congruent numbers. A positive integer f is called *congruent number* or HERON *number* if there is a right triangle with *rational* side lengths a, b, c and area f. We can assume without loss of generality that f is square-free. The problem of which numbers are congruent numbers in this sense was already dealt with by the Greeks and is (besides the problem of perfect numbers) the last major unsolved problem of antiquity (cf. L.E. DICKSON, *History of the theory of numbers II*, Chelsea, New York 1960, chap. XVI). Assuming that c is the hypotenuse of a right triangle, the following is true:

$$\frac{1}{2}ab = f. \tag{5.34}$$

The obvious idea of using the classical description of Pythagorean triples (cf. T.M. APOSTOL [6], p. 4) for the solution of the problem leads to the

5.20 Proposition. *A natural square-free number f is a congruent number if and only if there are an integer q and coprime integers u, v satisfying*

$$q^2 f = uv(u^2 - v^2) \quad \text{and} \quad u > v, u \not\equiv v \pmod{2}. \tag{5.35}$$

Proof For a congruent number f, choose rational a, b, c with (5.34) and $q \in \mathbb{N}$ such that qa, qb, qc is a primitive Pythagorean triple, where without loss of generality qb is even. Then there are coprime integers u, v with $u > v > 0$, $u \not\equiv v \pmod{2}$ and

$$qa = u^2 - v^2, \quad qb = 2uv, \quad qc = u^2 + v^2. \tag{*}$$

It follows that $q^2 f = \frac{1}{2} qa \cdot qb = uv(u^2 - v^2)$, thus (5.35).
Conversely, if q, u, v are given by (5.35), then a, b, c may be defined by (*). □

As $n^2 f$ is a congruent number whenever f is, we can restrict ourselves to the description of square-free congruent numbers. The list of square-free congruent numbers ≤ 50 is given by

$$5, 6, 7, 13, 14, 15, 21, 22, 23, 29, 30, 31, 34, 37, 38, 39, 41, 46, 47.$$

There are exactly 361 square-free congruent numbers ≤ 1000. The following small table shows that we get many congruent numbers this way, but we cannot decide whether a given number is a congruent number. Here, $q \in \mathbb{N}$ is chosen maximal satisfying $q^2 | uv(u^2 - v^2)$:

u	v	$uv(u^2-v^2)$	q	$a = (u^2-v^2)/q$	$b = 2uv/q$	$c = (u^2+v^2)/q$	f
2	1	6	1	3	4	5	6
3	2	30	1	5	12	13	30
4	1	60	2	15/2	4	17/2	15
4	3	84	2	7/2	12	25/2	21
5	4	180	6	3/2	20/3	41/6	5
7	2	630	3	15	28/3	53/3	70
8	1	505	6	21/2	8/3	65/6	14
9	4	2340	6	65/6	12	97/6	65
9	8	1224	6	17/6	24	145/6	34
13	12	3900	10	5/2	156/5	313/10	39
16	9	25200	60	35/12	24/5	337/60	7
25	16	147600	60	123/20	40/3	881/60	41
50	49	242550	105	33/35	140/3	4901/105	22
72	49	9818424	462	253/42	168/11	7585/462	46
325	36	1220649300	9690	323/30	780/323	106921/9690	13
1250	289	534281163750	118575	5301/425	1700/279	1646021/118575	38
1600	81	330925694400	103320	8897/360	720/287	2566561/103320	31
4901	4900	235370034900	90090	99/910	52780/99	48029801/90090	29

There are 3 square-free congruent numbers ≤ 50 missing in the table above. They are given by

$$f = 23, \quad u = 24\,336, \quad v = 17\,689,$$
$$f = 37, \quad u = 777\,925, \quad v = 1\,764,$$
$$f = 47, \quad u = 14\,561\,856, \quad v = 2\,289\,169.$$

An example with a large "minimal" solution was given by D. ZAGIER:

§ 5 Elliptic curves and the Addition Theorem of the ℘-Function

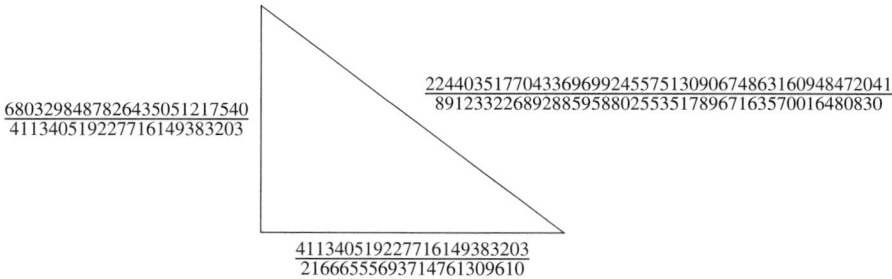

Figure 14: Congruent number 157

The ratio is about 16.6 : 19 : 25. Here $f = 157$. Surprisingly, there is a relation between the congruent numbers and certain rational points of elliptic curves: for abbreviation, in the remainder of this section let $v(x) \in \mathbb{N}$ denote the denominator of $0 \neq x \in \mathbb{Q}$ in reduced fraction notation.

5.21 Lemma. *For a natural square-free number f, the following assertions are equivalent:*

(i) *f is a congruent number.*
(ii) *There exists a rational point (x, y) on the elliptic curve $Y^2 = 4X^3 - 4f^2X$ such that x is a square in $\mathbb{Q}$ with even $v(x)$.*

Proof (i) $\Longrightarrow$ (ii): We choose $q \in \mathbb{N}$ and coprime integers $u, v, u \not\equiv v \pmod{2}, u > v$, such that the triangle with sides

$$a = (u^2 - v^2)/q, \quad b = 2uv/q, \quad c = (u^2 + v^2)/q$$

has area f:

$$a^2 + b^2 = c^2 \quad \text{and} \quad f = \frac{1}{2}ab. \tag{*}$$

Since qa, qb, qc is a primitive Pythagorean triple, $qc = u^2 + v^2$ is odd. If we now set $x := (c/2)^2$, then $v(x)$ is even. Because of $(*)$,

$$x \pm f = (c/2)^2 \pm f = ((a \pm b)/2)^2$$

holds, such that $4x^3 - 4f^2x = 4x(x^2 - f^2) =: y^2$ with $y = c(a^2 - b^2)/4 \in \mathbb{Q}$ follows.
(ii) $\Longrightarrow$ (i): We set $\alpha := \sqrt{x} \in \mathbb{Q}$, $\beta := y/\alpha$ and obtain

$$\beta^2 = 4x^2 - 4f^2, \quad \text{i.e.} \quad \beta^2 + 4f^2 = 4x^2.$$

By assumption, $t := v(\alpha)$ is even. Since f is an integer, β^2 and $4x^2$ both have denominator $v(4x^2) = t^4/4$. Thus $t^2\beta/2, t^2f, t^2x$ is a primitive Pythagorean triple. Therefore, there exist coprime natural numbers $u, v, u \not\equiv v \pmod{2}$, $u > v$ satisfying

$$t^2\beta/2 = u^2 - v^2, \quad t^2f = 2uv, \quad (t\alpha)^2 = u^2 + v^2,$$

as t^2f is even. Thus $2u/t, 2v/t, 2\alpha$ are the sides of a right triangle of area f. □

References: N. KOBLITZ [47]. G. KRAMARZ, *Math. Ann.* **273**, 337–340 (1986).

5.22 Exercises.
1) Express $\wp(2z)$ and $\wp(3z)$ rationally by $\wp(z)$.
2) Show that $4\wp'(z+w)$ is equal to

$$-2(\wp'(z) + \wp'(w)) + \frac{(\wp'(z) - \wp'(w))(\wp''(z) - \wp''(w))}{(\wp(z) - \wp(w))^2} - \left(\frac{\wp'(z) - \wp'(w)}{\wp(z) - \wp(w)}\right)^3$$

as well as

$$\wp'(2z) = -\wp'(z) + \frac{1}{4}\frac{\wp''(z)\wp'''(z)}{\wp'(z)^2} - \frac{1}{4}\left(\frac{\wp''(z)}{\wp'(z)}\right)^3$$

$$= -\wp'(z) + \frac{3\wp(z)\wp''(z)}{\wp'(z)} - \frac{1}{4}\left(\frac{\wp''(z)}{\wp'(z)}\right)^3$$

$$= \frac{28\wp^3(z) - g_2\wp(z) + 2g_3}{2\wp'(z)} - \frac{1}{4}\left(\frac{12\wp^2(z) - g_2}{2\wp'(z)}\right)^3.$$

3) Given $z, w \in \mathbb{C}$ with $z, w, z \pm w \notin \Omega$,

$$\wp(z+w) + \wp(z-w) = \frac{\left(2\wp(z)\wp(w) - \frac{1}{2}g_2\right)(\wp(z) + \wp(w)) - g_3}{(\wp(z) - \wp(w))^2},$$

$$\frac{\wp'(z + \omega/2)}{\wp'(z)} = -\frac{1}{2}\frac{\wp''(\omega/2)}{(\wp(z) - \wp(\omega/2))^2}, \quad \omega \in \Omega, \ \omega/2 \notin \Omega.$$

4) Show that

$$(\wp(\omega_1/4) - e_1)^2 = 2e_1^2 + e_2 e_3 = (e_1 - e_2)(e_1 - e_3).$$

5) Show that

$$X^4 - \frac{1}{2}g_2 X^2 - g_3 X - \frac{1}{48}g_2^2 =$$
$$\left(X - \wp\left(\frac{\omega_1}{3}\right)\right)\left(X - \wp\left(\frac{\omega_2}{3}\right)\right)\left(X - \wp\left(\frac{\omega_1 + \omega_2}{3}\right)\right)\left(X - \wp\left(\frac{\omega_1 - \omega_2}{3}\right)\right).$$

6) If $z_0 \in \mathbb{C}$ with $\wp(z_0) = 0$, then

$$\wp(2z_0) = -\frac{g_2^2}{16g_3} \quad \text{and} \quad \wp(3z_0) = \frac{8g_3}{g_2} - \frac{256g_3^3}{g_2^4}.$$

7) a) There are exactly 3 points of order 2 in $\overline{\mathbb{E}}$, namely $(e_1, 0), (e_2, 0)$ and $(e_3, 0)$.
b) For $N \in \mathbb{N}$, there exist exactly N^2 points in $\overline{\mathbb{E}}$ whose order is a divisor of N. How many points of order N are there?
8) Let $g_2, g_3 \in \mathbb{Q}$ and $N \in \mathbb{Z}, N > 1$. If $P = (x, y) \in \overline{\mathbb{E}}$ is of order N, then the coefficients x, y are algebraic over $\mathbb{Q}$ of degree $\leq N^2$.

§ 6 Product expansions

9) $\overline{\mathbb{E}}$ is always isomorphic to $(\mathbb{R}/\mathbb{Z}) \times (\mathbb{R}/\mathbb{Z})$.
10) Let Ω be a rectangular lattice. Then

$$\mathbb{E}_\mathbb{R} := \mathbb{E} \cap (\mathbb{R} \times \mathbb{R}) = \{(\wp(z), \wp'(z))\, ; \, z \in \left(\tfrac{\omega}{2} + \mathbb{R}\right) \setminus \Omega,\; \omega \in \Omega\}.$$

$\overline{\mathbb{E}}_\mathbb{R} := \mathbb{E}_\mathbb{R} \cup \{O\}$ is a subgroup of $\overline{\mathbb{E}}$ isomorphic to $(\mathbb{Z}/2\mathbb{Z}) \times (\mathbb{R}/\mathbb{Z})$.
11) Give a parametrization of the FERMAT curve $\mathbb{F}_3$ in sect.3.
12) Let f be a square-free congruent number and (x, y) be a rational point of the elliptic curve $Y^2 = 4X^3 - 4f^2 X$ according to Lemma 5.21, such that x is a square in $\mathbb{Q}$ with even $\nu(x)$. Then $\nu(y) \equiv 0 \pmod{4}$ holds.
13) Let $f \in \mathbb{N}$ be square-free. f is a congruent number if and only if there exists some x such that x, $x + f$, $x - f$ are squares in $\mathbb{Q}$.
14) Given $f \in \mathbb{N}$, determine the lattice Ω for the elliptic curve $Y^2 = 4X^3 - 4f^2 X$.
15) Each *tetrahedral number*, i.e. a number of the form $\binom{n+2}{3}$, $n \in \mathbb{N}$, which is simultaneously a square resp. a *triangular number*, i.e. a number of the form $\binom{m+1}{2}$, $m \in \mathbb{N}$, leads to an integer point of the elliptic curve $Y^2 = 4X^3 - 144X$ resp. $Y^2 = 4X^3 - 36X + 81$.
16) Each sum of squares $\sum_{k=1}^{n} k^2$, $n \in \mathbb{N}$, which is simultaneously a square, leads to an integer point of the elliptic curve $Y^2 = 4X^3 - 144X$.

§ 6 Product expansions

In addition to presenting the theory of the ζ- and σ-functions the aim of this paragraph is to derive the product formula for the discriminant $\Delta = g_2^3 - 27g_3^2$. The arguments are classical, we follow the well organized presentation of S. LANG [52], chap. 18. Furthermore, as an application, we consider the problem of multiplication of the $\wp$-function. Let Ω always be a lattice in $\mathbb{C}$ and (ω_1, ω_2) a basis of Ω.

1. The σ-function. According to the WEIERSTRASS Product Theorem A.10 and the Convergence Lemma 1.17, the product

$$\sigma(z) := \sigma(z; \Omega) := z \cdot \prod_{0 \neq \omega \in \Omega} \left(1 - \frac{z}{\omega}\right) \cdot e^{\frac{z}{\omega} + \frac{1}{2}\left(\frac{z}{\omega}\right)^2} \tag{6.1}$$

is absolutely and uniformly convergent on every compact subset of $\mathbb{C}$. Thus σ is an entire function that has 1st order roots exactly at $z \in \Omega$. Since $-\omega$ runs through Ω as ω does, σ *is odd*. We call $\sigma(z)$ the WEIERSTRASS *sigma function*. Following WEIERSTRASS, the logarithmic derivative of σ is denoted by ζ, i.e.

$$\zeta(z) := \zeta(z; \Omega) := \frac{\sigma'(z)}{\sigma(z)} = \frac{1}{z} + \sum_{0 \neq \omega \in \Omega} \left(\frac{1}{z - \omega} + \frac{1}{\omega} + \frac{z}{\omega^2}\right), \quad z \notin \Omega. \tag{6.2}$$

In order to avoid misunderstandings with the RIEMANN zeta function, we speak here of the WEIERSTRASS *zeta function*. Because of

$$\frac{1}{z-\omega} + \frac{1}{\omega} + \frac{z}{\omega^2} = \frac{z^2}{\omega^2(z-\omega)}$$

and the Convergence Lemma 1.17, the series (6.2) converges absolutely uniformly on every compact subset of $\mathbb{C}$ that does not contain a lattice point. *Obviously ζ is also odd.*
A comparison with (3.3) results in

$$\zeta' = \left(\frac{\sigma'}{\sigma}\right)' = -\wp. \tag{6.3}$$

Therefore, $\wp(z+\omega) = \wp(z)$ for $\omega \in \Omega$ shows that

$$\eta(\omega) := \eta(\omega; \Omega) := \zeta(z+\omega) - \zeta(z) \tag{6.4}$$

is independent of z. It follows that

$$\eta(\omega + \omega') = \eta(\omega) + \eta(\omega') \quad \text{for all} \ \ \omega, \omega' \in \Omega. \tag{6.5}$$

Hence $\eta : \Omega \to \mathbb{C}$ is a homomorphism of groups.

6.1 LEGENDRE Relation. *It holds that*

$$\eta(\omega_2) \cdot \omega_1 - \eta(\omega_1) \cdot \omega_2 = 2\pi i \quad \text{if} \ \ \operatorname{Im}(\omega_1/\omega_2) > 0.$$

Proof We integrate $\zeta(z)$ over the positively oriented boundary ∂P of a period parallelogram $P := \Diamond(u; \omega_1, \omega_2)$, where u is chosen such that 0 is an interior point of P. Using the Residue Theorem A.5, we obtain

$$\int_{\partial P} \zeta(z)dz = 2\pi i,$$

because ζ has exactly one 1st order pole with residue 1 in P. On the other hand,

$$\int_{\partial P} \zeta(z)dz = \int_u^{u+\omega_2} \zeta(z)dz + \int_{u+\omega_2}^{u+\omega_1+\omega_2} \zeta(z)dz + \int_{u+\omega_1+\omega_2}^{u+\omega_1} \zeta(z)dz + \int_{u+\omega_1}^{u} \zeta(z)dz$$

$$= \int_u^{u+\omega_2} (\zeta(z) - \zeta(z+\omega_1))dz + \int_{u+\omega_1}^{u} (\zeta(z) - \zeta(z+\omega_2))dz,$$

$$= -\eta(\omega_1)\omega_2 + \eta(\omega_2)\omega_1$$

follows, if we note that the parallelogram $(0, \omega_2, \omega_1 + \omega_2, \omega_1)$ is positively oriented by assumption. The assertion follows from (6.4). $\square$

§ 6 Product expansions

6.2 Corollary. *For any $\omega, \omega' \in \Omega$,*

$$\eta(\omega) \cdot \omega' - \eta(\omega') \cdot \omega \in 2\pi i \mathbb{Z}.$$

Proof Use (6.5) and the LEGENDRE Relation 6.1. □

Sometimes we simply write

$$\eta_1 := \eta(\omega_1), \quad \eta_2 := \eta(\omega_2), \tag{6.6}$$

but note that this definition depends on the choice of the basis. Finally, from (6.1), (6.2) and (6.4), we obtain for $0 \neq \lambda \in \mathbb{C}$:

$$\left.\begin{aligned}\sigma(\lambda z; \lambda\Omega) &= \lambda \cdot \sigma(z; \Omega), \\ \zeta(\lambda z; \lambda\Omega) &= \tfrac{1}{\lambda} \cdot \zeta(z; \Omega), \\ \eta(\lambda\omega; \lambda\Omega) &= \tfrac{1}{\lambda} \cdot \eta(\omega; \Omega).\end{aligned}\right\} \tag{6.7}$$

2. The behavior of σ under translations from Ω. Of course, σ, as an entire non-constant function, is not an elliptic function with respect to the lattice Ω. In order to describe the behavior of σ under the translations $z \mapsto z + \omega$, $\omega \in \Omega$, we define a mapping $\chi : \Omega \mapsto \{\pm 1\}$ by

$$\chi(\omega) := \begin{cases} 1, & \text{if } \omega/2 \in \Omega, \\ -1, & \text{if } \omega/2 \notin \Omega. \end{cases} \tag{6.8}$$

6.3 Theorem. *For all $\omega \in \Omega$ and $z \in \mathbb{C}$,*

$$\sigma(z + \omega) = \chi(\omega) \cdot e^{\eta(\omega)(z+\omega/2)} \cdot \sigma(z). \tag{6.9}$$

Proof For $z \in \Omega$, both sides of the equation are equal to 0. Because of (6.2), we have $\sigma'(z) = \sigma(z) \cdot \zeta(z)$ for $z \notin \Omega$, and (6.4) yields

$$\frac{d}{dz}\left(\frac{\sigma(z+\omega)}{\sigma(z)}\right) = \frac{\sigma(z+\omega)}{\sigma(z)} \cdot \eta(\omega) \quad \text{for } \omega \in \Omega.$$

Thus

$$\frac{\sigma(z+\omega)}{\sigma(z)} \cdot e^{-\eta(\omega)z}, \quad \text{hence} \quad \psi(\omega) := \frac{\sigma(z+\omega)}{\sigma(z)} \cdot e^{-\eta(\omega)(z+\omega/2)}$$

is independent of z. Since σ is odd, we first set $z = -\omega/2$ and obtain

$$\psi(\omega) = -1, \quad \text{if } \omega/2 \notin \Omega. \tag{*}$$

Then (6.5) leads to

$$\psi(2\omega) = \frac{\sigma(z+2\omega)}{\sigma(z+\omega)} \cdot \frac{\sigma(z+\omega)}{\sigma(z)} \cdot e^{-2\eta(\omega)(z+\omega)} = \psi^2(\omega). \qquad (**)$$

Now let $0 \neq \omega/2 \in \Omega$. Since Ω is discrete, there is a natural number $n \geq 1$ with the property $\omega' := 2^{-n}\omega \in \Omega$, but $\frac{1}{2}\omega' = 2^{-n-1}\omega \notin \Omega$. It follows that

$$\psi(\omega) = \psi(2^n\omega') = (\psi(\omega'))^{2^n}$$

according to $(**)$ and thus $\psi(\omega) = 1$ from $(*)$. In summary, we get $\psi = \chi$. $\square$

6.4 Corollary. *If we set $f(z) := \sigma(z-a)/\sigma(z-b)$ for $a, b \in \mathbb{C}$, then for all $\omega \in \Omega$ and $z \in \mathbb{C}, z \notin b + \Omega$,*

$$f(z + \omega) = e^{\eta(\omega)(b-a)} \cdot f(z).$$

3. Existence and representation of elliptic functions. Let f be a non-constant elliptic function with respect to the lattice Ω and P a period parallelogram. Repeating the zeros and poles of f in P according to their multiplicities, LIOUVILLE's Third Theorem 2.5 shows that there exist $a_1, \ldots, a_r$ and $b_1, \ldots, b_r$ in P with $r \geq 2$ such that f in P has roots exactly at $a_1, \ldots, a_r$ and poles at $b_1, \ldots, b_r$. According to ABEL's Relation 2.7 we have

$$a_1 + \ldots + a_r \equiv b_1 + \ldots + b_r \pmod{\Omega}. \qquad (6.10)$$

Up to now it is not clear whether this necessary condition is also sufficient!

6.5 Existence Theorem. *Let $a_1, \ldots, a_r$ and $b_1, \ldots, b_r$ be two finite sequences in $\mathbb{C}$, for which the sets $\{a_1 + \Omega, \ldots, a_r + \Omega\}$ and $\{b_1 + \Omega, \ldots, b_r + \Omega\}$ are disjoint and for which*

$$\omega_0 := b_1 + \ldots + b_r - (a_1 + \ldots + a_r) \in \Omega \qquad (6.11)$$

holds. Then

$$f(z) := e^{-\eta(\omega_0)z} \cdot \frac{\sigma(z-a_1) \cdot \ldots \cdot \sigma(z-a_r)}{\sigma(z-b_1) \cdot \ldots \cdot \sigma(z-b_r)} \qquad (6.12)$$

is an elliptic function that has roots exactly at the points $a_1 + \Omega, \ldots, a_r + \Omega$ and poles at the points $b_1 + \Omega, \ldots, b_r + \Omega$.

Proof Use Corollary 6.4, Corollary 6.2 and (6.10). $\square$

Two elliptic functions with the same roots and poles (including multiplicities) differ only by a constant factor according to LIOUVILLE's First Theorem 2.3. A direct consequence of the Existence Theorem 6.5 is therefore the

6.6 Representation Theorem. *If f is an elliptic function with respect to the lattice Ω and f has roots at $a_1, \ldots, a_r$ and poles at $b_1, \ldots, b_r$ (with repetition according to their multiplicities) in a period parallelogram P, then there is a constant C and*

$$\omega_0 := b_1 + \ldots + b_r - a_1 - \ldots - a_r \in \Omega$$

with the property

§ 6 Product expansions

$$f(z) = C \cdot \frac{\sigma(z - a_1 - \omega_0) \cdot \sigma(z - a_2) \cdot \ldots \cdot \sigma(z - a_r)}{\sigma(z - b_1) \cdot \ldots \cdot \sigma(z - b_r)}.$$

As an application to the $\wp$-function we obtain the

6.7 Corollary. *Given* $z, w \in \mathbb{C} \setminus \Omega$,

$$\wp(z) - \wp(w) = -\frac{\sigma(z+w) \cdot \sigma(z-w)}{\sigma^2(z) \cdot \sigma^2(w)}.$$

Proof For continuity reasons, we may assume $2w \notin \Omega$. Now let w be fixed. Then $\wp(z) - \wp(w)$ has a second order pole at $z = 0$ and a first order root at $z = \pm w$ due to (2.12). According to the Representation Theorem 6.6,

$$\wp(z) - \wp(w) = C \cdot f(z) \quad \text{with} \quad f(z) := \frac{\sigma(z+w) \cdot \sigma(z-w)}{\sigma^2(z) \cdot \sigma^2(w)}$$

holds with a constant C. Because of (6.1) and (3.4), one has

$$\lim_{z \to 0} z^2 \cdot f(z) = -1, \quad \lim_{z \to 0} z^2 \cdot (\wp(z) - \wp(w)) = 1, \quad \text{hence } C = -1. \qquad \square$$

6.8 Corollary. *Given* $z \in \mathbb{C} \setminus \Omega$,

$$\wp'(z) = -\frac{\sigma(2z)}{\sigma^4(z)}.$$

Proof Multiply the statement of Corollary 6.7 by $\frac{1}{z-w}$ and let $w \to z$. $\qquad \square$

6.9 Corollary. *Given* $z \in \mathbb{C} \setminus \Omega$ *and* $\omega \in \Omega$, $\omega/2 \notin \Omega$,

$$\wp(z) - \wp(\omega/2) = \left(e^{\eta(\omega)z/2} \cdot \frac{\sigma(z - \omega/2)}{\sigma(z) \cdot \sigma(\omega/2)} \right)^2.$$

Proof We insert $w = \omega/2$ into Corollary 6.7 and use Theorem 6.3 in the form

$$\sigma(z + \omega/2) = -e^{\eta(\omega)z} \cdot \sigma(z - \omega/2). \qquad \square$$

According to Corollary 6.9, $\wp(z) - \wp(\omega/2)$, $\omega \in \Omega \setminus 2\Omega$, is the square of a meromorphic function, hence we can define

$$\sqrt{\wp(z) - \wp(\omega/2)} := -e^{\eta(\omega)z/2} \cdot \frac{\sigma(z - \omega/2)}{\sigma(z) \cdot \sigma(\omega/2)} = \frac{1}{z} + \cdots, \quad \frac{\omega}{2} \notin \Omega. \quad (6.13)$$

Note that (6.13) is not an elliptic function with respect to the lattice Ω because of Liouville's Second Theorem 2.4, since it has only one simple pole in a period parallelogram. However, the following holds

6.10 Corollary. *Given* $\omega_0 \in \Omega \setminus 2\Omega$,

$$\sqrt{\wp(z) - \wp(\omega_0/2)} := -e^{\eta(\omega_0)z/2} \cdot \frac{\sigma(z - \omega_0/2)}{\sigma(z) \cdot \sigma(\omega_0/2)}$$

is an elliptic function of order 4 with respect to the lattice 2Ω. Its poles are all simple and lie in Ω. Its roots are also all simple and lie in $\omega_0/2 + \Omega$.

Proof The statements about the roots and poles follow directly from the properties of the σ-function in sect. 1. Applying Corollary 6.2 and (6.5), it follows for $\omega \in \Omega$ that

$$\sqrt{\wp(z + 2\omega) - \wp(\omega_0/2)} = -e^{\eta(\omega_0)(z+2\omega)/2} \cdot \frac{\sigma(z + 2\omega - \omega_0/2)}{\sigma(z + 2\omega) \cdot \sigma(\omega_0/2)}$$
$$= \sqrt{\wp(z) - \wp(\omega_0/2)} \cdot e^{\eta(\omega_0)\omega - \eta(\omega)\omega_0} = \sqrt{\wp(z) - \wp(\omega_0/2)}. \qquad \square$$

With the well-known notation

$$e_k := \wp(\omega_k/2), \quad k = 1, 2, 3, \quad \omega_3 = \omega_1 + \omega_2,$$

as in (2.10), Corollary 6.9 and Theorem 6.3 yield

$$\left.\begin{aligned}
e_2 - e_1 &= e^{-\eta(\omega_1)\omega_2/2} \cdot \left(\frac{\sigma(\omega_3/2)}{\sigma(\omega_1/2) \cdot \sigma(\omega_2/2)}\right)^2, \\
e_3 - e_1 &= e^{\eta(\omega_1)\omega_3/2} \cdot \left(\frac{\sigma(\omega_2/2)}{\sigma(\omega_1/2) \cdot \sigma(\omega_3/2)}\right)^2, \\
e_3 - e_2 &= e^{\eta(\omega_2)\omega_3/2} \cdot \left(\frac{\sigma(\omega_1/2)}{\sigma(\omega_2/2) \cdot \sigma(\omega_3/2)}\right)^2.
\end{aligned}\right\} \quad (6.14)$$

As another application of Corollary 6.7 we derive the

6.11 Sigma Relation. *For all $u, v, w, z \in \mathbb{C}$,*

$$\sigma(u - v)\sigma(u + v) \cdot \sigma(z - w)\sigma(z + w)$$
$$+ \sigma(v - w)\sigma(v + w) \cdot \sigma(z - u)\sigma(z + u)$$
$$+ \sigma(w - u)\sigma(w + u) \cdot \sigma(z - v)\sigma(z + v) = 0.$$

Proof For $u, v, w, z \in \mathbb{C} \setminus \Omega$, we insert $U := \wp(u)$, $V := \wp(v)$, $W := \wp(w)$ and $Z := \wp(z)$ in the identity

$$(U - V)(Z - W) + (V - W)(Z - U) + (W - U)(Z - V) = 0$$

and apply Corollary 6.7. $\qquad \square$

6.12 Remarks. a) A. HURWITZ ([39], vol. I, 722–730) determined all functions that are holomorphic in a neighborhood of 0 and satisfy the Sigma Relation 6.11.
b) If we especially define a meromorphic function f by

§ 6 Product expansions

$$f(z) := \sqrt{\frac{\wp(z) - e_1}{e_3 - e_1}},$$

then we can easily verify the differential equation

$$f'^2 = (e_2 - e_1) \cdot (1 - f^2)(1 - \chi^2 f^2) \quad \text{with} \quad \chi^2 := \frac{e_3 - e_1}{e_2 - e_1} \neq 0, 1.$$

This allows us to calculate the integrals in LEGENDRE normal form (cf. (0.5)).

4. Another product expansion. As in §4, we now restrict ourselves to lattices of the form

$$\Omega = \mathbb{Z}\tau + \mathbb{Z} \quad \text{with} \quad \tau \in \mathbb{C}, \quad \text{Im } \tau > 0, \tag{6.15}$$

thus to $\omega_1 = \tau$, $\omega_2 = 1$, and use the abbreviations

$$\sigma(z; \tau) := \sigma(z; \Omega), \quad \eta(\omega; \tau) := \eta(\omega; \Omega) \quad \text{for} \quad \omega \in \Omega = \mathbb{Z}\tau + \mathbb{Z} \tag{6.16}$$

and

$$\eta := \eta(1; \tau) = \eta(1; \Omega). \tag{6.17}$$

The LEGENDRE Relation 6.1 is then given by

$$\eta \cdot \tau - \eta(\tau; \tau) = 2\pi i. \tag{6.18}$$

We use the abbreviation

$$q := e^{2\pi i \tau} \quad \text{for} \quad \tau \in \mathbb{H}, \quad w := e^{2\pi i z} \quad \text{for} \quad z \in \mathbb{C} \tag{6.19}$$

In particular, let $\sqrt{q} := e^{\pi i \tau}$ and $\sqrt{w} := e^{\pi i z}$.

6.13 Lemma. *The function*

$$f(z) := e^{-\eta z^2/2} \cdot \sqrt{w} \cdot \sigma(z; \tau) \tag{6.20}$$

satisfies

$$f(z + 1) = f(z) \quad \text{and} \quad f(z + \tau) = -\frac{1}{w} \cdot f(z). \tag{6.21}$$

Proof (6.9) yields

$$\sigma(z + 1; \tau) = -e^{\eta \cdot (z + 1/2)} \cdot \sigma(z; \tau)$$

and

$$\sigma(z + \tau; \tau) = -e^{\eta(\tau;\tau) \cdot (z + \tau/2)} \cdot \sigma(z; \tau) = -\frac{1}{w\sqrt{q}} \cdot e^{\eta \cdot \tau \cdot (z + \tau/2)} \cdot \sigma(z; \tau),$$

if we use (6.16). The assertion follows from (6.20). □

6.14 Theorem. *If we define the function f by (6.20), then*

$$f(z) = \frac{1}{2\pi i}(w-1) \prod_{n=1}^{\infty} \frac{(1-wq^n)\left(1-\frac{1}{w}q^n\right)}{(1-q^n)^2}, \qquad (6.22)$$

i.e. thus

$$\sigma(z;\tau) = \frac{1}{2\pi i} e^{\eta z^2/2}\left(\sqrt{w}-\frac{1}{\sqrt{w}}\right) \prod_{n=1}^{\infty} \frac{(1-wq^n)\left(1-\frac{1}{w}q^n\right)}{(1-q^n)^2}. \qquad (6.23)$$

Proof Because $|q| < 1$, the product converges absolutely and uniformly in z on compact subsets. We denote the right-hand side of (6.22) by $g(z)$. Then g is an entire function and $g(z+1) = g(z)$ follows from (6.19). If $z \mapsto z+\tau$, then wq^n and $\frac{1}{w}q^n$ transform into wq^{n+1} and $\frac{1}{w}q^{n-1}$. Thus it follows that

$$g(z+\tau) = \frac{1}{2\pi i}(wq-1) \prod_{n=1}^{\infty} \frac{(1-wq^{n+1})\left(1-\frac{1}{w}q^{n-1}\right)}{(1-q^n)^2},$$

i.e.

$$\frac{g(z+\tau)}{g(z)} = \frac{wq-1}{w-1} \cdot \frac{1-1/w}{1-wq} = -\frac{1}{w}.$$

Thus we have $g(z+\tau) = -\frac{1}{w} \cdot g(z)$. A comparison with Lemma 6.13 shows that the meromorphic function $h := f/g$ has periods 1 and τ, so it is an elliptic function with respect to the lattice (6.15). As an absolutely convergent product, g has 1st order roots at the points z for which

$$w = e^{2\pi i z} = 1 \quad \text{or} \quad wq^m = e^{2\pi i(m\tau+z)} = 1 \quad \text{or} \quad \frac{1}{w}q^m = e^{2\pi i(m\tau-z)} = 1$$

for $m = 1, 2, \ldots$. But these are exactly the points $z \in \Omega = \mathbb{Z}\tau + \mathbb{Z}$. Since f has the same roots according to (6.20) and (6.1), h is constant. Because

$$\lim_{z \to 0} \frac{\sigma(z;\tau)}{z} = 1, \quad \text{so} \quad \lim_{z \to 0} \frac{f(z)}{z} = 1, \quad \text{and} \quad \lim_{z \to 0} \frac{g(z)}{z} = 1,$$

we finally have $h = 1$. □

6.15 Remark. Knowing the WEIERSTRASS Product Theorem A.10 the fundamental identity (6.23) can be interpreted as follows. Both sides of (6.22) have the same roots with equal multiplicities, so there is an entire function F without roots satisfying

$$e^{-\eta z^2/2} \cdot \sqrt{w} \cdot \sigma(z;\tau) = f(z) = F(z) \cdot (w-1) \prod_{n=1}^{\infty} \frac{(1-wq^n)\left(1-\frac{1}{w}q^n\right)}{(1-q^n)^2}.$$

§ 6 Product expansions

However, the subtle result of the previous sections, namely $F(z) = 1/2\pi i$, can probably not be expected from general principle. If, however, we take the HADAMARD theory of entire functions of finite order into account, then we can immediately conclude that $F(z) = e^{q(z)}$ holds with a polynomial q of degree at most 2. For the proof, we must show that the entire function denoted by g in the proof has order 2. Then σ is the associated canonical product and HADAMARD's Theorem (cf. J.B. CONWAY [13], XI.3.4) provides the assertion.

5. The Δ product. We are going to prove the

6.16 Delta Product Formula. *It holds that*

$$\Delta(\tau) = (2\pi)^{12} \cdot q \cdot \prod_{n=1}^{\infty}(1-q^n)^{24}, \quad q := e^{2\pi i\tau}, \quad \operatorname{Im}\tau > 0. \tag{6.24}$$

We just have to put the parts together. But whenever we follow the partly subtle calculations in detail, we are astonished how such a simple and yet complex expression can arise from such complicated formulas.

Proof We abbreviate

$$\left.\begin{array}{ll} P_0 := \prod_{n=1}^{\infty}(1-q^n), & P_2 := \prod_{n=1}^{\infty}(1+q^n), \\ P_1 := \prod_{n=1}^{\infty}\left(1-q^{n-1/2}\right), & P_3 := \prod_{n=1}^{\infty}\left(1+q^{n-1/2}\right). \end{array}\right\} \tag{6.25}$$

Note that all products converge absolutely if $|q| < 1$. Because

$$P_0 P_1 P_2 P_3 = \prod_{n=1}^{\infty}(1-q^{2n}) \cdot \prod_{n=1}^{\infty}(1-q^{2n-1}) = P_0$$

we have

$$P_1 P_2 P_3 = 1. \tag{6.26}$$

Now the product representation (6.23) of the σ-function is applied. Using the abbreviation $\eta := \eta(1;\tau)$, we calculate

$$\left.\begin{array}{ll} 2\pi i \cdot P_0^2 \cdot \sigma(1/2;\tau) & = 2ie^{\eta/8} \cdot P_2^2, \\ 2\pi i \cdot P_0^2 \cdot \sigma(\tau/2;\tau) & = -q^{-1/4}e^{\eta\cdot\tau^2/8} \cdot P_1^2, \\ 2\pi i \cdot P_0^2 \cdot \sigma((\tau+1)/2;\tau) & = iq^{-1/4}e^{\eta\cdot(\tau+1)^2/8} \cdot P_3^2. \end{array}\right\} \tag{6.27}$$

Observing the LEGENDRE Relation 6.1, (6.18) and (6.14) now yield

$$e_1 - e_2 = e^{-\eta \cdot \tau/2} \cdot \left(\frac{\sigma((\tau+1)/2; \tau)}{\sigma(1/2; \tau) \cdot \sigma(\tau/2; \tau)} \right)^2,$$

$$e_3 - e_1 = e^{(\eta \cdot \tau - 2\pi i)(\tau+1)/2} \cdot \left(\frac{\sigma(1/2; \tau)}{\sigma(\tau/2; \tau) \cdot \sigma((\tau+1)/2; \tau)} \right)^2, \qquad (6.28)$$

$$e_3 - e_2 = e^{\eta \cdot (\tau+1)/2} \cdot \left(\frac{\sigma(\tau/2; \tau)}{\sigma(1/2; \tau) \cdot \sigma((\tau+1)/2; \tau)} \right)^2.$$

The formulas (6.27), inserted in (6.28), while considering (6.25), lead to the surprising identities

$$e_2 - e_1 = \pi^2 P_0^4 P_3^8, \quad e_3 - e_1 = 16\pi^2 \sqrt{q} P_0^4 P_2^8, \quad e_3 - e_2 = -\pi^2 P_0^4 P_1^8.$$

Finally, Corollary 3.18 implies

$$\Delta(\tau) = 16(e_1 - e_2)^2 (e_2 - e_3)^2 (e_3 - e_1)^2 = (2\pi)^{12} \cdot q \cdot P_0^{24} \cdot (P_1 P_2 P_3)^{16}.$$

and the assertion (6.24) follows from (6.26).

6.17 Remark. Except for the factor $(2\pi)^{12}$, the Δ-function is the 24th power of the *eta function* usually named after R. DEDEKIND,

$$\eta(\tau) := e^{\pi i \tau / 12} \cdot \prod_{n=1}^{\infty} \left(1 - e^{2\pi i n \tau} \right). \qquad (6.29)$$

This product was already treated by L. EULER [22], vol. 8, *Introductio in analysin infinitorum*, Caput XVI: *De partitione numerorum*, and *Opera posthuma I*, 76–84, as well as by C.G.J. JACOBI [42], vol. I, 141–155. Compare also III, §6.

6. Application to the absolute invariant. According to (4.25), the absolute invariant was defined by

$$j(\tau) = (12g_2(\tau))^3 / \Delta(\tau), \quad \text{Im } \tau > 0.$$

Due to (4.20) the following holds

$$12g_2(\tau) = (2\pi)^4 \left(1 + 240 \cdot \sum_{m=1}^{\infty} \sigma_3(m) \cdot e^{2\pi i m \tau} \right),$$

and (6.24) says that

$$\Delta(\tau) = (2\pi)^{12} e^{2\pi i \tau} \cdot \prod_{m=1}^{\infty} (1 - e^{2\pi i m \tau})^{24}.$$

Therefore it follows that

$$e^{2\pi i \tau} \cdot j(\tau) = \left(1 + 240 \cdot \sum_{m=1}^{\infty} \sigma_3(m) \cdot e^{2\pi i m \tau} \right)^3 \cdot \prod_{m=1}^{\infty} \left(\sum_{n=0}^{\infty} e^{2\pi i m n \tau} \right)^{24}.$$

§6 Product expansions

Here all factors have positive FOURIER coefficients. Therefore substituting

$$j(\tau) = e^{-2\pi i \tau} + \sum_{m=0}^{\infty} j_m \cdot e^{2\pi i m \tau},$$

yields

6.18 Theorem. *The FOURIER coefficients j_m of j are positive integers.*

Further arithmetic statements about the j_m can be found in III, §2.

7. The JACOBI theta series with respect to the lattice $\Omega = \mathbb{Z}\tau + \mathbb{Z}$, $\tau \in \mathbb{H}$, is defined by

$$\vartheta(z; \tau) := \sum_{n \in \mathbb{Z}} e^{\pi i n^2 \tau + 2\pi i n z}. \tag{6.30}$$

Series of this type were used by C.G.J.JACOBI as early as 1829 in his *Fundamenta nova* ([42], vol. I) in order to construct elliptic functions.

6.19 Lemma. *The series (6.30) converges absolutely and uniformly on compact subsets of $\mathbb{C} \times \mathbb{H}$. Given $\tau \in \mathbb{H}$, $\vartheta(z; \tau)$ is an entire function in z that has roots at the points $\frac{\tau+1}{2} + \Omega$. Furthermore,*

$$\vartheta(z + 1; \tau) = \vartheta(z; \tau) \quad \text{and} \quad \vartheta(z + \tau; \tau) = e^{-\pi i \tau - 2\pi i z} \cdot \vartheta(z; \tau). \tag{6.31}$$

Proof If $C \subseteq \mathbb{C} \times \mathbb{H}$ is compact, there exists some $\varepsilon > 0$ with $\operatorname{Im} \tau \geq \varepsilon$ and $|\operatorname{Im} z| \leq 1/\varepsilon$ for all $(z, \tau) \in C$, i.e.

$$\sum_{n \in \mathbb{Z}} |e^{\pi i n^2 \tau + 2\pi i n z}| \leq 1 + 2 \cdot \sum_{n=1}^{\infty} e^{-\pi n^2 \varepsilon + 2\pi n/\varepsilon} < \infty.$$

Accordingly, the series converges uniformly on compact subsets of $\mathbb{C} \times \mathbb{H}$ and thus is holomorphic in z and τ. (6.30) directly yields $\vartheta(z + 1; \tau) = \vartheta(z; \tau)$ as well as

$$\vartheta(z + \tau; \tau) = e^{-\pi i \tau - 2\pi i z} \cdot \sum_{n \in \mathbb{Z}} e^{\pi i (n+1)^2 \tau + 2\pi i (n+1) z} = e^{-\pi i \tau - 2\pi i z} \cdot \vartheta(z; \tau).$$

Furthermore one has

$$\vartheta\left(\tfrac{\tau+1}{2}; \tau\right) = \sum_{n \in \mathbb{Z}} (-1)^n e^{\pi i n(n+1)\tau}$$

$$= \sum_{m \in \mathbb{Z}} (-1)^{-m-1} e^{\pi i (-m-1)(-m)\tau} = -\vartheta(\tfrac{\tau+1}{2}; \tau),$$

hence $\vartheta(\tfrac{\tau+1}{2}; \tau) = 0$. Thus (6.31) leads to

$$\vartheta\left(\tfrac{\tau+1}{2} + m\tau + n; \tau\right) = 0 \quad \text{for all} \quad m, n \in \mathbb{Z}. \qquad \square$$

As an immediate consequence, we note

6.20 Corollary. *For all $a, b, c, d \in \mathbb{C}$ with $a + b - (c + d) \in \mathbb{Z}$,*

$$f(z) := \frac{\vartheta(z - a; \tau) \cdot \vartheta(z - b; \tau)}{\vartheta(z - c; \tau) \cdot \vartheta(z - d; \tau)}$$

is an elliptic function with respect to the lattice $\Omega = \mathbb{Z}\tau + \mathbb{Z}$, $\tau \in \mathbb{H}$.

We now show the famous

6.21 Jacobi's Triple Product Identity. *For all $z \in \mathbb{C}, \tau \in \mathbb{H}$*

$$\vartheta(z; \tau) = \sum_{n \in \mathbb{Z}} e^{\pi i n^2 \tau + 2\pi i n z}$$
$$= \prod_{m=1}^{\infty} (1 - e^{2\pi i m \tau}) \cdot (1 + e^{\pi i (2m-1)\tau + 2\pi i z}) \cdot (1 + e^{\pi i (2m-1)\tau - 2\pi i z}). \tag{6.32}$$

Proof Let $g(z; \tau)$ denote the right side of (6.32). Then $g(z; \tau)$ is an entire function in z with simple roots exactly at the points

$$z = \tfrac{\tau+1}{2} + m\tau + n, \quad m, n \in \mathbb{Z}.$$

Furthermore, $g(z + 1; \tau) = g(z; \tau)$ follows, as well as

$$\frac{g(z + \tau; \tau)}{g(z; \tau)} = \frac{1 + e^{-\pi i \tau - 2\pi i z}}{1 + e^{\pi i \tau + 2\pi i z}} = e^{-\pi i \tau - 2\pi i z}.$$

According to Lemma 6.19, $\vartheta(z; \tau)/g(z; \tau)$ is an entire elliptic function for the lattice $\Omega = \mathbb{Z}\tau + \mathbb{Z}$. By Liouville's First Theorem 2.3 the quotient is constant. Hence there exists a holomorphic function $\varphi : \{q \in \mathbb{C} ; 0 < |q| < 1\} \to \mathbb{C}, q = e^{\pi i \tau}$, with the property

$$\varphi(q) = \frac{\vartheta(z, \tau)}{g(z; \tau)} = \frac{\sum_{n \in \mathbb{Z}} q^{n^2} \cdot e^{2\pi i n z}}{\prod_{m=1}^{\infty} (1 - q^{2m}) \cdot (1 + q^{2m-1} \cdot e^{2\pi i z}) \cdot (1 + q^{2m-1} \cdot e^{-2\pi i z})}.$$

Now one has

$$\varphi(q) = \frac{\vartheta(1/4; \tau)}{g(1/4; \tau)} = \frac{\sum_{n \in \mathbb{Z}} i^n q^{n^2}}{\prod_{m=1}^{\infty} (1 - q^{2m}) \cdot (1 + iq^{2m-1}) \cdot (1 - iq^{2m-1})}$$

$$= \frac{\sum_{n \in \mathbb{Z}} i^{2n} q^{(2n)^2}}{\prod_{m=1}^{\infty} (1 - q^{4m}) \cdot (1 - q^{4m-2}) \cdot (1 + q^{4m-2})}.$$

This yields

§ 6 Product expansions

$$\varphi(q) = \frac{\vartheta(1/2; 4\tau)}{\prod_{m=1}^{\infty}(1-q^{8m})\cdot(1-q^{8m-4})\cdot(1-q^{8m-4})} = \frac{\vartheta(1/2; 4\tau)}{g(1/2; 4\tau)} = \varphi(q^4).$$

φ is holomorphic at 0 with $\varphi(0) = 1$. By induction, we conclude $\varphi(q^{4k}) = \varphi(q)$ for all $k \in \mathbb{N}$. The Identity Theorem A.1 yields

$$\varphi \equiv \varphi(0) = 1, \quad \text{i.e.} \quad \vartheta = g. \qquad \square$$

Replacing τ by $3\tau/2$ and setting $z = (\tau + 2)/4$, with $q = e^{\pi i \tau}$ we directly obtain

6.22 Euler's Pentagonal Number Theorem. *For $q \in \mathbb{C}$, $|q| < 1$,*

$$\prod_{m=1}^{\infty}(1-q^m) = \sum_{n \in \mathbb{Z}}(-1)^n q^{(3n^2+n)/2}.$$

6.23 Remarks. a) The *partition number* of $m \in \mathbb{N}_0$ is the number of monotonically decreasing sequences of non-negative integers, whose sum is m. Let $p_e(m)$ resp. $p_o(m)$ denote the number of partitions of m with an even or odd number of summands. Then Euler's Pentagonal Number Theorem 6.22 is equivalent to the statement

$$p_e(m) - p_o(m) = \begin{cases} (-1)^n, & \text{if } n \in \mathbb{Z} \text{ with } (3n^2+n)/2 = m, \\ 0, & \text{otherwise}. \end{cases}$$

This identity was proved by induction by L. Euler in 1750 (cf. *Introductio in analysin infinitorum*, Caput XVI, in [22], vol. 8). The proof given here goes back to C.G.J. Jacobi ([42], vol. I, 49–239). The *pentagonal numbers* $1, 5, 12, 22, \ldots$ arise in the form $(3n^2 + n)/2$ for $n = -1, -2, -3, -4, \ldots$ and graphically by regularly placing pentagons, whose edge length increases by 1, inside each other and counting their vertices.

Historical notes can be found in T.M. Apostol [5]. A general approach is described by E. Neher, *Jahresber. Dtsch. Math.–Ver.* **87**, 164–181 (1987). A formula for the partition function $p(n)$ as a finite sum of algebraic numbers was found by J.H. Bruinier and K. Ono, *Adv. Math.* **246**, 198-219 (2013).

b) The theta series satisfies the so-called *heat equation*

$$\frac{\partial}{\partial \tau} \vartheta(z; \tau) = \frac{1}{4\pi i} \frac{\partial^2}{\partial z^2} \vartheta(z; \tau).$$

8. Multiplication of the $\wp$-function. For each $n \in \mathbb{N}$, following to Weierstrass ([88], vol. V, 212) a mapping $\psi_n : \mathbb{C} \setminus \Omega \to \mathbb{C}$ is defined by

$$\psi_n(z) := \sigma(nz)/\sigma^{n^2}(z), \quad \psi_1 = 1. \tag{6.33}$$

Obviously, ψ_n is even if n is odd, and ψ_n is odd, if n is even.

6.24 Theorem. *The function ψ_n is an elliptic function with respect to the lattice Ω.*

It has poles of order $n^2 - 1$ only in the points of Ω. It is
a) *a polynomial in $\wp$ of degree $(n^2 - 1)/2$ with highest coefficient n if n is odd, and*
b) *a product of $\wp'$ with a polynomial in $\wp$ of degree $(n^2 - 4)/2$ with highest coefficient $-n/2$ if n is even.*

Proof Theorem 6.3 yields

$$\psi_n(z+\omega) = \psi_n(z) \cdot \chi(n\omega)/\chi^{n^2}(\omega) = \psi_n(z), \quad \text{hence} \quad \psi_n \in \mathcal{E}(\Omega).$$

The remaining assertions follow from Corollary 3.10 as well as $\sigma(z) = z + O(z^3)$, if we note that ψ_n is an even function for odd n and an odd function for even n. □

6.25 Corollary. *If $n \geq 2$, then*

$$\wp(nz) - \wp(z) = -\psi_{n-1}(z) \cdot \psi_{n+1}(z)/\psi_n^2(z).$$

Proof We use Corollary 6.7 and (6.33). □

6.26 Corollary. *We have*

$$\psi_2 = -\wp', \tag{6.34}$$

$$\psi_3(z) = (\wp(z) - \wp(2z))\wp'^2(z) = 3\wp^4(z) - \frac{3}{2}g_2\wp^2(z) - 3g_3\wp(z) - \frac{1}{16}g_2^2, \tag{6.35}$$

$$\psi_4(z)/\wp'(z) = -\wp'(2z) \cdot \wp'^3(z) = \tag{6.36}$$

$$-2\wp^6(z) + \frac{5}{2}g_2\wp^4(z) + 10g_3\wp^3(z) + \frac{5}{8}g_2^2\wp^2(z) + \frac{1}{2}g_2g_3\wp(z) - \frac{1}{32}g_2^3 + g_3^2.$$

Proof (6.34) follows from Theorem 6.24 with $n = 2$. Thus Corollary 6.25 implies

$$\psi_3(z) = (\wp(z) - \wp(2z)) \cdot \wp'^2(z).$$

(6.35) is a consequence of Corollary 5.3, (3.7), and Corollary 3.8. We have

$$\psi_4(z) = \frac{\sigma(4z)}{\sigma^{16}(z)} = \frac{\sigma(4z)}{\sigma^4(2z)} \cdot \left(\frac{\sigma(2z)}{\sigma^4(z)}\right)^4$$

and can insert (6.34). The further transformations are dealt with in an analogous way. □

Due to Theorem 6.24, there are polynomials A_n and B_n, $n \geq 2$, such that

$$A_n(\wp(z)) = \psi_{n-1}(z) \cdot \psi_{n+1}(z) \quad \text{and} \quad B_n(\wp(z)) = \psi_n^2(z). \tag{6.37}$$

Here,

$$A_n(\wp) = (n^2 - 1)\wp^{n^2} + \ldots \quad \text{and} \quad B_n(\wp) = n^2\wp^{n^2-1} + \ldots. \tag{6.38}$$

Thus Corollary 6.25 can be written in the form

§ 6 Product expansions

$$\wp(nz) - \wp(z) = -\frac{A_n(\wp(z))}{B_n(\wp(z))}, \quad n \geq 2. \tag{6.39}$$

The problem of *multiplication of the $\wp$-function* now consists of the explicit determination of the functions ψ_n or the polynomials A_n and B_n. Compare the "explicit" formula in Corollary 7.9.

6.27 Lemma. *The following recursion formulas hold:*

$$\psi_{2n+1} = A_{n+1}(\wp)B_n(\wp) - A_n(\wp)B_{n+1}(\wp), \tag{6.40}$$

$$\wp' \cdot \psi_{2n} = A_{n-1}(\wp)B_{n+1}(\wp) - A_{n+1}(\wp)B_{n-1}(\wp). \tag{6.41}$$

Proof In the Sigma Relation 6.11, we set $z = 0$ and get

$$\sigma^2(u) \cdot \sigma(v-w) \cdot \sigma(v+w) + \sigma^2(v) \cdot \sigma(w-u) \cdot \sigma(w+u) + \sigma^2(w) \cdot \sigma(u-v) \cdot \sigma(u+v) = 0.$$

If $u = nz$, $v = mz$ and $w = z$ with $m > n$, this yields

$$\psi_{m+n} \cdot \psi_{m-n} = \psi_n^2 \cdot \psi_{m-1} \cdot \psi_{m+1} - \psi_m^2 \cdot \psi_{n-1} \cdot \psi_{n+1}.$$

Now set $m = n + 1$ and on the other hand replace n by $n - 1$ and m by $n + 1$. We obtain (6.40) and (6.41) by means of (6.37) and (6.34). □

For the moment we call a polynomial P of degree $m \geq 2$ a *lacunary polynomial* if its subbading coefficient is zero, i.e. if there exists some $0 \neq \alpha_m \in \mathbb{C}$ such that

$$P(X) = \alpha_m X^m + \alpha_{m-2} X^{m-2} + \ldots + \alpha_0.$$

Note that the following rules apply to lacunary polynomials P and Q:
a) If P, Q and $P + Q$ have the same degree, then $P + Q$ is also a lacunary polynomial.
b) $P \cdot Q$ is a lacunary polynomial.
c) If P/Q is a polynomial, then P/Q is a lacunary polynomial.

6.28 Proposition. *The polynomials A_n and B_n are lacunary polynomials.*

Proof By Corollary 6.26, ψ_2, ψ_3, ψ_4 and because of (6.37) also A_2, A_3, B_2, B_3 are lacunary polynomials. Now let A_ν, B_ν be lacunary polynomials for all $\nu \leq n$. Then, because of (6.40) and (6.41), $\psi_1, \ldots, \psi_{2n-1}$ are also lacunary polynomials and, according to (6.37), also A_ν, B_ν for $\nu \leq n + 1 \leq 2n - 2$. □

6.29 Remarks. a) The first formulas in (6.35) and (6.36) suggest that there may be a simple closed form for the ψ_n if higher derivatives are allowed. In fact,

$$\psi_n = \frac{(-1)^{n-1}}{(2! \cdot \ldots \cdot (n-1)!)^2} \cdot \det \begin{pmatrix} \wp' & \wp'' & \cdots & \wp^{(n-1)} \\ \wp'' & \wp''' & & \wp^{(n)} \\ \vdots & \vdots & & \vdots \\ \wp^{(n-1)} & \wp^{(n)} & \cdots & \wp^{(2n-3)} \end{pmatrix}.$$

This and other formulas for ψ_n can be found in R. FRICKE [28], 184–196.

b) Another "explicit" formula is given in Corollary 7.9.

9*. Multiplicative invariant functions. In this section, we will briefly present another approach to elliptic functions. For this purpose, let $q \in \mathbb{C}$ be chosen arbitrarily but fixed with $0 < |q| < 1$. We denote the set of all functions g meromorphic in $\mathbb{C}^\times := \mathbb{C} \setminus \{0\}$ that satisfy

$$g(qw) = g(w) \quad \text{for } w \in \mathbb{C}^\times \tag{6.42}$$

by $\mathcal{F}_q$. Obviously, $\mathcal{F}_q$ is a subfield of the field of all meromorphic functions on $\mathbb{C}^\times$ containing $\mathbb{C}$. Obviously, $g(q^n w) = g(w)$ holds for all $n \in \mathbb{Z}$. If $r > 0$ let

$$B := B_r := \{w \in \mathbb{C}^\times \, ; \, r < |w| \leq r/|q|\}$$

be a semi-open annulus. Each $g \in \mathcal{F}_q$ then attains all its values in B.

6.30 Lemma. *If $g \neq 0$ is holomorphic in $\mathbb{C}^\times$ and $g(qw) = c \cdot g(w)$ holds for some $c \in \mathbb{C}$ and all $w \in \mathbb{C}^\times$, then there exists $k \in \mathbb{Z}$ such that $c = q^k$ and $g(w) = g(1) \cdot w^k$.*

Proof We develop g into a LAURENT series around 0,

$$g(w) = \sum_{\nu=-\infty}^{\infty} a_\nu w^\nu,$$

and obtain $a_\nu q^\nu = c \cdot a_\nu$ for all $\nu \in \mathbb{Z}$ by the Identity Theorem A.1. We now choose $k \in \mathbb{Z}$ with $a_k \neq 0$ and get $c = q^k$. It follows that $a_\nu = 0$ for $\nu \neq k$. □

For $c = 1$ we obtain the

6.31 Corollary. *Any holomorphic function $g \in \mathcal{F}_q$ is constant.*

The double product (cf. Theorem 6.14)

$$\begin{aligned} p(w) := p_q(w) : &= \prod_{n=1}^{\infty}(1 - wq^n) \cdot \prod_{n=0}^{\infty}\left(1 - \frac{1}{w}q^n\right) \\ &= \prod_{n=1}^{\infty}(1 - wq^n)\left(1 - \frac{1}{w}q^{n-1}\right), \quad w \in \mathbb{C}^\times, \end{aligned} \tag{6.43}$$

is absolutely and uniformly convergent an every compact subset of $\mathbb{C}^\times$, thus is holomorphic on $\mathbb{C}^\times$. We obviously have

$$p(qw) = -\frac{1}{qw} \cdot p(w) \quad \text{for } w \in \mathbb{C}^\times. \tag{6.44}$$

Moreover, p has 1st order roots exactly at the points $w = q^n$, $n \in \mathbb{Z}$.

Each function $g \in \mathcal{F}_q$ has only finitely many zeros and poles in B. These points cannot be chosen arbitrarily:

6.32 Existence and Representation Theorem. *Let $a_1, \ldots, a_m$ and $b_1, \ldots, b_n$ be two finite sequences from B such that the sets $\{a_1, \ldots, a_m\}$ and $\{b_1, \ldots, b_n\}$ are disjoint. Then the following assertions are equivalent:*

(i) *There exists some $g \in \mathcal{F}_q$ which has roots in B, exactly at points a_l, $1 \leq l \leq m$, and poles at the points b_l, $1 \leq l \leq n$, where the orders are indicated by the number of repetitions.*

(ii) *$m = n \geq 2$ and $b_1 \cdot \ldots \cdot b_m = a_1 \cdot \ldots \cdot a_m \cdot q^k$ with some $k \in \mathbb{Z}$.*

In this case, any function g for which (i) holds is uniquely representable in the form

$$g(w) = c \cdot w^k \cdot \frac{p(w/a_1) \cdot \ldots \cdot p(w/a_m)}{p(w/b_1) \cdot \ldots \cdot p(w/b_m)} \quad \text{with} \quad c \in \mathbb{C}^\times. \tag{6.45}$$

Proof After a potential transition from g to $1/g$, we may assume $m \leq n$. We set

$$h(w) := \frac{p(w/a_1) \cdot \ldots \cdot p(w/a_m)}{p(w/b_1) \cdot \ldots \cdot p(w/b_m)}, \quad \lambda := \frac{a_1 \cdot \ldots \cdot a_m}{b_1 \cdot \ldots \cdot b_m} \neq 0.$$

Then h is meromorphic on $\mathbb{C}^\times$ and (6.44) yields $h(qw) = \lambda \cdot h(w)$.
(i) $\Longrightarrow$ (ii) : Since $p(w/a)$ has simple roots exactly at the points $w = aq^l$, $l \in \mathbb{Z}$, $f := g/h$ has no poles in B. Because $f(qw) = \frac{1}{\lambda} f(w)$ and $\mathbb{C}^\times = \bigcup_{\nu \in \mathbb{Z}} q^\nu B$, the function f is holomorphic on $\mathbb{C}^\times$. According to Lemma 6.30, there is therefore some $k \in \mathbb{Z}$ with $\frac{1}{\lambda} = q^k$ as well as $f(w) = f(1) \cdot w^k$. Then (6.45) follows with $c = f(1)$ and $n = m$ holds. In the case $m = n = 1$, we have $b_1 = a_1 q^k$, hence $k = 0$ because of $a_1 \in B$ and $b_1 \in B$.
(ii) $\Longrightarrow$ (i): Any function given by (6.45) satisfies (i). □

6.33 Remarks a) To develop a p-adic theory of elliptic functions, then we should not try to generalize the $\wp$-function, instead, we should follow the "multiplicative path" outlined here.
b) If $g \in \mathcal{F}_q$ is not constant, then for $h \in \mathcal{F}_q$ there always exists a polynomial $0 \neq P \in \mathbb{C}[g][X]$ with $P(g, h) = 0$. A proof can be obtained by counting the coefficients appearing in P and comparing them with the number of coefficients appearing in the principal parts of $P(g, h)$.
c) Choose $\tau \in \mathbb{H}$ and $q = e^{2\pi i \tau}$. Given $g \in \mathcal{F}_q$, we now define a function $\hat{g}$, meromorphic on $\mathbb{C}$, by $\hat{g}(z) := g(e^{2\pi i z})$, $z \in \mathbb{C}$. Obviously, $\hat{g}(z+1) = \hat{g}(z)$ holds. However, it also follows from (6.42) that $\hat{g}(z + \tau) = \hat{g}(z)$. Thus $g \mapsto \hat{g}$ defines an injective homomorphism of $\mathcal{F}_q$ into the field $\mathcal{E}(\Omega)$ of elliptic functions for the lattice $\Omega := \mathbb{Z}\tau + \mathbb{Z}$. A comparison of (6.43) with Theorem 6.14 yields

$$\sigma(z; \tau) = \frac{1}{2\pi i P} e^{\frac{1}{2}\eta z^2 + \pi i z} p(e^{2\pi i z}) \quad \text{with} \quad P := \prod_{n=1}^{\infty} (1 - q^n)^2.$$

Thus, the Existence and Representation Theorem 6.32 corresponds to the results of sect. 3. □

6.34 Exercises.

1) Show that
$$\wp'(z) = 2 \cdot \frac{\sigma(z+w_1/2) \cdot \sigma(z+w_2/2) \cdot \sigma(z-w_3/2)}{\sigma^3(z) \cdot \sigma(w_1/2) \cdot \sigma(w_2/2) \cdot \sigma(w_3/2)}.$$

2) Show that
$$-2 \cdot \frac{\sigma(z+w_1/2) \cdot \sigma(z+w_2/2) \cdot \sigma(z-w_3/2)}{\sigma(w_1/2) \cdot \sigma(w_2/2) \cdot \sigma(w_3/2)} = \frac{\sigma(2z)}{\sigma(z)}.$$

3) If $u, v \in \mathbb{C}$ with $u \not\equiv v \pmod{\Omega}$, then $\zeta(z-u) - \zeta(z-v)$ is an elliptic function with respect to the lattice Ω with two poles of first order in each period parallelogram.

4) If $f(z; w)$ is defined as in (5.2), then $f(z; w) = \zeta(z+w) - \zeta(z) - \zeta(w)$.

5) $\zeta(z+w) + \zeta(z-w) - 2\zeta(z) = \wp'(z)/(\wp(z) - \wp(w))$.

6) $2\zeta(2z) = \zeta(z) + \zeta(z+\omega_1/2) + \zeta(z+\omega_2/2) + \zeta(z-(\omega_1+\omega_2)/2)$.

7) $\psi_3(z) = 3\wp(z)\wp'^2(z) - \frac{1}{4}\wp''(z)^2$.

8) $\wp''(z)/\wp'(z) = 2\zeta(2z) - 4\zeta(z)$.

9) Let $f \in M$ with $f(z+\omega) = c(\omega) \cdot f(z)$ for all $\omega \in \Omega$. In the period parallelogram P, the function may have roots in $a_1, \ldots, a_r$ and poles in $b_1, \ldots, b_r$ (cf. Exercise 2.17 5)). Then there are $\alpha, \beta \in \mathbb{C}$ such that
$$f(z) = \alpha \cdot \frac{\sigma(z-a_1) \cdot \ldots \cdot \sigma(z-a_r)}{\sigma(z-b_1) \cdot \ldots \cdot \sigma(z-b_r)} \cdot e^{\beta z}. \tag{$*$}$$

Conversely, any such function $(*)$ also satisfies $f(z+\omega) = c(\omega) \cdot f(z)$, where
$$c(\omega) = e^{\beta\omega + \eta(\omega)(b_1 + \cdots + b_r - a_1 - \cdots - a_r)}.$$

10) Suppose $\begin{pmatrix}\omega\\\omega'\end{pmatrix} = U\begin{pmatrix}\omega_1\\\omega_2\end{pmatrix}$, $U \in \text{Mat}(2; \mathbb{Z})$, $\text{Im}(\omega_1/\omega_2) > 0$. Then
$$\eta(\omega')\omega - \eta(\omega)\omega' = 2\pi i \cdot \det U.$$

11) For $a \in \mathbb{C}$, $a \neq 0, 1$, there exists an elliptic function f satisfying the differential equation $f'^2 = (1-f^2)(1-af^2)$.

12) For mappings $\lambda, \mu : \Omega \to \mathbb{C}$, we denote by $\Theta[\lambda, \mu]$ the set of entire functions f with the property
$$f(z+\omega) = e^{-\pi i(\lambda(\omega)z + \mu(\omega))} \cdot f(z) \quad \text{for all } z \in \mathbb{C}, \omega \in \Omega.$$

a) $\Theta[\lambda, \mu]$ is a $\mathbb{C}$-vector space.
b) $\Theta[\lambda, \mu] \cdot \Theta[\lambda^*, \mu^*] \subseteq \Theta[\lambda + \lambda^*, \mu + \mu^*]$.
c) If $\Theta[\lambda, \mu] \neq \{0\}$, then $\lambda : \Omega \to \mathbb{C}$ is a group homomorphism and one has for all $\omega, \omega' \in \Omega$
$$\lambda(\omega)\omega' - \lambda(\omega')\omega \in 2\mathbb{Z}, \quad \mu(\omega+\omega') - \mu(\omega) - \mu(\omega') - \lambda(\omega')\omega \in 2\mathbb{Z}.$$

d) $\Theta[\lambda, \mu]$ contains a function without roots if and only if there are $a, b \in \mathbb{C}$ such that
$$\lambda(\omega) = 2a\omega \quad \text{and} \quad \mu(\omega) = a\omega^2 + b\omega.$$

In this case,
$$\Theta[\lambda, \mu] = \mathbb{C} \cdot f_{a,b}, \quad f_{a,b}(z) := e^{-\pi i(az^2 + bz)}.$$

e) If $\Omega = \mathbb{Z}\tau + \mathbb{Z}$ with $\tau \in \mathbb{H}$, then the JACOBI theta series satisfies
$$\vartheta(\cdot; \tau) \in \Theta[\lambda, \mu] \quad \text{for } \lambda(m\tau + n) = 2m, \ \mu(m\tau + n) = m^2\tau.$$

13) Let $0 \neq f \in \mathcal{E}(\Omega)$ and let $a_1, \ldots, a_l$ represent the poles of f modulo Ω, i.e., each pole of f is congruent to one of the a_ν modulo Ω and the a_ν are not congruent to each other. Let the principal part of f around a_ν be given by

$$\frac{c_{\nu,1}}{z - a_\nu} + \cdots + \frac{c_{\nu, l_\nu}}{(z - a_\nu)^{l_\nu}}$$

with $l_\nu \geq 1$. Demonstrate the existence of a constant C with the property

$$f(z) = \sum_{\nu=1}^{l} \left(c_{\nu,1} \zeta(z - a_\nu) + \sum_{k=2}^{l_\nu} \frac{(-1)^k}{(k-1)!} c_{\nu, k} \wp^{(k-2)}(z - a_\nu) \right) + C.$$

Is this representation unique?

14) Let $a_1, \ldots, a_l \in \mathbb{C}$ be pairwise incongruent modulo Ω. Furthermore, let any principal parts in the a_ν be given. Determine a (necessary and sufficient) criterion for the existence of an elliptic function with respect to Ω with poles exactly in $a_1, \ldots, a_l \in \mathbb{C}$ modulo Ω and the given principal parts around a_ν, $1 \leq \nu \leq l$.

§ 7* $\wp$-partial values, algebraic dependence and complex multiplication

Parts of this paragraph require elementary knowledge of imaginary–quadratic fields. Let Ω always be an arbitrary lattice in $\mathbb{C}$.

1. Sublattices. A subgroup Ω^* of a lattice Ω in $\mathbb{C}$ is called a *sublattice*, if Ω^* is itself a lattice (cf. §1), so if Ω^* contains an $\mathbb{R}$-basis of $\mathbb{C}$. If Ω^* is a sublattice of Ω, then a subset $\mathcal{R}$ of Ω is called a *system of representatives of Ω modulo Ω^** or a *system of representatives of the coset classes Ω/Ω^**, if:

(RS.1) Given $\omega \in \Omega$ there is $v \in \mathcal{R}$ with $\omega \in v + \Omega^*$.
(RS.2) If $v_1, v_2 \in \mathcal{R}$ and $\omega^* \in \Omega^*$ with $v_1 = v_2 + \omega^*$, then $\omega^* = 0$ holds.

The cardinality of any system of representatives $\mathcal{R}$ of Ω modulo Ω^* is equal to the index $[\Omega : \Omega^*]$. Furthermore, each $\omega \in \Omega$ can uniquely be written as

$$\omega = v + \omega^* \quad \text{with} \quad v \in \mathcal{R} \text{ and } \omega^* \in \Omega^*. \tag{7.1}$$

If $\mathcal{R}$ and $\mathcal{R}'$ are two systems of representatives of Ω modulo Ω^* and $v \in \mathcal{R}$, (7.2) there are uniquely determined $v' \in \mathcal{R}'$ and $\omega^* \in \Omega^*$ such that $v = v' + \omega^*$.

Conversely, this way we obtain all such systems of representatives from one system of representatives. Directly clear is the following

7.1 Proposition. *For a positive integer n, $n\Omega$ is a sublattice of Ω and*

$$\mathcal{R}(n) := \{m_1\omega_1 + m_2\omega_2 \, ; \, m_1, m_2 \in \mathbb{Z}, \, 0 \leq m_1 < n, \, 0 \leq m_2 < n\}$$

is a system of representatives of Ω modulo $n\Omega$. It holds that

$$[\Omega : n\Omega] = \sharp \mathcal{R}(n) = n^2.$$

Furthermore, of course, we have the

7.2 Lemma. *If Ω^* is a sublattice of Ω, then $\mathcal{E}(\Omega) \subseteq \mathcal{E}(\Omega^*)$.*

We state a purely algebraic assertion.

7.3 Equivalence Theorem for Sublattices. *Given a subgroup Ω^* of the lattice $\Omega = \mathbb{Z}\omega_1 + \mathbb{Z}\omega_2$ the following assertions are equivalent:*

(i) *Ω^* is a sublattice of Ω.*
(ii) *There exist $a, b, c, d \in \mathbb{Z}$ with $ad - bc \neq 0$ and*

$$\Omega^* = \mathbb{Z}(a\omega_1 + b\omega_2) + \mathbb{Z}(c\omega_1 + d\omega_2).$$

(iii) *There exists $0 \neq n \in \mathbb{N}$ such that $n\Omega \subseteq \Omega^*$.*
(iv) *The index $[\Omega : \Omega^*]$ is finite.*

Proof (i) $\implies$ (ii): By assumption, there is a basis ω_1^*, ω_2^* of the lattice Ω^*, so $\Omega^* = \mathbb{Z}\omega_1^* + \mathbb{Z}\omega_2^*$. Because $\Omega^* \subseteq \Omega$, there is some $M \in \text{Mat}(2; \mathbb{Z})$ satisfying

$$\begin{pmatrix} \omega_1^* \\ \omega_2^* \end{pmatrix} = M \begin{pmatrix} \omega_1 \\ \omega_2 \end{pmatrix}.$$

But since ω_1^*, ω_2^* is also an $\mathbb{R}$-basis of $\mathbb{C}$, there exists an $N \in \text{GL}(2; \mathbb{R})$ such that

$$\begin{pmatrix} \omega_1 \\ \omega_2 \end{pmatrix} = N \begin{pmatrix} \omega_1^* \\ \omega_2^* \end{pmatrix} = NM \begin{pmatrix} \omega_1 \\ \omega_2 \end{pmatrix},$$

thus $NM = I$. It follows that $\det M \neq 0$.

(ii) $\implies$ (iii): Let $n := |ad - bc|$. Then we have

$$G := n \begin{pmatrix} a & b \\ c & d \end{pmatrix}^{-1} = \pm \begin{pmatrix} d & -b \\ -c & a \end{pmatrix} \in \text{Mat}(2; \mathbb{Z})$$

and
$$n\begin{pmatrix}\omega_1\\\omega_2\end{pmatrix} = G\begin{pmatrix}a & b\\c & d\end{pmatrix}\begin{pmatrix}\omega_1\\\omega_2\end{pmatrix} = G\begin{pmatrix}a\omega_1 + b\omega_2\\c\omega_1 + d\omega_2\end{pmatrix},$$

i.e. $n\omega_1$ and $n\omega_2$ belong to Ω^*.

(iii) $\Longrightarrow$ (iv): The index $[\Omega : \Omega^*]$ divides $[\Omega : n\Omega] = n^2$.

(iv) $\Longrightarrow$ (i): Let $l := [\Omega : \Omega^*]$ and $\omega \in \Omega$. Then, among the $l + 1$ coset classes $m\omega + \Omega^*, 0 \leq m \leq l$, at least two coincide, i.e. there exists $0 \neq k \in \mathbb{N}$ with $k\omega \in \Omega^*$. In particular, there is some $k \neq 0$ with $k\omega_1 \in \Omega^*$ and $k\omega_2 \in \Omega^*$. Thus Ω^* contains an $\mathbb{R}$-basis of $\mathbb{C}$. $\square$

If Ω^* is a sublattice of Ω, then in the notation of (ii),
$$[\Omega : \Omega^*] = |ad - bc|. \tag{7.3}$$

Using the SMITH normal form (cf. Lemma IV.1.17) we can show that there exist $\omega_1, \omega_2 \in \mathbb{C}$ and positive integers a, d with $a | d$ such that (ω_1, ω_2) is a basis of Ω and $(a\omega_1, d\omega_2)$ is a basis of Ω^*. Thus we can immediately see that (7.3) is true and that
$$m_1\omega_1 + m_2\omega_2, \ 0 \leq m_1 < a, \ 0 \leq m_2 < b,$$
is a system of representatives of the cosets classes Ω/Ω^* in this case.

2. The field extension $\mathcal{E}(\Omega^*)$ over $\mathcal{E}(\Omega)$. Let Ω^* again be a sublattice of Ω and $\mathcal{R}$ a system of representatives of Ω modulo Ω^* in the sense of sect. 1. Given $v \in \Omega$ and $f \in \mathcal{E}(\Omega^*)$, we define the elliptic function $f_v \in \mathcal{E}(\Omega^*)$ by
$$f_v(z) = f(z + v). \tag{7.4}$$

Obviously $f_v = f$ holds for $f \in \mathcal{E}(\Omega)$. Specifically, let
$$\wp_v^*(z) = \wp_v(z; \Omega^*) = \wp(z + v; \Omega^*), \tag{7.5}$$

where $\wp^* := \wp_0^*$ is the $\wp$-function with respect to the lattice Ω^*. Obviously, $\mathbb{C}(\wp)$ is the field of even elliptic functions with respect to Ω and it is a subfield of $\mathbb{C}(\wp^*)$. Given an indeterminate X, we now define the polynomial
$$P(X) := P_{\Omega,\Omega^*}(X) := \prod_{v \in \mathcal{R}}(X - \wp_v^*) \tag{7.6}$$

in $\mathcal{E}(\Omega^*)[X]$ of degree $[\Omega : \Omega^*]$. By (7.2), P does not depend on the choice of the system of representatives $\mathcal{R}$, because $\wp_{v+\omega^*}^* = \wp_v^*$, $\omega^* \in \Omega^*$, according to (7.5). The translations $z \mapsto z + u$, $u \in \Omega$, permute the roots $\wp_v^*$ of the polynomial P. The coefficients of P, i.e. the elementary symmetric polynomials in the roots $\wp_v^*$, are therefore even elliptic functions with respect to the lattice Ω. It follows that
$$P_{\Omega,\Omega^*}(X) \in \mathbb{C}(\wp)[X]. \tag{7.7}$$

7.4 Proposition. *If Ω^* is a sublattice of Ω the polynomial*

$$P(X) = P_{\Omega,\Omega^*}(X) \in \mathcal{E}(\Omega)[X]$$

in (7.6) is irreducible and has $\mathcal{E}(\Omega^*)$ as a splitting field. The degree of the field extension of $\mathcal{E}(\Omega^*)$ over $\mathcal{E}(\Omega)$ is $[\Omega : \Omega^*]$.

Proof Any two roots of P can be mapped by a translation $z \mapsto z + u$, $u \in \Omega$, onto each other. Since these mappings are automorphisms of the field $\mathcal{E}(\Omega^*)$ that keep $\mathcal{E}(\Omega)$ elementwise fixed, P is irreducible, because the $\wp_v^*$ are pairwise distinct. The splitting field $\mathcal{F} \subseteq \mathcal{E}(\Omega^*)$ of $P(X)$ contains $\mathbb{C}(\wp^*)$. Since $\wp' \in \mathcal{E}(\Omega)$ is odd, it follows that $\mathcal{F} = \mathcal{E}(\Omega^*)$. □

This yields the first main result of this paragraph.

7.5 Theorem. *If Ω^* is a sublattice of the lattice Ω, then $\mathcal{E}(\Omega^*)$ is a GALOIS extension of the field $\mathcal{E}(\Omega)$ of degree $[\Omega : \Omega^*]$ and GALOIS group is isomorphic to the factor group Ω/Ω^*.*

Proof The mappings $f \mapsto f_v$, $v \in \mathcal{R}$, defined in (7.4) are $[\Omega : \Omega^*]$ pairwise distinct elements of the GALOIS group. The assertion now follows from the Main Theorem of GALOIS theory (cf. S. LANG [53], VI). □

As a special example, we consider the $\wp$–function.

7.6 Lemma. *If Ω^* is a sublattice of Ω, then one has for all $z \in \mathbb{C} \setminus \Omega$*

$$\wp(z; \Omega) = \sum_{v \in \mathcal{R}} \wp(z + v; \Omega^*) - \sum_{v \in \mathcal{R}, v \notin \Omega^*} \wp(v; \Omega^*).$$

Proof According to (7.1), the assertion follows from

$$\wp(z; \Omega) = z^{-2} + \sum_{0 \neq \omega \in \Omega} \left((z + \omega)^{-2} - \omega^{-2}\right)$$

$$= z^{-2} + \sum_{v \in \mathcal{R}, v \notin \Omega^*, \omega' \in \Omega^*} \left((z + v + \omega')^{-2} - (v + \omega')^{-2}\right)$$

$$+ \sum_{0 \neq \omega' \in \Omega^*} \left((z + \omega')^{-2} - \omega'^{-2}\right)$$

$$= \wp(z; \Omega^*) + \sum_{v \in \mathcal{R}, v \notin \Omega^*} \left((z + v)^{-2} - v^{-2}\right)$$

$$+ \sum_{\substack{v \in \mathcal{R}, v \notin \Omega^* \\ 0 \neq \omega' \in \Omega^*}} \left((z + v + \omega')^{-2} - \omega'^{-2} - \left((v + \omega')^{-2} - \omega'^{-2}\right)\right)$$

$$= \wp(z; \Omega^*) + \sum_{v \in \mathcal{R}, v \notin \Omega^*} \left(\wp(z + v; \Omega^*) - \wp(v; \Omega^*)\right). \quad \square$$

7.7 Remarks. a) Theorem 7.5 fits into the general theory of compact RIEMANN surfaces as follows: consider $\mathcal{R} := \mathbb{C}/\Omega$ and $\mathcal{R}^* := \mathbb{C}/\Omega^*$ as RIEMANN surfaces. Then $\mathcal{R}^*$ is a $[\Omega : \Omega^*]$-fold unramified covering of $\mathcal{R}$ by virtue of the mapping $a + \Omega^* \mapsto a + \Omega$. The field $\mathcal{F}^*$ of all meromorphic functions on $\mathcal{R}^*$ is then an extension field of the field $\mathcal{F}$ of all meromorphic functions on $\mathcal{R}$ of degree $[\Omega : \Omega^*]$. Compare O. FORSTER [25].

b) Any $f \in \mathcal{E}(\Omega^*)$ is a root of the polynomial

$$P(X) = \prod_{v \in \mathcal{R}}(X - f_v) \in \mathcal{E}(\Omega)[X]. \tag{7.6'}$$

Of course, $\mathbb{C}(\wp^*)$ over $\mathbb{C}(\wp)$ also has degree $[\Omega : \Omega^*]$.

3. $\wp$-partial values. Let $\wp$ always denote the $\wp$-function for the fixed lattice Ω. For a fixed natural number n, consider the case $\Omega^* := n\Omega$. Because of (4.1), we have

$$E_k(n\Omega) = n^{-k} \cdot E_k(\Omega), \quad k \geq 3, \quad \wp(z; n\Omega) = n^{-2} \cdot \wp(z/n; \Omega). \tag{7.8}$$

As in Proposition 7.1, we obtain a system of representatives $\mathcal{R}(n)$ of Ω modulo $n\Omega$ in the form

$$v = m_1\omega_1 + m_2\omega_2, \quad m_1, m_2 \in \mathbb{Z}, \quad 0 \leq m_1, m_2 < n \tag{7.9}$$

Now let $n \geq 2$. Then by (6.39) there exist polynomials A_n and B_n in $\mathbb{C}[X]$ satisfying

$$\wp(nz) - \wp(z) = -A_n(\wp(z))/B_n(\wp(z)) \tag{7.10}$$

According to (6.37) as well as Theorem 6.24, one has

$$A_n(X) = (n^2 - 1)X^{n^2} + O(X^{n^2-2}) \quad \text{and} \quad B_n(X) = n^2 X^{n^2-1} + O(X^{n^2-3}), \tag{7.11}$$

where $O(X^r)$ stands for a polynomial of degree $\leq r$. If we define a polynomial $P_n \in \mathbb{C}[X, Y]$ by

$$P_n(X, Y) := (X - Y) \cdot B_n(X) - A_n(X), \quad n \geq 2, \tag{7.12}$$

then (7.10) immediately yields

$$P_n(\wp(z), \wp(nz)) = 0 \tag{7.13}$$

and (7.11) leads to

$$P_n(X, Y) = X^{n^2} - n^2 \cdot Y \cdot X^{n^2-1} + O(X^{n^2-2}) + YO(X^{n^2-3}). \tag{7.14}$$

Note that P_n has degree 1 in Y.

7.8 Theorem. *If $n \geq 2$ then*

$$P_n(X, \wp(nz)) = \prod_{v \in \mathcal{R}(n)} (X - \wp(z + v/n)). \tag{7.15}$$

Proof If $\mathcal{R}$ is an arbitrary system of representatives of Ω modulo $n\Omega$, consider the polynomial

$$\prod_{v \in \mathcal{R}} (X - \wp(z + v/n)). \qquad (*)$$

Because of (7.2), the product $(*)$ does not depend on the choice of the system of representatives $\mathcal{R}$. So we can replace $\mathcal{R}$ by the special system of representatives $\mathcal{R}(n)$. By (7.13), $P_n(\wp(z + v/n), \wp(nz)) = 0$ holds. Thus the two monic polynomials $(*)$ and $P_n(X, \wp(nz))$ of degree n^2 have the n^2 distinct roots $\wp(z + v/n)$, $v \in \mathcal{R}(n)$. So they coincide. □

A comparison of the coefficients of the second highest power of X in (7.14) and (7.15) yields

7.9 Corollary. *If $n \geq 2$ then*

$$n^2 \wp(nz) = \sum_{v \in \mathcal{R}(n)} \wp(z + v/n).$$

This is evidently another "explicit" formula for the multiplication of the $\wp$-function. A comparison of the constant terms in the LAURENT series expansion around zero in Corollary 7.9, yields

7.10 Corollary. *If $n \geq 2$ then*

$$\sum_{0 \neq v \in \mathcal{R}(n)} \wp(v/n) = 0.$$

In the case $n = 2$, we get the well known relation $e_1 + e_2 + e_3 = 0$ (cf. Corollary 3.17). Finally, we replace z by z/n in (7.15) and obtain

7.11 Corollary. *Given $n \geq 2$,*

$$P_n(X, \wp(z)) = \prod_{v \in \mathcal{R}(n)} (X - \wp((z + v)/n)).$$

Thus, the so-called $\wp$-*partial values*

$$\wp((z + v)/n), \quad v \in \mathcal{R}(n),$$

are algebraic over the field $\mathbb{C}(\wp(z))$. They all belong to a field extension of the field $\mathbb{C}(\wp(z))$ of degree n^2. We call $P_n(X, \wp(z)) = 0$ the *n-division equation* and P_n the *n–division polynomial* of the $\wp$-function.

4. Lattices with complex multiplication. Early on, the question arose for which $\lambda \in \mathbb{C}$ and $f \in \mathcal{E}(\Omega)$, the function f_λ, $f_\lambda(z) := f(\lambda z)$, is also an elliptic function with respect to the same lattice. As a simple example, consider the square lattice $\Omega = \mathbb{Z}i + \mathbb{Z}$ and the corresponding $\wp$-function: because $i\Omega = \Omega$, we can immediately see that

$$\wp_i = -\wp \quad \text{and} \quad (\wp')_i = i \cdot \wp'$$

§ 7* $\wp$-partial values, algebraic dependence and complex multiplication

from the definition (3.1). Given $f \in \mathcal{E}(\Omega)$, we have $f = R(\wp) + Q(\wp) \cdot \wp'$ by (2.14) with rational functions P and Q. It then follows that $f_i = R(-\wp) + Q(-\wp) \cdot i\wp'$ and we obtain a complete description of the functions f_i for $f \in \mathcal{E}(\Omega)$.

In the general situation, we proceed as follows: given $f \in \mathcal{E}(\Omega)$, for $0 \neq \lambda \in \mathbb{C}$, the elliptic function f_λ is contained in $\mathcal{E}(\Omega)$ if

$$\lambda \Omega \subseteq \Omega \tag{7.16}$$

holds, because then

$$f_\lambda(z + \omega) = f(\lambda z + \lambda \omega) = f(\lambda z) = f_\lambda(z) \quad \text{for } \omega \in \Omega.$$

Of course, (7.16) is satisfied for all $\lambda \in \mathbb{Z}$. Remember here that the Addition Theorem 5.1 gives explicit formulas for the functions $\wp_m$ and $(\wp')_m$, $m \in \mathbb{Z}$, $m \neq 0$. Due to (2.14) we then understand all f_m, $m \in \mathbb{Z}$, $m \neq 0$, for $f \in \mathcal{E}(\Omega)$. Now let

$$\Omega = \mathbb{Z}\omega_1 + \mathbb{Z}\omega_2 \quad \text{with} \quad \omega_1 = \omega_2 \cdot \tau, \tau \in \mathbb{H}, i.e. \ \Omega = \omega_2 \cdot (\mathbb{Z}\tau + \mathbb{Z}) \tag{7.17}$$

be an arbitrary lattice in $\mathbb{C}$. The condition (7.16) is obviously only a constraint on τ. Let $\mathcal{R}(\tau)$ denote the set of $\lambda \in \mathbb{C}$ for which (7.16) holds. Obviously, $\mathcal{R}(\tau)$ is a subring of $\mathbb{C}$ containing $\mathbb{Z}$. A lattice (7.17) is called a *lattice with complex multiplication* if $\mathcal{R}(\tau) \neq \mathbb{Z}$ holds. Then $\mathcal{R}(\tau)$ is called the *multiplier ring* of Ω. According to the example above, $\mathbb{Z}i + \mathbb{Z}$ is a lattice with complex multiplication; the same is true for $\mathbb{Z}\rho + \mathbb{Z}$ with $\rho := \frac{1}{2}\left(1 + i\sqrt{3}\right)$.

7.12 Proposition. *Let $\tau \in \mathbb{H}$ and $\lambda \in \mathbb{C}$. Then the following assertions are equivalent:*
(i) $\lambda \in \mathcal{R}(\tau)$.
(ii) *There exist $a, b, c, d \in \mathbb{Z}$ such that $\lambda \tau = a\tau + b$ and $\lambda = c\tau + d$.*
(iii) *If $w := (\tau, 1)^{tr}$ then $\lambda \cdot w = Mw$ with some $M = \begin{pmatrix} a & b \\ c & d \end{pmatrix} \in \text{Mat}(2; \mathbb{Z})$.*

Proof (i) $\iff$ (ii): $\lambda \in \mathcal{R}(\tau)$ is equivalent to $\lambda \cdot 1 \in \mathbb{Z}\tau + \mathbb{Z}$ and $\lambda \cdot \tau \in \mathbb{Z}\tau + \mathbb{Z}$.
(ii) $\iff$ (iii): Clear. □

The name *complex multiplication* is justified by

7.13 Theorem. *If (7.17) is a lattice with complex multiplication, i.e. $\mathcal{R}(\tau) \neq \mathbb{Z}$, then τ belongs to an imaginary-quadratic number field $\mathbb{Q}[\sqrt{-D}]$, $D \in \mathbb{Z}$, $D > 0$ square-free, and $\mathcal{R}(\tau)$ is a subring of the ring of integers of $\mathbb{Q}[\sqrt{-D}]$.*

As a submodule of the free $\mathbb{Z}$-module of the integers in $\mathbb{Q}[\sqrt{-D}]$, $\mathcal{R}(\tau)$ is also free. Because $\mathcal{R}(\tau) \neq \mathbb{Z}$, the multiplier ring $\mathcal{R}(\tau)$ has rank 2, i.e. $\mathcal{R}(\tau)$ is a lattice in $\mathbb{C}$.

Proof As an eigenvalue of a matrix M of $\text{Mat}(2; \mathbb{Z})$ according to Proposition 7.12, λ is a root of a monic quadratic equation over $\mathbb{Z}$. Thus λ is an algebraic integer of degree 1 or 2. In the case $\lambda \notin \mathbb{Z}$, λ is quadratic. Then $c \neq 0$ holds in (ii). Thus τ and λ are imaginary-quadratic numbers and contained in some field $\mathbb{Q}\left[\sqrt{-D}\right]$, $D \in \mathbb{Z}$, $D > 0$. □

5. The corresponding subring of Mat(2; $\mathbb{Z}$).

We define

$$\mathcal{M}(\tau) := \{M \in \text{Mat}(2;\mathbb{Z}) \; ; \; c\tau^2 + (d-a)\tau - b = 0\}. \tag{7.18}$$

In Proposition 7.12, the matrix $M = \begin{pmatrix} a & b \\ c & d \end{pmatrix}$ is uniquely determined by $\lambda w = Mw$ and $M \in \mathcal{M}(\tau)$. However, if $M \in \mathcal{M}(\tau)$, then $\lambda := c\tau + d$ is an element of $\mathcal{R}(\tau)$ according to Proposition 7.12.

7.14 Proposition. $\mathcal{M}(\tau)$ *is a subring of* Mat(2; $\mathbb{Z}$) *that is isomorphic to the ring* $\mathcal{R}(\tau)$ *by virtue of* $M \mapsto \lambda, \lambda w = Mw, w = (\tau, 1)^{tr}$.

We now proceed inversely from an imaginary-quadratic number field

$$\mathbb{Q}\left[\sqrt{-D}\right], \; D \in \mathbb{Z}, \; D > 0 \quad \text{square-free},$$

and choose

$$\tau = r + s\sqrt{-D} \in \mathbb{Q}[\sqrt{-D}] \quad \text{with} \quad \text{Im } \tau > 0, \tag{7.19}$$

so especially $r, s \in \mathbb{Q}, s \neq 0$. Let $q = q(\tau)$ be the least common denominator of $2r$ and $r^2 + Ds^2$, hence

$$q := \text{lcm (denominator of tr } \tau, \text{ denominator of norm } \tau). \tag{7.20}$$

Here, of course, tr $\tau := \tau + \bar{\tau} = 2r$ and norm $\tau := \tau \cdot \bar{\tau} = r^2 + Ds^2$. Now we define $\mathcal{M}(\tau)$ according to (7.18) and observe that $\mathcal{M}(\tau)$ is a subring of Mat(2; $\mathbb{Z}$).

7.15 Lemma. *If τ is given by (7.19), q by (7.20) and* $M = \begin{pmatrix} a & b \\ c & d \end{pmatrix} \in \text{Mat}(2;\mathbb{Z})$, *then the following assertions are equivalent:*
(i) $M \in \mathcal{M}(\tau)$.
(ii) $b = -c \cdot \text{norm } \tau, \; d = a - c \cdot \text{tr } \tau$ *with* $a, c \in \mathbb{Z}$ *and* $c \equiv 0 \pmod{q}$.

Proof Substituting (7.19) into (7.18), we get $a - d = 2rc$ and thus moreover $b = -c \cdot \text{norm } \tau$. □

Due to (ii), we can write

$$\mathcal{M}(\tau) = \mathbb{Z} \cdot I + \mathbb{Z}q \cdot T \quad \text{with} \quad T := \begin{pmatrix} 0 & \text{norm } \tau \\ -1 & \text{tr } \tau \end{pmatrix}. \tag{7.21}$$

Using Proposition 7.12 and Theorem 7.13 we get

7.16 Theorem. *For a lattice (7.17), the following assertions are equivalent:*
(i) Ω *is a lattice with complex multiplication.*
(ii) τ *belongs to an imaginary-quadratic number field.*
In this case, in the notation (7.20), the multiplier ring is given by

$$\mathcal{R}(\tau) = \mathbb{Z}q\tau + \mathbb{Z}. \tag{7.22}$$

In particular, $\mathcal{R}(\tau) \subseteq \mathbb{Z}\tau + \mathbb{Z}$.

Proof Just (7.22) remains is to be shown. According to Proposition 7.14 and part (ii) of Lemma 7.15, however,

$$\mathcal{R}(\tau) = \{c\tau + a - c(\tau + \overline{\tau}) \,;\, a, c \in \mathbb{Z},\, c \equiv 0 \pmod{q}\} = \mathbb{Z}q\overline{\tau} + \mathbb{Z}$$

holds. Because $\overline{\tau} = \operatorname{tr} \tau - \tau$ and $q \operatorname{tr} \tau \in \mathbb{Z}$, the assertion follows. □

7.17 Corollary. τ *is an algebraic integer if and only if* $\mathcal{R}(\tau) = \mathbb{Z}\tau + \mathbb{Z}$ *holds.*

A statement about the values of the j-function for lattices with complex multiplication, called *singular values*, will be derived in IV, §1, sect. 7.

6. Example. Consider the lattice

$$\Omega := \mathbb{Z}\tau + \mathbb{Z} \quad \text{with} \quad \tau := \sqrt{-2} = i\sqrt{2}. \tag{7.23}$$

By Proposition 7.12 or Corollary 7.17, $\mathcal{R}(\tau) = \Omega = \mathbb{Z}\tau + \mathbb{Z}$ holds. So, for example, $\wp(\sqrt{-2}z; \Omega)$ is rationally expressible as an even function in $\wp(z; \Omega)$. In order to explicitly determine $\wp(\sqrt{-2}z; \Omega)$, we consider the sublattice

$$\Omega^* := \sqrt{-2} \cdot \Omega = \mathbb{Z}2 + \mathbb{Z}\tau, \quad [\Omega : \Omega^*] = 2,$$

and abbreviate $\wp(z) := \wp(z; \Omega)$ as well as $\wp^*(z) := \wp(z; \Omega^*)$. By Lemma 7.6, then,

$$\wp(z) = \wp^*(z) + \wp^*(z+1) - \wp^*(1).$$

Substitute $z \mapsto \sqrt{-2}z$, then (4.1') leads to

$$\wp(\sqrt{-2}z) = -\frac{1}{2}\left(\wp(z) + \wp\left(z - \frac{1}{2}\sqrt{-2}\right)\right) - \wp^*(1).$$

As usual, let

$$e_1 := \wp\left(\sqrt{-2}/2\right),\ e_2 := \wp(1/2),\ e_3 := \wp\left((1 + \sqrt{-2})/2\right).$$

Since $\wp'$ vanishes at the corresponding points because of Lemma 2.10, the Addition Theorem 5.1 yields

$$\wp(\sqrt{-2}z) = \frac{1}{2}e_1 - \frac{1}{2} \cdot \frac{1}{4}\left(\frac{\wp'(z)}{\wp(z) - e_1}\right)^2 - \wp^*(1).$$

Now we have to use the differential equation of Theorem 2.12 and obtain

$$\wp(\sqrt{-2}z) = \frac{1}{2}e_1 - \frac{1}{2}\frac{(\wp(z) - e_2) \cdot (\wp(z) - e_3)}{\wp(z) - e_1} - \wp^*(1).$$

For $z = \frac{1}{2}$ we get $e_1 = \frac{1}{2}e_1 - \wp^*(1)$, i.e.

7.18 Theorem. *If* $\Omega = \mathbb{Z}\sqrt{-2} + \mathbb{Z}$, *then*

$$\wp(\sqrt{-2}z) = e_1 - \frac{1}{2} \frac{(\wp(z) - e_2) \cdot (\wp(z) - e_3)}{\wp(z) - e_1}.$$

7.19 Exercises.
1) Calculate the polynomial $P_2(X, Y)$ in sect.3.
2) In the notation of (7.5), $\wp_v^*$ belongs to $\mathbb{C}(\wp^*)$ if and only if $2v \in \Omega^*$ holds.
3) Let $\Omega = \mathbb{Z}\frac{1}{2}(1 + i\sqrt{7}) + \mathbb{Z}$. Represent $\wp\left(\frac{1}{2}(1 + i\sqrt{7})z; \Omega\right)$ rationally by $\wp(z; \Omega)$.
4) Let $\Omega = \mathbb{Z}i\sqrt{3} + \mathbb{Z}$. Represent $\wp(i\sqrt{3}z; \Omega)$ rationally by $\wp(z; \Omega)$.
5) Let $\Omega = \mathbb{Z}\frac{1}{2}(1 + i\sqrt{3}) + \mathbb{Z}$. Then $\wp\left(\frac{1}{2}(1 + i\sqrt{3})z; \Omega\right) = -\frac{1}{2}(1 + i\sqrt{3}) \cdot \wp(z; \Omega)$.
6) Let $\Omega = \mathbb{Z}\tau + \mathbb{Z}$, $\operatorname{Im} \tau > 0$, be a lattice with complex multiplication. Then $\wp(\tau z; \Omega)$ is a polynomial in $\wp(z; \Omega)$ if and only if $\tau = i$ or $\tau = \frac{1}{2}\left(\pm 1 + i\sqrt{3}\right)$.
7) Given two lattices Ω and Ω' in $\mathbb{C}$, demonstrate the equivalence of the following statements:

   (i) Ω and Ω' are commensurable, i.e., the intersection $\Omega \cap \Omega'$ has finite index in Ω and Ω'.
   (ii) The intersection $\Omega \cap \Omega'$ is a lattice in $\mathbb{C}$.
   (iii) The sum $\Omega + \Omega'$ is a lattice in $\mathbb{C}$.
   (iv) $\mathcal{E}(\Omega) \cap \mathcal{E}(\Omega') \neq \mathbb{C}$.

8) Let Ω be a lattice in $\mathbb{C}$ and $n \in \mathbb{N}$. Show that Ω possesses exactly $\sigma_1(n)$ sublattices Ω' with $[\Omega : \Omega'] = n$.
9) Show that (7.19) and (7.20) imply

$$\{M^\sharp;\ M \in \mathcal{M}(\tau)\} = \mathcal{M}(\tau)$$

in (7.21), where $M^\sharp$ is the adjugate of M.
10) Determine the units of the ring $\mathcal{M}(\tau)$ in Proposition 7.14.
11) Show that $\mathcal{M}(\tau) = \mathbb{Z}I$ in (7.18), whenever $\tau \in \mathbb{H}$ does not belong to an imaginary-quadratic number field.
12) Show that any $M \neq 0$ in (7.18) satisfies $\det M \neq 0$.

§ 8* Miscellaneous remarks

1. The zeros of the $\wp$-function. After more than a hundred years of theory of elliptic functions, M. EICHLER and D. ZAGIER (*Math. Ann.* **258**, 399–407) only in 1982 noticed that the zeros of the $\wp$-function can be given explicitly. We quote the result without proof:

8.1 Theorem. *Given $\tau \in \mathbb{H}$, consider the lattice $\Omega = \mathbb{Z}\tau + \mathbb{Z}$ in $\mathbb{C}$ and the corresponding $\wp$-function $\wp(z; \tau) = \wp_\Omega(z)$. In the notation of §4, the points $z \in \mathbb{C}$ with $\wp(z; \tau) = 0$ are given by*

$$z = m\tau + n + \frac{1}{2} \pm \left(\frac{\log(5 + 2\sqrt{6})}{2\pi i} + \frac{i}{72\pi^2} \cdot \int_0^\infty \frac{\xi \cdot \Delta(\tau + i\xi)}{[g_3(\tau + i\xi)]^{3/2}} d\xi \right), \quad m, n \in \mathbb{Z}.$$

A generalization of this formula, which is the solution of an equation of the form $\wp(z; \tau) = \varphi(\tau)$, is also given there. Later W. DUKE and Ö. IMAMOGLU (*Math. Ann.* **340**, 897-905 (2008)) took up the approach and gave a power series representation in g_2 and g_3 for the roots of the $\wp$-function.

2. Theta series. C.G.J. JACOBI systematically used theta series for the construction of elliptic functions in 1829 in his *Fundamenta nova* ([42], vol. I, 49–239). Already in 1822 these series appear in a different context in *Théorie Analytique de la Chaleur* by J. FOURIER (1768–1830; [26], vol. I, §241). Compare the Introduction of Chapter III.

Following JACOBI, we consider four different *theta series*, which are (in this or similar notation) given by:

$$\vartheta_0(z; \tau) := i \cdot \sum_{n \in \mathbb{Z}} e^{\pi i (n - 1/2)^2 \tau + 2\pi i (n - 1/2) z + \pi i n},$$

$$\vartheta_1(z; \tau) := \sum_{n \in \mathbb{Z}} e^{\pi i n^2 \tau + 2\pi i n z + \pi i n},$$

$$\vartheta_2(z; \tau) := \sum_{n \in \mathbb{Z}} e^{\pi i (n - 1/2)^2 \tau + 2\pi i (n - 1/2) z},$$

$$\vartheta_3(z; \tau) := \sum_{n \in \mathbb{Z}} e^{\pi i n^2 \tau + 2\pi i n z}$$

with $z \in \mathbb{C}$ and $\tau \in \mathbb{H}$. The theta series ϑ_3 was already treated in (6.30). All series converge absolutely and uniformly on compact subsets of $\mathbb{C} \times \mathbb{H}$. The roots of $\vartheta_k(\cdot, \tau)$ are contained in

$$\tfrac{1}{2} \omega_k + \mathbb{Z}\tau + \mathbb{Z}, \quad \omega_0 = 0, \ \omega_1 = \tau, \ \omega_2 = 1, \ \omega_3 = \tau + 1, \quad k = 0, 1, 2, 3.$$

The importance of these theta series comes from their behavior under the translations $z \mapsto z + \omega$ for $\omega \in \Omega := \mathbb{Z}\tau + \mathbb{Z}$

ν	$\vartheta_\nu(z + 1/2; \tau)$	$\vartheta_\nu(z + \tau/2; \tau)$	$\vartheta_\nu(z + 1; \tau)$	$\vartheta_\nu(z + \tau; \tau)$
0	ϑ_2	$i\xi \cdot \vartheta_1$	$-\vartheta_0$	$-\zeta \cdot \vartheta_0$
1	ϑ_3	$i\xi \cdot \vartheta_0$	ϑ_1	$-\zeta \cdot \vartheta_1$
2	$-\vartheta_0$	$\xi \cdot \vartheta_3$	$-\vartheta_2$	$\zeta \cdot \vartheta_2$
3	ϑ_1	$\xi \cdot \vartheta_2$	ϑ_3	$\zeta \cdot \vartheta_3$

with the abbreviations $\xi := e^{-\frac{1}{4}\pi i \tau - \pi i z}$ and $\zeta := e^{-\pi i \tau - 2\pi i z}$. The roots given here are all of 1st order. A *proof* of these results (and of the other results quoted without proof) can be found, e.g., in A. HURWITZ and R. COURANT [40], 188–212. Compare also K. CHANDRASEKHARAN [10], Chap. V. From this table, we can see that each function of the form

$$\frac{\vartheta(z+a;\tau)\cdot\vartheta(z+b;\tau)}{\vartheta(z+c;\tau)\cdot\vartheta(z+d;\tau)}, \quad \text{where } \vartheta \text{ represents } \vartheta_0, \vartheta_1, \vartheta_2 \text{ or } \vartheta_3$$

and $a, b, c, d \in \mathbb{C}$ with $a + b - c - d \in \mathbb{Z}$, is an elliptic function with respect to the lattice $\mathbb{Z}\tau + \mathbb{Z}$. The exact knowledge of the roots leads to the explicit formulas

$$\sigma(z;\tau) = e^{z^2 \cdot \eta/2} \cdot \frac{\vartheta_1(z;\tau)}{\vartheta_1'(0;\tau)}$$

and

$$\sqrt{\wp(z;\tau) - e_k} = \frac{\vartheta_0'(0;\tau)}{\vartheta_0(z;\tau)} \cdot \frac{\vartheta_k(z;\tau)}{\vartheta_k(0;\tau)} \quad \text{for } k = 1, 2, 3.$$

Following Ch. HERMITE the four theta series can be regarded as special cases of the general theta series

$$\Theta_{\mu,\nu}(z;\tau) := \sum_{n \in \mathbb{Z}} e^{\pi i \tau (n+\mu/2)^2 + 2\pi i z(n+\mu/2) + \pi i n \nu} \quad \text{with } \nu, \mu \in \mathbb{Z}.$$

3. The arithmetic-geometric mean. For real numbers $a \geq b \geq 0$, we recursively define sequences $f_n = f_n(a, b)$ and $g_n = g_n(a, b)$, $n \in \mathbb{N}$, by

$$f_{n+1} := \frac{1}{2}(f_n + g_n), \quad g_{n+1} := \sqrt{f_n g_n}, \quad f_0 := a, \quad g_0 := b. \tag{8.1}$$

This is a mixture of the arithmetic mean and the geometric mean.

8.2 Proposition. *For all $n \in \mathbb{N}$,*

$$0 \leq f_{n+1} - g_{n+1} \leq \frac{1}{2}(f_n - g_n) \tag{8.2}$$

and

$$g_n \leq g_{n+1} \leq f_{n+1} \leq f_n. \tag{8.3}$$

Proof We use induction on n in order to get

$$(f_{n+1} - g_{n+1})(f_{n+1} + g_{n+1}) = \frac{1}{4}(f_n - g_n)^2$$

as well as

$$2(f_{n+1} + g_{n+1}) \geq f_n + g_n. \qquad \square$$

(8.2) leads to

$$0 \leq f_n - g_n \leq 2^{-n} \cdot (a - b). \tag{8.4}$$

By (8.3), the sequences $(f_n)_n$ and $(g_n)_n$ therefore converge to the same limit:

$$M(a, b) := \lim_{n \to \infty} f_n(a, b) = \lim_{n \to \infty} g_n(a, b). \tag{8.5}$$

(8.1) implies

§ 8* Miscellaneous remarks

$$M\left(\frac{a+b}{2}, \sqrt{ab}\right) = M(a,b), \tag{8.6}$$

$$M(\alpha a, \alpha b) = \alpha \cdot M(a,b) \quad \text{for } \alpha > 0, \tag{8.7}$$

$$M(a,b) = a \cdot M(1, b/a) = b \cdot M(a/b, 1), \text{ if } a \neq 0, b \neq 0. \tag{8.8}$$

This fact was already known to J.-L. LAGRANGE (1736–1813) [51], vol. I, 272, 304f.. C.F. GAUSS (1777–1857; *Ostwald's Klassiker der exakten Wissenschaften* **225**, 1927) dealt with it independently at the age of 14. Following his example, we shall now show that $M(a,b)$ can be expressed by an elliptic integral: for real $a, b, a \geq b > 0$ let

$$\mu(a,b;x) := \int_0^x \frac{d\varphi}{\sqrt{a^2 \cos^2 \varphi + b^2 \sin^2 \varphi}}, \quad 0 \leq x \leq \frac{\pi}{2}. \tag{8.9}$$

Replacing $\cos^2 \varphi$ here by $1 - \sin^2 \varphi$, we see that $\mu(a,b;x)$ is essentially the LEGENDRE integral. This requires

8.3 Lemma. *There is a continuously differentiable, bijective and strictly monotonically increasing function*

$$\varphi : [0, \pi/2] \to [0, \pi/2], \quad t \mapsto \varphi(t),$$

with the following properties:

$$\left. \begin{array}{l} \sin \varphi = \dfrac{2a \sin t}{(a+b)\cos^2 t + 2a \sin^2 t}, \\[2mm] \cos \varphi = \dfrac{\sqrt{(a+b)^2 \cos^2 t + 4ab \sin^2 t}}{(a+b)\cos^2 t + 2a \sin^2 t} \cos t, \end{array} \right\} \tag{8.10}$$

and

$$\frac{d\varphi}{\sqrt{a^2 \cos^2 \varphi + b^2 \sin^2 \varphi}} = \frac{dt}{\sqrt{\left(\frac{a+b}{2}\right)^2 \cos^2 t + \left(\sqrt{ab}\right)^2 \sin^2 t}}. \tag{8.11}$$

Proof We set

$$A := \frac{2a \sin t}{(a+b)\cos^2 t + 2a \sin^2 t}, \quad B := \frac{\sqrt{(a+b)^2 \cos^2 t + 4ab \sin^2 t}}{(a+b)\cos^2 t + 2a \sin^2 t} \cos t$$

and verify $A^2 + B^2 = 1$. From this we obtain the existence of a continuously differentiable mapping $\varphi : [0, \frac{\pi}{2}] \to [0, \frac{\pi}{2}]$ satisfying (8.10), and

$$\sin \varphi \cdot \frac{d\varphi}{dt} = 2a \cdot \frac{(a+b)\cos^2 t + 2b \sin^2 t}{((a+b)\cos^2 t + 2a \sin^2 t)^2} \cos t$$

follows. Therefore, φ is strictly monotonically increasing, and we obtain (8.11). □

Now we use this transformation into the integral (8.9) and get

8.4 Theorem. *If $0 \leq x, y \leq \frac{\pi}{2}$ such that*

$$\sin x = \frac{2a \sin y}{(a+b)\cos^2 y + 2a \sin^2 y},$$

then

$$\mu(a, b; x) = \mu\left(\frac{a+b}{2}, \sqrt{ab}; y\right).$$

In particular, $\mu(a, b) = \mu\left(\frac{a+b}{2}, \sqrt{ab}\right)$ holds for $\mu(a, b) := \mu\left(a, b; \frac{\pi}{2}\right)$. Therefore, a comparison with (8.1) and an induction yields $\mu(a, b) = \mu(f_n, g_n)$, and for $n \to \infty$ it follows that $\mu(a, b) = \mu(M, M) = \frac{1}{M} \cdot \mu(1, 1)$ with $M := M(a, b)$ from (8.5). Because $\mu(1, 1) = \frac{\pi}{2}$, we obtain

8.5 GAUSS's Formula. *We have*

$$\frac{1}{M(a,b)} = \frac{2}{\pi} \int_0^{\pi/2} \frac{d\varphi}{\sqrt{a^2 \cos^2 \varphi + b^2 \sin^2 \varphi}}.$$

A fast iteration procedure has been developed from this formula for the calculation of π.

References: F. KLEIN and M. BRENDEL [44]. K. KOMMERELL: *Das Grenzgebiet der elementaren und höheren Mathematik.* Koehler, Leipzig 1936. J.M. and P.B. BORWEIN: *Pi and the AGM.* J. Wiley, New York 1987. D.A. COX, L'Enseign. Math. **30**, 275–330 (1984).

4. JACOBI forms. Let $\mathbb{H}$ denote the upper half-plane in $\mathbb{C}$. If k, m are natural numbers, then a holomorphic function $\Phi : \mathbb{C} \times \mathbb{H} \to \mathbb{C}$ is called a JACOBI *form of weight k and index m* if

(JF.1) $\quad \Phi\left(\dfrac{z}{c\tau+d}, \dfrac{a\tau+b}{c\tau+d}\right) = (c\tau+d)^k \cdot e^{2\pi i m c z^2/(c\tau+d)} \cdot \Phi(z, \tau)$

holds for all $\begin{pmatrix} a & b \\ c & d \end{pmatrix} \in \mathrm{SL}(2; \mathbb{Z})$,

(JF.2) $\quad \Phi(z + \lambda\tau + \mu, \tau) = e^{-2\pi i m(\lambda^2 \tau + 2\lambda z)} \cdot \Phi(z, \tau)$ holds for all $\lambda, \mu \in \mathbb{Z}$,

(JF.3) $\quad \Phi$ has a FOURIER series expansion of the form

$$\Phi(z, \tau) = \sum_{n=0}^{\infty} \sum_{\substack{r \in \mathbb{Z} \\ r^2 \leq 4nm}} c(n, r) e^{2\pi i(n\tau + rz)}.$$

A detailed exposition of this theory can be found in M. EICHLER and D. ZAGIER [19]. With respect to (4.9), the WEIERSTRASS $\wp$-function is a *meromorphic* JACOBI form of weight 2 and index 0. It can be represented as a quotient of two (holomorphic)

§ 8* Miscellaneous remarks

JACOBI forms of weight 12 and 10 and index 1. JACOBI forms can be seen as a cross between elliptic functions and modular forms (see Chapter III).

5. Transcendence statements. C.L. SIEGEL [78], vol. I, 267–274, started a new development in the theory of transcendental numbers in 1932 with the first statements on the transcendence of values of the $\wp$-function. This was continued in 1937 by his student Th. SCHNEIDER. These are statements of the following form:

8.6 Theorem. *Given* $\tau \in \mathbb{H}$, $a, b \in \mathbb{C}$ *not both zero and* $z \in \mathbb{C}$, $z \notin \Omega := \mathbb{Z}\tau + \mathbb{Z}$, *the six quantities*

$$a, \ b, \ g_2(\tau), \ g_3(\tau), \ \wp(z;\tau) \ and \ az + b\zeta(z;\Omega)$$

are not all algebraic over $\mathbb{Q}$.

Thus, if we specifically choose $a = 1$, $b = 0$ and assume that $g_2(\tau)$, $g_3(\tau)$ and $z \notin \Omega$ are algebraic over $\mathbb{Q}$, then $\wp(z;\tau)$ is transcendent.

References: T. SCHNEIDER: *Einführung in die transzendenten Zahlen.* Springer-Verlag, Berlin–Heidelberg–New York 1957. G.V. CHUDNOVSKY, *Proc. Int. Congr. Helsinki I*, 339–350 (1978). J. WOLFART, *Invent. Math.* **92**, 187–216 (1988).

6. Example applications of elliptic functions. The main examples treated in the literature are the following:
a) Arc length of ellipse and hyperbola.
b) Closed figures in the plane.
c) The mathematical and the spherical pendulum.
d) Surface area of a triaxial ellipsoid.
e) Geodesics on rotational ellipsoids.
f) Elastica theory.
g) Conformal mapping in aerodynamics.
h) Theory of the gyroscope.

References: F. TRICOMI and M. KRAFFT [84]. G.H. HALPHÉN [34], vol. II. K.T.W. WEIERSTRASS [88], vol. VI. C.G.J. JACOBI [42], vol. I. A.G. GREENHILL [33]. E. GRAESER [32]. R. BULIRSCH, *Mitteilungen der DMV*, 21–36 (1996).

8.7 Exercises.
1) Derive conditions on μ, ν under which $\Theta_{\mu,\nu}(0;\tau)$ vanishes identically.
2) Relate the arithmetic geometric mean $M(a, b)$ in (8.5) to the FAGNANO integral (0.8).
3) Let $\Phi(z, \tau)$ be a JACOBI form of weight k and index m. Let $\tau \in \mathbb{H}$ be fixed such that $\Phi(\cdot;\tau) \not\equiv 0$. Show that $z \mapsto \Phi(z, \tau)$ has exactly $2m$ roots counted with multiplicities in every fundamental parallelogram of the lattice $\mathbb{Z}\tau + \mathbb{Z}$.
4) Show that a holomorphic function $\Phi : \mathbb{C} \times \mathbb{H} \to \mathbb{C}$ satisfying (JF.3) is a JACOBI form of weight k and index m if and only if

$$\tau^{-k} e^{-2\pi i m z^2/\tau} \cdot \Phi\left(\frac{z}{\tau}, -\frac{1}{\tau}\right) = \Phi(z, \tau) = e^{2\pi i m(\tau + 2z)} \cdot \Phi(z + \tau, \tau)$$

for all $(z, \tau) \in \mathbb{C} \times \mathbb{H}$.

5) What does Theorem 8.6 say for $\tau = i, a = 1, b = 0, z = (1+i)/2$?

6) Consider the four theta series in section 2. Prove the assertion on their zeros and the associated orders.

7) Derive product expansions of the four theta series in section 2.

8) Show that
$$\frac{\log(5 + 2\sqrt{6})}{\pi} + \frac{1}{36\pi^2} \int_0^\infty \frac{\xi \Delta(i(1+\xi))}{(g_3(i(1+\xi)))^{3/2}} d\xi$$
is always an odd integer.

Chapter II
Geometry in the upper half-plane and the action of the modular group

Introduction

1. Preliminaries. The elliptic functions discussed in Chapter 1 are, by definition, exactly those meromorphic functions on $\mathbb{C}$ which are invariant under a class of distinguished groups of biholomorphic automorphisms of $\mathbb{C}$: the group of translations $z \mapsto z + \omega$, $\omega \in \Omega$, where Ω is a given lattice in $\mathbb{C}$. During our considerations, it was often useful to assume the lattice Ω given in the form

$$\Omega = \mathbb{Z}\tau + \mathbb{Z} \quad \text{with} \quad \operatorname{Im} \tau > 0.$$

According to the Basis Lemma for Lattices I.1.12 and the considerations in I, §4, τ is only determined up to a mapping of the form

$$\tau \mapsto M\tau := \frac{a\tau + b}{c\tau + d} \quad \text{with} \quad M = \begin{pmatrix} a & b \\ c & d \end{pmatrix} \in \operatorname{SL}(2; \mathbb{Z}). \tag{0.1}$$

This is the historical basis for the interest that those functions which are invariant or at least show a clear behavior under the substitutions (0.1) have found since the beginning of the theory of elliptic functions.

2. Modular substitutions. The center of all the following considerations is therefore the so-called *modular group*

$$\Gamma := \operatorname{SL}(2; \mathbb{Z}) = \{M \in \operatorname{Mat}(2; \mathbb{Z}) \, ; \, \det M = 1\}.$$

The group Γ acts as a group of biholomorphic automorphisms (0.1), the so-called *modular substitutions*, on the *upper half-plane*

$$\mathbb{H} := \{\tau \in \mathbb{C} \, ; \, \operatorname{Im} \tau > 0\}.$$

Before we discuss functions that are invariant or at least "relatively invariant" under modular substitutions in Chapter III, we will present the geometric foundations for such investigations.
On one hand this is a review of the basic properties of the transformations (0.1) for

real matrices M of determinant 1, i.e. if $M \in \mathrm{SL}(2;\mathbb{R})$ is allowed. On the other hand, in §2 the geometry of modular substitutions is investigated, and in particular, a canonical exact fundamental domain for Γ in $\mathbb{H}$ is constructed. The following paragraphs of this chapter can be skipped on a first reading.

3. The absolute invariant and the moduli space. In I(3.18), we assigned the so-called *absolute invariant*

$$j = j(\Omega) := (12g_2)^3/(g_2^3 - 27g_3^2)$$

to each lattice Ω with the help of the WEIERSTRASS invariants g_2 and g_3. Theorem I.4.1 states that

$$j(\Omega') = j(\Omega) \quad \text{is equivalent to} \quad \Omega' = \lambda\Omega \quad \text{for some} \quad 0 \neq \lambda \in \mathbb{C}.$$

Therefore, the image of j is already determined on lattices of the form

$$\Omega = \mathbb{Z}\tau + \mathbb{Z} \quad \text{with} \quad \tau \in \mathbb{H}.$$

Using the abbreviation $j(\tau) := j(\mathbb{Z}\tau + \mathbb{Z})$, then (0.1) shows that j is invariant under all modular substitutions. This was already stated in I, §4. In Theorem I.4.9 it was shown that the converse of this result is valid as well:

If $j(\tau') = j(\tau)$ holds for $\tau', \tau \in \mathbb{H}$, then there is some $M \in \Gamma$ with $\tau' = M\tau$.

Therefore, if we consider the space

$$\Gamma\backslash\mathbb{H} := \{\Gamma\tau \,;\, \tau \in \mathbb{H}\} \quad \text{with} \quad \Gamma\tau := \{M\tau \,;\, M \in \Gamma\}$$

of all *orbits* of Γ, then j induces a bijection

$$j^* : \Gamma\backslash\mathbb{H} \longrightarrow \mathbb{C}, \; j^*(\Gamma\tau) := j(\tau).$$

In this sense, $\Gamma\backslash\mathbb{H}$ is a *moduli space* for the function j. A first main task is therefore to find a simple realization of this moduli space in the upper half-plane. This is done by specifying an exact fundamental domain for Γ. Compare §2.

4. Generalizations. The upper half-plane with its geometry on the one hand and the group of modular substitutions on the other hand can be generalized in many ways: we can stick with $\mathbb{H}$ and use subgroups of Γ (§3) or more generally so-called discontinuous subgroups of $\mathrm{SL}(2;\mathbb{R})$ (§4). On the other hand, we can also move from the upper half-plane $\mathbb{H}$ to higher dimensions and put the focus on

a) holomorphy of the mappings or
b) birationality of the mappings or
c) preservation of metric properties

etc. None of these generalizations will be discussed in this book in detail. However, some aspects of generalizations will be discussed in sections that can be omitted at first glance.

§ 1 The upper half-plane

In this paragraph, the upper half-plane $\mathbb{H}$ in $\mathbb{C}$ is studied in more detail. The automorphism group of $\mathbb{H}$ is described. Then the hyperbolic geometry is developed.

1. Fractional linear transformations. We usually write 2×2 matrices in the form

$$M = \begin{pmatrix} a & b \\ c & d \end{pmatrix}$$

and use the usual abbreviations for *determinant* and *trace*

$$\det M := ad - bc, \quad \text{tr } M := a + d,$$

and

$$M^{tr} = \begin{pmatrix} a & c \\ b & d \end{pmatrix} \quad \text{resp.} \quad M^\sharp = \begin{pmatrix} d & -b \\ -c & a \end{pmatrix}$$

for the *transposed* resp. *adjugate* matrix. Let $\text{GL}(2;\mathbb{C})$ denote the group of the invertible complex 2×2 matrices,

$$\text{GL}(2;\mathbb{C}) := \{M \in \text{Mat}(2;\mathbb{C}) \ ; \ \det M \neq 0\}.$$

We use I for the identity matrix. As is well-known, the inverse of a 2×2 matrix $M \in \text{GL}(2;\mathbb{C})$ is given by

$$M^{-1} = \frac{1}{\det M} M^\sharp = \frac{1}{ad - bc} \begin{pmatrix} d & -b \\ -c & a \end{pmatrix}.$$

For $M \in \text{GL}(2;\mathbb{C})$, under obvious conditions on $\tau \in \mathbb{C}$, the complex number

$$M\tau := \frac{a\tau + b}{c\tau + d} \tag{1.1}$$

is well-defined. Thus

$$\Phi_M : \tau \mapsto M\tau \tag{1.2}$$

is a meromorphic function on $\mathbb{C}$, which is also called a *fractional linear transformation* or a MÖBIUS *transformation*. If $c = 0$, then Φ_M is an entire function. If $c \neq 0$, then Φ_M has exactly one pole of 1st order at $\tau = -d/c$.

Warning: The notation $M\tau$ must not be mixed up with the scalar multiplication τM of τ with M! If misunderstandings are possible, we write $M\langle\tau\rangle$ instead of $M\tau$.

From (1.1), we conclude for $L, M \in \text{GL}(2;\mathbb{C})$ and $0 \neq \lambda \in \mathbb{C}$ under obvious conditions on τ and τ':

$$I\tau = \tau, \quad \text{i.e..} \quad \Phi_I = \text{id}, \tag{1.3}$$

$$(\lambda M)\tau = M\tau, \quad \text{i.e.} \quad \Phi_{\lambda M} = \Phi_M, \tag{1.4}$$

$$(LM)\tau = L\langle M\tau\rangle, \quad \text{i.e.} \quad \Phi_{LM} = \Phi_L \circ \Phi_M, \tag{1.5}$$

$$M\tau - M\tau' = \frac{\det M}{(c\tau + d)(c\tau' + d)} \cdot (\tau - \tau'), \tag{1.6}$$

$$\frac{d}{d\tau}\Phi_M(\tau) = \Phi'_M(\tau) = \frac{dM\tau}{d\tau} = \frac{\det M}{(c\tau + d)^2}. \tag{1.7}$$

The last statement follows if we divide (1.6) by $\tau - \tau'$ and let $\tau' \to \tau$. As a converse of (1.4) we have the

1.1 Proposition. *Given $L, M \in \mathrm{GL}(2; \mathbb{C})$, the following assertions are equivalent*:

(i) $M\tau = L\tau$ holds for at least three different $\tau \in \mathbb{C}$.
(ii) There is some $0 \neq \lambda \in \mathbb{C}$ satisfying $M = \lambda L$.

Proof Because of (1.5) and (1.3), $M\tau = L\tau$ and $(L^{-1}M)\tau = \tau$ are equivalent. Therefore we may assume $L = I$ in both cases. Since $M\tau = \tau$ can now be expressed as
$$c\tau^2 + (d-a)\tau - b = 0,$$
the assertion follows, because a polynomial of degree 2 has at most 2 roots. □

Each *circle* in $\mathbb{C}$ can be represented by an equation of the form

$$A\tau\bar{\tau} + B\tau + \overline{B\tau} + C = 0 \quad \text{with} \quad A, C \in \mathbb{R}, \ A \neq 0 \text{ and } B \in \mathbb{C}. \tag{1.8}$$

More precisely, then $|B|^2 > AC$ and the *center* m and the *radius* $r > 0$ are given by

$$m = -\bar{B}/A \quad \text{resp.} \quad r^2 = (|B|^2 - AC)/A^2. \tag{1.9}$$

Conversely, (1.8) always describes a circle as long as $A \neq 0$ and $|B|^2 > AC$. In the case $A = 0$, $B \neq 0$ we obtain exactly all lines in $\mathbb{C}$ from (1.8).
In order to study the images of lines and circles under the fractional linear transformations (1.2), we have to consider $z = M\tau$, $M \in \mathrm{GL}(2; \mathbb{C})$. Now we insert $\tau = M^{-1}z$ into (1.8), multiply the denominators up and get an equation of the form

$$\alpha z\bar{z} + \beta z + \overline{\beta z} + \gamma = 0 \quad \text{with} \quad \alpha, \gamma \in \mathbb{R} \text{ and } \beta \in \mathbb{C}.$$

When interpreted correctly, we get the

1.2 Theorem. *The set of circles and lines in $\mathbb{C}$ is mapped into itself under fractional linear transformations.*

Finally, let us consider the obvious rules for the calculus with infinity. If M is given in the standard form, then we define

$$M\infty := \begin{cases} \infty, & \text{if } c = 0, \\ a/c, & \text{if } c \neq 0. \end{cases} \tag{1.10}$$

Clearly, Proposition 1.1 also holds if one of the points is ∞, because $M\infty = \infty$ is equivalent to $c = 0$ and we are left with a linear transformation in the proof.

1.3 Remarks. a) Equations (1.3), (1.5) and (1.10) state that (1.1) defines a group action of $\mathrm{GL}(2; \mathbb{C})$ on $\mathbb{P} = \mathbb{C} \cup \{\infty\}$. By Proposition 1.1, $\Phi_M = \Phi_L$ for matrices $M, L \in \mathrm{SL}(2; \mathbb{C})$ holds if and only if $M = \pm L$. By means of (1.2), we obtain a natural identification of the group $\mathrm{Aut}\,\mathbb{P}$ of the biholomorphic self-mappings of $\mathbb{P}$ with the

group $\text{PSL}(2;\mathbb{C}) := \text{SL}(2;\mathbb{C})/\{\pm I\}$ (cf. W. FISCHER and I. LIEB [24], IX.3.1).
b) If $f : D \to \mathbb{C}$ is a non-constant holomorphic function on a domain $D \subseteq \mathbb{C}$, then the SCHWARZian *derivative* Σf of f is defined by

$$(\Sigma f)(z) := \frac{f'''}{f'} - \frac{3}{2}\left(\frac{f''}{f'}\right)^2, \quad z \in D.$$

Using (1.7), we can verify $\Sigma(\Phi_M \circ f) = \Sigma f$ for all $M \in \text{GL}(2;\mathbb{C})$. Thus: *If f is a solution of the differential equation $\Sigma f = \varphi$, then $\Phi_M \circ f$ for $M \in \text{GL}(2;\mathbb{C})$ is a solution as well.*

2. The upper half-plane and the unit disk. As in Chapter I, the *upper half-plane* in $\mathbb{C}$ is denoted by $\mathbb{H}$, thus

$$\mathbb{H} := \{\tau \in \mathbb{C}\,;\, \text{Im}\,\tau > 0\}.$$

Moreover, we denote the *unit disk* by $\mathbb{D}$,

$$\mathbb{D} := \{z \in \mathbb{C}\,;\, |z| < 1\}.$$

If D is an arbitrary domain in $\mathbb{C}$, then we write

$$\text{Aut}\, D := \{\varphi : D \to D\,;\, \varphi \text{ biholomorphic}\}$$

for the *group of biholomorphic automorphisms of D*.

1.4 Theorem. *The mapping*

$$\Phi_C : \mathbb{H} \to \mathbb{D},\quad \tau \mapsto C\tau = \frac{\tau - i}{\tau + i},\quad C = \begin{pmatrix} 1 & -i \\ 1 & i \end{pmatrix}, \tag{1.11}$$

is biholomorphic with the inverse

$$\Phi_{C^{-1}} : \mathbb{D} \to \mathbb{H},\quad z \mapsto C^{-1}z = i \cdot \frac{1+z}{1-z}.$$

We have

$$\text{Aut}\,\mathbb{H} = \Phi_{C^{-1}} \circ \text{Aut}\,\mathbb{D} \circ \Phi_C. \tag{1.12}$$

We call (1.11) the CAYLEY *transformation*.

Proof The mapping $\varphi : \mathbb{H} \to \mathbb{C}$, $\tau \mapsto C\tau$, is holomorphic with

$$|C\tau|^2 = \left|\frac{\tau-i}{\tau+i}\right|^2 = \frac{x^2 + (y-1)^2}{x^2 + (y+1)^2} < 1 \quad \text{for } \tau \in \mathbb{H},$$

because $y > 0$ implies $|y - 1| < y + 1$. Thus $\varphi(\mathbb{H}) \subseteq \mathbb{D}$ follows. The mapping

$$\psi : \mathbb{D} \to \mathbb{C},\quad z \mapsto C^{-1}z = i \cdot \frac{z+1}{-z+1},\quad C^{-1} = \frac{1}{2i}\begin{pmatrix} i & i \\ -1 & 1 \end{pmatrix},$$

is holomorphic and satisfies

$$\operatorname{Im}\psi(z) = \operatorname{Re}\left(\frac{z+1}{-z+1}\right) = \frac{1-|z|^2}{|1-z|^2} > 0 \quad \text{for } z \in \mathbb{D},$$

thus $\psi(\mathbb{D}) \subseteq \mathbb{H}$. From (1.5) and (1.3) we get

$$\psi(\varphi(\tau)) = C^{-1}C\tau = \tau, \quad \varphi(\psi(z)) = CC^{-1}z = z$$

for all $\tau \in \mathbb{H}$, $z \in \mathbb{D}$. Thus Φ_C is biholomorphic with inverse $\Phi_{C^{-1}}$. Finally (1.12) follows by a calculation. □

For further investigations, we need the following tool.

1.5 Schwarz's Lemma. *If $\varphi : \mathbb{D} \to \mathbb{D}$ is holomorphic with $\varphi(0) = 0$, then*

$$|\varphi(z)| \le |z| \quad \text{for all } z \in \mathbb{D} \quad \text{and} \quad |\varphi'(0)| \le 1.$$

If there exists some $a \in \mathbb{D}\setminus\{0\}$ with $|\varphi(a)| = |a|$ or if $|\varphi'(0)| = 1$, then there exists some $\lambda \in \mathbb{R}$ such that

$$\varphi(z) = e^{i\lambda} \cdot z \quad \text{for all } z \in \mathbb{D}.$$

Proof We consider

$$F : \mathbb{D} \to \mathbb{C}, \quad z \mapsto \begin{cases} \varphi(z)/z, & \text{if } z \ne 0, \\ \varphi'(0), & \text{if } z = 0. \end{cases}$$

Then F is holomorphic by Riemann's Theorem on Removable Singularities A.2. Because $\varphi(\mathbb{D}) \subseteq \mathbb{D}$, the Maximum Principle A.15 for $0 < |z| \le r < 1$ leads to

$$|F(z)| \le \max_{|\zeta|=r} \frac{|\varphi(\zeta)|}{|\zeta|} \le \frac{1}{r}.$$

Letting $r \uparrow 1$, we get $|F(z)| \le 1$ for all $z \in \mathbb{D}$, i.e.

$$|\varphi(z)| \le |z| \quad \text{for all } z \in \mathbb{D} \quad \text{and} \quad |\varphi'(0)| = |F(0)| \le 1.$$

If $|\varphi'(0)| = 1$ or $|\varphi(a)| = |a|$ holds for some $0 \ne a \in \mathbb{D}$, then $|F(0)| = 1$ or $|F(a)| = 1$ follows. According to the Maximum Principle A.15, F is a constant of absolute value 1, i.e. $F(z) = e^{i\lambda}$, $\lambda \in \mathbb{R}$, and hence

$$\varphi(z) = e^{i\lambda}z \quad \text{for all } z \in \mathbb{D}. \qquad \square$$

Thus we can describe all automorphisms of $\mathbb{D}$ with fixed point 0.

1.6 Lemma. *If $\varphi \in \operatorname{Aut} \mathbb{D}$, then $\varphi(0) = 0$ if and only if there exists some $\lambda \in \mathbb{R}$ satisfying*

$$\varphi(z) = e^{i\lambda} \cdot z \quad \text{for all } z \in \mathbb{D}. \tag{1.13}$$

Proof If φ is of the form (1.13), then $\varphi \in \operatorname{Aut} \mathbb{D}$ holds with $\varphi(0) = 0$. Now let $\varphi \in \operatorname{Aut} \mathbb{D}$ be arbitrary with $\varphi(0) = 0$. Then $|\varphi(z)| \le |z|$ for all $z \in \mathbb{D}$ by Schwarz's Lemma 1.5. The same conclusion for φ^{-1} yields $|\varphi^{-1}(z)| \le |z|$. Thus we finally

§ 1 The upper half-plane

obtain
$$|z| = |\varphi(\varphi^{-1}(z))| \le |\varphi^{-1}(z)| \le |z|,$$
so
$$|\varphi(z)| = |z| \quad \text{for all} \quad z \in \mathbb{D}.$$
By the SCHWARZ's Lemma 1.5, φ has the form (1.13). □

3. Aut $\mathbb{H}$. Instead of $\mathrm{GL}(2;\mathbb{C})$, consider the subgroup $\mathrm{GL}(2;\mathbb{R})$ of real invertible matrices. If $M \in \mathrm{GL}(2;\mathbb{R})$ and $\operatorname{Im}\tau \ne 0$, then $M\tau \in \mathbb{C}$. In (1.6), we set $\tau' = \overline{\tau}$, observe that $\overline{M\tau} = M\overline{\tau}$ and obtain

$$\operatorname{Im} M\tau = \frac{\det M}{|c\tau + d|^2} \cdot \operatorname{Im}\tau. \tag{1.14}$$

Thus $\mathbb{H}$ is mapped into itself by

$$\Phi_M : \tau \longmapsto M\tau, \quad M \in \mathrm{GL}(2;\mathbb{R}), \quad \det M > 0. \tag{1.15}$$

Because of (1.4), we may assume without loss of generality that M belongs to the subgroup $\quad \mathrm{SL}(2;\mathbb{R}) := \{M \in \mathrm{GL}(2;\mathbb{R}) \,;\, \det M = 1\},$

the so-called *special linear group over* $\mathbb{R}$.

1.7 Theorem. *The biholomorphic automorphisms of $\mathbb{H}$ are exactly the transformations*

$$\Phi_M : \mathbb{H} \to \mathbb{H}, \quad \tau \mapsto M\tau, \quad M \in \mathrm{SL}(2;\mathbb{R}), \tag{1.16}$$

with inverse $\Phi_{M^{-1}}$. The group Aut $\mathbb{H}$ *acts transitively on $\mathbb{H}$ and is isomorphic to the factor group* $\mathrm{PSL}(2;\mathbb{R}) := \mathrm{SL}(2;\mathbb{R})/\{\pm I\}$.

Proof (1.14) immediately leads to $\Phi_M(\mathbb{H}) \subseteq \mathbb{H}$ for $M \in \mathrm{SL}(2;\mathbb{R})$. Because of (1.5) and (1.3) we also have

$$\Phi_M \circ \Phi_{M^{-1}} = \Phi_{M^{-1}} \circ \Phi_M = \Phi_I = id.$$

Accordingly, all mappings

$$\Phi_M : \mathbb{H} \to \mathbb{H}, \quad M \in \mathrm{SL}(2;\mathbb{R}),$$

are biholomorphic.

$$\Phi : \mathrm{SL}(2;\mathbb{R}) \to \mathrm{Aut}\,\mathbb{H}, \quad M \mapsto \Phi_M,$$

is a homomorphism of groups because of (1.5), and it satisfies

$$\text{kernel}\,\Phi = \{M \in \mathrm{SL}(2;\mathbb{R}) \,;\, \Phi_M = id\} = \{\pm I\}$$

according to Proposition 1.1. The First Isomorphism Theorem for Groups then immediately yields $\mathrm{SL}(2;\mathbb{R})/\{\pm I\} \cong \Phi(\mathrm{SL}(2;\mathbb{R}))$.
If $\tau = x + iy \in \mathbb{H}$, let

$$M := \begin{pmatrix} 1/\sqrt{y} & 0 \\ 0 & \sqrt{y} \end{pmatrix} \begin{pmatrix} 1 & -x \\ 0 & 1 \end{pmatrix} \in \mathrm{SL}(2;\mathbb{R}).$$

Then $M\tau = i$, such that $\mathrm{SL}(2;\mathbb{R})$ acts transitively on $\mathbb{H}$.
Now let $\varphi \in \mathrm{Aut}\,\mathbb{H}$ be arbitrary. Then there exists a matrix $M \in \mathrm{SL}(2;\mathbb{R})$ such that $\Psi := \Phi_M \circ \varphi$ already satisfies $\Psi(i) = i$. Accordingly,

$$\Psi^* := \Phi_C \circ \Psi \circ \Phi_{C^{-1}} \in \mathrm{Aut}\,\mathbb{D} \quad \text{with} \quad \Psi^*(0) = 0.$$

Then Lemma 1.6 shows that there exists some $\lambda \in \mathbb{R}$ satisfying

$$\Psi^*(z) = e^{2i\lambda} \cdot z = \Phi_K(z), \quad K = \begin{pmatrix} e^{i\lambda} & 0 \\ 0 & e^{-i\lambda} \end{pmatrix} \in \mathrm{SL}(2;\mathbb{C}).$$

Hence

$$\varphi = \Phi_{M^{-1}} \circ \Psi = \Phi_{M^{-1}} \circ \Phi_{C^{-1}} \circ \Psi^* \circ \Phi_C = \Phi_L,$$

$$L = M^{-1} \cdot C^{-1} \cdot K \cdot C = M^{-1} \cdot \begin{pmatrix} \cos\lambda & \sin\lambda \\ -\sin\lambda & \cos\lambda \end{pmatrix} \in \mathrm{SL}(2;\mathbb{R}).$$

Accordingly, $\Phi(\mathrm{SL}(2;\mathbb{R})) = \mathrm{Aut}\,\mathbb{H}$. $\square$

The decomposition

$$M\tau = \begin{cases} \dfrac{\tau}{d^2} + \dfrac{b}{d} & \text{if } c = 0, \\ \dfrac{a}{c} - \dfrac{1}{c^2} \cdot \dfrac{1}{\tau + d/c} & \text{if } c \neq 0, \end{cases}$$

for $M \in \mathrm{SL}(2;\mathbb{R})$ yields

1.8 Lemma. *The group $\mathrm{Aut}\,\mathbb{H}$ is generated by the mappings*

$$\tau \longmapsto -1/\tau, \quad \tau \longmapsto \tau + \alpha,\ \alpha \in \mathbb{R}, \quad \tau \longmapsto \lambda^2 \tau,\ 0 \neq \lambda \in \mathbb{R}.$$

The preimages of these transformations under Φ therefore generate the group $\mathrm{SL}(2;\mathbb{R})$. Theorem 1.7 and Lemma 1.8 yield

1.9 Corollary. *The group $\mathrm{SL}(2;\mathbb{R})$ is generated by the matrices*

$$\begin{pmatrix} 0 & -1 \\ 1 & 0 \end{pmatrix}, \quad \begin{pmatrix} 1 & \alpha \\ 0 & 1 \end{pmatrix},\ \alpha \in \mathbb{R}, \quad \begin{pmatrix} \lambda & 0 \\ 0 & 1/\lambda \end{pmatrix},\ 0 \neq \lambda \in \mathbb{R}.$$

We define an *orthogonal circle* of $\mathbb{H}$ as any arc in $\mathbb{H}$ of a circle or any half straight line in $\mathbb{H}$ which is perpendicular to the real axis. In the case of a circular arc, it is a semicircle with center on the real axis.

1.10 Proposition. *The automorphisms of $\mathbb{H}$ map orthogonal circles onto orthogonal circles.*

Proof According to Lemma 1.8, it is sufficient to consider the case of the mapping $\tau \mapsto -1/\tau$. According to (1.8) and (1.9), an orthogonal circle is given by an equation

$$A\tau\bar{\tau} + B(\tau + \bar{\tau}) + C = 0 \quad \text{with} \quad A, B, C \in \mathbb{R} \text{ and } B^2 > AC.$$

Under the mapping $\tau \mapsto z = -1/\tau$ however, we obtain such an equation again. □

Using the geometry of $\mathbb{P} = \mathbb{C} \cup \{\infty\}$, we can also conclude as follows: according to Theorem 1.2, the mappings in question on $\mathbb{P}$ take "circles" onto "circles" and these mappings preserve angles.

1.11 Remark. The mapping $\tau \mapsto \lambda\tau$, $\lambda > 0$, can be obtained from the maps $\tau \mapsto \tau + \alpha$ and $\tau \mapsto -1/\tau$. We easily verify the so-called HUA *identity*

$$\mu^2 \tau = -\left(\lambda - [\mu - (\tau + \lambda)^{-1}]^{-1}\right)^{-1}, \quad \text{if } \lambda\mu = 1.$$

4. $\mathbb{H}$ as a homogeneous space. According to Theorem 1.7, for each $\tau \in \mathbb{H}$, there is a matrix $M \in \mathrm{SL}(2;\mathbb{R})$ that satisfies $\tau = Mi$. If two matrices $M, L \in \mathrm{SL}(2;\mathbb{R})$ satisfy $\tau = Mi = Li$, then $(L^{-1}M)i = i$ due to (1.5) and (1.3), i.e. there is a matrix $K \in \mathrm{SL}(2;\mathbb{R})$ with $M = LK$ and $Ki = i$.

1.12 Proposition. *For $K \in \mathrm{SL}(2;\mathbb{R})$, the following assertions are equivalent*:
(i) $Ki = i$.
(ii) $K = \begin{pmatrix} \alpha & \beta \\ -\beta & \alpha \end{pmatrix}$ *with* $\alpha^2 + \beta^2 = 1$.
(iii) $K \in \mathrm{SO}(2) := \{M \in \mathrm{SL}(2;\mathbb{R});\, M \text{ orthogonal}\} = \left\{\begin{pmatrix} \cos\varphi & \sin\varphi \\ -\sin\varphi & \cos\varphi \end{pmatrix}; \varphi \in \mathbb{R}\right\}$.
(iv) *If* $C := \begin{pmatrix} 1 & -i \\ 1 & i \end{pmatrix}$, *then* $C \cdot K \cdot C^{-1} = \begin{pmatrix} e^{i\varphi} & 0 \\ 0 & e^{-i\varphi} \end{pmatrix}$ *with some* $\varphi \in \mathbb{R}$.

Proof If $K = \begin{pmatrix} \alpha & \beta \\ \gamma & \delta \end{pmatrix}$, then $Ki = i$ is equivalent to $\alpha i + \beta = i(\gamma i + \delta)$, thus to $\delta = \alpha$ and $\gamma = -\beta$. The equivalence of (ii) with (iii) and (iv) is a calculation. □

Using the *special orthogonal subgroup* $\mathrm{SO}(2)$ of $\mathrm{SL}(2;\mathbb{R})$, we can form the quotient space

$$\mathrm{SL}(2;\mathbb{R})/\mathrm{SO}(2) := \{L \cdot \mathrm{SO}(2)\,;\, L \in \mathrm{SL}(2;\mathbb{R})\}. \tag{1.17}$$

Since $\mathrm{SO}(2)$ is not a normal subgroup of $\mathrm{SL}(2;\mathbb{R})$, this is not a factor group, but only a so-called *homogeneous space*. The group $\mathrm{SL}(2;\mathbb{R})$ acts in an obvious way from the left on the quotient space (1.17):

$$(M, L \cdot \mathrm{SO}(2)) \longmapsto (ML) \cdot \mathrm{SO}(2) \quad \text{for } M \in \mathrm{SL}(2;\mathbb{R}). \tag{1.18}$$

Proposition 1.12 shows that the map

$$\mathrm{SL}(2;\mathbb{R})/\mathrm{SO}(2) \longrightarrow \mathbb{H},\ L \cdot \mathrm{SO}(2) \longmapsto Li,$$

is a bijection compatible with the action (1.18). Therefore, we also write

$$\mathbb{H} \cong \mathrm{SL}(2;\mathbb{R})/\mathrm{SO}(2).$$

A stronger version of the transitivity statement is given in

1.13 Theorem. *Let $\tau, \tau' \in \mathbb{H}$, $\tau \neq \tau'$. Then there is a unique $\lambda > 1$ and a unique transformation Φ_M, $M \in \mathrm{SL}(2;\mathbb{R})$, such that*

$$M\tau = i \quad \text{and} \quad M\tau' = \lambda i.$$

Proof To demonstrate the *existence*, apply Theorem 1.7. We only have to show that for $\tau' = u + iv \in \mathbb{H}$ with $\tau' \neq i$ there is some $K \in \mathrm{SL}(2;\mathbb{R})$ such that

$$Ki = i \quad \text{and} \quad \tau' = \lambda i, \; \lambda > 1. \qquad (*)$$

Because of Theorem 1.4 and Theorem 1.7, it is sufficient to find some $\varphi \in \mathrm{Aut}\,\mathbb{D}$ with $\varphi(0) = 0$ and $\varphi(\Phi_C(\tau')) \in\,]0,1[$. But this follows directly from Lemma 1.6. For *uniqueness*, consider an $M' \in \mathrm{SL}(2;\mathbb{R})$ and $\lambda' > 1$ satisfying $M'\tau = i$ and $M'\tau' = \lambda'i$. This leads to

$$Ki = i \quad \text{and} \quad K\lambda i = \lambda' i \quad \text{for} \quad K = M'M^{-1} \in \mathrm{SL}(2;\mathbb{R}).$$

Proposition 1.12 yields

$$K = \begin{pmatrix} \alpha & \beta \\ -\beta & \alpha \end{pmatrix}, \; \alpha^2 + \beta^2 = 1 \quad \text{and} \quad \alpha\lambda i + \beta = i\lambda'(-\beta\lambda i + \alpha),$$

hence $\lambda' = \lambda$ and $K = \pm I$. Therefore, the transformation $\tau \mapsto M\tau$ is uniquely determined. $\square$

5. Fixed points. Let $M = \begin{pmatrix} a & b \\ c & d \end{pmatrix} \in \mathrm{GL}(2;\mathbb{C})$. We call $\tau \in \mathbb{P} = \mathbb{C} \cup \{\infty\}$ *a fixed point* of M if $M\tau = \tau$ holds, where we use (1.10) for $\tau = \infty$. If $\tau \in \mathbb{C}$, then $M\tau = \tau$ is equivalent to

$$c\tau^2 + (d-a)\tau - b = 0. \qquad (1.19)$$

If M is real, then $\bar{\tau}$ also satisfies this equation whenever τ does. In this case, for $M \neq \pm I$, there is at most one $\tau \in \mathbb{H}$ satisfying $M\tau = \tau$.

1.14 Proposition. *For $M \in \mathrm{SL}(2;\mathbb{R})$, the following assertions are equivalent*:

  (i) *M has exactly one fixed point τ in $\mathbb{H}$.*
  (ii) *M has no real eigenvalue.*
  (iii) *$|\mathrm{tr}\,M| < 2$.*

In this case $c \neq 0$ holds, and the fixed point is

$$\tau = \frac{a-d}{2c} + \frac{i}{2|c|}\sqrt{4 - (\mathrm{tr}\,M)^2} \quad \text{for} \quad M = \begin{pmatrix} a & b \\ c & d \end{pmatrix}. \qquad (1.20)$$

Matrices M with this property are called *elliptic*.

Proof Obviously, (i), i.e. a solution $\tau \in \mathbb{H}$ of (1.19), is equivalent to (1.20). However,

§ 1 The upper half-plane

there is only one solution $\tau \in \mathbb{H}$ if (iii) holds. The eigenvalues of M are given by

$$\lambda_{1,2} = \frac{\operatorname{tr} M}{2} \pm \frac{i}{2}\sqrt{4 - (\operatorname{tr} M)^2}. \tag{1.21}$$

Therefore, (ii) and (iii) are also equivalent. □

Now we examine the matrices M with fixed points on the real axis. Analogous arguments then yield the

1.15 Proposition. *For $M \in \operatorname{SL}(2;\mathbb{R})$ the following assertions are equivalent:*

(i) *M has exactly two different fixed points in $\mathbb{R} \cup \{\infty\}$.*
(ii) *M has two different real eigenvalues.*
(iii) *$|\operatorname{tr} M| > 2$.*

In the case $c \neq 0$, the fixed points are given by

$$\tau = \frac{a-d}{2c} \pm \frac{1}{2c}\sqrt{(\operatorname{tr} M)^2 - 4} \quad \text{for} \quad M = \begin{pmatrix} a & b \\ c & d \end{pmatrix}.$$

In the case $c = 0$, $a \neq d$ holds and the fixed points are ∞ as well as $\tau = b/(d-a)$.

Matrices M with this property are called *hyperbolic*. Finally, we call a matrix $M \in \operatorname{SL}(2;\mathbb{R})$, $M \neq \pm I$ with $|\operatorname{tr} M| = 2$ *parabolic*. Parabolic matrices have 1 or -1 as a double eigenvalue. The corresponding linear transformations have exactly one fixed point. The fixed point is real in the case $c \neq 0$, namely $\tau = (a-d)/2c$, and in the case $c = 0$ it is $\tau = \infty$. Thus $\operatorname{SL}(2;\mathbb{R})$ has a disjoint decomposition into $\{\pm I\}$ and the sets of elliptic, hyperbolic and parabolic matrices.

From linear algebra (cf. S. LANG [53], XV, §3), it is known that hyperbolic matrices in $\operatorname{SL}(2;\mathbb{R})$ are diagonalizable. Parabolic matrices are conjugate in $\operatorname{SL}(2;\mathbb{R})$ to the matrices $\pm \begin{pmatrix} 1 & \pm 1 \\ 0 & 1 \end{pmatrix}$.

6*. An invariant metric on $\mathbb{H}$. A function $\gamma : [\alpha, \beta] \to \mathbb{H}$ of a real interval into $\mathbb{H}$ is called a *path* from $\gamma(\alpha)$ to $\gamma(\beta)$ if γ is continuous and piecewise continuously differentiable. In contrast to the Euclidean length of γ, we call

$$L(\gamma) := \int_\gamma \frac{|d\tau|}{\operatorname{Im} \tau} = \int_\alpha^\beta \frac{|\gamma'(\xi)|}{\operatorname{Im} \gamma(\xi)} d\xi \tag{1.22}$$

the $\mathbb{H}$–*length* or *hyperbolic length* of γ. As is well-known, an integral of the form (1.22) does not depend on the parametrization. We abbreviate $M\gamma := \Phi_M \circ \gamma$ for a matrix $M \in \operatorname{SL}(2;\mathbb{R})$. Then (1.7) and (1.14) yield

$$\frac{|(M\gamma)'|}{\operatorname{Im} M\gamma} = \frac{|\gamma'|}{\operatorname{Im} \gamma}.$$

Thus (1.22) implies

$$L(M\gamma) = L(\gamma) \quad \text{for all } M \in \operatorname{SL}(2;\mathbb{R}). \tag{1.23}$$

For $z, w \in \mathbb{H}$, the quantity
$$|z, w| := \inf\{L(\gamma) \; ; \; \gamma \text{ path in } \mathbb{H} \text{ from } z \text{ to } w\} \tag{1.24}$$
is called the $\mathbb{H}$–*distance* or *hyperbolic distance* between z and w. It is easy to see that (1.24) defines a metric on $\mathbb{H}$, the $\mathbb{H}$–*metric* or *hyperbolic metric*, and (1.23) yields the invariance property
$$|Mz, Mw| = |z, w| \quad \text{for} \quad M \in \mathrm{SL}(2; \mathbb{R}). \tag{1.25}$$
Therefore, (1.24) is also called the *invariant metric*. Now we first need the

1.16 Proposition. *For two different points z, w in $\mathbb{H}$, there exists exactly one orthogonal circle passing through z and w.*

Proof According to Theorem 1.7 and Theorem 1.13, it is sufficient to prove the claim for the points $z = i$ and $w = i\lambda$, $\lambda > 1$, but in this case the ray $\{iy \; ; \; y > 0\}$ is the only orthogonal circle passing through the two points. $\square$

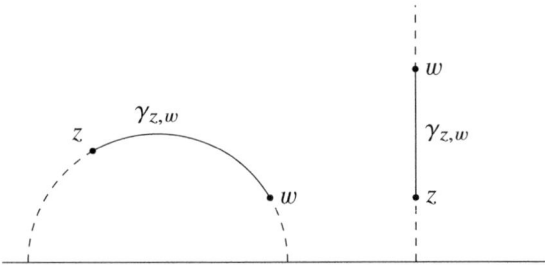

Figure 15: Orthogonal circles

Let $\gamma_{z,w}$ denote the arc of the orthogonal circle from z to w. This is the shortest connection between z and w in the $\mathbb{H}$-metric, as shown by the following Theorem. Thus the orthogonal circles correspond to the lines in Euclidean geometry.

1.17 Theorem. a) *Given $z, w \in \mathbb{H}$, then $|z, w| = L(\gamma_{z,w})$ and $\gamma_{z,w}$ is the only path in $\mathbb{H}$ with this property up to reparametrization.*
b) $|i, \lambda i| = \log \lambda$ *holds for $\lambda \geq 1$.*

Proof Let $z \neq w$. We choose M according to Theorem 1.13 such that $Mz = i$ and $Mw = \lambda i$, $\lambda > 1$. Then we obtain $M\gamma_{z,w} = \gamma_{i,\lambda i}$. So we only need to prove
$$|i, \lambda i| = L(\gamma_{i,\lambda i}) = \int_1^\lambda \frac{1}{\xi} d\xi = \log \lambda. \tag{$*$}$$

Now let γ be an arbitrary path in $\mathbb{H}$ connecting i and λi with the parametrization $\gamma = \gamma(\xi)$, $\alpha \leq \xi \leq \beta$, $\gamma(\alpha) = i$, $\gamma(\beta) = \lambda i$. If $\gamma = x + iy$, it follows that

§ 1 The upper half-plane 121

$$L(\gamma) = \int_\alpha^\beta \frac{|\gamma'(\xi)|}{\operatorname{Im} \gamma(\xi)} d\xi \geq \int_\alpha^\beta \frac{|y'(\xi)|}{y(\xi)} d\xi \geq \left| \int_\alpha^\beta \frac{y'(\xi)}{y(\xi)} d\xi \right| = \left| [\log y(\xi)]_\alpha^\beta \right|$$
$$= \log \lambda = L(\gamma_{i,\lambda i}).$$

Therefore, (∗) is a consequence of (1.24). From $L(\gamma) = \log \lambda$ we get $x'(\xi) = 0$, hence $x(\xi) = 0$ because $\gamma(\alpha) = i$, and $y'(\xi) \geq 0$ for almost all ξ. Thus $\gamma_{i,i\lambda}$ is the only path γ up to reparametrization satisfying $L(\gamma) = \log \lambda$. □

7*. Explicit formulas for the $\mathbb{H}$-distance can be obtained as follows: we start from $z, w \in \mathbb{H}$ and define the *cross-ratio*

$$D(z, w) := \left| \frac{z - w}{z - \overline{w}} \right|. \tag{1.26}$$

Because of (1.6), we get

$$D(Mz, Mw) = D(z, w) \quad \text{for all} \quad M \in \operatorname{SL}(2; \mathbb{R}). \tag{1.27}$$

Then

$$D(i, \lambda i) = \frac{\lambda - 1}{\lambda + 1} \quad \text{for } \lambda \geq 1 \tag{1.28}$$

and Theorem 1.13 leads to

$$0 \leq D(z, w) < 1. \tag{1.29}$$

The announced explicit formula is now:

1.18 Proposition. *For all $z, w \in \mathbb{H}$ we have*

$$|z, w| = \log \frac{1 + D(z, w)}{1 - D(z, w)} = \log \frac{|z - \overline{w}| + |z - w|}{|z - \overline{w}| - |z - w|}.$$

Proof We may again assume $z = i$ and $w = \lambda i$ with $\lambda > 1$. Now the assertion follows from (1.28) and Theorem 1.17 b). □

1.19 Corollary. *Every compact subset C of $\mathbb{H}$ is contained in an $\mathbb{H}$–circular disk with $\mathbb{H}$-center i*
$$\mathcal{D}_{i,r} := \{z \in \mathbb{H} \,;\, |z, i| \leq r\}, \quad r > 0.$$

Proof The mapping $z \mapsto |z, i|$ is continuous according to Proposition 1.18 and it takes its maximum r on the compact set C. □

We can improve the invariance relation (1.25).

1.20 Theorem. *For two pairs of points (z, w) and (z', w') in $\mathbb{H} \times \mathbb{H}$, the following assertions are equivalent*:

(i) $|z, w| = |z', w'|$.
(ii) $D(z, w) = D(z', w')$.
(iii) *There exists a matrix $M \in \operatorname{SL}(2; \mathbb{R})$ satisfying $Mz = z'$ and $Mw = w'$.*

Proof (i) $\Longleftrightarrow$ (ii): Use Proposition 1.18 and (1.29).
(i) $\Longleftrightarrow$ (iii): Because of Theorem 1.13 and (1.25), we may assume $z = z' = i$ and $w = i\lambda$, $w' = i\lambda'$ with $\lambda, \lambda' \geq 1$. Now apply Theorem 1.17. $\square$

$$C_{z,r} := \{w \in \mathbb{H} \, ; \, |z,w| = r\} \quad \text{with } z \in \mathbb{H} \text{ and } r > 0 \tag{1.30}$$

is called an $\mathbb{H}$–*circle* with $\mathbb{H}$–*center* z and $\mathbb{H}$–*radius* r. (1.27) and Theorem 1.20 imply

$$MC_{z,r} := \{Mw \, ; \, w \in C_{z,r}\} = C_{Mz,r}. \tag{1.31}$$

1.21 Lemma. *Every $\mathbb{H}$-circle is a Euclidean circle in $\mathbb{H}$ and vice versa. More precisely, if $z = x + iy \in \mathbb{H}$, then $w = u + iv \in C_{z,r}$, $r > 0$, is equivalent to*

$$(u-x)^2 + \left(v - \frac{1+s^2}{1-s^2}y\right)^2 = \left(\frac{2sy}{1-s^2}\right)^2, \quad s = \frac{e^r - 1}{e^r + 1}. \tag{1.32}$$

Proof First, $w \in C_{z,r}$ holds, if and only if $D(z,w) = s$ due to Proposition 1.18. This proves (1.32). Because

$$0 < \frac{2sy}{1-s^2} < \frac{1+s^2}{1-s^2}y,$$

an $\mathbb{H}$-circle is a Euclidean circle in $\mathbb{H}$. If we are given a Euclidean circle with center $a + ib \in \mathbb{H}$ and radius ρ, $0 < \rho < b$, we choose

$$z = a + i\frac{1-s^2}{1+s^2}b$$

and then determine $r > 0$ by

$$\frac{2s}{1+s^2} = \frac{\rho}{b}, \quad s = \frac{e^r - 1}{e^r + 1}. \qquad \square$$

Lemma 1.21 immediately implies that the topology induced by the $\mathbb{H}$-metric coincides with the natural topology on $\mathbb{H}$. Moreover, (1.32) says that the $\mathbb{H}$-center and the Euclidean center for $r > 0$ *do not* coincide, but have the same real part.

1.22 Corollary. *Let $(w_k)_{k\geq 1}$ be a sequence in $\mathbb{H}$ and $z \in \mathbb{H}$ such that the sequence $(|z,w_k|)_{k\geq 1}$ is bounded. Then the sequence $(w_k)_{k\geq 1}$ has a limit point in $\mathbb{H}$.*

Proof According to Lemma 1.21 $\mathcal{D}_{z,r} = \{w \in \mathbb{H}; |z,w| \leq r\}$ is a closed Euclidean disk in $\mathbb{H}$. Hence the BOLZANO-WEIERSTRASS Theorem yields the assertion. $\square$

That the perpendicular bisectors in Euclidean geometry correspond to the orthogonal circles in hyperbolic geometry is the content of

1.23 Corollary. *Let $z, w \in \mathbb{H}$, $z \neq w$. Then*

$$\{\tau \in \mathbb{H} \, ; \, |z, \tau| = |w, \tau|\}$$

is an orthogonal circle.

§ 1 The upper half-plane

Proof Due to Proposition 1.10 and Theorem 1.13, we may assume without loss of generality that $z = i$ and $w = \lambda i$ with $\lambda > 1$. According to Theorem 1.20 and (1.26) the set

$$\mathcal{M} := \{\tau \in \mathbb{H}; \ |i, \tau| = |\lambda i, \tau|\} = \{\tau \in \mathbb{H}; \ D(i, \tau) = D(\lambda i, \tau)\}$$

consists of the points $\tau = x + iy \in \mathbb{H}$ satisfying

$$0 = |i - \tau|^2 \cdot |\lambda i - \overline{\tau}|^2 - |i - \overline{\tau}|^2 \cdot |\lambda i - \tau|^2 = 4y(\lambda - 1) \cdot [x^2 + y^2 - \lambda].$$

Hence $\mathcal{M}$ describes the orthogonal circle with center 0 and radius $\sqrt{\lambda}$. □

8*. The action of $\mathrm{GL}(2;\mathbb{R})$ on $\mathbb{H}$. In analogy with (1.1), for $M \in \mathrm{GL}(2;\mathbb{R})$ and $\tau \in \mathbb{H}$, we define a complex number

$$M\langle\tau\rangle := \begin{cases} M\tau, & \text{if } \det M > 0, \\ M\overline{\tau}, & \text{if } \det M < 0. \end{cases} \quad (1.33)$$

Because of (1.14), we get

$$\mathrm{Im}\, M\langle\tau\rangle = |\det M| \cdot \frac{\mathrm{Im}\,\tau}{|c\tau + d|^2},$$

thus

$$M\langle\tau\rangle \in \mathbb{H} \quad \text{for} \quad \tau \in \mathbb{H} \quad \text{and} \quad M \in \mathrm{GL}(2;\mathbb{R}). \quad (1.34)$$

We can immediately verify that

$$(LM)\langle\tau\rangle = L\langle M\langle\tau\rangle\rangle \quad \text{for} \quad L, M \in \mathrm{GL}(2;\mathbb{R}). \quad (1.35)$$

Thus $\mathrm{GL}(2;\mathbb{R})$ acts on $\mathbb{H}$. Note, however, that this transformation is not holomorphic in general. The operation of the group $\mathrm{GL}(2;\mathbb{R})$ on $\mathbb{H}$ is generated by the transformations $\tau \mapsto M\tau$, $M \in \mathrm{SL}(2;\mathbb{R})$, and $\tau \mapsto -\overline{\tau}$.

9*. A parametrization of positive definite 2×2 matrices. A real symmetric 2×2 matrix

$$S = \begin{pmatrix} \alpha & \beta \\ \beta & \gamma \end{pmatrix}$$

is known to be positive definite if and only if $\alpha > 0$ and $\det S > 0$ (cf. S. LANG [53], XV, §5). Let $\mathrm{Pos}(2;\mathbb{R})$ denote the set of real symmetric and positive definite 2×2 matrices. Consider the map

$$F : \mathbb{H} \longrightarrow \mathrm{Pos}(2;\mathbb{R}), \quad \tau \longmapsto F_\tau := \frac{1}{y}\begin{pmatrix} 1 & -x \\ -x & x^2 + y^2 \end{pmatrix}, \quad \tau = x + iy. \quad (1.36)$$

Obviously, $\det F_\tau = 1$ holds for all $\tau \in \mathbb{H}$. Let

$$w : \mathrm{Pos}(2;\mathbb{R}) \longrightarrow \mathbb{H}, \quad S \longmapsto w(S) := \frac{1}{\alpha}\left(-\beta + i\sqrt{\det S}\right). \quad (1.37)$$

A calculation yields

1.24 Proposition. *If $S \in \text{Pos}(2;\mathbb{R})$, then $w = w(S)$ is the unique solution $w \in \mathbb{H}$ of the equation*
$$\alpha w^2 + 2\beta w + \gamma = 0. \tag{1.38}$$
The other solution of (1.38) is given by $\overline{w} = \overline{w(S)}$.

Furthermore we easily demonstrate the

1.25 Lemma. a) $w(F_\tau) = \tau$ *holds for all* $\tau \in \mathbb{H}$.
b) $S = \sqrt{\det S} \cdot F_{w(S)}$ *holds for all* $S \in \text{Pos}(2;\mathbb{R})$.
c) *The mapping*
$$]0, \infty[\times \mathbb{H} \longrightarrow \text{Pos}(2;\mathbb{R}), \quad (t, \tau) \longmapsto t \cdot F_\tau,$$
is a bijection with inverse mapping $S \longmapsto \left(\sqrt{\det S}, w(S)\right)$.

Hence $t \cdot F_\tau$ with $t > 0$ and $\tau \in \mathbb{H}$ is a parametrization of $\text{Pos}(2;\mathbb{R})$. We obtain a bijection $F : \mathbb{H} \longrightarrow \{S \in \text{Pos}(2;\mathbb{R}) \; ; \; \det S = 1\}$. As is well-known, the group $\text{GL}(2;\mathbb{R})$ acts on $\text{Pos}(2;\mathbb{R})$ by

$$(M, S) \longmapsto M * S := \left(M^{-1}\right)^{tr} S M^{-1}. \tag{1.39}$$

This action is related to the action of $\text{GL}(2;\mathbb{R})$ on $\mathbb{H}$. According to sect. 8 we get:

1.26 Theorem. *Given $L, M \in \text{GL}(2;\mathbb{R})$, $S \in \text{Pos}(2;\mathbb{R})$ and $\tau \in \mathbb{H}$, we have:*
a) $(LM) * S = L * (M * S)$.
b) $w(M * S) = M\langle w(S) \rangle$.
c) $F_{M\langle \tau \rangle} = |\det M| \cdot (M * F_\tau)$.

Proof a) A calculation yields the assertion.
b) Let $\tilde{w} = w(M * S)$, $w = w(S)$ and $M * S = \begin{pmatrix} \tilde{\alpha} & \tilde{\beta} \\ \tilde{\beta} & \tilde{\gamma} \end{pmatrix}$. According to Proposition 1.24, $\tilde{w}$ is the unique solution in $\mathbb{H}$ of the equation

$$\begin{aligned} 0 &= \tilde{\alpha}\tilde{w}^2 + 2\tilde{\beta}\tilde{w} + \tilde{\gamma} = (\tilde{w} \; 1)(M^{-1})^{tr} S M^{-1} \begin{pmatrix} \tilde{w} \\ 1 \end{pmatrix} \\ &= \frac{(-c\tilde{w} + a)^2}{(\det M)^2} \cdot (z \; 1) S \begin{pmatrix} z \\ 1 \end{pmatrix} = \frac{(-c\tilde{w} + a)^2}{(\det M)^2} \cdot (\alpha z^2 + 2\beta z + \gamma) \end{aligned}$$

with $z = M^{-1}\langle \tilde{w} \rangle$ and $M = \begin{pmatrix} a & b \\ c & d \end{pmatrix}$. It follows that $z = w$, hence $\tilde{w} = M\langle w \rangle$ from (1.35).
c) We write $\tau = w(S)$ and use b) as well as part b) of Lemma 1.25. □

1.27 Exercises.
1) The group $\text{GL}(2;\mathbb{C})$ is generated by the matrices

§ 1 The upper half-plane   125

$$\begin{pmatrix} 1 & 1 \\ 0 & 1 \end{pmatrix}, \quad \begin{pmatrix} 0 & -1 \\ 1 & 0 \end{pmatrix}, \quad \begin{pmatrix} \lambda & 0 \\ 0 & 1 \end{pmatrix}, \quad 0 \neq \lambda \in \mathbb{C}.$$

2) If $M = \begin{pmatrix} a & b \\ c & d \end{pmatrix} \in \mathrm{SL}(2;\mathbb{R})$ and $\tau \in \mathbb{H}$, then $\frac{1}{\mathrm{Im}\, M\tau} = g^{tr} F_\tau g$ with $g = \begin{pmatrix} -d \\ c \end{pmatrix}$.

3) Let $M \in \mathrm{SL}(2;\mathbb{R})$ and $n \in \mathbb{Z}$ with $M^n \neq \pm I$. The matrix M^n is elliptic, or hyperbolic, or parabolic if M is.

4) $M \in \mathrm{SL}(2;\mathbb{R})$, $M \neq \pm I$, has finite order if and only if there is some $q \in \mathbb{Q}$ with $0 < q < 1$ such that $\mathrm{tr}\, M = 2\cos(\pi q)$. In this case, M is elliptic.

5) Every $M \in \mathrm{SL}(2;\mathbb{R})$ is a product of 2 hyperbolic matrices.

6) Let $M \in \mathrm{SL}(2;\mathbb{R})$ be parabolic with $\varepsilon = \frac{1}{2}\mathrm{tr}\, M = \pm 1$. Then

$$M^n = n\varepsilon^{n-1} M - (n-1)\varepsilon^n I \quad \text{for all} \quad n \in \mathbb{Z}.$$

7) $M \in \mathrm{SL}(2;\mathbb{R})$, $M \neq \pm I$, is conjugate in $\mathrm{SL}(2;\mathbb{R})$ to $\begin{pmatrix} 0 & \mp 1 \\ \pm 1 & t \end{pmatrix}$, $t = \mathrm{tr}\, M$. For which M are these two matrices themselves conjugate?

8) If w is the $\mathbb{H}$-center of an $\mathbb{H}$-circle, m its Euclidean center and y^+ and y^- denote the imaginary parts of the points of intersection of the circle with the line through w and m, then the following is true:

$$\mathrm{Im}\, m = \tfrac{1}{2}(y^+ + y^-) \quad \text{and} \quad \mathrm{Im}\, w = \sqrt{y^+ y^-}.$$

9) Describe the orthogonal circle passing through the points $z = 4+5i$ and $w = 2+8i$.

10) Determine the hyperbolic distance between the points $z = (73 + i)/130$ and $w = (157 + 2i)/277$.

11) For $z, w \in \mathbb{H}$, let $d(z,w) := \mathrm{tr}(F_z F_w^{-1}) - 2$ (cf. (1.36)). Then, all $M \in \mathrm{GL}(2;\mathbb{R})$ satisfy

$$d(M\langle z\rangle, M\langle w\rangle) = d(z,w) \quad \text{and} \quad d(z,w) > 0 \quad \text{for} \quad z \neq w.$$

12) Let $\mathcal{K} = \{Z \in \mathrm{Mat}(2;\mathbb{R}) \,;\, Z + Z^{tr} \in \mathrm{Pos}(2;\mathbb{R})\}$ and $J = \begin{pmatrix} 0 & -1 \\ 1 & 0 \end{pmatrix}$. The mapping

$$\varphi : \mathbb{H} \times \mathbb{H} \longrightarrow \mathcal{K}, \quad (\tau, w) \longmapsto Z := uJ + vF_\tau, \quad w = u + iv,$$

is a bijection and one has for $M \in \mathrm{SL}(2;\mathbb{R})$, $\lambda > 0$

$$\varphi((M\tau, w)) = (M^{-1})^{tr} ZM^{-1}, \quad \varphi((\tau, w+1)) = Z + J,$$
$$\varphi((\tau, \lambda w)) = \lambda Z, \quad \varphi((-1/\tau, -1/w)) = Z^{-1},$$
$$\varphi((\tau, -\overline{w})) = Z^{tr}, \quad \varphi((-\overline{\tau}, w)) = D^{tr} Z^{tr} D, \quad D = \begin{pmatrix} 1 & 0 \\ 0 & -1 \end{pmatrix}.$$

13) Describe the images of orthogonal circles in $\mathbb{H}$ under the CAYLEY transformation.

14) Derive the connection between the CAYLEY transformation and the stereographic projection, i.e. show that i, $t \in \mathbb{R}$ and $i\phi_C(it)$ lie on a line.

15) Determine all $z, w \in \mathbb{H}$, $z \neq w$, such that

$$\{z \in \mathbb{H}; \ |z,w| = |z,\tau|\} = \{it; t > 0\}.$$

16) Determine all pairs of points $(z, w) \in \mathbb{H} \times \mathbb{H}$ for which the perpendicular bisectors in the hyperbolic geometry and the Euclidean geometry in $\mathbb{H}$ coincide.

17) Determine the LAURENT series expansion of Φ_M in (1.1) at the pole if $c \neq 0$.

§ 2 The modular group

The modular group is studied in this paragraph. A canonical exact fundamental domain of the modular group in $\mathbb{H}$ is determined.

1. Generators. The *modular group* (or the *elliptic modular group*) as in the Introduction is defined by
$$\Gamma := \mathrm{SL}(2; \mathbb{Z}) := \{M \in \mathrm{Mat}(2; \mathbb{Z}) \ ; \ \det M = 1\}.$$

Two matrices from Γ are specially denoted:
$$J := \begin{pmatrix} 0 & -1 \\ 1 & 0 \end{pmatrix} \quad \text{and} \quad T := \begin{pmatrix} 1 & 1 \\ 0 & 1 \end{pmatrix}. \tag{2.1}$$

You can immediately convince yourself that
$$J^2 = -I \quad \text{and} \quad T^m = \begin{pmatrix} 1 & m \\ 0 & 1 \end{pmatrix} \quad \text{for} \ m \in \mathbb{Z}. \tag{2.2}$$

A verification yields
$$(JT)^3 = (TJ)^3 = -I. \tag{2.3}$$

The transformations of the upper half-plane corresponding to the matrices (2.1) are
$$\tau \longmapsto -1/\tau \quad \text{and} \quad \tau \longmapsto \tau + 1. \tag{2.4}$$

We rephrase the Completion Lemma I.1.11 in the form

2.1 Completion Lemma. *For given coprime integers c, d there is a matrix $M \in \Gamma$ with $M = \begin{pmatrix} * & * \\ c & d \end{pmatrix}$. Any two matrices with this property differ only by a left-hand factor of the form T^m with $m \in \mathbb{Z}$.*

An algebraic description of the group Γ is given by the

2.2 Theorem. textitThe modular group Γ is generated by the matrices J and T.

Proof Let Δ be the subgroup of Γ generated by J and T. Because of (2.2), $-I \in \Delta$ holds. Now let $M = \begin{pmatrix} a & b \\ c & d \end{pmatrix} \in \Gamma$. We will show by induction on $|c|$ that $M \in \Delta$. In the case $c = 0$, we already have $M = \pm T^m \in \Delta$ according to (2.2). Now let $c \neq 0$. For suitable $m \in \mathbb{Z}$,
$$M' = JT^m M = \begin{pmatrix} * & * \\ c' & * \end{pmatrix} \quad \text{with} \quad 0 \leq c' = a + mc < |c| \ .$$

§ 2 The modular group

By the induction hypothesis, we obtain $M' \in \Delta$ and then also

$$M = T^{-m}J^{-1}M' \in \Delta,$$

hence $\Delta = \Gamma$. □

Thus, each element of Γ is a finite product of the form

$$J^\varepsilon T^{m_1} J T^{m_2} J \cdot \ldots \cdot J T^{m_k} J^\delta \tag{2.5}$$

with $m_1, \ldots, m_k \in \mathbb{Z}$ and ε, δ equal to 0 or 1. Because of (2.3), we can even assume $\varepsilon = \delta = 0$. However, this representation is not unique.

2.3 Corollary. *The group of modular substitutions $\tau \mapsto M\tau$, $M \in \Gamma$, is generated by the mappings $\tau \mapsto \tau + 1$ and $\tau \mapsto -1/\tau$.*

If we set (cf. (2.3))

$$U := -TJ = \begin{pmatrix} -1 & 1 \\ -1 & 0 \end{pmatrix}, \quad \text{hence} \quad U^3 = I, \tag{2.6}$$

then we obtain generators of Γ consisting of matrices of finite order.

2.4 Corollary. *The modular group Γ is generated by the matrices J and U of order 4 and 3, respectively.*

2.5 Remarks. a) The identity $M^{tr}JM = \det M \cdot J$, valid for 2×2 matrices M over any field, shows that the modular group can also be described as

$$\Gamma = \{M \in \text{Mat}(2; \mathbb{Z}) \,;\, M^{tr}JM = J\}.$$

b) Following Corollary 2.4, we can show that Γ can be described as the group with two generators J and U and the defining relations $J^4 = U^3 = I$ as well as $J^2U = UJ^2$. Compare H. MAASS [56], 54–55.

c) The group $\text{PSL}(2; \mathbb{Z}) := \text{SL}(2; \mathbb{Z})/\{\pm I\}$ is canonically isomorphic to the group of modular substitutions because of Proposition 1.1 and Theorem 1.7. It is generated by the modular substitutions $\tau \mapsto J\tau = -1/\tau$ and $\tau \mapsto U\tau = 1 - 1/\tau$ of order 2 and 3. Thus $\text{PSL}(2; \mathbb{Z})$ is the free product of two cyclic groups of order 2 and 3.

d) The notation of the matrices (2.1) with J and T (because of "involution" and "translation") is not universal. H. PETERSSON and his students used the notation T and U, respectively.

2. The exact fundamental domain $\mathbb{F}$. Let

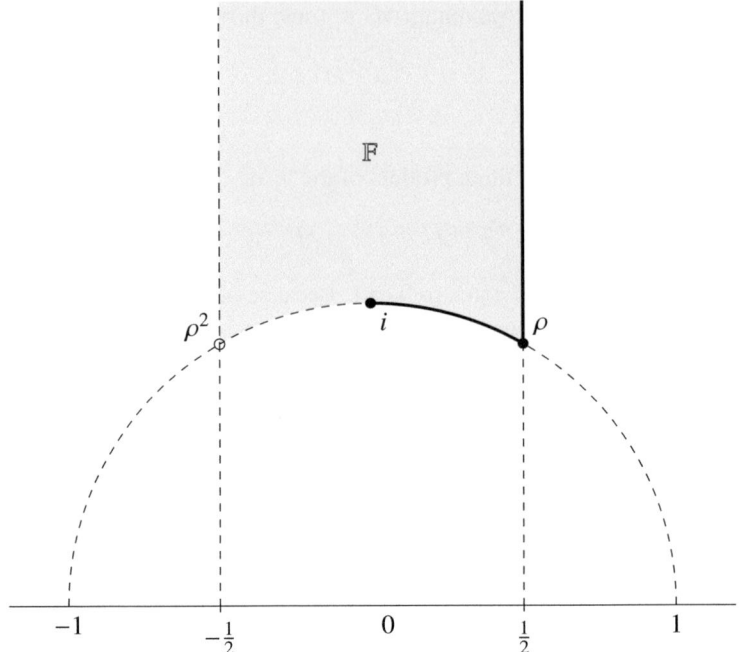

Figure 16: The exact fundamental domain

$$\mathbb{F} := \left\{ \tau \in \mathbb{H}; \ -\frac{1}{2} < \operatorname{Re} \tau \leq \frac{1}{2}, \ |\tau| \geq 1 \text{ and } |\tau| > 1 \text{ for } -\frac{1}{2} < \operatorname{Re} \tau < 0 \right\} \quad (2.7)$$

be given by the figure above. Obviously $\mathbb{F}$ is bounded by parts of the lines $\operatorname{Re} \tau = \pm \frac{1}{2}$ and an arc of the unit circle, i.e. by parts of orthogonal circles. Let

$$\begin{aligned}
\overline{\mathbb{F}} &= \{\tau \in \mathbb{H}; \ |\operatorname{Re} \tau| \leq \tfrac{1}{2}, |\tau| \geq 1\}, \\
\overset{\circ}{\mathbb{F}} &= \{\tau \in \mathbb{H}; \ |\operatorname{Re} \tau| < \tfrac{1}{2}, |\tau| > 1\}, \\
\partial \mathbb{F} &= \{\tau \in \mathbb{H}; \ |\operatorname{Re} \tau| = \tfrac{1}{2}, |\tau| \geq 1\} \cup \{\tau \in \mathbb{H}; \ |\operatorname{Re} \tau| \leq \tfrac{1}{2}, |\tau| = 1\},
\end{aligned} \quad (2.8)$$

denote the closure, the interior and the boundary of $\mathbb{F}$. The boundary points i and

$$\rho := \tfrac{1}{2} + \tfrac{1}{2} i \sqrt{3} \quad \text{with} \quad \rho^3 = -1$$

belong to $\mathbb{F}$, whereas

$$\rho^2 = \rho - 1 = -\overline{\rho} = -\tfrac{1}{2} + \tfrac{1}{2} i \sqrt{3}$$

belongs to $\overline{\mathbb{F}}$ but not to $\mathbb{F}$. Obviously,

$$\operatorname{Im} \tau \geq \tfrac{1}{2} \sqrt{3} \quad \text{for all } \tau \in \overline{\mathbb{F}}. \quad (2.9)$$

Using (2.1) and (2.6) we obtain

2.6 Theorem. a) *For any $\tau \in \mathbb{H}$, there is a matrix $M \in \Gamma$ with $M\tau \in \mathbb{F}$.*

§ 2 The modular group

b) *If τ and $M\tau$, $M \in \Gamma$, belong to $\mathbb{F}$, then $\tau = M\tau$. If $M \neq \pm I$, then*
either

$$\tau = M\tau = i \quad \text{and} \quad M = \pm J \tag{2.10}$$

or

$$\tau = M\tau = \rho \quad \text{and} \quad M = \pm U, \pm U^2. \tag{2.11}$$

c) *If $\tau \in \overline{\mathbb{F}}$ and $M\tau \in \overset{\circ}{\mathbb{F}}$, $M \in \Gamma$, then $M = \pm I$.*

Because of these properties, $\mathbb{F}$ is called an *exact fundamental domain* for the modular group Γ.

Proof a) By Proposition I.1.5, the lattice $\mathbb{Z}\tau + \mathbb{Z}$ is discrete in $\mathbb{C}$. Thus there is a pair $(c, d) \in \mathbb{Z} \times \mathbb{Z}$ with $(c, d) \neq (0, 0)$ such that

$$|m\tau + n| \geq |c\tau + d| \quad \text{for all} \quad (m, n) \in \mathbb{Z} \times \mathbb{Z}, (m, n) \neq (0, 0). \tag{2.12}$$

Of course, c and d are coprime. According to the Completion Lemma 2.1, there is a matrix $L = \begin{pmatrix} a & b \\ c & d \end{pmatrix} \in \Gamma$, and (2.12) implies $|T^m L\tau| \geq 1$ for all $m \in \mathbb{Z}$. Selecting m suitably and setting $\tau' = T^m L\tau = m + L\tau$, we get

$$-\frac{1}{2} < \operatorname{Re} \tau' \leq \frac{1}{2} \quad \text{and} \quad |\tau'| \geq 1.$$

In the case $-\frac{1}{2} < \operatorname{Re} \tau' < 0$ and $|\tau'| = 1$, we replace τ' by $J\tau' = -1/\tau' = -\overline{\tau'}$.
b) In the case $c = 0$, $M\tau = \tau + m$ holds for some $m \in \mathbb{Z}$. Then (2.7) implies $m = 0$ and $M = \pm I$. Hence we may assume $c > 0$ without loss of generality. If $\tau = x + iy$, then (1.14) and (2.9) immediately yield

$$\frac{1}{2}\sqrt{3} \leq \operatorname{Im} M\tau = \frac{y}{|c\tau + d|^2} \quad \text{and} \quad \frac{1}{2}\sqrt{3} \leq y. \tag{$*$}$$

Assertion. $c = 1$ and $d \in \{0, \pm 1\}$.

Proof We have $|c\tau + d|^2 \geq c^2 y^2$ and $(*)$ yields $\frac{3}{4}c^2 \leq 1$, thus $c = 1$. But now also

$$|\tau + d|^2 \geq 2|x + d|y, \quad \text{hence} \quad \sqrt{3}|x + d| \leq 1,$$

holds due to $(*)$. Because $|x| \leq \frac{1}{2}$, we get $|d| \leq 1$. □

Now the cases $d = 0$ and $d = \pm 1$ are treated separately:
The case $d = 0$: it follows $M = T^m J$, thus $M\tau = m - 1/\tau$. For $|\tau| > 1$, we get

$$|\operatorname{Re}(-1/\tau)| = |\operatorname{Re}(\tau)|/|\tau|^2 < 1/2$$

and therefore the immediate contradiction $m = 0$ and $|M\tau| = |-1/\tau| < 1$. Thus $|\tau| = 1$ follows. Because $M\tau = m - \overline{\tau} = (m - x) + iy$ a look at the fundamental domain shows that only the cases

$$m = 0, \quad M = J \quad \text{and} \quad \tau = M\tau = i$$

and

$$m = 1, \quad M = TJ = -U \quad \text{and} \quad \tau = M\tau = \rho$$

are possible.

The case $d = \pm 1$: We have $|\tau + d|^2 = (x \pm 1)^2 + y^2 \geq y^2 + 1/4$, i.e.

$$\frac{1}{2}\sqrt{3} \leq \operatorname{Im} M\tau \leq f(y), \quad \text{where} \quad f(y) := \frac{y}{y^2 + 1/4} \qquad (**)$$

according to (∗). Since $f(y)$ for $y \geq \frac{1}{2}$ is strictly monotonically decreasing and $f\left(\frac{1}{2}\sqrt{3}\right) = \frac{1}{2}\sqrt{3}$,

$$\operatorname{Im} M\tau = y = \frac{1}{2}\sqrt{3} \qquad (***)$$

follows from (∗∗). But then $x = \frac{1}{2}$ and $\tau = M\tau = \rho$. Using (∗ ∗ ∗), we get

$$\left|d + \frac{1}{2}\right| = \frac{1}{2}, \quad \text{hence} \quad d = -1 \quad \text{and} \quad M = \begin{pmatrix} * & * \\ 1 & -1 \end{pmatrix}.$$

The Completion Lemma 2.1 yields

$$M = T^m (TJ)^2 \quad \text{for some} \quad m \in \mathbb{Z}.$$

As $TJ\rho = \rho$ and $M\rho = \rho$, we find $m = 0$ and hence $M = U^2$.

c) As τ or $T\tau$ or $J\tau$ belongs to $\mathbb{F}$, this is an immediate consequence of b). □

For $M \in \Gamma$ and $\tau \in \overline{\mathbb{F}}$,

$$|c\tau + d|^2 = c^2|\tau|^2 + 2cd \cdot \operatorname{Re} \tau + d^2 \geq c^2 - |cd| + d^2 = (|c| - |d|)^2 + |cd| \geq 1.$$

holds. Hence (1.14) yields

2.7 Lemma. *If $\tau \in \overline{\mathbb{F}}$, then all $M \in \Gamma$ satisfy*

$$\operatorname{Im} M\tau \leq \operatorname{Im} \tau.$$

Due to Theorem 2.6, the images

$$M\mathbb{F} := \{M\tau \; ; \; \tau \in \mathbb{F}\}, \quad M \in \Gamma,$$

cover the upper half plane $\mathbb{H}$ without gaps, producing the famous *modular tesselation*. In this context $\overline{\mathbb{F}}$ is also called the *modular triangle*.

When successively constructing of the images $M\overline{\mathbb{F}}$ of $\overline{\mathbb{F}}$, note that the boundary of $\overline{\mathbb{F}}$ consists of parts of orthogonal circles in the sense of §1. Thus, according to Proposition 1.10, the images of the boundary again consist of parts of orthogonal

§ 2 The modular group

circles. Finally, we use the fact that orthogonal circles are uniquely determined by two points in $\mathbb{H}$. Thus it suffices to compute the images of $i, \rho, -\overline{\rho}$ and ∞.

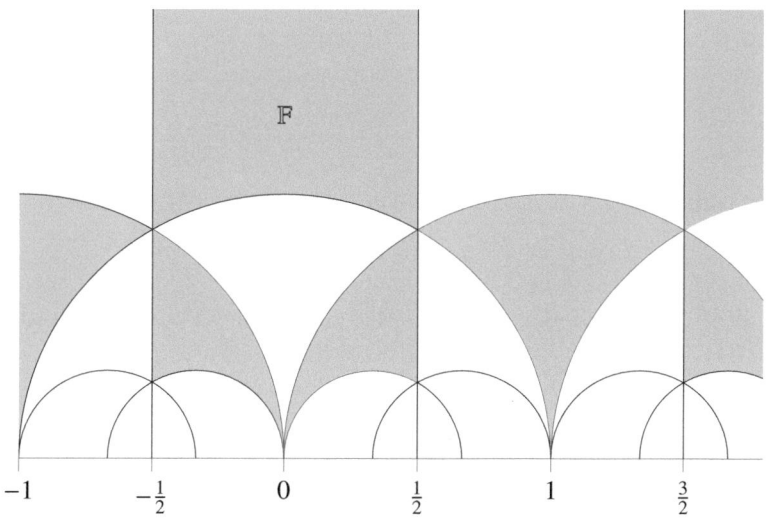

Figure 17: Images of $\mathbb{F}$

3. Fixed points and neighbors. Given $\tau \in \mathbb{H}$, we define
$$\Gamma_\tau := \{M \in \Gamma \; ; \; M\tau = \tau\}. \tag{2.13}$$

Obviously, Γ_τ is a subgroup of Γ, the so-called *stabilizer subgroup of* τ. Theorem 2.6 states that

$$\Gamma_i = \{\pm I, \pm J\}\,, \; \Gamma_\rho = \{\pm I, \pm U, \pm U^2\}\,, \; \Gamma_\tau = \{\pm I\}, \; \tau \in \mathbb{F}, \tau \neq i, \rho\,. \tag{2.14}$$

$\tau \in \mathbb{H}$ is called a *fixed point of* Γ if $\Gamma_\tau \neq \{\pm I\}$. Thus i and ρ are the fixed points of Γ in $\mathbb{F}$. If τ is a fixed point, then there is a matrix $M \in \Gamma$ with $M\tau \in \mathbb{F}$. Since the stabilizer subgroups
$$\Gamma_{M\tau} = M\Gamma_\tau M^{-1} \tag{2.15}$$

and Γ_τ are conjugate in Γ, $M\tau$ is also a fixed point. Thus Theorem 2.6 resp. (2.14) lead to the

2.8 Corollary. *The fixed points of Γ are exactly the points Mi and $M\rho$ with $M \in \Gamma$.*

Given $\varepsilon > 0$, we call
$$\mathcal{V}_\varepsilon := \{\tau \in \mathbb{H};\ y \geq \varepsilon,\ |x| \leq \frac{1}{\varepsilon}\} \qquad (2.16)$$

a *vertical strip of height ε* in $\mathbb{H}$. According to (2.7) and (2.9),

$$\mathbb{F} \subset \overline{\mathbb{F}} \subset \mathcal{V}_\varepsilon \text{ for all } \varepsilon \leq \frac{1}{2}\sqrt{3}. \qquad (2.17)$$

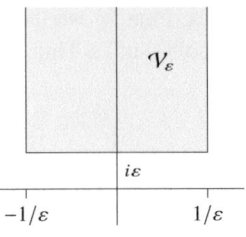

Figure 18: Vertical strips

2.9 Proposition. *If $\varepsilon > 0$, then there are only finitely many $M \in \Gamma$ satisfying*
$$M\mathcal{V}_\varepsilon \cap \mathcal{V}_\varepsilon \neq \emptyset.$$

Proof If $M \in \Gamma$ and $\tau \in \mathcal{V}_\varepsilon$ with $M\tau \in \mathcal{V}_\varepsilon$, then it follows from (1.14) that

$$\frac{1}{\varepsilon} \geq (\operatorname{Im} M\tau)^{-1} = \frac{|c\tau + d|^2}{y} \geq c^2 y \geq \varepsilon c^2, \quad \text{hence} \quad |c| \leq \frac{1}{\varepsilon}. \qquad (*)$$

Thus there are only finitely many c. If $c = 0$, then $M\tau = \tau + m$, $m \in \mathbb{Z}$ and we have only finitely many $M \in \Gamma$. If $|c| \geq 1$, then $(*)$ implies

$$\frac{1}{\varepsilon} \geq \frac{|cy(cx+d)|}{y} \geq |cx + d|, \quad \text{hence} \quad |d| \leq \frac{1}{\varepsilon} + |cx| \leq \frac{1}{\varepsilon} + \frac{1}{\varepsilon^2}.$$

This shows that there are only finitely many possibilities for the second row of M. Then the Completion Lemma 2.1 yields the assertion. □

A set $M\mathbb{F}$ with $M \in \Gamma$ and $M\mathbb{F} \cap \mathbb{F} \neq \emptyset$ is called a *neighbor of $\mathbb{F}$*. Thus Proposition 2.9 implies

2.10 Corollary. $\mathbb{F}$ *has only finitely many neighbors.*

By Theorem 2.6, $\mathbb{F}$, $J\mathbb{F}$, $U\mathbb{F}$ and $U^2\mathbb{F}$ are exactly the neighbors of $\mathbb{F}$.

2.11 Corollary. *If $C \subset \mathbb{H}$ is compact, then $M\overline{\mathbb{F}} \cap C \neq \emptyset$ for only finitely many $M \in \Gamma$.*

Proof Choose $\varepsilon \leq \frac{1}{2}\sqrt{3}$ with $C \subseteq \mathcal{V}_\varepsilon$ and use Proposition 2.9. □

4*. The quotient space $\Gamma\backslash\mathbb{H}$ is defined by
$$\Gamma\backslash\mathbb{H} := \{\Gamma\tau;\ \tau \in \mathbb{H}\} \qquad (2.18)$$

and thus consists of all the *orbits*
$$\Gamma\tau := \{M\tau;\ M \in \Gamma\} \quad \text{for} \quad \tau \in \mathbb{H} \qquad (2.19)$$

under Γ. As usual, we have a canonical surjection
$$\pi : \mathbb{H} \longrightarrow \Gamma\backslash\mathbb{H},\ \tau \longmapsto \pi(\tau) := \Gamma\tau. \qquad (2.20)$$

The importance of Theorem 2.6 is

2.12 Proposition. *The restriction of π to $\mathbb{F}$, $\pi : \mathbb{F} \to \Gamma\backslash\mathbb{H}$, is bijective.*

Proof According to Theorem 2.6, $\mathbb{F} \cap \Gamma\tau \neq \emptyset$ for any $\tau \in \mathbb{H}$, i.e. the restriction in question is surjective. Now let $\tau, \tau' \in \mathbb{F}$ be given with $\pi(\tau) = \pi(\tau')$, i.e. with $\Gamma\tau = \Gamma\tau'$. Then, by (2.19), there is a matrix $M \in \Gamma$ with $\tau' = M\tau$. As τ and $M\tau$ belong to $\mathbb{F}$, Theorem 2.6 yields $\tau' = M\tau = \tau$. □

Intuitively, $\Gamma\backslash\mathbb{H}$ is obtained from $\mathbb{F}$ by identifying the parts of the boundary related by Γ. By looking at the figure, it immediately becomes clear that the quotient space $\Gamma\backslash\mathbb{H}$ can be compactified by adding one point, namely the orbit $\Gamma\infty = \mathbb{Q} \cup \{\infty\}$.

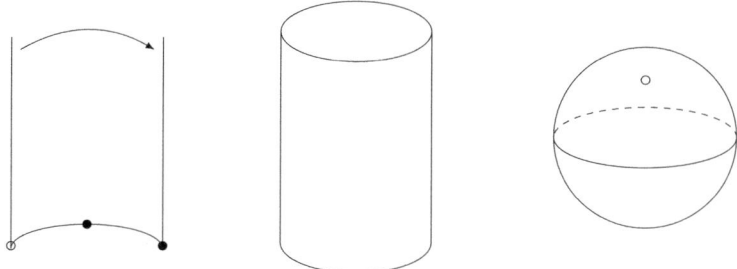

Figure 19: Compactification of $\Gamma\backslash\mathbb{H}$

5*. Reduction theory of positive definite 2 × 2 matrices. According to sect. 8 in §1 not only SL $(2; \mathbb{R})$ but even GL $(2; \mathbb{R})$ acts on $\mathbb{H}$ by virtue of

$$\tau \mapsto M\langle\tau\rangle := \begin{cases} M\tau, & \text{if } \det M > 0, \\ M\overline{\tau}, & \text{if } \det M < 0. \end{cases} \quad (2.21)$$

A fundamental domain of $\mathbb{H}$ with respect to the action of

$$\text{GL}(2; \mathbb{Z}) = \Gamma \cup \Gamma \cdot \begin{pmatrix} 1 & 0 \\ 0 & -1 \end{pmatrix} \quad (2.22)$$

can be determined as follows. We define the "left half" of $\overline{\mathbb{F}}$ to be

$$\mathbb{L} := \{\tau \in \mathbb{H}; -\frac{1}{2} \leq \text{Re}\,\tau \leq 0 \text{ and } |\tau| \geq 1\}, \quad (2.23)$$

and easily obtain the following assertion from Theorem 2.6.

2.13 Proposition. a) *Given $\tau \in \mathbb{H}$, there is a matrix $M \in \text{GL}(2; \mathbb{Z})$ with $M\langle\tau\rangle \in \mathbb{L}$.*
b) *If τ and $M\langle\tau\rangle$, $M \in \text{GL}(2; \mathbb{Z})$, belong to the interior of $\mathbb{L}$, then $M = \pm I$.*
c) *If τ and $M\langle\tau\rangle$, $M \in \text{GL}(2; \mathbb{Z})$, belong to $\mathbb{L}$, then $\tau = M\langle\tau\rangle$.*
d) *The boundary of $\mathbb{L}$ consists only of fixed points of $\text{GL}(2; \mathbb{Z})$.*

For the boundary points τ of $\mathbb{L}$, for example $-\overline{\tau} = \tau$ if $\tau = iy$ and $1/\overline{\tau} = \tau$ if $|\tau| = 1$ and $-\overline{(\tau + 1)} = \tau$ if $\text{Re}\,\tau = -1/2$.

According to Lemma 1.25, the mapping

$$]0, \infty[\times \mathbb{H} \longrightarrow \text{Pos}(2; \mathbb{R}), \quad (t, \tau) \longmapsto t \cdot F_\tau, \quad F_\tau := \frac{1}{y} \begin{pmatrix} 1 & -x \\ -x & x^2 + y^2 \end{pmatrix}, \quad (2.24)$$

is a bijection which, according to Theorem 1.26, is compatible with the action

$$(M, S) \longmapsto M * S := \left(M^{-1}\right)^{tr} S M^{-1}$$

of $\text{GL}(2; \mathbb{R})$ on $\text{Pos}(2; \mathbb{R})$. If we denote the image of $]0, \infty[\times \mathbb{L}$ under the map (2.24) by $\mathbb{P}$, then (2.23) immediately yields

$$\mathbb{P} = \left\{ S = \begin{pmatrix} \alpha & \beta \\ \beta & \gamma \end{pmatrix} \in \text{Pos}(2; \mathbb{R}) \ ; \ 0 \leq 2\beta \leq \alpha \leq \gamma \right\}. \quad (2.25)$$

Therefore, Theorem 1.26 and the Proposition 2.13 imply

2.14 Corollary. a) *For every $S \in \text{Pos}(2; \mathbb{R})$ there is a matrix $M \in \text{GL}(2; \mathbb{Z})$ with the property $M^{tr} S M \in \mathbb{P}$.*
b) *If S and $M^{tr} S M$, $M \in \text{GL}(2; \mathbb{Z})$, belong to the interior of $\mathbb{P}$, then $M = \pm I$.*
c) *If S and $M^{tr} S M$, $M \in \text{GL}(2; \mathbb{Z})$, belong to $\mathbb{P}$, then $S = M^{tr} S M$.*

This so-called *reduction theory* of positive definite *binary quadratic forms* goes back to A.-M.LEGENDRE and C.F. GAUSS. GAUSS was aware of the connection between the determination of a fundamental domain for Γ and the reduction theory formulated in Corollary 2.14 ([31], vol. VIII, 100–105).

2.15 Exercises.
1) Show that $\{M\infty \ ; M \in \Gamma\} = \mathbb{Q} \cup \{\infty\}$.
2) The order of $M \in \Gamma$ can be characterized as follows:

(i) $\text{ord } M = 2 \iff M = -I$,
(ii) $\text{ord } M = 3 \iff \text{tr } M = -1$,
(iii) $\text{ord } M = 4 \iff \text{tr } M = 0$,
(iv) $\text{ord } M = 6 \iff \text{tr } M = 1$.

3) Γ contains infinitely many elliptic, hyperbolic and parabolic matrices.
4) If n is a positive integer, then find an explicit $M \in \Gamma$ with $M\tau \in \mathbb{F}$ for $\tau = (1+i)/n$ and $\tau = (n+i)/2n$.
5) If $\tau \in \mathbb{H}$ is a fixed point of Γ, then τ belongs to the imaginary-quadratic number field $\mathbb{Q}[i]$ or $\mathbb{Q}[i\sqrt{3}]$, and there is some $n \in \mathbb{N}$ with the property

$$y = 1/n, \ n \equiv 1, 2 \ (\text{mod } 4) \quad \text{or} \quad y = \sqrt{3}/n, \ n \equiv 2 \ (\text{mod } 4).$$

6) We have

$$\overline{\mathbb{F}} = \{\tau \in \mathbb{H} \ ; \ |x| \leq \tfrac{1}{2} \text{ and } |m\tau + n| \geq 1 \text{ for all } (m, n) \in \mathbb{Z} \times \mathbb{Z}, (m, n) \neq (0, 0)\},$$

$$\overset{\circ}{\mathbb{F}} = \{\tau \in \mathbb{H} ; \ |x| < \tfrac{1}{2} \text{ and } |m\tau + n| > 1 \text{ for all } (m, n) \in \mathbb{Z} \times \mathbb{Z}, \ m \neq 0\},$$

$$\partial \mathbb{F} = \{\tau \in \overline{\mathbb{F}}; \ |x| \leq \tfrac{1}{2} \text{ and } |m\tau + n| = 1 \text{ for some } (m, n) \in \mathbb{Z} \times \mathbb{Z}, \ m \neq 0\}.$$

7) $\mathbb{F}$ is a "star domain with center λi", $\lambda \geq 2/\sqrt{3}$, i.e. for each $\tau \in \mathbb{F}$, the line between τ and λi belongs to $\mathbb{F}$ entirely.

§ 2 The modular group   135

8) $\mathbb{F}$ is $\mathbb{H}$–convex, i.e. if $\tau, \tau' \in \mathbb{F}$, then the orthogonal circle between τ and τ' is entirely contained in $\mathbb{F}$.

9) For which $\tau \in \overline{\mathbb{F}}$ does an $M \in \Gamma$ exist with $M\tau = -\overline{\tau}$?

10) Let $\mathcal{R} := \left\{ S = \begin{pmatrix} \alpha & \beta \\ \beta & \gamma \end{pmatrix} \in \text{Pos}(2; \mathbb{R}) \, ; \, -\alpha \leq 2\beta < \alpha \leq \gamma \right\}$.

a) If $S \in \text{Pos}(2; \mathbb{R})$, then there exists $M \in \Gamma$ with $M^{tr}SM \in \mathcal{R}$.

b) S and $T = M^{tr}SM$ with $M \in \Gamma$, $M \neq \pm I$, belong to $\mathcal{R}$ if and only if

$$S = T = \lambda I, \ \lambda > 0, \text{ and } M = \pm J \textbf{ or}$$

$$S = T = \lambda \begin{pmatrix} 2 & -1 \\ -1 & 2 \end{pmatrix}, \ \lambda > 0, \text{ and } M = \pm U \text{ or } M = \pm U^2.$$

11) Describe the image of $\mathbb{F}$ in the unit disk $\mathbb{D}$ under the CAYLEY transformation (1.11).

12) Determine the stabilizer group $\{M \in \text{GL}(2; \mathbb{Z}); \, M\langle w \rangle = w\}$ for $w \in \mathbb{L}$.

13) If C_1 and C_2 are compact subsets of $\mathbb{H}$, then there are only finitely many $M \in \Gamma$ with the property $MC_1 \cap C_2 \neq \emptyset$.

14) Determine all $M \in \Gamma$ satisfying $M\mathcal{V}_{1/2} \cap \mathcal{V}_{1/2} \neq \emptyset$.

15) Describe the analogon of the Completion Lemma 2.1 for the first row as well as the first and second column of $M \in \Gamma$.

16) Let $F_\tau \in \text{Pos}(2; \mathbb{R})$ be defined as in (1.36). For $\tau \in \mathbb{F}$ the eigenvalues of F_τ are greater than or equal to $1/2y$, and $1/2y$ occurs as an eigenvalue if and only if $\tau = \rho$.

17) Review the definition of φ and $\mathcal{K}$ from Exercise 1.27 12). Then show that

$$\varphi(\overline{\mathbb{F}} \times \overline{\mathbb{F}}) = \left\{ Z = \begin{pmatrix} a & b \\ c & d \end{pmatrix} \in \mathcal{K} \, ; \, |b+c| \leq a \leq d, \, |b-c| \leq 1, \, \det Z \geq 1 \right\}.$$

18) Let $||\cdot||$ denote the maximum norm and $n \in \mathbb{N}$. Show that

$$\#\{M \in \Gamma; \, ||M|| = n\} = \begin{cases} 20, & \text{if } n = 1, \\ 32\varphi(n), & \text{if } n > 1, \end{cases}$$

where φ denotes EULER's totient function.

19) Determine all $M \in \Gamma$ satisfying $M\overline{\mathbb{F}} \cap \overline{\mathbb{F}} \neq \emptyset$.

20) Let $S = \begin{pmatrix} 0 & J \\ -J & 0 \end{pmatrix}$ and $\Sigma = \{U \in \text{SL}(4; \mathbb{Z}); \, U^{tr}SU = S\}$. Show that, for $M \in \Gamma$,

$$M^\uparrow = \begin{pmatrix} M & 0 \\ 0 & M \end{pmatrix}, \quad M^\downarrow = \begin{pmatrix} aI & bI \\ cI & dI \end{pmatrix} \in \Sigma$$

always hold, as well as

$$\Sigma = \{\text{diag}(1, \varepsilon, \varepsilon, 1) M_1^\uparrow M_2^\downarrow; \, \varepsilon = \pm 1, \, M_1, M_2 \in \Gamma\}.$$

Show that

$$\{M \in \Sigma; \, Ue_1 = e_1\} \cong \mathbb{Z} \times \mathbb{Z}.$$

§ 3 Subgroups of the modular group

This paragraph describes a procedure to construct fundamental domains with respect to subgroups of the modular group. Let $\Gamma = \mathrm{SL}(2;\mathbb{Z})$ again be the modular group and Λ be a subgroup of Γ.

1. Fundamental domains for subgroups. Let Δ be any subgroup of $\mathrm{SL}(2;\mathbb{R})$. A subset $\mathcal{F}$ of $\mathbb{H}$ is called a *fundamental domain* for Δ if the following holds:

(FD.0) $\mathcal{F}$ is (relatively) closed in $\mathbb{H}$ and the boundary $\partial\mathcal{F}$ has LEBESGUE measure 0.
(FD.1) For every $\tau \in \mathbb{H}$, there is a matrix $M \in \Delta$ with $M\tau \in \mathcal{F}$.
(FD.2) If τ and $M\tau$, $M \in \Delta$, belong to the interior of $\mathcal{F}$, then $M = \pm I$ holds.

Let $M\mathcal{F} := \{M\tau\;;\;\tau \in \mathcal{F}\}$ be the image of $\mathcal{F}$ under the transformation $\tau \mapsto M\tau$. A (relatively) closed subset $\mathcal{F} \subseteq \mathbb{H}$ whose boundary has LEBESGUE measure 0 is a fundamental domain for Δ if and only if

(FD.1*) $\mathbb{H} = \bigcup_{M \in \Delta} M\mathcal{F}$.
(FD.2*) $\overset{\circ}{\mathcal{F}} \cap M\overset{\circ}{\mathcal{F}} \neq \emptyset$, $M \in \Delta \;\Rightarrow\; M = \pm I$.

If $\Delta = \Lambda$ is a subgroup of Γ, then Λ is countable. In this case, it follows from (FD.1*) and BAIRE's Category Theorem (cf. B. v. QUERENBURG [65], § 13D) that every fundamental domain possesses interior points.

Let $\mathbb{F}$ be defined as in (2.7). Due to Theorem 2.6, $\overline{\mathbb{F}}$ is a fundamental domain of Γ in the above sense. Now let $\Lambda' = \{\pm M;\; M \in \Lambda\}$, and let

$$\Gamma = \bigcup_{1 \leq \nu \leq [\Gamma:\Lambda']} \Lambda' M_\nu \tag{3.1}$$

be a disjoint decomposition of Γ into right cosets of Λ'. Then Γ in (3.1) is a finite or infinite union, depending on whether or not the index $[\Gamma : \Lambda']$ is finite. In (3.1) the representatives M_ν are, of course, only unique up to left-hand factors from Λ' and up to a permutation. Nevertheless we set

$$\mathcal{F}(\Lambda) := \bigcup_{1 \leq \nu \leq [\Gamma:\Lambda']} M_\nu \overline{\mathbb{F}}. \tag{3.2}$$

3.1 Theorem. *Let Λ be a subgroup of Γ. Then $\mathcal{F}(\Lambda)$ in (3.2) is a fundamental domain for Λ.*

Proof (FD.0): Let $\tau \in \mathbb{H}$, $\tau \notin \mathcal{F}(\Lambda)$. We choose $\varepsilon > 0$ such that $\overline{\mathbb{F}} \subseteq \mathcal{V}_\varepsilon$ and τ is an interior point of $\mathcal{V}_\varepsilon$. By Proposition 2.9, there are only finitely many $M \in \Gamma$ satisfying $M\mathcal{V}_\varepsilon \cap \mathcal{V}_\varepsilon \neq \emptyset$, and therefore also only finitely many ν with $M_\nu \overline{\mathbb{F}} \cap \mathcal{V}_\varepsilon \neq \emptyset$. Hence τ is not a limit point of $\mathcal{F}(\Lambda)$, and $\mathcal{F}(\Lambda)$ is therefore relatively closed in $\mathbb{H}$. As the boundary of $\mathcal{F}(\Lambda)$ is contained in $\bigcup_{1 \leq \nu \leq [\Gamma:\Lambda']} M_\nu \langle \partial \mathbb{F} \rangle$ and the latter union is finite or countable, it has LEBESGUE measure 0.

(FD.1): Given $\tau \in \mathbb{H}$, we choose $L \in \Gamma$ with $L\tau \in \mathbb{F}$ by Theorem 2.6. Due to (3.1), there is some $M \in \Lambda$ and ν with $L^{-1} = \pm M^{-1} M_\nu$. Thus $\pm M = M_\nu L$, and therefore

$$M\tau = M_\nu \langle L\tau \rangle \in M_\nu \overline{\mathbb{F}} \subseteq \mathcal{F}(\Lambda).$$

§ 3 Subgroups of the modular group

(FD.2): Let τ and $M\tau$, $M \in \Lambda$, be interior points of $\mathcal{F}(\Lambda)$ and let $\mathcal{U}$ be an open neighborhood of τ in $\mathcal{F}(\Lambda)$. After making $\mathcal{U}$ smaller we may immediately assume without loss of generality that $M\mathcal{U} = \{Mz\,;\, z \in \mathcal{U}\} \subseteq \mathcal{F}(\Lambda)$. We have

$$\mathcal{U} = \bigcup_\nu \mathcal{U}_\nu \quad \text{with} \quad \mathcal{U}_\nu := \mathcal{U} \cap M_\nu \overline{\mathbb{F}},$$

and at least one set $\mathcal{U}_\nu$ has interior points according to BAIRE's Category Theorem. Without loss of generality suppose $\overset{\circ}{\mathcal{U}_1} \neq \emptyset$ and $z \in \overset{\circ}{\mathcal{U}_1}$. Because $\mathcal{U}_1 \subseteq M_1 \overline{\mathbb{F}}$, $w := M_1^{-1} z$ is an interior point of $\mathbb{F}$. Furthermore, there is an index ν with $Mz \in M_\nu \overline{\mathbb{F}}$. Then $w' := M_\nu^{-1} M M_1 w$ lies in $\overline{\mathbb{F}}$ and w is an interior point of $\mathbb{F}$. We conclude from Theorem 2.6 that $w' = w$ and $M_\nu^{-1} M M_1 = \pm I$, hence $M_\nu = \pm M M_1$. Because $\pm M \in \Lambda'$, it follows that $\nu = 1$ and $M = \pm I$. $\square$

The *cusps* of $\mathcal{F}(\Lambda)$ are the images $M_\nu \infty$ of ∞. This definition depends on the choice of M_ν, whereas the *cusp classes* $\Lambda M_\nu \infty$ are uniquely determined by Λ. Using the Completion Lemma 2.1, we conclude that $\mathbb{Q} \cup \{\infty\}$ is the cusp class of Γ. Analogously to §2, a point $\tau \in \mathbb{H}$ is a called *fixed point of* Λ, if there is $M \in \Lambda$ with $M\tau = \tau$ and $M \neq \pm I$. Each fixed point of Λ is a fixed point of Γ, but not vice versa in general.

3.2 Remarks. a) In the past, the phrase *discontinuity region* was sometimes used instead of fundamental domain. R. DEDEKIND used the expression *Hauptfeld* for it ([16], vol. I, 174–201).
b) The notion of a fundamental domain is not uniform in the literature. But J. EL-STRODT (*What is a fundamental domain?*, arXiv:2308.11997, 2023) pointed out that the different definitions may lead to subtle mistakes, when calculating the area.

2. Principal congruence groups. Let $n \geq 1$ be a fixed natural number. The *congruence* $L \equiv M \pmod n$ for $L, M \in \text{Mat}(2; \mathbb{Z})$ is defined componentwise, i.e. there is a matrix $X \in \text{Mat}(2; \mathbb{Z})$ satisfying $L = M + nX$. A look at the components of a matrix product immediately shows that

$$LM \equiv L'M' \pmod n, \quad \text{if} \quad L \equiv L' \pmod n \text{ and } M \equiv M' \pmod n. \tag{3.3}$$

We define
$$\Gamma[n] := \{M \in \Gamma \,;\, M \equiv I \pmod n\}. \tag{3.4}$$

Let us consider the canonical projection

$$\mathbb{Z} \to \mathbb{Z}/n\mathbb{Z}, \quad x \mapsto \overline{x} := x + n\mathbb{Z}, \tag{3.5}$$

and extend it to 2×2 matrices by

$$\text{Mat}(2; \mathbb{Z}) \to \text{Mat}(2; \mathbb{Z}/n\mathbb{Z}), \quad M = \begin{pmatrix} a & b \\ c & d \end{pmatrix} \mapsto \overline{M} = \begin{pmatrix} \overline{a} & \overline{b} \\ \overline{c} & \overline{d} \end{pmatrix}. \tag{3.6}$$

To prove the next theorem, we need the following number-theoretic result.

3.3 Lemma. *Let $a, b, c \in \mathbb{Z}$ with $c \neq 0$ and $\gcd(a, b, c) = 1$. Then there exists some $x \in \mathbb{Z}$ satisfying*

$$\gcd(a + xb, c) = 1.$$

Proof Let x be the product of all primes p which divide c but not a, with the empty product equal to 1. Suppose there is a prime q with $q|(a + xb)$ and $q|c$.
If $q|a$, then $q \nmid x$. From $q|(a + xb)$, $q|a$, $q \nmid x$ it follows that $q|b$, contradicting to $\gcd(a, b, c) = 1$.
If $q \nmid a$, then $q|x$ holds and $q|(a + xb)$ implies $q|a$, which is a contradiction.
Hence $a + xb$ and c are coprime. □

This yields

3.4 Theorem. *The map*

$$\phi : \Gamma \to \mathrm{SL}(2; \mathbb{Z}/n\mathbb{Z}), \quad M \mapsto \overline{M},$$

is a surjective homomorphism of groups whose kernel is $\Gamma[n]$. Thus $\Gamma[n]$ is a normal subgroup of finite index in Γ and the factor subgroup $\Gamma/\Gamma[n]$ is isomorphic to $\mathrm{SL}(2; \mathbb{Z}/n\mathbb{Z})$.

The subgroup $\Gamma[n]$ of Γ is called the *principal congruence group* (mod n). The number n is also called the *level* of $\Gamma[n]$.

Proof Because $\det \overline{M} = \overline{\det M}$ for $M \in \mathrm{Mat}(2; \mathbb{Z})$, the image of ϕ is contained in $\mathrm{SL}(2; \mathbb{Z}/n\mathbb{Z})$. By (3.3), ϕ is a homomorphism of groups, whose kernel is just $\Gamma[n]$ due to (3.4), and thus a normal subgroup of Γ.

Let $K = \begin{pmatrix} \alpha & \beta \\ \gamma & \delta \end{pmatrix} \in \mathrm{Mat}(2; \mathbb{Z})$ with $\overline{K} \in \mathrm{SL}(2; \mathbb{Z}/n\mathbb{Z})$ and suppose $\gamma \neq 0$, because γ can otherwise be replaced by $\gamma + n$. Because $\alpha\delta - \beta\gamma \equiv 1 \pmod{n}$, $\gcd(\gamma, \delta, n) = 1$ holds. According to Lemma 3.3, there exists an $r \in \mathbb{Z}$ such that $\gcd(\gamma, d) = 1$ for $d = \delta + rn$. Now we have

$$\alpha d - \beta\gamma = \alpha\delta - \beta\gamma + \alpha rn = 1 + sn \quad \text{for some } s \in \mathbb{Z}.$$

Because $\gcd(\gamma, d) = 1$, we can find $x, y \in \mathbb{Z}$ such that

$$\gamma y - dx = s.$$

Then

$$M := \begin{pmatrix} \alpha + xn & \beta + yn \\ \gamma & \delta + rn \end{pmatrix} \in \mathrm{SL}(2; \mathbb{Z}) \quad \text{satisfies} \quad \overline{M} = \overline{K},$$

because

$$\det M = (\alpha + xn) \cdot d - (\beta + yn)\gamma = 1 + sn + n(xd - y\gamma) = 1.$$

Hence ϕ is also surjective and the First Isomorphism Theorem for Groups yields an isomorphism between $\Gamma/\Gamma[n]$ and $\mathrm{SL}(2; \mathbb{Z}/n\mathbb{Z})$. Thus

$$[\Gamma : \Gamma[n]] = \sharp\mathrm{SL}(2; \mathbb{Z}/n\mathbb{Z}) < \infty.$$

□

§ 3 Subgroups of the modular group

If $M \equiv L \pmod{n}$ holds for $M, L \in \Gamma$, then $ML^{-1} \equiv I \pmod{n}$ follows from the explicit formula of L^{-1}. Thus M and L belong to the same coset modulo $\Gamma[n]$ and we obtain the

3.5 Corollary. *Two matrices L and M in Γ belong to the same (left or right) coset with respect to $\Gamma[n]$ if and only if $L \equiv M \pmod{n}$.*

The index can be calculated using Theorem 3.4:

$$[\Gamma : \Gamma[n]] = \#\mathrm{SL}\,(2; \mathbb{Z}/n\mathbb{Z}) = n^3 \prod_{p|n} \left(1 - p^{-2}\right) \quad \text{for} \quad n \geq 2 \,. \tag{3.7}$$

A *proof* can be found in H. MAASS [56], 63–65, J. LEHNER [55], 356, R.A. RANKIN [66], 21–23.

3.6 Proposition. *For $n \geq 2$, $\Gamma[n]$ has no fixed points, i.e. $M\tau = \tau$ for $\tau \in \mathbb{H}$ and $M \in \Gamma[n]$ imply $M = \pm I$ if $n = 2$ and $M = I$ if $n > 2$.*

Proof If $M\tau = \tau$ for some $M \neq \pm I$ in $\Gamma[n]$, then τ is also a fixed point of Γ. From Corollary 2.8, we conclude that $\tau = Li$ or $\tau = L\rho$ with a suitable $L \in \Gamma$. We obtain $L^{-1}MLi = i$ or $L^{-1}ML\rho = \rho$. Now we can apply Theorem 2.6 to get

$$L^{-1}ML = \pm J \quad \text{or} \quad L^{-1}ML = \pm U, \; \pm U^2 \,.$$

But since $\Gamma[n]$ is a normal subgroup of Γ, $\pm J$ or $\pm U$ or $\pm U^2$ would belong to $\Gamma[n]$ as well. This is not the case. □

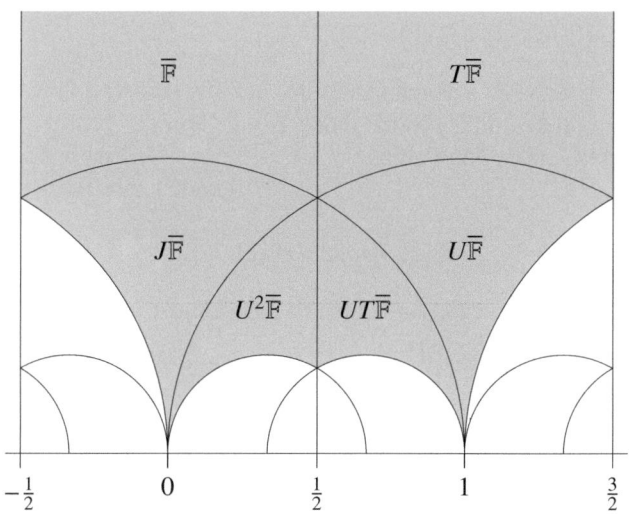

Figure 20: Fundamental domain for $\Gamma[2]$

As an example, consider the case $n = 2$. According to Corollary 3.5, we easily conclude that

$$\begin{pmatrix} 1 & 0 \\ 0 & 1 \end{pmatrix} = I, \quad \begin{pmatrix} 1 & 1 \\ 0 & 1 \end{pmatrix} = T, \quad \begin{pmatrix} 1 & 0 \\ 1 & 1 \end{pmatrix} = -UT,$$
$$\begin{pmatrix} 0 & -1 \\ 1 & 0 \end{pmatrix} = J, \quad \begin{pmatrix} -1 & 1 \\ -1 & 0 \end{pmatrix} = U, \quad \begin{pmatrix} 0 & -1 \\ 1 & -1 \end{pmatrix} = U^2 \tag{3.8}$$

is a system of representatives of the cosets of $\Gamma[2]$ in Γ. Theorem 3.1 produces a fundamental domain for $\Gamma[2]$ in the form above.

For a group equivalent to the group $\Gamma[2]$, C.F. GAUSS – as only became known from his estate – constructed an analogous fundamental domain ([31], vol. VIII, 100–105) as early as 1808.

3. Congruence subgroups. A subgroup Λ of Γ is called a *congruence subgroup* if there exists an $n \geq 1$ with $\Gamma[n] \subseteq \Lambda$. The smallest such n is called the *level* of Λ. Obviously, every congruence subgroup of Γ has finite index in Γ. An important class of examples is given by the groups

$$\Gamma_0[n] := \{M \in \Gamma ; c \equiv 0 \pmod{n}\}. \tag{3.9}$$

We again consider the case $n = 2$: a system of representatives of the right cosets of Γ modulo $\Gamma_0[2]$ is I, J and U^2, i.e.

$$\Gamma = \Gamma_0[2] \cup (\Gamma_0[2] \cdot J) \cup (\Gamma_0[2] \cdot U^2).$$

By Theorem 3.1, there is a fundamental domain of the form above. $\Gamma[2]$ has the index 2 in $\Gamma_0[2]$, which has index 3 in Γ, thus

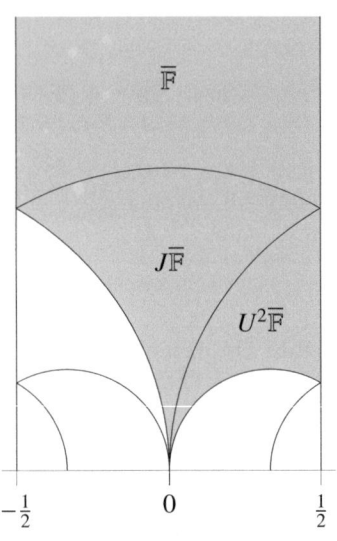

Figure 21:
Fundamental domain for $\Gamma_0[2]$

$$\Gamma_0[2] = \Gamma[2] \cup (\Gamma[2] \cdot T). \tag{3.10}$$

In analogy with $\Gamma_0[n]$, we can also define the subgroups

$$\left. \begin{aligned} \Gamma^0[n] &:= \left\{ M = \begin{pmatrix} a & b \\ c & d \end{pmatrix} \in \Gamma; \ b \equiv 0 \pmod{n} \right\}, \\ \Gamma_1[n] &:= \left\{ M = \begin{pmatrix} a & b \\ c & d \end{pmatrix} \in \Gamma_0[n]; \ a \equiv d \equiv 1 \pmod{n} \right\}. \end{aligned} \right\} \tag{3.11}$$

It is obvious that $\Gamma^0[n] = J \cdot \Gamma_0[n] \cdot J^{-1}$. A detailed discussion of the groups $\Gamma_0[n]$ and $\Gamma^0[n]$ can be found in H. PETERSSON [62], 241–251. By means of Theorem 3.4, we can easily show that

§ 3 Subgroups of the modular group

$$[\Gamma : \Gamma_0[n]] = [\Gamma : \Gamma^0[n]] = n \prod_{p|n} \left(1 + \frac{1}{p}\right).$$

$$[\Gamma : \Gamma_1[n]] = n^2 \prod_{p|n} \left(1 - \frac{1}{p^2}\right) = [\Gamma : \Gamma_0[n]] \cdot \varphi(n).$$

These and other subgroups of the modular group are discussed in F. KLEIN and R. FRICKE [45], section 2, M. NEWMAN [60], Chap. VIII, J. LEHNER [55], Chap. XI.3, R.A. RANKIN [66], Chap. 1.

3.7 Remarks. a) A subgroup Λ of Γ of finite index does not need to be a congruence subgroup. The first examples were found simultaneously by R. FRICKE and G. PICK (*Math. Ann.* **28**, 99–118 and 119–124 (1887)). Further examples can be found in H. MAASS [56], 77–79, and H. PETERSSON (*J. Reine Angew. Math.* **250**, 182–212 (1971); **268/9**, 94–109 (1974)).
b) $\Gamma_0[n]$, $\Gamma^0[n]$ and $\Gamma_1[n]$ are not normal subgroups of Γ whenever $n > 1$. But $\Gamma_1[n]$ is normal in $\Gamma_0[n]$.

4. The congruence groups of level 2 are exactly the proper subgroups of Γ which contain $\Gamma[2]$.

3.8 Theorem. a) $\Gamma/\Gamma[2]$ *is isomorphic to the permutation group* S_3.
b) $\Gamma[2]$ *is generated by the matrices*

$$-I, \quad T^2 \quad \text{and} \quad JT^2J^{-1}. \tag{3.12}$$

Proof a) By Theorem 3.4, $\Gamma/\Gamma[2]$ is isomorphic to $\mathrm{SL}(2;\mathbb{Z}/2\mathbb{Z})$, and this group permutes the set

$$\left\{\begin{pmatrix}\overline{1}\\\overline{0}\end{pmatrix}, \begin{pmatrix}\overline{0}\\\overline{1}\end{pmatrix}, \begin{pmatrix}\overline{1}\\\overline{1}\end{pmatrix}\right\} \subset (\mathbb{Z}/2\mathbb{Z})^2$$

via left multiplication. As $\mathrm{SL}(2;\mathbb{Z}/2\mathbb{Z})$ and S_3 consist of 6 elements, the claim follows.
b) Let Λ denote the subgroup of $\Gamma[2]$ generated by the matrices (3.12) and $M \in \Gamma[2]$. Because

$$T^{2m}M = \begin{pmatrix} a+2mc & * \\ c & * \end{pmatrix} \quad \text{and} \quad (JT^2J^{-1})^m M = \begin{pmatrix} a & * \\ c-2ma & * \end{pmatrix},$$

there is some $L_1 \in \Lambda$ such that

$$L_1 M = \begin{pmatrix} a_1 & b_1 \\ c_1 & d_1 \end{pmatrix} \quad \text{and} \quad |a_1| \leq |c_1| \quad \text{if} \quad c \neq 0.$$

Since a_1 is odd and c_1 is even, it even follows that $|a_1| < |c_1|$. Analogously, we determine some $L_2 \in \Lambda$ satisfying

$$L_2 M = \begin{pmatrix} a_2 & b_2 \\ c_2 & d_2 \end{pmatrix} \quad \text{and} \quad |c_2| < |a_2|, \quad \text{as } a_2 \neq 0.$$

After finitely many steps, we find a matrix $L \in \Lambda$ with

$$LM = \pm \begin{pmatrix} 1 & r \\ 0 & 1 \end{pmatrix}, \quad r \in \mathbb{Z}.$$

Because $LM \in \Gamma[2]$, r is even and $LM = \pm T^r \in \Lambda$, hence $M \in \Lambda$. □

Thus the subgroups of Γ containing $\Gamma[2]$ are in bijection with the subgroups of $\Gamma/\Gamma[2]$, i.e. of S_3. The trivial subgroups of S_3 thus correspond to $\Gamma[2]$ and Γ. The group S_3 has 3 subgroups of order 2. These correspond to the congruence groups

$$\Gamma_0[2], \quad \Gamma^0[2] \quad \text{and} \quad \Gamma_\vartheta := \Gamma[2] \cup \Gamma[2] \cdot J. \tag{3.13}$$

Γ_ϑ is called the *theta group*. Given $M \in \Gamma$, we have $M \in \Gamma_\vartheta$ if and only if

$$M \equiv I (\text{mod } 2) \quad \text{or} \quad M \equiv J (\text{mod } 2). \tag{3.14}$$

Equivalently, we can also write this as

$$a + b + c + d \equiv 0 (\text{mod } 2) \tag{3.15}$$

From the definition, Theorem 3.8, (3.13) and (3.8) we obtain

3.9 Corollary. *Γ_ϑ is a subgroup of Γ of index 3 and*

$$\Gamma = \Gamma_\vartheta \cup (T \cdot \Gamma_\vartheta) \cup (U^2 \cdot \Gamma_\vartheta) = \Gamma_\vartheta \cup (\Gamma_\vartheta \cdot T) \cup (\Gamma_\vartheta \cdot U)$$

holds. The group Γ_ϑ is generated by the matrices J and T^2.

According to Theorem 3.1,

$$\mathcal{F}(\Gamma_\vartheta) := \overline{\mathbb{F}} \cup T\overline{\mathbb{F}} \cup U\overline{\mathbb{F}}$$

is a fundamental domain of Γ_ϑ. We can also specify an exact fundamental domain $\mathbb{F}_\vartheta$ (cf. Exercise 3.18 4)).

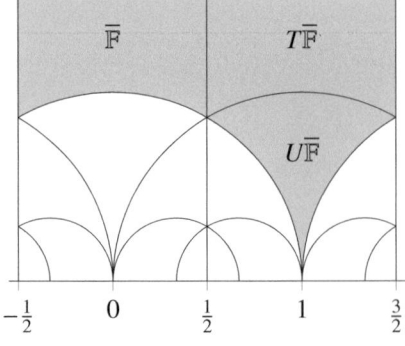 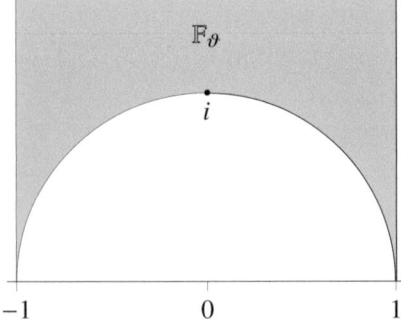

Figure 22: Fundamental domains for Γ_ϑ

§ 3 Subgroups of the modular group 143

Furthermore, S_3 has exactly one subgroup of order 3, and it corresponds to the congruence subgroup

$$\Gamma_N[2] := \Gamma[2] \cup (\Gamma[2] \cdot U) \cup \left(\Gamma[2] \cdot U^2\right). \tag{3.16}$$

3.10 Proposition. $\Gamma_N[2]$ *is the only normal subgroup of* Γ *of index* 2. *The group* $\Gamma_N[2]$ *is generated by the matrices* T^2 *and* $-U$ *and*

$$\Gamma = \Gamma_N[2] \cup \Gamma_N[2] \cdot T \tag{3.17}$$

holds. The map

$$\chi : \Gamma \to \{\pm 1\}, \quad M \mapsto (-1)^{ac+bc+bd},$$

is an abelian character of Γ *with kernel* $\chi = \Gamma_N[2]$.

Proof (3.17) follows from (3.16) and (3.8). Thus $\Gamma_N[2]$, as a subgroup of index 2, is a normal subgroup. According to Theorem 3.8 and (3.16), $\Gamma_N[2]$ is generated by T^2, $JT^2J^{-1} = U^{-1}T^2U$, $-I$ and U, thus because of $U^3 = I$ by T^2 and $-U$.
If Γ^* is any subgroup of Γ of index 2, then $M^2 \in \Gamma^*$ holds for all $M \in \Gamma$, since otherwise $[\Gamma : \Gamma^*] \geq 3$ follows. Therefore T^2, $J^2 = -I$, $U^4 = U$ belong to Γ^*. Hence $\Gamma_N[2] \subseteq \Gamma^*$ and thus equality holds.

For $M \in \Gamma$, we easily verify

$$\chi(MJ) = (-1)^{bd-ad+ac} = -\chi(M) = \chi(M) \cdot \chi(J),$$
$$\chi(MT) = (-1)^{ac+(a+b)c+(a+b)(c+d)} = -\chi(M) = \chi(M) \cdot \chi(T).$$

Since Γ is generated by J and T, according to Theorem 2.2, it follows that

$$\chi(ML) = \chi(M) \cdot \chi(L) \quad \text{for all} \quad M, L \in \Gamma.$$

$\chi(T^2) = \chi(-U) = 1$ leads to $\Gamma_N[2] \subseteq \text{kernel } \chi$. Because of $\chi(T) = -1$ we obtain the equality. □

3.11 Remarks. a) We call Γ_ϑ the theta group because it is the invariance group (cf. Introduction of Chapter III) of the *theta zero value* $\vartheta(\tau) := \vartheta(0; \tau)$ (cf. I(6.30)), which is (even in the English language literature) called *Theta Nullwert*.
b) The congruence subgroups $\Gamma_0[2], \Gamma^0[2]$ and Γ_ϑ are conjugate in Γ:

$$\Gamma^0[2] = J \cdot \Gamma_0[2] \cdot J^{-1}, \quad \Gamma_\vartheta = U \cdot \Gamma_0[2] \cdot U^{-1}.$$

5*. Generators of subgroups. We need the following purely algebraic result.

3.12 NIELSEN-SCHREIER Theorem. *Let S be a subgroup of a group G with a system R of representative of the right cosets, i.e.* $G = \overset{\bullet}{\bigcup}_{r \in R} Sr$, *and the identity element* $e \in R$. *Given* $g \in G$, *let* $\overline{g} \in R$ *be such that* $g \in S\overline{g}$. *If G is generated by a subset H, then S is generated by*

$$\{rh(\overline{rh})^{-1}; \ r \in R, \ h \in H\}. \tag{3.18}$$

Proof Clearly, (3.18) is a subset of S. Note that

$$\left(rh(\overline{rh})^{-1}\right)^{-1} = (\overline{rh})h^{-1}r^{-1} = \left((\overline{rh})h^{-1}\right)\left(\overline{(rh)}h^{-1}\right)^{-1}.$$

Given $s \in S$, we obtain a representation

$$s = h_1 \cdot \ldots \cdot h_n, \quad h_i \in H \text{ or } h_i^{-1} \in H.$$

Then we successively get

$$s = \left(eh_1\overline{h_1}^{-1}\right)\overline{h_1}h_2h_3 \cdot \ldots \cdot h_n$$
$$= \left(eh_1\overline{h_1}^{-1}\right)\left(\overline{h_1}h_2\left(\overline{h_1h_2}\right)^{-1}\right)\left(\left(\overline{h_1h_2}\right)h_3 \cdot \ldots \cdot h_n\right)$$
$$= u_1 \cdot \ldots \cdot u_n r.$$

for some $r \in R$, where the u_i or u_i^{-1} are of the form (3.18). Thus $s \in S$ leads to $r \in S$, i.e. $r = e$. □

Then Theorem 2.2 leads to

3.13 Corollary. *Let Λ be a subgroup of Γ of finite index r. Then Λ is finitely generated. If $\Lambda = \bigcup_{\nu=1}^{r} \Lambda M_\nu$, $M_1 = I$, then Λ is generated by*

$$M_\nu J(\overline{M_\nu J})^{-1}, \quad M_\nu T(\overline{M_\nu T})^{-1}, \quad \nu = 1, \ldots, r.$$

We deal with $\Lambda = \Gamma_0[p]$, p prime, as a concrete example. If $c, d \in \mathbb{Z}$, $d \not\equiv 0 \pmod{p}$, there exists $1 \le j \le p-1$ such that $c + jd \equiv 0 \pmod{p}$. Thus we may choose

$$J, \quad \begin{pmatrix} 1 & 0 \\ j & 1 \end{pmatrix} = J^{-1}T^{-j}J, \quad j = 0, \ldots, p-1,$$

as a system of representatives of the right cosets. Note that Γ is also generated by J and $J^{-1}T^{-1}J$. Then the procedure of Corollary 3.13 yields that $\Gamma_0[p]$ is generated by the matrices

$$-I, \quad \begin{pmatrix} 1 & 0 \\ p & 1 \end{pmatrix}, \quad \begin{pmatrix} j' & 1 \\ rp & j \end{pmatrix}, \quad j, j' \in \{1, \ldots, p-1\}, \quad jj' = 1 + rp, \quad r \in \mathbb{N}_0,$$

where we omit the matrices which can already be generated by these.

3.14 Corollary. *Let Λ be a subgroup of Γ of finite index. Then Λ is generated by the matrices.*

$$L \in \Lambda \quad \text{such that} \quad L\mathcal{F}(\Lambda) \cap \mathcal{F}(\Lambda) \ne \emptyset.$$

Proof According to (3.2), the points

§ 3 Subgroups of the modular group 145

$$\tau = \overline{M_\nu J}\langle i\rangle \quad \text{and} \quad M_\nu J(\overline{M_\nu J})^{-1}\langle \tau\rangle = M_\nu i,$$
$$\tau' = \overline{M_\nu U}\langle \rho\rangle \quad \text{and} \quad M_\nu U(\overline{M_\nu U})^{-1}\langle \tau'\rangle = M_\nu \rho$$

belong to $\mathcal{F}(\Lambda)$. As J and U generate Γ, the claim follows from the NIELSEN-SCHREIER Theorem 3.12. □

6*. Finite subgroups. First let us consider those $M \in \Gamma$, which have finite order.

3.15 Theorem. *For $\pm I \neq M \in \Gamma$, the following assertions are equivalent:*

(i) *M has order 3, 4 or 6.*
(ii) *M has finite order.*
(iii) *$|\operatorname{tr} M| < 2$.*
(iv) *There is some $\tau \in \mathbb{H}$ such that $M\tau = \tau$.*
(v) *There is some $L \in \Gamma$ such that $L^{-1}ML \in \{\pm J, \pm U, \pm U^2\}$.*

Proof (i) $\Longrightarrow$ (ii): Trivial.
(ii) $\Longrightarrow$ (iii): If M is of finite order, then the eigenvalues of M have absolute value 1. With $t := \operatorname{tr} M$, the eigenvalues have the form

$$\lambda_{1,2} = \frac{1}{2}\left(t \pm \sqrt{t^2 - 4}\right).$$

In the case $t > 2$, an eigenvalue would be greater than 1; in the case $t < -2$, one eigenvalue would be less than -1. Thus, after passing to $-M$ if necessary it suffices to consider $t = 2$. From the CAYLEY–HAMILTON Theorem, it follows that $M^2 = 2M - I$ and by induction

$$M^n = nM - (n-1)I, \quad n \in \mathbb{N}.$$

Because $M \neq \pm I$, M has infinite order in this case. Thus $|t| < 2$ follows.
(iii) $\Longrightarrow$ (iv): Apply Proposition 1.14.
(iv) $\Longrightarrow$ (v) $\Longrightarrow$ (i): We use Theorem 2.6, (1.15) and Corollary 2.8 □

3.16 Corollary. *If Λ is a finite subgroup of Γ, then Λ is conjugate in Γ to one of the groups*

$$\{I\}, \{\pm I\}, \{\pm I, \pm J\}, \{I, U, U^2\} \text{ or } \{\pm I, \pm U, \pm U^2\}.$$

Proof We have $S := \sum_{L \in \Lambda} L^{tr}L \in \operatorname{Pos}(2;\mathbb{R})$. Since for fixed $M \in \Lambda$, LM runs through all of elements Λ as L does, it follows that $M^{tr}SM = S$. By Lemma 1.25, there exists some $\tau = w(S) \in \mathbb{H}$ with $S = \sqrt{\det S} \cdot F_\tau$. Then it follows from Theorem 1.26 that Λ is a subgroup of Γ_τ. After a suitable conjugation in Γ, we may assume $\tau \in \mathbb{F}$. Then the assertion follows from (2.14). □

3.17 Corollary. *Every finite subgroup of Γ has the order 1, 2, 3, 4 or 6.*

3.18 Exercises.
1) a) Let $M \in \Gamma$, Λ be a subgroup of Γ and $\Lambda^* = M\Lambda M^{-1}$. If $\mathcal{F}$ is a fundamental domain for Λ, then $M\mathcal{F} := \{M\tau \ ; \ \tau \in \mathcal{F}\}$ is a fundamental domain for Λ^*.
b) Let $\mathcal{F}$ be a fundamental domain for $\Gamma[n]$, $n \geq 1$, and $M \in \Gamma$. Then $M\mathcal{F}$ is also a fundamental domain for $\Gamma[n]$.
2) a) The intersection of finitely many congruence groups is a congruence group.

b) The intersection of all congruence groups is $\{I\}$.
c) If $n > 2$ then $\{I\}$ is the only finite subgroup of $\Gamma[n]$.
3) The set $\{\tau \in \mathbb{H}; |x| \leq 1, |\tau - 1/2| \geq 1/2, |\tau + 1/2| \geq 1/2\}$ is a fundamental domain for $\Gamma[2]$.
4) Let $\mathbb{F}_\vartheta := \{\tau \in \mathbb{H}; |\tau| \geq 1, -1 < x \leq 1, |\tau| > 1 \text{ for } -1 < x < 0\}$.
a) $\overline{\mathbb{F}}_\vartheta$ is a fundamental domain for Γ_ϑ.
b) If τ and $M\tau$, $M \in \Gamma_\vartheta$, $M \neq \pm I$, belong to $\mathbb{F}_\vartheta$, then $\tau = M\tau = i$ and $M = \pm J$.
5) Let Λ be a finite subgroup of Γ_ϑ. Then Λ is conjugate to $\{I\}$, $\{\pm I\}$ or $\{\pm I, \pm J\}$ in Γ_ϑ.
6) Let $n \in \mathbb{Z}$, $n \geq 2$. The group $\Gamma_0[n]$ contains an element of order 4 if and only if -1 is a quadratic residue (mod n). If n is even, then $\Gamma_0[n]$ contains no element of order 3 or 6.
7) $\Gamma_0[n]$ is generated by the matrices $-I, T$ and $JT^n J^{-1}$, if and only if $n = 1, 2, 3, 4$.
8) Let p be a prime number. Then

$$M_j, J, J^{-1}T^k J, 0 \leq k < p, M_j = \begin{pmatrix} j & 1 \\ rp & j' \end{pmatrix}, 1 \leq j, j' < p, jj' = 1 + rp, r \in \mathbb{N}_0,$$

is a representation system of the right cosets of Γ for $\Gamma_1[p]$.
9) Let p be a prime. Then

$$\{\tau \in \mathbb{H}; |x| \leq p/2, |\tau - n| \geq 1 \text{ for all } 0 \neq n \in \mathbb{Z}, |n| \leq p/2\}$$

is a fundamental domain for $\Gamma^0[p]$.
10) a) $\Gamma_N[2]$ is generated by the matrices M^2, $M \in \Gamma$.
b) $\Gamma_N[2]$ is generated by the matrices $LML^{-1}M^{-1}$, $L, M \in \mathrm{GL}(2; \mathbb{Z})$.
11) The principal congruence subgroup $\Gamma[n]$, $n \in \mathbb{N}$, is conjugate in $\mathrm{GL}(2; \mathbb{Q})$ and in $\mathrm{SL}(2; \mathbb{R})$ to the subgroup $\Gamma_1[n^2]$.
12) The set $\{\tau \in \mathbb{H}; |\tau - 1| \geq 1, |\tau + 1| \geq 1\}$ is a fundamental domain for the group
$\left\{\begin{pmatrix} 1 & 0 \\ n & 1 \end{pmatrix} ; n \in \mathbb{Z}\right\}$.
13) For a subgroup Λ of Γ, let $\Lambda' := \Lambda \cup (-I)\Lambda$.
a) Describe a *transversal* of $\Gamma[3]'$ in Γ, i.e. a system of representatives of the right cosets.
b) Show that:

$$\Gamma/\Gamma[3]' \cong A_4 = \{\pi \in S_4; \mathrm{sgn}\,\pi = 1\}$$

c) Determine a fundamental domain for $\Gamma[3]$ and draw it.
d) Determine explicitly all subgroups of Γ which contain $\Gamma[3]'$.
e) Determine all congruence subgroups of level 3.
f) Determine a system of representatives of the inequivalent cusps of $\Gamma[3]$.
g) Show that $\Gamma[3]$ is generated by the matrices

$$\begin{pmatrix} 1 & 3 \\ 0 & 1 \end{pmatrix}, \begin{pmatrix} 1 & 0 \\ 3 & 1 \end{pmatrix}, \begin{pmatrix} 4 & -3 \\ 3 & -2 \end{pmatrix}, \begin{pmatrix} -2 & -3 \\ 3 & 4 \end{pmatrix}.$$

14) Show that $\Gamma[4]$ is generated by the matrices

$$\begin{pmatrix} 1 & 4 \\ 0 & 1 \end{pmatrix}, \begin{pmatrix} 1 & 0 \\ 4 & 1 \end{pmatrix}, \begin{pmatrix} 5 & -4 \\ 4 & -3 \end{pmatrix}, \begin{pmatrix} -3 & -4 \\ 4 & 5 \end{pmatrix}, \begin{pmatrix} 9 & 20 \\ 4 & 9 \end{pmatrix}.$$

15) Let Λ be a subgroup of Γ such that $-I \notin \Lambda$. Then $[\Gamma : \Lambda] \geq 4$. If $[\Gamma : \Lambda] = 4$, then Λ is uniquely determined, a normal subgroup of Γ, a congruence subgroup of level 4 and generated by U and T^2.

16) If Λ is a subgroup of finite index in Γ and $\mathcal{F}(\Lambda)$ is given by (3.2), then there are only finitely many $M \in \Gamma$ satisfying $M\langle \mathcal{F}(\Lambda)\rangle \cap \mathcal{F}(\Lambda) \neq \emptyset$. Is this property true for an arbitrary fundamental domain of Λ ?

17) Show that

$$\Lambda = \{M \in \Gamma;\ M \equiv \pm I \pmod{3} \text{ or } \operatorname{tr} M \equiv 0 \pmod{3}\}$$

is a congruence subgroup of level 3. Λ is a normal subgroup of index 3 in Γ and $T^{-1}\overline{\mathbb{F}} \cup \overline{\mathbb{F}} \cup T\overline{\mathbb{F}}$ is a fundamental domain for Λ.

18) Show that $\Gamma[m] \cdot \Gamma[n] = \Gamma$ holds for coprime $m, n \in \mathbb{N}$.

§ 4* Discontinuous groups

In this paragraph, we consider more general discontinuous subgroups of $\mathrm{SL}(2;\mathbb{R})$. For these groups, a fundamental domain is constructed according to a method of H. POINCARÉ (1854–1912).

1. Discrete and discontinuous. The modular group Γ is a "discrete" subgroup of the "topological" group $\mathrm{SL}(2;\mathbb{R})$. In order to define these terms ad hoc for arbitrary subgroups of $\mathrm{SL}(2;\mathbb{R})$, we define a *norm* for the elements of $\mathrm{Mat}(2;\mathbb{R})$ by

$$|M| := \sqrt{\operatorname{tr}(M^{tr}M)} = \sqrt{a^2 + b^2 + c^2 + d^2}, \quad \text{if} \quad M = \begin{pmatrix} a & b \\ c & d \end{pmatrix}, \tag{4.1}$$

and thus we can use (4.1) to transfer all the metric terms of $\mathbb{R}^4$ to $\mathrm{Mat}(2;\mathbb{R})$. Furthermore, we have

$$|LM| \leq |L| \cdot |M| \quad \text{for all} \quad L, M \in \mathrm{Mat}(2;\mathbb{R}). \tag{4.2}$$

Let Λ be a subgroup of $\mathrm{SL}(2;\mathbb{R})$. If Λ is a discrete subset of $\mathrm{Mat}(2;\mathbb{R})$, then we call Λ a *discrete* subgroup of $\mathrm{SL}(2;\mathbb{R})$. The subgroup Λ of $\mathrm{SL}(2;\mathbb{R})$ is called *discontinuous* if for every $\tau \in \mathbb{H}$ and every sequence $(M_k)_{k \geq 1}$ of mutually distinct matrices M_k in Λ, the sequence $(M_k\tau)_{k \geq 1}$ has no limit point in $\mathbb{H}$. This obviously implies the validity of the following statements for every $\tau \in \mathbb{H}$:

(D.1) The set $\{M\tau\ ;\ M \in \Lambda\}$ is discrete in $\mathbb{H}$, i.e. has no limit point in $\mathbb{H}$.
(D.2) The set $\{M \in \Lambda\ ;\ M\tau = \tau\}$ is finite.

Often, it is not easy to decide whether a given group is discontinuous.

4.1 Theorem. *For a subgroup Λ of $\mathrm{SL}(2;\mathbb{R})$, the following assertions are equivalent*:

(i) Λ *is discrete*.
(ii) Λ *is discontinuous*.
(iii) *For any two compact subsets C_1 and C_2 of $\mathbb{H}$, there are only finitely many $M \in \Lambda$ satisfying $(MC_1) \cap C_2 \neq \emptyset$.*

Proof We first demonstrate

$$\text{If } \tau \in \mathbb{H} \text{ and } (L_k)_{k\geq 1} \text{ is a sequence in } \mathrm{SL}(2;\mathbb{R}) \text{ such that} \qquad (*)$$
$$L_k\tau \to \tau \text{ for } k \to \infty, \text{ then the sequence } (L_k)_{k\geq 1} \text{ is bounded}.$$

Proof. By Theorem 1.7, we may choose $N \in \mathrm{SL}(2;\mathbb{R})$ satisfying $\tau = Ni$. Defining $P_k := N^{-1}L_k N$ and $Q_k := J^{-1}P_k J$, we have $P_k i \to i$ and $Q_k i \to i$, hence also $\mathrm{Im}(P_k i) \to 1$ and $\mathrm{Im}(Q_k i) \to 1$. Because

$$Q_k = \begin{pmatrix} * & * \\ -b_k & a_k \end{pmatrix} \quad \text{for} \quad P_k = \begin{pmatrix} a_k & b_k \\ c_k & d_k \end{pmatrix}$$

and (1.14), it follows that

$$c_k^2 + d_k^2 \to 1 \quad \text{and} \quad b_k^2 + a_k^2 \to 1.$$

Hence the matrices P_k are bounded and, because of (4.2), so are the matrices L_k. □

(i) $\Longrightarrow$ (ii): We assume that Λ is not discontinuous. Then there exist $\tau, \tau' \in \mathbb{H}$ and a sequence $(M_k)_{k\geq 1}$ of pairwise distinct matrices in Λ with $M_k\tau \to \tau'$ for $k \to \infty$. For $N_k := M_k^{-1}M_{k+1} \in \Lambda$, the invariant $\mathbb{H}$–metric (1.24) yields

$$|\tau, N_k\tau| = |M_k\tau, M_{k+1}\tau| \longrightarrow |\tau', \tau'| = 0 \quad \text{for } k \longrightarrow \infty$$

according to (1.25). Because the $\mathbb{H}$-topology coincides with the natural topology, it follows that

$$N_k\tau \longrightarrow \tau \quad \text{for } k \longrightarrow \infty. \qquad (**)$$

According to $(*)$, the sequence $(N_k)_{k\geq 1}$ is bounded. Since Λ is discrete, among the N_k there are only finitely many different matrices. So, because of $(**)$, there is some k_0 with $N_k\tau = \tau$ for $k \geq k_0$. Thus $M_{k+1}\tau = M_k N_k\tau = M_k\tau = \cdots = M_{k_0}\tau$ for $k \geq k_0$, i.e. $M_k^{-1}M_{k_0}\tau = \tau$. Because of $(*)$ and the discreteness of Λ there are only finitely many different matrices among the matrices $M_k^{-1}M_{k_0}$, thus also only finitely many among the M_k in contradiction to the hypothesis.

(ii) $\Longrightarrow$ (iii): Assume that there are compact subsets C_1 and C_2 of $\mathbb{H}$ with the property $(MC_1) \cap C_2 \neq \emptyset$ for infinitely many $M \in \Lambda$. By Corollary 1.19, C_1 and C_2 may be assumed to be given by an $\mathbb{H}$-circular disk $\mathcal{D} = \{w \in \mathbb{H} \,;\, |w,i| \leq r\}$. If $w_k \in (M_k\mathcal{D}) \cap \mathcal{D}$ for infinitely many $M_k \in \Lambda$, then the points $\tau_k = M_k^{-1}w_k$ belong to $\mathcal{D}$. By (1.25), we conclude

$$|M_k i, i| \leq |M_k i, M_k \tau_k| + |M_k \tau_k, i| = |i, \tau_k| + |w_k, i| \leq 2r$$

§4* Discontinuous groups

and the sequence $(M_k i)_{k \geq 1}$ would have a limit point in $\mathbb{H}$. This is a contradiction.
(iii) $\Longrightarrow$ (i): We assume that Λ is not discrete. Then there is a matrix $M \in \mathrm{Mat}\,(2;\mathbb{R})$ and a sequence of pairwise distinct $M_k \in \Lambda$ with the property

$$\lim_{k \to \infty} M_k = M \quad \text{(componentwise)}.$$

Since $\mathrm{SL}\,(2;\mathbb{R})$ is closed in $\mathrm{Mat}\,(2;\mathbb{R})$, it follows that $M \in \mathrm{SL}\,(2;\mathbb{R})$. For $\tau \in \mathbb{H}$,

$$\lim_{k \to \infty} (M_k \tau) = M\tau =: w.$$

If $\mathcal{D} := \{\tau\}$ and C is a compact neighborhood of w in $\mathbb{H}$, then $M_k \tau \in C$ holds for all sufficiently large k, hence $(M_k \mathcal{D}) \cap C \neq \emptyset$ for infinitely many k. This contradicts (iii). $\square$

2. Fixed points. For $\varepsilon > 0$ and $z \in \mathbb{H}$, the $\mathbb{H}$-circular disk

$$\mathcal{D}_{z,\varepsilon} := \{w \in \mathbb{H}\,;\, |z,w| < \varepsilon\} \tag{4.3}$$

is a neighborhood of z by Corollary 1.19. Because of (1.25), we obtain

$$M \mathcal{D}_{z,\varepsilon} = \mathcal{D}_{Mz,\varepsilon} \quad \text{for } M \in \mathrm{SL}\,(2;\mathbb{R}). \tag{4.4}$$

Now let Λ be a discontinuous subgroup of $\mathrm{SL}\,(2;\mathbb{R})$.

4.2 Lemma. *Given $z \in \mathbb{H}$, there is some $\varepsilon = \varepsilon_\Lambda(z) > 0$ such that $Mz \in \mathcal{D}_{z,\varepsilon}$ for any $M \in \Lambda$ if and only if $Mz = z$.*

Proof Otherwise there would be a sequence of mutually distinct $M_k \in \Lambda$ satisfying $M_k z \to z$. Since Λ is discontinuous we have a contradiction. $\square$

4.3 Corollary. *For every $z \in \mathbb{H}$, there is some $\delta = \delta_\Lambda(z) > 0$ such that*

$$\mathcal{D}_{z,\delta} \cap M \mathcal{D}_{z,\delta} \neq \emptyset \quad \text{and} \quad M \in \Lambda \tag{4.5}$$

always implies $Mz = z$.

Proof We set $\delta = \varepsilon/2$ in Lemma 4.2. If w belongs to the intersection in question, then $|w,z| < \delta$ and $|Mz,w| = |z, M^{-1}w| < \delta$, thus $|Mz,z| \leq |Mz,w| + |w,z| < \varepsilon$ and consequently $Mz = z$. $\square$

The neighborhood $\mathcal{D}_{z,\delta}$ does not contain fixed points of Λ except for possibly z. Hence we get

4.4 Theorem. *The fixed points of Λ are a discrete subset of $\mathbb{H}$.*

3. The normal polygon. Let Λ again be a discontinuous subgroup of $\mathrm{SL}\,(2;\mathbb{R})$. For $z \in \mathbb{H}$, following H. POINCARÉ (1854–1912), we define the so-called *normal polygon*

$$\mathcal{P} = \mathcal{P}_\Lambda(z) := \{w \in \mathbb{H}\,;\, |z,w| \leq |z,Mw| \text{ for all } M \in \Lambda\}.$$

In analogy with Theorem 2.6, we get

4.5 Theorem. *If z is not a fixed point of Λ, then $\mathcal{P}_\Lambda(z)$ is a fundamental domain for Λ, i.e.*
a) *$\mathcal{P}_\Lambda(z)$ is relatively closed in $\mathbb{H}$ and its boundary has LEBESGUE measure 0.*
b) *Given $\tau \in \mathbb{H}$, there is some $M \in \Lambda$ with $M\tau \in \mathcal{P}_\Lambda(z)$.*
c) *If τ and $M\tau$, $M \in \Lambda$, are interior points of $\mathcal{P}_\Lambda(z)$, then $M = \pm I$.*

Proof a) For $M \in \Lambda$, the set $\{w \in \mathbb{H} \,;\, |z,w| \leq |M^{-1}z,w|\}$ is relatively closed in $\mathbb{H}$ by Proposition 1.18. So $\mathcal{P}$ is also relatively closed in $\mathbb{H}$.
Due to (1.25) we have

$$\partial \mathcal{P} \subseteq \bigcup_{M \in \Lambda,\, M \neq \pm I} \{w \in \mathbb{H};\, |z,w| = |M^{-1}z,w|\}.$$

As Λ is countable, this set has LEBESGUE measure 0 by Corollary 1.23.
b) Given $\tau \in \mathbb{H}$, consider $\mu := \inf\{|z, M\tau| \,;\, M \in \Lambda\}$ and choose a sequence $(M_k)_{k \geq 1}$ in Λ with the property

$$|z, M_k \tau| \longrightarrow \mu. \tag{$*$}$$

If $\mu = 0$, then the sequence $(M_k\tau)_{k \geq 1}$ converges to z. Since Λ is discontinuous, there exists some $L \in \Lambda$ with $M_k\tau = L\tau$ for all sufficiently large k. Then $L\tau = z \in \mathcal{P}$ follows.
So let $\mu > 0$. Because of Lemma 1.18,

$$\overline{\mathcal{D}_{z,2\mu}(z)} = \{w \in \mathbb{H}\,;\, |z,w| \leq 2\mu\}$$

is a compact neighborhood of z in $\mathbb{H}$. According to Corollary 1.22, we may assume without loss of generality that the sequence $(M_k\tau)_{k \geq 1}$ is convergent. Since Λ is discontinuous, there is some $L \in \Lambda$ with $M_k\tau = L\tau$ for all sufficiently large k. From $(*)$ we obtain $\mu = |z, L\tau|$ and consequently $L\tau \in \mathcal{P}$.
c) According to the assumption, there is an open neighborhood $\mathcal{U}$ of τ in $\mathbb{H}$ with $\mathcal{U} \subseteq \mathcal{P}$. Since $M\tau$ is an interior point of $\mathcal{P}$, by using a subset of $\mathcal{U}$ if necessary we may also assume $M\mathcal{U} := \{Mw\,;\, w \in \mathcal{U}\} \subseteq \mathcal{P}$. For all $w \in \mathcal{U}$, therefore,

$$|z,w| \leq |z, Lw| \quad \text{and} \quad |z, Mw| \leq |z, LMw| \quad \text{for all} \quad L \in \Lambda.$$

If we set $L = M$ in the first and $L = M^{-1}$ in the second inequality, we have

$$|z,w| = |z, Mw| = |M^{-1}z, w| \quad \text{for all} \quad w \in \mathcal{U}.$$

Since $\mathcal{U}$ is a non-empty open set, $z = M^{-1}z$ follows from Corollary 1.23. But since z is not a fixed point of Λ, we get $M = \pm I$. □

The fundamental domain $\mathcal{P}$ of Λ is the intersection of countably many $\mathbb{H}$–half planes, thus convex in the $\mathbb{H}$-geometry. The interior of $\mathcal{P}$ is not empty, according to BAIRE's Category Theorem (B. V. QUERENBURG [65], §13 D). We can show that it is given by

$$\overset{\circ}{\mathcal{P}} = \{w \in \mathbb{H}\,;\, |z, w| < |z, Mw| \text{ for all } \pm I \neq M \in \Lambda\}.$$

4.6 Remarks. a) The considerations of this section transfer mutatis mutandis to discontinuous groups of isometries of complete metric spaces. Compare M. KOECHER and W. ROELCKE (*Math. Z.* **71**, 258–267 (1959)).
b) A detailed exposition of discontinuous groups and related questions can be found in H. MAASS [56], Chap. I, J. LEHNER [55], Chap. III, IV, VII, and C.L. SIEGEL [77].
c) To give another example, let $G(\lambda)$, $\lambda > 0$, be the subgroup of $\mathrm{SL}(2;\mathbb{R})$ generated by the matrices

$$J = \begin{pmatrix} 0 & -1 \\ 1 & 0 \end{pmatrix} \quad \text{and} \quad \begin{pmatrix} 1 & \lambda \\ 0 & 1 \end{pmatrix}$$

the so-called HECKE *group* with parameter λ. In particular $G(1) = \Gamma$ by Theorem 2.2 and $G(2) = \Gamma_\vartheta$ by Corollary 3.9. E. HECKE ([36], 599–616) showed that the group $G(\lambda)$ is discrete if and only if

$$\lambda \geq 2 \quad \text{or} \quad \lambda = 2\cos(\pi/q), \; q \in \mathbb{Z}, \; q \geq 3 \;.$$

R. EVANS (*J. Number Th.* **5**, 108–115) gave an elementary proof of this result in 1973.

4.7 Exercises.
1) For $p = 2, 3$, show that $G(\sqrt{p}) = \Delta \cup J\Delta$, where

$$\Delta = \left\{ M = \begin{pmatrix} a & b\sqrt{p} \\ c\sqrt{p} & d \end{pmatrix} ; \; a, b, c, d \in \mathbb{Z}, \; \det M = 1 \right\} \;.$$

$G(\sqrt{p})$ is a discontinuous subgroup of $\mathrm{SL}(2;\mathbb{R})$ and the subgroup Δ is conjugate in $\mathrm{SL}(2;\mathbb{R})$ to $\Gamma_0[p]$.
2) If $p = 2, 3$ then $\mathcal{F} := \{\tau \in \mathbb{H} \,;\, |x| \leq \frac{1}{2}\sqrt{p}, \; |\tau| \geq 1\}$ is a fundamental domain for $G(\sqrt{p})$.
3) The fixed points of $G(\sqrt{p})$ are Mi and $M\tau$, $\tau = \frac{1}{2}\sqrt{p} + i\frac{1}{2}\sqrt{4-p}$, with $M \in G(\sqrt{p})$ for $p = 2, 3$. Determine the stabilizer groups explicitly.
4) Let o be the ring of integers of a real–quadratic number field. Then the group $\mathrm{SL}(2;o)$ is not discontinuous.
5) Let $\mathcal{M}$ be the set of fixed points of a discontinuous group Λ. Then $L\mathcal{M} = \mathcal{M}$ holds for all $L \in \Lambda$.
6) Let Λ be a subgroup of $\mathrm{SL}(2;\mathbb{R})$ and $M \in \mathrm{SL}(2;\mathbb{R})$. The group Λ is discontinuous if and only if $\Lambda^* = M\Lambda M^{-1}$ is discontinuous. If $z \in \mathbb{H}$, then $M\mathcal{P}_\Lambda(z) = \mathcal{P}_{\Lambda^*}(Mz)$.
7) Let $M \in \mathrm{SL}(2;\mathbb{R})$, $M \neq \pm I$, and let Λ be be the subgroup of $\mathrm{SL}(2;\mathbb{R})$ generated by M. The group Λ is discontinuous if and only if M is parabolic or hyperbolic or elliptic of finite order.
8) Let $\lambda \in \mathbb{R}$, $\lambda > 1$. Then $\mathcal{P}_\Gamma(\lambda i) = \overline{\mathbb{F}}$ holds.
9) Let $\lambda \in \mathbb{R}$, $\lambda > 1$. Then $\mathcal{P}_{\Gamma_\vartheta}(\lambda i) = \overline{\mathbb{F}_\vartheta}$ (cf. Exercise 3.18 4)) holds.
10) Let $\Gamma_\infty = \{\pm T^n \,;\, n \in \mathbb{Z}\}$. Then $\mathcal{P}_{\Gamma_\infty}(z) = \{\tau \in \mathbb{H} \,;\, |\mathrm{Re}(\tau - z)| \leq 1/2\}$ holds for each $z \in \mathbb{H}$.
11) A discontinuous subgroup of $\mathrm{SL}(2;\mathbb{R})$ is countable.
12) Let G_n, $n > 2$, be the subgroup $\mathrm{SL}(2;\mathbb{R})$ generated by $\begin{pmatrix} \cos(\pi/n) & \sin(\pi/n) \\ -\sin(\pi/n) & \cos(\pi/n) \end{pmatrix}$. Then

$$\mathcal{F}_n := \{\tau \in \mathbb{H} \,;\, x \leq 0, \; \tau\bar{\tau} + 2x\cot(2\pi/n) \geq 1\}$$

is a fundamental domain for G_n.

13) If Λ is a discontinuous subgroup of $\mathrm{SL}(2;\mathbb{R})$, then $\mathcal{P}_\Lambda(z)$ is an $\mathbb{H}$-star with the center z.

14) Describe the analogon of the normal polygon $\mathcal{P}$ for the group of translations with respect to a lattice $\Omega = \mathbb{Z}\tau + \mathbb{Z}$, $\tau \in \mathbb{F}$,

$$\mathcal{P} = \{z \in \mathbb{C};\ |z| \leq |z+\omega|\ \text{for all}\ \omega \in \Omega\}$$

by a minimal set of linear inequalities. Show that $\mathcal{P}$ is the closure of a fundamental parallelotope.

15) Prove the following stronger version of (4.2)

$$|LM|^2 = |L|^2|M|^2 + |M^\# L|^2 \quad \text{for all}\ L, M \in \mathrm{SL}(2;\mathbb{R}).$$

Chapter III
Modular forms

Introduction

1. Preface. As the elliptic functions are invariant under certain self-mappings of $\mathbb{C}$, namely the translations of a lattice, so are the *modular functions* under suitable self-mappings of the upper half-plane $\mathbb{H}$, namely the *modular substitutions*

$$\tau \longmapsto M\tau := \frac{a\tau + b}{c\tau + d} \quad \text{with} \quad M = \begin{pmatrix} a & b \\ c & d \end{pmatrix} \in \Gamma := \mathrm{SL}(2; \mathbb{Z}).$$

The most important example of such a function, which is moreover holomorphic on $\mathbb{H}$, is the *absolute invariant* $j = j(\tau)$, which we have already met in I, §4 and in the introduction of Chapter II. It will turn out that j can be used to describe all modular functions.

2. Possible transformation behavior. In addition to functions which remain invariant under modular substitutions functions $f : \mathbb{H} \to \mathbb{C}$ are interesting which at least still show a clear behavior under modular substitutions:

$$f(M\tau) = \gamma_M(\tau) \cdot f(\tau) \quad \text{for all} \quad M \in \Gamma. \tag{0.1}$$

Let $\gamma_M(\tau)$ be an "elementary" factor, which will have to be defined more precisely. If we write (0.1) for MN instead of M and use $(MN)\tau = M\langle N\tau\rangle$, we get (in the case $f(\tau) \neq 0$) the condition

$$\gamma_{MN}(\tau) = \gamma_M(N\tau) \cdot \gamma_N(\tau) \quad \text{for} \quad M, N \in \Gamma \quad \text{and} \quad \tau \in \mathbb{H} \tag{0.2}$$

on γ. This "Cocycle Condition" resembles the chain rule of the differentiation. In fact,

$$\gamma_M(\tau) := \frac{dM\tau}{d\tau} = (c\tau + d)^{-2} \tag{0.3}$$

and its powers satisfy (0.2). The transformation formula thus obtained for any even number k is

$$f(M\tau) = (c\tau + d)^k \cdot f(\tau) \quad \text{for} \quad M \in \Gamma \tag{0.4}$$

and characterizes of the so-called *modular forms*. Since the group of modular substitutions is generated by the mappings

$$\tau \longmapsto \tau + 1 \quad \text{and} \quad \tau \longmapsto -1/\tau \tag{0.5}$$

according to Corollary II.2.3 (0.4) can be replaced by the two conditions

$$f(\tau + 1) = f(\tau) \quad \text{and} \quad f(-1/\tau) = \tau^k \cdot f(\tau). \tag{0.6}$$

In I, §4 we saw that the EISENSTEIN series E_k are examples of such functions. We shall sketch another example which has a transformation behavior analogously to (0.6).

3. The classical theta series. In his letters to GOLDBACH on May 5,1748 and August 17,1750 L. EULER already treated the *theta series*

$$\vartheta(\tau) := \sum_{n \in \mathbb{Z}} e^{\pi i n^2 \tau} = 1 + 2 \cdot \sum_{n=1}^{\infty} q^{n^2} \quad \text{with} \quad q := e^{\pi i \tau} \quad \text{and} \quad \tau \in \mathbb{H}. \tag{0.7}$$

In connection with the heat equation, the theta series appears in J. FOURIER's *Théorie Analytique de la Chaleur* (Paris 1822) (I, §6). In the legacy of C.F. GAUSS ([31], vol. III, 436–445) a note from about 1808 was found, in which a more general series (namely the one in I(6.30)) is considered and a transformation formula is already proved for it. Then, in the *Fundamenta nova* by C.G.J. JACOBI ([42], vol. I, 198–239), the general series

$$\sum_{n \in \mathbb{Z}} (-1)^n q^{n^2} \cos(2nx)$$

is introduced under the letter Θ and is used to construct the elliptic functions. In the notation of I, §6 the function $\vartheta(\tau)$ is equal to the *zero value* $\vartheta(0; \tau)$, also called *Theta Nullwert*.

Obviously, $\vartheta(\tau)$ is absolutely and uniformly convergent on compact subsets of $\mathbb{H}$, such that $\vartheta : \mathbb{H} \to \mathbb{C}$ is holomorphic. Furthermore,

$$\vartheta(\tau + 2) = \vartheta(\tau) \quad \text{for} \quad \tau \in \mathbb{H} \tag{0.8}$$

holds. The importance and interest that the theta series has always found are due to the so-called

0.1 Theta Transformation Formula:

$$\vartheta(-1/\tau) = \sqrt{\tau/i} \cdot \vartheta(\tau) \quad \text{for all} \quad \tau \in \mathbb{H}. \tag{0.9}$$

Here, choose the branch of the root that is positive for positive arguments.

III Modular forms

Proof Consider $y > 0$ and the function

$$\varphi : \mathbb{C} \to \mathbb{C}, \quad w \mapsto \sum_{n \in \mathbb{Z}} e^{-\pi(n+w)^2 y},$$

which is holomorphic and periodic modulo 1. Due to Theorem A.4, φ possesses a FOURIER series expansion

$$\vartheta(iy) = \varphi(0) = \sum_{m \in \mathbb{Z}} \alpha_m,$$

$$\alpha_m := \int_0^1 \varphi(t) \cdot e^{-2\pi i m t} \, dt = \int_{-\infty}^{\infty} e^{-\pi t^2 y} \cdot e^{-2\pi i m t} \, dt$$

$$= e^{-\pi m^2/y} \int_{-\infty}^{\infty} e^{-\pi(t\sqrt{y} + im/\sqrt{y})^2} \, dt = \frac{1}{\sqrt{y}} e^{-\pi m^2/y},$$

if we apply Corollary A.14. Thus we get

$$\vartheta(iy) = \frac{1}{\sqrt{y}} \vartheta(i/y) \quad \text{for} \quad y > 0.$$

The Transformation Formula (0.9) is obtained by analytic continuation. □

If we now consider $f(\tau) := \vartheta^8(\tau)$, (0.8) and (0.9) lead to

$$f(\tau + 2) = f(\tau) \quad \text{and} \quad f(-1/\tau) = \tau^4 \cdot f(\tau). \tag{0.10}$$

Thus we know the transformation behavior of $f = \vartheta^8$ under the group generated by the modular substitutions $\tau \mapsto \tau + 2$ and $\tau \mapsto -1/\tau$. With the *theta group* Γ_ϑ as in II, §3 we get

$$f(M\tau) = (c\tau + d)^4 \cdot f(\tau) \quad \text{for all} \quad M \in \Gamma_\vartheta.$$

Another approach – also suggested by II, §3 – yields a function, whose transformation behavior under all modular substitutions is known: we set

$$g(\tau) := \frac{1}{\tau} \cdot \vartheta^2(\tau) \cdot \vartheta^2(\tau + 1) \cdot \vartheta^2 (1 - 1/\tau) \tag{0.11}$$

and verify

$$g(\tau + 1) = i \cdot g(\tau) \quad \text{and} \quad g(-1/\tau) = i \cdot \tau^3 \cdot g(\tau),$$

using (0.8) and (0.9). Therefore, the fourth power of g is a modular form of weight 12 in the sense of (0.6) with $k = 12$. Compare Proposition 4.17.

4. Generating functions. One can interpret modular forms as generating functions of particular arithmetic functions, i.e. sequences. Given a complex sequence $(\alpha_n)_{n \geq 0}$ consider

$$f(\tau) = \sum_{n=0}^{\infty} \alpha_n q^n, \quad q = e^{2\pi i \tau}, \quad \tau \in \mathbb{H}. \tag{0.12}$$

This series converges absolutely if for instance $(\alpha_n)_{n\geq 0}$ is of polynomial growth at most. Clearly $f(\tau + 1) = f(\tau)$ holds by construction.

The classical theta series $\vartheta(2\tau)$ fits into this framework when considering

$$\alpha_n = \#\{x \in \mathbb{Z}; x^2 = n\}.$$

The EISENSTEIN series provide another example via

$$\alpha_n = \sigma_{k-1}(n) = \sum_{d\mid n} d^{k-1}, \quad n \geq 1,$$

(cf. (2.3)).

The behavior of f under $\tau \mapsto -1/\tau$ is only valid for special sequences. This question will be solved in IV, §4.

The theory of modular forms yields arithmetic applications. First of all we show that the vector space of modular forms of fixed weight k has finite dimension. Hence we find examples where a modular form has two different representations (e.g. (4.8)). If we compare the coefficients of the generating functions we obtain arithmetical identities (cf. RAMANUJAN Congruence 4.9, Proposition 7.13, Four-Squares-Theorem Corollary 7.14).

§ 1 The elementary theory

The goal of this paragraph is to develop the elementary theory of modular forms and to derive initial structural statements.

We agree to reserve the letter M for 2×2 matrices and to always write $M = \begin{pmatrix} a & b \\ c & d \end{pmatrix}$. A point τ from the upper half plane $\mathbb{H}$ is always given in the form $\tau = x + iy$.

1. Modularity. First we define an operation of $\mathrm{SL}(2;\mathbb{R})$ on the space of meromorphic functions on $\mathbb{H}$. Let $k \in \mathbb{Z}$ and f be meromorphic on $\mathbb{H}$. Then there exists a discrete and relatively closed subset $D_f \subseteq \mathbb{H}$ such that f is holomorphic on $\mathbb{H} \setminus D_f$. For $M \in \mathrm{SL}(2;\mathbb{R})$, we define a function $f|M = f|_k M$, meromorphic on $\mathbb{H}$ by

$$(f \mid_k M)(\tau) = (f\mid M)(\tau) := (c\tau + d)^{-k} \cdot f(M\tau) \quad \text{for} \quad \tau \in \mathbb{H} \setminus D_{f \circ M}. \tag{1.1}$$

The letter $k \in \mathbb{Z}$ is usually omitted here. Together with II(1.5) a verification yields

$$(f\mid M)\mid N = f\mid(MN) \quad \text{for} \quad M, N \in \mathrm{SL}(2;\mathbb{R}). \tag{1.2}$$

Thus $(M, f) \mapsto f\mid M$ defines an operation of $\mathrm{SL}(2;\mathbb{R})$ on the space of meromorphic functions on $\mathbb{H}$, which is called the *slash operator*.

We now call f *modular of weight k*, if:

(M.1) f is meromorphic on $\mathbb{H}$.
(M.2) $f\mid_k M = f$ for all $M \in \Gamma$.

Obviously, the functions which are modular of weight k form a vector space over $\mathbb{C}$.

§ 1 The elementary theory

If we enter $M = -I$ into (M.2), we obtain the

1.1 Proposition. *Any function which is modular of odd weight is 0.*

Therefore, in the following it is always assumed that k is even. Because of (1.2) and Theorem II.2.2, we can replace (M.2) by

(M.2*) $\qquad f(\tau + 1) = f(\tau) \quad \text{and} \quad f(-1/\tau) = \tau^k \cdot f(\tau).$

Finally, II(1.7) shows that (1.1) can also be written in the form

$$(f|M)(\tau) = \left(\frac{dM\tau}{d\tau}\right)^{k/2} \cdot f(M\tau) \qquad (1.3)$$

2. Periodic functions. Let

$$\mathbb{D} := \{z \in \mathbb{C} \, ; \, |z| < 1\}$$

again be the open unit disk in $\mathbb{C}$. As is well-known,

$$\mathbb{H} \to \mathbb{D} \setminus \{0\}, \quad \tau \mapsto q := e^{2\pi i \tau},$$

is a surjective holomorphic function that is periodic mod 1. If f is meromorphic on $\mathbb{H}$ with pole set D_f and periodic with period 1 (cf. sect. I.1.2), then it is well-known (cf. Theorem on the FOURIER Series Development A.4) that there exists a function $\hat{f}$, meromorphic on $\mathbb{D} \setminus \{0\}$, satisfying

$$f(\tau) = \hat{f}\left(e^{2\pi i \tau}\right) \quad \text{for} \quad \tau \in \mathbb{H} \setminus D_f. \qquad (1.4)$$

Conversely, if $\hat{f}$ is a function which is meromorphic on $\mathbb{D} \setminus \{0\}$, then the corresponding function f of the form (1.4) is meromorphic on $\mathbb{H}$ and periodic with period 1. Note here that the poles of $\hat{f}$ can have 0 as a limit point. To avoid this, we say that f *has at most a pole* at ∞ if $\hat{f}$ has a meromorphic continuation on $\mathbb{D}$.

Now if f has at most a *pole at* ∞ in this sense, then 0 is an isolated singularity of $\hat{f}$ and there exists a LAURENT series development of $\hat{f}$ around 0 with finite principal part (cf. A.3):

$$\hat{f}(q) = \sum_{m \geq m_0} \alpha_f(m) \cdot q^m, \qquad (1.5)$$

which converges absolutely and uniformly on compact subsets of a punctured neighborhood of 0. Then, due to (1.4), f possesses a FOURIER series development

$$f(\tau) = \sum_{m \geq m_0} \alpha_f(m) \cdot e^{2\pi i m \tau} \qquad (1.6)$$

for $\operatorname{Im} \tau > \gamma$ with suitable $\gamma > 0$. The integral formula for the coefficients of the LAURENT series translates into

$$\alpha_f(m) = \int_w^{w+1} f(\tau) \cdot e^{-2\pi i m \tau} d\tau, \tag{1.7}$$

where the integration for $\operatorname{Im} w > \gamma$ is carried out along the line from w to $w + 1$.

1.2 Lemma. *Given a function $f \not\equiv 0$, meromorphic on $\mathbb{H}$ and periodic modulo 1 the following assertions are equivalent:*

(i) *f has at most a pole at ∞.*
(ii) *There exists some $\gamma > 0$ with the following properties:*

    (a) *f is holomorphic on the domain $\{\tau \in \mathbb{H} \,;\, \operatorname{Im} \tau > \gamma\}$.*
    (b) *There is some $m_0 \in \mathbb{Z}$ such that, for every $\varepsilon > 0$, there is a constant C with*

$$|f(\tau)| \le C \cdot e^{-2\pi m_0 \cdot \operatorname{Im} \tau} \quad \text{for all} \quad \tau \quad \text{with} \quad \operatorname{Im} \tau \ge \gamma + \varepsilon,$$

where m_0 is the minimum of $m \in \mathbb{Z}$ with $\alpha_f(m) \ne 0$.

If this is the case and f is holomorphic on $\mathbb{H}$, then (b) holds for all $\gamma > 0$.

Proof Everything depends on the behavior of $\hat{f}$ in a neighborhood of 0: $\hat{f}$ has a LAURENT series development of the form (1.5) if and only if an estimate of the form $|\hat{f}(q)| \le C \cdot |q|^{m_0}$ holds. □

If m_0 is chosen as in Lemma 1.2, then according to the behavior of $\hat{f}$ at 0, we say that f

$$\begin{aligned} \text{has a pole of order } -m_0 &\quad \text{at} \quad \infty, \quad \text{if} \quad m_0 < 0, \\ \text{is holomorphic} &\quad \text{at} \quad \infty, \quad \text{if} \quad m_0 \ge 0, \\ \text{has a root of order } m_0 &\quad \text{at} \quad \infty, \quad \text{if} \quad m_0 > 0. \end{aligned}$$

We naturally set

$$\operatorname{ord}_\infty f := m_0. \tag{1.8}$$

3. Meromorphic modular forms. A function f on $\mathbb{H}$ is called a *meromorphic modular form of the weight k* if f is meromorphic and modular of weight k and has at most a pole at ∞. This means

(M.1) f is meromorphic on $\mathbb{H}$.
(M.2) $f|_k M = f$ for all $M \in \Gamma$.
(M.3) f has at most a pole at ∞.

Because of Lemma 1.2, (M.3) can be replaced here by

(M.3*) f has a FOURIER series expansion of the form:

$$f(\tau) = \sum_{m \ge m_0} \alpha_f(m) \cdot e^{2\pi i m \tau},$$

§ 1 The elementary theory

which, for suitable $\gamma > 0$, converges absolutely and uniformly on compact subsets of $\{\tau \in \mathbb{H};\ \operatorname{Im}\tau > \gamma\}$.

Since $\alpha f + \beta g$, $\alpha, \beta \in \mathbb{C}$, and $f \cdot g$ have at most a pole at ∞ if f and g do, a look at (1.2) shows that *the set of meromorphic modular forms of weight k forms a vector space $\mathbb{V}_k$ over $\mathbb{C}$*. It also holds that

$$\mathbb{V}_k \cdot \mathbb{V}_\ell \subseteq \mathbb{V}_{k+\ell} \qquad \text{for} \quad k, \ell \in \mathbb{Z}. \tag{1.9}$$

Because of Proposition 1.1, of course,

$$\mathbb{V}_k = \{0\} \quad \text{for odd} \quad k. \tag{1.10}$$

Finally,

$$\frac{1}{f} \in \mathbb{V}_{-k} \qquad \text{for} \quad 0 \neq f \in \mathbb{V}_k. \tag{1.11}$$

A meromorphic modular form of weight 0 is called a *modular function*. Because of (1.9) and (1.11) the set

$$\mathbb{K} := \mathbb{V}_0 \tag{1.12}$$

of all modular functions is a field containing the constant functions and all quotients of modular forms of the same weight.

1.3 Remarks. a) The name *modular function* is due to R. DEDEKIND [16], vol. I, 159–172. The first occurrence of a modular function is the *modulus* of elliptic integrals in LEGENDRE's normal form. However, this is a function, which does not belong to the modular group Γ, but to the principal congruence group $\Gamma[2]$ (cf. Introduction of Chapter I, Corollary I.3.20, Exercise I. 4.18 7)). Modular forms first appear systematically by WEIERSTRASS in his lectures on elliptic functions as the invariants g_2 and g_3 and as Δ, respectively (cf. I, §4), which were considered previously by G. EISENSTEIN (cf. I, §7).

The actual founders of the theory of modular functions are F. KLEIN and H. POINCARÉ. The main works of POINCARÉ [64] on this subject can be found in the first volumes of *Acta Mathematica* (namely in vol. **1, 3, 4, 5**) and in his *Œuvres II*. Important works of KLEIN can be found in [43], vol. III. The theory was then developed by R. FRICKE, in the detailed survey [28].

b) R. DEDEKIND used the abbreviated notation $1^\tau := e^{2\pi i\tau}$ ([16], vol. I, 174–201), which has sometimes been used later, but has not become generally accepted.

4. Modular forms. A function f, which is modular of weight k, is called a *modular form of weight k*, if f is holomorphic on $\mathbb{H}$ and at ∞. Thus f is a modular form of weight k if and only if:

(M.1') $f : \mathbb{H} \to \mathbb{C}$ is holomorphic.
(M.2') $f|_k M = f$ for all $M \in \Gamma$.
(M.3') f is bounded on $\{\tau \in \mathbb{H};\ \operatorname{Im}\tau \geq \gamma\}$ for all $\gamma > 0$.

Because of Lemma 1.2, the condition (M.3') can be replaced by

(M.3") f has a FOURIER series expansion of the form

$$f(\tau) = \sum_{m \geq 0} \alpha_f(m) \cdot e^{2\pi i m \tau},$$

which converges absolutely and uniformly on each set $\{\tau \in \mathbb{H} ;\ \mathrm{Im}\,\tau \geq \gamma\}$, $\gamma > 0$.

The set $\mathbb{M}_k$ of all modular forms of weight k is obviously a subspace of the vector space $\mathbb{V}_k$. A modular form f is called a *cusp form* (German: Spitzenform) if f has a root at ∞, i.e. $\alpha_f(0) = 0$. The vector space of cusp forms of weight k is denoted by $\mathbb{S}_k$. We obviously have

$$\mathbb{S}_k \subseteq \mathbb{M}_k \subseteq \mathbb{V}_k \quad \text{for all } k \in \mathbb{Z}.$$

From (M.3") it follows immediately that

$$\alpha_f(0) = \lim_{y \to \infty} f(iy) \quad \text{for} \quad f \in \mathbb{M}_k. \tag{1.13}$$

In view of (1.9), we obtain

$$\mathbb{M}_k \cdot \mathbb{M}_\ell \subseteq \mathbb{M}_{k+\ell}, \quad \mathbb{S}_k \cdot \mathbb{M}_\ell \subseteq \mathbb{S}_{k+\ell} \quad \text{for} \quad k, \ell \in \mathbb{Z}. \tag{1.14}$$

Because of its importance, we repeat a special case of Lemma 1.2 as

1.4 Lemma. *If $f \in \mathbb{M}_k$ and $\gamma > 0$, then there exists some $C = C(f, \gamma) > 0$ such that*

$$|f(\tau) - \alpha_f(0)| \leq C \cdot e^{-2\pi \mathrm{Im}\,\tau} \quad \text{as well as} \quad |f'(\tau)| \leq C \cdot e^{-2\pi \mathrm{Im}\,\tau}$$

on the set $\{\tau \in \mathbb{H} ;\ \mathrm{Im}\,\tau \geq \gamma\}$.

In the rest of the book we will only study the field $\mathbb{K} = \mathbb{V}_0$ of the modular functions and the vector spaces $\mathbb{M}_k$ of the modular forms of weight k.

5. Negative weight. Given $f \in \mathbb{M}_k$, let $\tilde{f} : \mathbb{H} \to \mathbb{R}$ be defined by

$$\tilde{f}(\tau) := (\mathrm{Im}\,\tau)^{k/2} \cdot |f(\tau)|, \quad \tau \in \mathbb{H}. \tag{1.15}$$

II(1.14) and (1.1) imply

$$\tilde{f}(M\tau) = \tilde{f}(\tau) \quad \text{for all} \quad M \in \Gamma, \tag{1.16}$$

such that $\tilde{f}$ is a function invariant under Γ. The precise knowledge of the exact fundamental domain $\mathbb{F}$ in II(2.7) leads to the

1.5 Proposition. *If $\varphi : \mathbb{H} \to \mathbb{R}$ is bounded for $\mathrm{Im}\,\tau \geq \frac{1}{2}\sqrt{3}$ and satisfies*

$$\varphi(M\tau) = \varphi(\tau) \quad \text{for all} \quad M \in \Gamma,$$

then φ is bounded on $\mathbb{H}$.

Proof According to II(2.9), φ is bounded on the fundamental domain $\mathbb{F}$. Now we can apply Theorem II.2.6 and see that φ is bounded on $\mathbb{H}$. □

As the first structural result, we get the

§1 The elementary theory

1.6 Proposition. $\mathbb{M}_k = \{0\}$ *holds for* $k < 0$.

Proof According to (M.3"), f is bounded for all τ with $\mathrm{Im}\,\tau \geq \frac{1}{2}\sqrt{3}$. Since k is negative, this is also true for $\tilde{f}$. Now we can apply Proposition 1.5 to $\varphi = \tilde{f}$ and conclude that $\tilde{f}$ is bounded on $\mathbb{H}$.

We develop f into a FOURIER series and obtain

$$\alpha_f(m) = e^{2\pi my} \cdot \int_0^1 f(x + iy) \cdot e^{-2\pi imx}\,dx,\ m \geq 0$$

for the FOURIER coefficients according to (1.7). It follows from (1.15) that

$$|\alpha_f(m)| \leq y^{-k/2} \cdot e^{2\pi my} \cdot \int_0^1 \tilde{f}(x + iy)dx \leq C \cdot y^{-k/2} \cdot e^{2\pi my}$$

with a constant C independent of y. Since the left-hand side also does not depend on y, we can take the limit $y \to 0$ and get $\alpha_f(m) = 0$ for all $m \geq 0$, i.e. $f \equiv 0$. □

6. Growth of FOURIER coefficients. Given $f \in \mathbb{M}_k$, let $\tilde{f}$ be defined as in (1.15).

1.7 Proposition. *Let* $k > 0$ *and* $f \in \mathbb{M}_k$.
a) $\tilde{f}$ *is bounded on* $\mathbb{H}$ *if and only if* $f \in \mathbb{S}_k$. *In this case, there exists some* $w \in \mathbb{H}$ *with the property*

$$\tilde{f}(\tau) \leq \tilde{f}(w)\quad \text{for all}\ \tau \in \mathbb{H}.$$

b) *If* $f \in \mathbb{S}_k$, *then*

$$\alpha_f(m) = O(m^{k/2})\quad \text{for all}\ m \in \mathbb{N}.$$

The last assertion means that there is some $C > 0$ such that

$$|\alpha_f(m)| \leq Cm^{k/2}\quad \text{for all}\quad m \in \mathbb{N}.$$

Proof a) If f is a cusp form, $\tilde{f}$ is bounded for $\mathrm{Im}\,\tau \geq \frac{1}{2}\sqrt{3}$ due to Lemma 1.4. The assertion now follows from Proposition 1.5. Conversely, if $\tilde{f}$ is bounded, then because of Lemma 1.4 $\alpha_f(0) \cdot y^{k/2}$ is also bounded in $\mathbb{F}$. From $k > 0$ we get $\alpha_f(0) = 0$, hence $f \in \mathbb{S}_k$. The existence of w follows from $\lim_{y \to \infty} \tilde{f}(\tau) = 0$ according to Lemma 1.4.
b) We develop f into a FOURIER series and

$$|\alpha_f(m)| \leq C \cdot y^{-k/2} \cdot e^{2\pi my}\quad \text{for}\quad y > 0$$

follows from the proof of Proposition 1.6. Since the left-hand side does not depend on y, we may set $y = 1/m$, $m > 0$, on the right-hand side and get

$$|\alpha_f(m)| \leq C \cdot e^{2\pi} \cdot m^{k/2}.$$

□

1.8 Exercises.
Let $A(\mathbb{H})$ denote the set of all holomorphic functions $f : \mathbb{H} \to \mathbb{C}$ which are periodic

modulo 1 and possess a FOURIER series of the form

$$f(\tau) = \sum_{m=0}^{\infty} \alpha_f(m) \cdot e^{2\pi i m \tau}, \quad \tau \in \mathbb{H},$$

as in Lemma 1.2. $A(\mathbb{H})$ is obviously a $\mathbb{C}$-algebra which contains all modular forms. Denote by $B(\mathbb{H})$ the subset of those $f \in A(\mathbb{H})$ for which there is some $\ell \geq 0$ with the property

$$\alpha_f(m) = O(m^\ell) \quad \text{for} \quad m \geq 1.$$

Given $f \in B(\mathbb{H})$, we set

$$\kappa(f) := \inf \left\{ \ell \geq 0 \,;\, \alpha_f(m) = O\left(m^\ell\right) \quad \text{for} \quad m \geq 1 \right\}.$$

1) $B(\mathbb{H})$ is a subalgebra of $A(\mathbb{H})$. In particular, for $f, g \in B(\mathbb{H})$:
a) $\kappa(f + g) \leq \max\{\kappa(f), \kappa(g)\}$,
b) $\kappa(f \cdot g) \leq \kappa(f) + \kappa(g) + 1$.

2) If $f \in A(\mathbb{H})$ and $(\operatorname{Im} \tau)^\kappa \cdot f(\tau)$ is bounded on $\mathbb{H}$ for some $\kappa \geq 0$, then $f \in B(\mathbb{H})$ with $\kappa(f) \leq \kappa$.

3) If $f \in \mathbb{S}_k$, then f belongs to $B(\mathbb{H})$ with $\kappa(f) \leq k/2$.

4) If $f \in B(\mathbb{H})$, then the DIRICHLET series

$$D_f(s) := \sum_{m=1}^{\infty} \alpha_f(m) \cdot m^{-s}$$

converges absolutely for $s \in \mathbb{C}$ with $\operatorname{Re} s > \kappa(f) + 1$.

5) The function $f(\tau) = \vartheta(2\tau)$ (cf. (0.7)) belongs to $B(\mathbb{H})$ with $\kappa(f) = 0$ and $D_f(s) = 2\zeta(2s)$.

6) $B(\mathbb{H})$ is a proper subalgebra of $A(\mathbb{H})$.

7) Given $f \in \mathbb{M}_k$, we define a function

$$f^* : \mathbb{H} \longrightarrow \mathbb{C}, \quad \tau \longmapsto \overline{f(-\bar{\tau})}.$$

Then $f^* \in \mathbb{M}_k$ and $(f^*)^* = f$. All the FOURIER coefficients of f are real if and only if $f^* = f$. All the FOURIER coefficients of f are purely imaginary if and only if $f^* = -f$.

8) $\mathbb{M}_k$ has a basis of modular forms with real FOURIER coefficients.

9) $f \in \mathbb{M}_k$ is a cusp form if and only if, for every $\gamma > 0$, there are positive constants α and β such that

$$|f(\tau)| \leq \alpha \cdot e^{-\beta y} \quad \text{for all} \quad \tau \in \mathbb{H} \quad \text{with} \quad y \geq \gamma.$$

10) Let $f \in \mathbb{M}_k$ and $\rho = \frac{1}{2}(1 + i\sqrt{3})$. We have $f(i) = 0$ if $k \not\equiv 0 \pmod{4}$ and $f(\rho) = 0$ if $k \not\equiv 0 \pmod{6}$.

11) If $f \in \mathbb{S}_k$ and $r \in \mathbb{Q}$, then

$$\lim_{y \downarrow 0} f(r + iy) = 0.$$

If $f \in \mathbb{M}_k$, $f \notin \mathbb{S}_k$, $k > 0$ and $r \in \mathbb{Q}$, then

$$\lim_{y \downarrow 0} |f(r + iy)| = \infty.$$

§ 2 Examples

1. The EISENSTEIN series. The classical examples of modular forms are the EISENSTEIN series

$$E_k(\tau) := {\sum_{m,n}}' (m\tau + n)^{-k} \quad \text{for } k \in \mathbb{Z}, \, k \geq 3 \tag{2.1}$$

as in I(1.18) and I(3.5). Here, the prime in the sum shall mean that the sum is to be extended over all pairs $(m, n) \in \mathbb{Z} \times \mathbb{Z}$ with $(m, n) \neq (0, 0)$. Although the essential properties of E_k have already been derived in I, §4, some of them shall be proved again in a different way.

2.1 Proposition. *For every compact subset $C \subseteq \mathbb{H}$, there are positive constants γ and δ satisfying*

$$\gamma \cdot |mi + n| \leq |m\tau + n| \leq \delta \cdot |mi + n|$$

for all $m, n \in \mathbb{R}$ and all $\tau \in C$.

Proof For homogeneity reasons, we may assume $m^2 + n^2 = 1$, thus $|mi + n| = 1$. Then the continuous function

$$(\tau, m, n) \mapsto |m\tau + n|$$

on the compact set $C \times \{(m, n) \in \mathbb{R} \times \mathbb{R} \, ; \, m^2 + n^2 = 1\}$ attains its minimum γ and its maximum δ. Since $\mathrm{Im}\,\tau$, $\tau \in C$, is bounded from below by a positive constant, $\gamma > 0$ holds. $\square$

In analogy with the Convergence Lemma I.1.17, we now obtain the

2.2 Convergence Lemma. *For $k \in \mathbb{Z}$, $k \geq 3$, E_k is absolutely and uniformly convergent on compact subsets of $\mathbb{H}$. Thus E_k is a holomorphic function on $\mathbb{H}$.*

Proof In view of Proposition 2.1, only the convergence of

$$ {\sum_{m,n}}' (m^2 + n^2)^{-\alpha} \quad \text{for } \alpha > 1 $$

needs to be proved. Probably the simplest proof of this fact was already given by WEIERSTRASS [88], vol. V, 117: using of the inequality $m^2 + n^2 \geq |mn|$, we get

$$\sum_S (m^2 + n^2)^{-\alpha} \leq 4 \sum_{m \geq 1} m^{-2\alpha} + 4\left(\sum_{m \geq 1} m^{-\alpha}\right)^2 = 4(\zeta(2\alpha) + \zeta^2(\alpha)) < \infty$$

for $\alpha > 1$ and any finite subset S of $(\mathbb{Z} \times \mathbb{Z}) \setminus \{(0,0)\}$. □

A simple consequence (cf. I(4.15)) is the following transformational behavior of the EISENSTEIN series.

2.3 Transformation Lemma. *For $k \in \mathbb{Z}$, $k \geq 3$ and $M \in \Gamma$, we have $E_k|_k M = E_k$, i.e.*

$$E_k(M\tau) = (c\tau + d)^k \cdot E_k(\tau), \quad \tau \in \mathbb{H}. \tag{2.2}$$

Proof Note that

$$(m \cdot M\tau + n) \cdot (c\tau + d) = m'\tau + n' \quad \text{with} \quad (m', n') = (m, n) \cdot M$$

According to the Equivalence Theorem I.1.10, (m', n') also runs through all pairs of integers exactly once, as (m, n) does. □

With $M = -I$ we get the

2.4 Corollary. $E_k = 0$ *holds for odd $k \geq 3$.*

We quote Theorem I.4.3, the proof of which has no direct reference to the theory of elliptic functions: for even $k \geq 4$,

$$E_k(\tau) = 2\zeta(k) + 2 \cdot \frac{(2\pi i)^k}{(k-1)!} \cdot \sum_{m=1}^{\infty} \sigma_{k-1}(m) \cdot e^{2\pi i m\tau}. \tag{2.3}$$

In addition to the series E_k it is convenient to consider the *normalized* EISENSTEIN series

$$E_k^* := \frac{1}{2\zeta(k)} \cdot E_k, \quad k \geq 4 \text{ even.} \tag{2.4}$$

Since every pair of integers $(m, n) \in (\mathbb{Z} \times \mathbb{Z}) \setminus \{(0,0)\}$ can be uniquely expressed in the form $(m, n) = (t\mu, t\nu)$ with $t \in \mathbb{N}$ and coprime $\mu, \nu \in \mathbb{Z}$, we also have

$$E_k^*(\tau) = \frac{1}{2} \cdot \sum_{\gcd(m,n)=1} (m\tau + n)^{-k}, \quad k \geq 4 \text{ even} \tag{2.4'}$$

Let us derive another representation of E_k^*. For this, let

$$\Gamma_\infty := \{\pm T^n \ ; \ n \in \mathbb{Z}\} = \{M \in \Gamma \ ; \ c = 0\}. \tag{2.5}$$

"$M : \Gamma_\infty \setminus \Gamma$" means that M runs through a system of representatives $\mathcal{R}$ of the right cosets of Γ_∞, i.e.

$$\bigcup_{M \in \mathcal{R}} \Gamma_\infty M = \Gamma \quad \text{and} \quad \Gamma_\infty M \neq \Gamma_\infty N \quad \text{for } M, N \in \mathcal{R} \text{ with } M \neq N.$$

Using the Completion Lemma II.2.1, it follows that

§ 2 Examples

$$E_k^*(\tau) = \sum_{M:\Gamma_\infty \backslash \Gamma} 1|_k M(\tau) = \sum_{M:\Gamma_\infty \backslash \Gamma} (c\tau + d)^{-k} \quad \text{for even } k \geq 4. \tag{2.6}$$

We now use EULER's *formula* (cf. B.5)

$$2\zeta(k) = -\frac{(2\pi i)^k}{k!} \cdot B_k, \quad k \geq 2 \text{ even}, \tag{2.7}$$

where the first BERNOULLI *numbers* B_k are given by (cf. B.4)

k	2	4	6	8	10	12	14
B_k	1/6	−1/30	1/42	−1/30	5/66	−691/2730	7/6
$-2k/B_k$	−24	240	−504	480	−264	65520/691	−24

(2.8)

Then (2.3) leads to

$$E_k^*(\tau) = 1 - \frac{2k}{B_k} \cdot \sum_{m=1}^{\infty} \sigma_{k-1}(m) \cdot e^{2\pi i m \tau}, \quad k \geq 4 \text{ even}. \tag{2.9}$$

This means that the FOURIER coefficients of E_k^* are rational numbers with bounded denominators. The least common denominator is the denominator of $2k/B_k$. In particular we have

$$E_4^*(\tau) = 1 + 240 \cdot \sum_{m=1}^{\infty} \sigma_3(m) \cdot e^{2\pi i m \tau}, \tag{2.10}$$

$$E_6^*(\tau) = 1 - 504 \cdot \sum_{m=1}^{\infty} \sigma_5(m) \cdot e^{2\pi i m \tau}. \tag{2.11}$$

Having made these preparations we see out that the vector space of modular forms for even weight $k \geq 4$ is not trivial:

2.5 Proposition. *For even $k \geq 4$ we have*:
a) $E_k \in \mathbb{M}_k$.
b) $\mathbb{M}_k = \mathbb{C} \cdot E_k \oplus \mathbb{S}_k = \mathbb{C} \cdot E_k^* \oplus \mathbb{S}_k$.

Proof a) We use the Convergence Lemma 2.2 as well as (2.2) and (2.3).
b) If $f \in \mathbb{M}_k$, then $f - \alpha_f(0) \cdot E_k^*$ is a cusp form. □

2.6 Corollary. *For even $k \geq 4$ and $f \in \mathbb{M}_k$, we have*

$$\alpha_f(m) = -\alpha_f(0) \cdot \frac{2k}{B_k} \cdot \sigma_{k-1}(m) + O\left(m^{k/2}\right), \quad m \in \mathbb{N}. \tag{2.12}$$

Proof We use part b) of Proposition 2.5, (2.9) and part b) of Proposition 1.7. □

Because of

$$m^r \leq \sigma_r(m) = \sum_{d|m} d^r = m^r \sum_{d|m} d^{-r} \leq \zeta(r) \cdot m^r$$

for $r > 1$, we have

$$\alpha_f(m) = O(m^{k-1}) \quad \text{for all} \quad f \in \mathbb{M}_k, \ k \geq 4 \text{ even}. \tag{2.13}$$

This estimate cannot be improved for $f \notin \mathbb{S}_k$.

The EISENSTEIN series have certain forced roots.

2.7 Roots Lemma. Let $k \geq 4$ be even.
a) $E_k(i) = 0$ holds for $k \not\equiv 0 \pmod 4$. In particular we have $E_6(i) = 0$.
b) $E_k(\rho) = 0$ holds for $k \not\equiv 0 \pmod 6$. In particular $E_4(\rho) = 0$, $\rho := \frac{1}{2}(1 + i\sqrt{3})$.

Proof We have $i(\mathbb{Z}i + \mathbb{Z}) = \mathbb{Z}i + \mathbb{Z}$ and $\rho(\mathbb{Z}\rho + \mathbb{Z}) = \mathbb{Z}\rho + \mathbb{Z}$ because $\rho^2 = \rho - 1$. Thus $E_k(i) = i^{-k} \cdot E_k(i)$ and analogously $E_k(\rho) = \rho^{-k} \cdot E_k(\rho)$. □

We can describe the location of the roots little bit more detailed. According to a result of F.K.C. RANKIN and H.P.F. SWINNERTON–DYER (*Bull. Lond. Math. Soc.* **2**, 169–170 (1970)), for even $k > 2$ the roots of E_k in $\mathbb{F}$ lie on the unit circle (cf. Exercise 4.27 15)). Also compare B. SCHOENEBERG [71], III.1.6.

2. The discriminant Δ has already been introduced in I(3.17) and is explained in I, §4 in more detail. According to I(4.19), it is given by

$$\Delta := (60E_4)^3 - 27(140E_6)^2. \tag{2.14}$$

The Transformation Lemma 2.3 immediately yields

$$\Delta(M\tau) = (c\tau + d)^{12} \cdot \Delta(\tau) \quad \text{for all} \quad \tau \in \mathbb{H} \quad \text{and} \quad M \in \Gamma \tag{2.15}$$

and Theorem I.4.6 shows that Δ has a FOURIER series expansion of the form

$$\Delta(\tau) = (2\pi)^{12} \cdot \sum_{m=1}^{\infty} \tau(m) \cdot e^{2\pi i m \tau}, \ \tau \in \mathbb{H}, \tag{2.16}$$

where all coefficients $\tau(m)$ are integers and $\tau(1) = 1$. We have given a short table of $\tau(m)$ in I, §4.

From (2.14), (2.15) and (2.16), we get the important

2.8 Theorem. $\Delta \in \mathbb{S}_{12}$.

As the most important application of the theory of elliptic functions, we note that $\Delta(\tau) \neq 0$ according to Corollary I.3.18 as well as the product expansion of I(6.24). We will not use these facts here, but give new proofs in Proposition 4.3 and Corollary 6.6. It is again convenient to denote a normalized version separately. In view of (2.16), we define the *normalized discriminant* by

$$\Delta^*(\tau) := (2\pi)^{-12} \cdot \Delta(\tau) = \sum_{m=1}^{\infty} \tau(m) \cdot e^{2\pi i m \tau}, \tau \in \mathbb{H}. \tag{2.17}$$

Using (2.4), (2.10) and (2.11), we easily verify

$$\Delta^* = \frac{1}{1728}(E_4^{*3} - E_6^{*2}). \tag{2.18}$$

If we formally multiply out the FOURIER series in (2.18), the coefficients are evidently integers. We are led to ask, whether a "reasonably" simple arithmetic description of the coefficients $\tau(m)$ exists. However, up to now, such a description has not been found. After extensive numerical calculations (beyond $m = 10^{15}$), D.H. LEHMER (*Duke Math. J.* **14**, 429–433 (1947)) expressed the conjecture that $\tau(m) \neq 0$ for all m. So far, no proof of this LEHMER *conjecture* is known. Recent approaches concentrate on an analog of the LEHMER conjecture for powers of the DEDEKIND η-function (cf. B. HEIM, M. NEUHAUSER, F. RUPP, *Ramanujan J.* **48**, 1–11 (2019)).

3. Srinivasa RAMANUJAN (1887–1920) was born in a small town near Chennai (the former Madras) in southern India. According to G.H. HARDY, he was only "half–educated" and remained without a better education. His scientific development, however, was so unique that it cannot be compared with that of any other mathematician: he learned the basics of calculus from textbooks. Around 1903 he started to write down his discoveries (without proofs) in his *notebooks*, until about 1910 he lived completely alone except for his mathematics.

In a letter to G.H. HARDY he shared some of his results. Gradually, he transmitted about 120 theorems, mostly formal identities and (partly false or incorrectly formulated) statements on prime number theory to HARDY. Some of the formulas he gave were complicated and appeared profound. On the invitation of HARDY, RAMANUJAN visited Cambridge from 1914 to 1917, and an extremely fruitful collaboration between the two began. RAMANUJAN died in his hometown at the age of 32.

On the basis of numerical calculations (he gives the values of $\tau(m)$ for $m \leq 30$), RAMANUJAN conjectured that the number-theoretic function $\tau(m)$ is *multiplicative*, i.e. that
$$\tau(mn) = \tau(m) \cdot \tau(n) \quad \text{for all coprime } m, n \in \mathbb{N}.$$

This was confirmed shortly afterwards by L.J. MORDELL (*Proc. Cambridge Phil. Soc.* **19**, 117–124 (1920)). We give a proof by E. HECKE in Theorem IV.1.10.

RAMANUJAN discovered or rather conjectured a series of congruence properties of the coefficients $\tau(m)$. For example,
$$\tau(m) \equiv \sigma_{11}(m) \pmod{691} \text{ for all } m \in \mathbb{N}$$

(cf. 4.9). He conjectured congruences of the form
$$\tau(7m + 3) \equiv 0 \pmod{7} \quad \text{and} \quad \tau(23m + k) \equiv 0 \pmod{23} \quad \text{for } m \in \mathbb{N},$$
if k is a so-called quadratic non-residue modulo 23.

The most spectacular conjecture of RAMANUJAN still carries his name, although it has now been proved by P. DELIGNE (*Publ. Math. IHES* **43**, 273–307 (1974)) as a special case of a much more general statement:

2.9 RAMANUJAN Conjecture: *All primes p satisfy*

$$|\tau(p)| \leq 2p^{11/2}.$$

The arithmetic function $\tau(m)$ is often called the RAMANUJAN *tau function*.

References: G. ANDREWS et al: *Ramanujan revisited.* Academic Press, Boston 1988. B. C. BERNDT: *Ramanujan's notebooks I – IV.* Springer-Verlag, New York 1985–1994. B. C. BERNDT, *Math. Intel.* **10**, No. 3, 24–29 (1988). G.H. HARDY: *Ramanujan.* 3rd ed., Chelsea, New York 1978. K. G. RAMANATHAN, *J. Indian Math. Soc.* **51**, 1–25 (1987). S. RAMANUJAN: *Collected Papers.* Chelsea, New York 1962. S. RAMANUJAN: *Notebook I and II*, Tata Institute of Fundamental Research, Bombay 1957.

4. The absolute invariant j is defined according to I(3.18) or I(4.25) as the quotient of two modular forms of weight 12:

$$j := (720\, E_4)^3/\Delta = E_4^{*3}/\Delta^*. \tag{2.19}$$

j is certainly meromorphic (and in view of $\Delta(\tau) \neq 0$, even holomorphic) on $\mathbb{H}$. The Transformation Lemma 2.3, together with (2.15), leads to

$$j(M\tau) = j(\tau) \quad \text{for all } M \in \Gamma \text{ and } \tau \in \mathbb{H}. \tag{2.20}$$

As stated in Theorem I.4.8, (2.19) leads to a FOURIER series expansion of the form

$$j(\tau) = e^{-2\pi i \tau} + \sum_{m=0}^{\infty} j_m \cdot e^{2\pi i m \tau}, \quad \tau \in \mathbb{H}, \tag{2.21}$$

with $j_m \in \mathbb{Z}$. Thus we also have

$$\overline{j(-\bar{\tau})} = j(\tau).$$

In summary we get the

2.10 Theorem. *j is a modular function, i.e.*

$$j \in \mathbb{K} = \mathbb{V}_0.$$

The Roots Lemma 2.7 yields

$$j(i) = 12^3 = 1728, \quad j(\rho) = 0. \tag{2.22}$$

By Theorem I.6.18, the j_m are positive integers. We give a short table:

§ 2 Examples

m	j_m	prime factorization
0	744	$2^3 \cdot 3 \cdot 31$
1	196 884	$2^2 \cdot 3^3 \cdot 1\,823$
2	21 493 760	$2^{11} \cdot 5 \cdot 2\,099$
3	864 299 970	$2 \cdot 3^5 \cdot 5 \cdot 355\,679$
4	20 245 856 256	$2^{14} \cdot 3^3 \cdot 45\,767$
5	333 202 640 600	$2^3 \cdot 5^2 \cdot 2\,143 \cdot 777\,421$
6	4 252 023 300 096	$2^{13} \cdot 3^6 \cdot 11 \cdot 13^2 \cdot 383$
7	44 656 994 071 935	$3^3 \cdot 5 \cdot 7 \cdot 271 \cdot 174\,376\,673$
8	401 490 886 656 000	$2^{17} \cdot 3 \cdot 5^3 \cdot 199 \cdot 41\,047$
9	3 176 440 229 784 420	$2^2 \cdot 3^7 \cdot 5 \cdot 4\,723 \cdot 15\,376\,021$
10	22 567 393 309 593 600	$2^{12} \cdot 3^5 \cdot 5^2 \cdot 13^2 \cdot 5\,366\,467$

A. van Wijngaarden has calculated j_m up to $m = 100$ (*Indag. Math.* **15**, 389–400 (1953)). The obviously rapid growth is confirmed by a result independently proved by H. Petersson (*Acta Math.* **58**, 169–215 (1932)) and H. Rademacher (*Amer. J. Math.* **61**, 237–248 (1939)): the asymptotic formula is

$$j_m \sim \frac{e^{4\pi\sqrt{m}}}{\sqrt{2} \cdot m^{3/4}} \quad \text{for} \quad m \to \infty.$$

The numbers j_m satisfy a number of interesting congruences: D.H. Lehmer discovered the congruence

$$(m+1) \cdot j_m \equiv 0 \pmod{24} \quad \text{for } m \geq 1.$$

in 1942 (*Amer. J. Math.* **64**, 488–502). Furthermore, we have

2.11 Lehner's Theorem. *If* $m \equiv 0 \pmod{2^a 3^b 5^c 7^d}$, *then*

$$j_m \equiv 0 \pmod{2^{3a+8} 3^{2b+3} 5^{c+1} 7^d}.$$

For a *proof*, see J. Lehner [55], chap. XI.4, or T.M. Apostol [5], chap. 4.

2.12 Remark. The function j was introduced by R. Dedekind [16], vol. I, 174–201, in 1877 under the name *Valenz*. Dedekind used the Riemann Mapping Theorem in order to prove that there is a uniquely determined "analytic" function $v\,[=j]$, which bijectively maps the exact fundamental domain $\mathbb{F}$ for the modular group to $\mathbb{C}$. Compare Corollary 5.4. The name *absolute invariant* and the now usual definition (2.19) go back to F. Klein (*Math. Ann.* **14**, 111–172 (1879)) The Fourier coefficients of modular forms are of polynomial growth due to Corollary 2.6, which is not true for the j-function.

5*. Monstrous Moonshine. In connection with the classification of finite simple groups, B. Fischer and R.L. Griess Jr. independently conjectured the existence of the largest sporadic group $\mathcal{F}$ of the order

$$2^{46} \cdot 3^{20} \cdot 5^9 \cdot 7^6 \cdot 11^2 \cdot 13^3 \cdot 17 \cdot 19 \cdot 23 \cdot 29 \cdot 31 \cdot 41 \cdot 47 \cdot 59 \cdot 71 \sim 808 \cdot 10^{51}$$

in 1973. This group quickly became known as the *monster*. A calculation of the dimensions χ_ν of the irreducible representations gave

ν	χ_ν
1	1
2	196 883
3	21 296 876
4	842 609 326
5	18 538 750 076

J. MCKAY noted the first of the following curious connections with the coefficients j_m of the absolute invariant:

$$j_1 = \chi_2 + \chi_1,$$
$$j_2 = \chi_3 + \chi_2 + \chi_1,$$
$$j_3 = \chi_4 + \chi_3 + 2\chi_2 + 2\chi_1.$$

J.G. THOMPSON (*Bull. Lond. Math. Soc.* **11**, 352–353) asked in 1979 whether there is a an analytic description of j_m as a dimension of a vector space V_m for which homomorphisms $\mathcal{F} \to \mathrm{GL}(V_m)$ exist, which would explain this fact. R.L. GRIESS JR. gave a first construction of the *Friendly Giant* $\mathcal{F}$ as a subgroup of the orthogonal group of a 196 883–dimensional euclidean space (*Invent. Math.* **69**, 1–102) in 1983. I. FRENKEL, J. LEPOWSKY, and A. MEURMAN constructed a representation module of the monster with so-called vertex operators.

References: J. CONWAY and S. NORTON, *Bull. London Math. Soc.* **11**, 303–339 (1979). I. FRENKEL, J. LEPOWSKY and A. MEURMAN: *Vertex Operator Algebras and the Monster*. Academic Press, Boston 1988. G. HISS, *Jahresber. Deutsch. Math.-Ver.* **105**, 169-194 (2003). G. MASON, *Math. Ann.* **283**, 381–409 (1989).

2.13 Exercises.
For $\varepsilon > 0$, denote by

$$\mathcal{V}_\varepsilon := \{\tau \in \mathbb{H} \,;\, |x| \leq 1/\varepsilon,\, y \geq \varepsilon\}$$

the *vertical strip of height* ε in $\mathbb{H}$ (cf. II(2.16)).
1) a) If $\tau \in \mathbb{H}$ and $u \in \mathbb{R}$, we have $\frac{1}{\gamma} \cdot |\tau| \leq |\tau + u| \leq \gamma \cdot |\tau|$ with $\gamma = 1 + |u|/y$.
b) If $\varepsilon > 0$, there is some $\gamma > 0$ with the property

$$|m\tau + n| \geq \gamma \cdot |mi + n| \quad \text{for all } m, n \in \mathbb{R} \text{ and } \tau \in \mathcal{V}_\varepsilon.$$

2) The EISENSTEIN series E_k in (2.1) converges absolutely and uniformly in every vertical strip for even $k \geq 4$. Conclude that $\lim_{y \to \infty} E_k(iy) = 2\zeta(k)$.
3) Let $m \in \mathbb{Z}$ and $k \geq 4$ be even. Then the POINCARÉ *series of index m*,

$$Q_{k,m}(\tau) := \sum_{M:\Gamma_\infty \backslash \Gamma} e^{2\pi i m \cdot M\tau} \cdot (c\tau + d)^{-k} = \sum_{M:\Gamma_\infty \backslash \Gamma} \varphi|_k M(\tau),$$

§ 2 Examples   171

where $\varphi(\tau) := e^{2\pi i n m \tau}$, converges absolutely and uniformly in each vertical strip. We have $Q_{k,m} \in \mathbb{S}_k$. All FOURIER coefficients of $Q_{k,m}$ are real.

4) a) For $\tau \in \mathbb{H}$ and real $k > 1$, we have

$$\int_{-\infty}^{\infty} |\tau + u|^{-k} du = \sqrt{\pi} \cdot \frac{\Gamma((k-1)/2)}{\Gamma(k/2)} \cdot y^{1-k},$$

where $\Gamma(s)$ denotes the gamma function.

b) For $\tau \in \mathbb{H}$ and real $k > 1$ we have

$$\sum_{m \in \mathbb{Z}} |\tau + m|^{-k} \leq \gamma^k \cdot \int_{-\infty}^{\infty} |\tau + u|^{-k} du \quad \text{with } \gamma = 1 + 1/y.$$

5) For $M = \begin{pmatrix} a & b \\ c & d \end{pmatrix} \in \mathrm{SL}(2;\mathbb{R})$ and $\tau, w \in \mathbb{H}$, let

$$M\{\tau, w\} := (w + M\tau) \cdot (c\tau + d), \quad \widehat{M} := \begin{pmatrix} d & b \\ c & a \end{pmatrix} \in \mathrm{SL}(2;\mathbb{R}).$$

a) $M\{\tau, w\} = \widehat{M}\{w, \tau\}$.

b) For $\varepsilon > 0$ and even $k \geq 4$, the POINCARÉ *series in τ and w*,

$$P_k(\tau, w) := \sum_{M \in \Gamma} (M\{\tau, w\})^{-k},$$

converges absolutely and uniformly for $(\tau, w) \in \mathcal{V}_\varepsilon \times \mathcal{V}_\varepsilon$. For fixed $w \in \mathbb{H}$, we have $P_k(\cdot, w) \in \mathbb{S}_k$.

c) For even $k \geq 4$, we have

$$\overline{P_k(-\overline{\tau}, w)} = P_k(\tau, -\overline{w}) \quad \text{and} \quad P_k(\tau, w) = P_k(w, \tau).$$

All FOURIER coefficients of $P_k(\tau, iv)$ are real.

d) Let $k > 2$ and $w \in \mathbb{H}$. If k is not a multiple of the order of the stabilizer subgroup Γ_w (cf. II(2.13)), then $P_k(\cdot, w)$ vanishes identically.

6) If $k \geq 4$ is even, then a relation between the two types of POINCARÉ series can be described by the FOURIER series expansion of $P_k(\tau, w)$ with respect to w:

$$P_k(\tau, w) = 2 \cdot \frac{(2\pi i)^k}{(k-1)!} \cdot \sum_{m=1}^{\infty} m^{k-1} \cdot Q_{k,m}(\tau) \cdot e^{2\pi i m w}, \quad \tau, w \in \mathbb{H}.$$

7) For $m \geq 1$ and $k > 1$ the series

$$R_{k,m}(\tau) := \sum_{S \in \mathrm{Sym}(2;\mathbb{Z}), \det S = -m} (\alpha\tau^2 + 2\beta\tau + \gamma)^{-k}, \quad S = \begin{pmatrix} \alpha & \beta \\ \beta & \gamma \end{pmatrix}, \quad \tau \in \mathbb{H},$$

converges absolutely and uniformly on compact subsets of $\mathbb{H}$. We have $R_{k,m} \in \mathbb{S}_{2k}$. The FOURIER coefficients of $R_{k,m}(\tau)$ are all real.

8) For indefinite $S \in \mathrm{Sym}(2; \mathbb{R})$, let $\langle S \rangle := \{M^{tr} S M \; ; \; M \in \Gamma\}$. For $k > 1$, the series
$$R_k(\tau; \langle S \rangle) := \sum_{\binom{\alpha\ \beta}{\beta\ \gamma} \in \langle S \rangle} (\alpha \tau^2 + 2\beta\tau + \gamma)^{-k} \quad , \tau \in \mathbb{H},$$
converges absolutely and uniformly on all compact subsets of $\mathbb{H}$ and it satisfies $R_k(\cdot, \langle S \rangle) \in \mathbb{S}_{2k}$.

9) Let $\varphi : \mathbb{H} \to \mathbb{C}$ be holomorphic, bounded and periodic modulo 1. Show for even $k > 2$ that the series
$$P_k(\tau, \varphi) := \sum_{M : \Gamma_\infty \backslash \Gamma} \varphi |_k M(\tau)$$
converges absolutely and uniformly on vertical strips and belongs to $\mathbb{M}_k$. How do $P_k(\cdot, w)$ and $R_k(\cdot, \langle S \rangle)$, $S = \begin{pmatrix} 0 & 1 \\ 1 & 0 \end{pmatrix}$, from Exercise 5) and 8) fit under this construction principle?

10) For even $k \in \mathbb{Z}$, $\mathbb{V}_k = \mathbb{K}(j')^{k/2}$.

11) The FOURIER coefficients $\alpha_k(m)$ of E_k^*, $k \geq 4$ even, satisfy
$$\frac{(2\pi)^k}{(k-1)!} \frac{1}{\zeta(k)} m^{k-1} \leq |\alpha_k(m)| \leq \frac{(2\pi)^k}{(k-1)!} \frac{\zeta(k-1)}{\zeta(k)} m^{k-1}, \; m \in \mathbb{N}.$$

12) Let $\Phi(z, \tau)$ be a JACOBI form of weight k and index m (cf. sect. 4 in I, §8). Show that $\Phi(0, \tau)$ belongs to $\mathbb{M}_k$.

We can obtain a direct proof of the dimension formula for $\mathbb{M}_k$ from the results of §1 and §2 without the use of the weight formula in §3. This proof only uses $\Delta(\tau) \neq 0$ and is sketched in the following exercises:

13) $\Delta(\tau) \neq 0$ for all $\tau \in \mathbb{H}$ implies $\mathbb{S}_k = \Delta \cdot \mathbb{M}_{k-12}$ for even $k \geq 0$.

14) $\mathbb{S}_k = \{0\}$ for $k < 12$, $\mathbb{M}_0 = \mathbb{C}$, $\mathbb{M}_2 = \{0\}$ and $\mathbb{M}_k = \mathbb{C} \cdot E_k = \mathbb{C} \cdot E_k^*$ for $k = 4, 6, 8, 10$.

15) For even $k \geq 0$, we have
$$\dim_\mathbb{C} \mathbb{M}_k = \begin{cases} [\frac{k}{12}], & \text{if } k \equiv 2 \pmod{12}, \\ [\frac{k}{12}] + 1, & \text{if } k \not\equiv 2 \pmod{12}. \end{cases}$$

§3 The weight formula

1. Orders. Let $f \neq 0$ be a meromorphic modular form of weight k in the sense of §1, i.e. $f \in \mathbb{V}_k$. Then for $w \in \mathbb{H}$ there exists a LAURENT series expansion
$$f(\tau) = \sum_{m \geq r} \gamma(m) \cdot (\tau - w)^m, \; \gamma(r) \neq 0, \tag{3.1}$$

§ 3 The weight formula

which converges absolutely and uniformly on compact subsets of a punctured neighborhood of w. As in I, §2, we define the *order* of f at w by

$$\text{ord}_w f := r. \tag{3.2}$$

Thus, the function f has a root or a pole at w, depending on whether (3.2) is positive or negative.

3.1 Proposition. *If $f \not\equiv 0$ is meromorphic on $\mathbb{H}$, then*

$$\text{ord}_z f|_k M = \text{ord}_{Mz} f \quad \text{for all} \quad z \in \mathbb{H} \text{ and } M \in \text{SL}(2;\mathbb{R}).$$

In particular, for $0 \neq f \in \mathbb{V}_k$ we have

$$\text{ord}_z f = \text{ord}_{Mz} f \quad \text{for all} \quad z \in \mathbb{H} \text{ and } M \in \text{SL}(2;\mathbb{R}).$$

Proof As is well-known, (3.2) holds if there is a function g, holomorphic in a neighborhood of w, satisfying

$$f(\tau) = (\tau - w)^r \cdot g(\tau) \quad \text{and} \quad g(w) \neq 0.$$

Then, for $M \in \text{SL}(2;\mathbb{R})$ and $z := M^{-1}w$, the identities (1.1) and II(1.6) lead to

$$(f|_k M)(\tau) = (c\tau + d)^{-k} \cdot (M\tau - Mz)^r \cdot g(M\tau) = (\tau - z)^r \cdot h(\tau),$$
$$h(\tau) = (c\tau + d)^{-k-r} \cdot (cz + d)^{-r} \cdot g(M\tau).$$

But here, h is holomorphic in z with $h(z) = (cz + d)^{-k-2r} g(w) \neq 0$. □

The order of f at ∞ is defined in (1.8): according to (M.3*), f has a Fourier series expansion of the form

$$f(\tau) = \sum_{m \geq m_0} \alpha_f(m) \cdot e^{2\pi i m \tau}, \quad \alpha_f(m_0) \neq 0, \tag{3.3}$$

and we set

$$\text{ord}_\infty f := m_0. \tag{3.4}$$

Thus, $\text{ord}_w f$ is defined for all w in

$$\mathbb{F}^* := \mathbb{F} \cup \{\infty\}. \tag{3.5}$$

In addition for $w \in \mathbb{F}^*$, we define

$$\text{ord } w := \begin{cases} 2 & \text{for } w = i, \\ 3 & \text{for } w = \rho, \\ 1 & \text{otherwise.} \end{cases} \tag{3.6}$$

Thus, $2 \cdot \text{ord } w$ for $w \in \mathbb{F}$ is equal to the order of the stabilizer subgroup of w,

$$\Gamma_w = \{M \in \Gamma \ ; \ Mw = w\}$$

(cf. II(2.13)). Most non–trivial results from now on will be based directly or indirectly on the

3.2 Weight Formula. *Given* $0 \neq f \in \mathbb{V}_k$, $k \in \mathbb{Z}$ *even, then*

$$\sum_{w \in \mathbb{F}^*} \frac{1}{\operatorname{ord} w} \cdot \operatorname{ord}_w f = \frac{k}{12}.$$

Of course, this is a finite sum, since the roots and poles of f do not have a limit point in $\mathbb{F}^*$ because of (3.3). The *proof* will be given in the following sections of this paragraph.

2. The contour integral. Let $0 \neq f \in \mathbb{V}_k$. We want to calculate the poles and roots of f using the argument principle

$$\int_\gamma F(\tau) d\tau \quad \text{for } F := f'/f \tag{3.7}$$

along a path γ which is a modified boundary of the exact fundamental domain $\mathbb{F}$.

Given $w \in \mathbb{H}$, it is well-known that

$$\operatorname{res}_w F = \operatorname{ord}_w f. \tag{3.8}$$

If, on the other hand, f is given by (3.3) for a sufficiently large imaginary part of τ then the FOURIER series of F has the form

$$F(\tau) = 2\pi i \cdot \operatorname{ord}_\infty f + \sum_{m \geq 1} \alpha_F(m) \cdot e^{2\pi i m \tau}, \tag{3.9}$$

if we apply the FOURIER series of f and f'. Now the path γ is the contour of $\mathbb{F}$, modified as shown in Figure 23.

In the fixed points and in those boundary points of $\mathbb{F}$, where f has a zero or a pole, the path of integration on the boundary is replaced by an arc κ resp. κ' of radius $\varepsilon > 0$ into the interior of $\mathbb{F}$. There ε is to be chosen small enough that, apart from possibly the center, neither a root nor a pole of f lies in the disk of the radius ε. The path γ now consists of the line and the arc pieces γ_ν, γ'_ν respectively κ_μ, κ'_μ in the given sense. Here, y_0 is to be chosen such that f has neither poles nor roots in the set $\{\tau \in \mathbb{H}; \operatorname{Im} \tau \geq y_0\}$, $y_0 = \operatorname{Im} \gamma_0(t)$.

After choosing γ because of (3.8) the Residue Theorem A.5 now yields

$$2\pi i \cdot \sum_{w \in \overset{\circ}{\mathbb{F}}} \operatorname{ord}_w f = \int_\gamma F(\tau) d\tau. \tag{3.10}$$

§ 3 The weight formula

We note here that the left-hand side of (3.10) does not depend on the choice of ε, if it is sufficiently small. Therefore, on the right-hand side, taking the limit $\varepsilon \to 0$ is allowed.

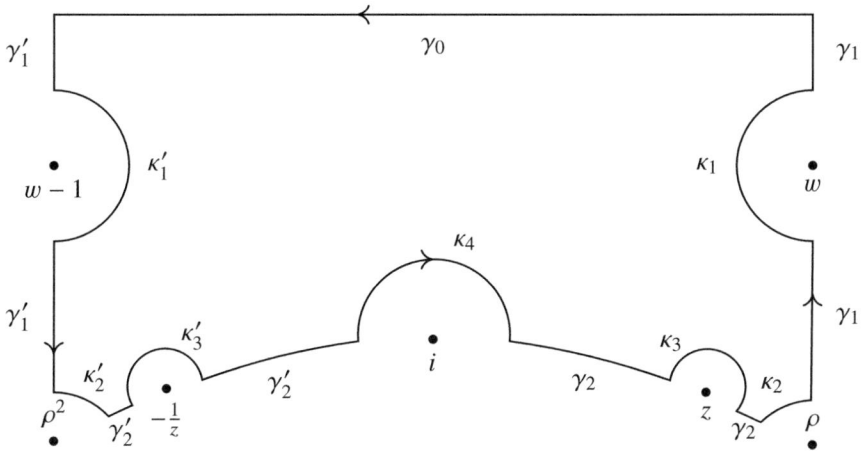

Figure 23: Integration path

Since here certain parts of γ can be related to other parts by modular substitutions we need to describe the behavior of F under modular substitutions. Note that logarithmic differentiation of $f(M\tau) = (c\tau + d)^k \cdot f(\tau)$ yields

$$F(M\tau) \cdot \frac{dM\tau}{d\tau} = \frac{kc}{c\tau + d} + F(\tau) \quad \text{for } M \in \Gamma. \tag{3.11}$$

3. The paths γ'_ν and γ_ν. According to (3.9) and the coefficient representation (1.7) we have

$$\int_{\gamma_0} F(\tau)d\tau = -2\pi i \cdot \operatorname{ord}_\infty f. \tag{3.12}$$

In order to see that

$$\int_{\gamma'_1} F(\tau)d\tau + \int_{\gamma_1} F(\tau)d\tau = 0 \quad \text{independently of } \varepsilon \tag{3.13}$$

we first observe that $F(\tau + 1) = F(\tau)$ due to (3.11). Now (3.13) follows, since the mapping $\tau \mapsto \tau + 1$ maps the path γ'_1 to the path $-\gamma_1$, where the minus sign indicates that the orientation has been reversed.

If $M = J = \begin{pmatrix} 0 & -1 \\ 1 & 0 \end{pmatrix}$, then (3.11) yields

$$F(J\tau) \cdot \frac{dJ\tau}{d\tau} = \frac{k}{\tau} + F(\tau), \tag{3.14}$$

thus

$$\int_{J\sigma} F(\tau)d\tau = \int_\sigma F(J\tau) \cdot \frac{dJ\tau}{d\tau} d\tau = k \cdot \int_\sigma \frac{d\tau}{\tau} + \int_\sigma F(\tau)d\tau \tag{3.15}$$

for any path σ in $\mathbb{H}$ which does not hit any pole of F.

Because $\gamma_2' = -(J\gamma_2)$, we obtain

$$\int_{\gamma_2'} F(\tau)d\tau + \int_{\gamma_2} F(\tau)d\tau = -k \cdot \int_{\gamma_2} \frac{d\tau}{\tau}$$

from (3.15). Thus we get

$$\lim_{\varepsilon \to 0} \left\{ \int_{\gamma_2'} F(\tau)d\tau + \int_{\gamma_2} F(\tau)d\tau \right\} = -k \cdot \int_i^\rho \frac{d\tau}{\tau} = 2\pi i \cdot \frac{k}{12}. \tag{3.16}$$

4. The paths κ_μ' and κ_μ. The paths κ_3' and κ_3 are arcs of the circles with center $-1/z$ and z, respectively. Thus

$$\lim_{\varepsilon \to 0} \left\{ \int_{\kappa_3'} F(\tau)d\tau + \int_{\kappa_3} F(\tau)d\tau \right\} = -2\pi i \cdot \mathrm{ord}_z f \tag{3.17}$$

follows. Because of (3.15), the left-hand side is equal to

$$\lim_{\varepsilon \to 0} \left\{ \int_{J\kappa_3'} F(\tau)d\tau + \int_{\kappa_3} F(\tau)d\tau \right\}.$$

According to the CAUCHY's Integral Theorem A.6, the two integrals may be replaced by an integral over the negatively oriented circle $|\tau - z| = \varepsilon$. Thus, (3.17) is equal to $-2\pi i \cdot \mathrm{res}_z F$. By (3.8), the assertion (3.17) follows.
In order to prove

$$\lim_{\varepsilon \to 0} \int_{\kappa_2} F(\tau)d\tau = -\frac{2\pi i}{6} \cdot \mathrm{ord}_\rho f = \lim_{\varepsilon \to 0} \int_{\kappa_2'} F(\tau)d\tau \tag{3.18}$$

we express for the left-hand side as

$$-\mathrm{res}_\rho F \cdot \lim_{\varepsilon \to 0} \int_{\kappa_2} \frac{d\tau}{\tau - \rho}. \tag{*}$$

§ 4 Modular forms

Here, we choose $\alpha = \alpha(\varepsilon) \in \mathbb{R}$ with

$$|e^{i\alpha} - \rho| = \varepsilon, \quad \frac{\pi}{3} < \alpha < \frac{\pi}{2},$$

and denote by $\varphi = \varphi(\varepsilon)$ the angle between $e^{i\alpha}$ and $\rho + i\varepsilon$ (see Fig. 24). In $(*)$, we substitute $\tau = \rho + \varepsilon e^{it}$, $\frac{\pi}{2} + \varphi \leq t \leq \frac{\pi}{2}$ and obtain

$$i \cdot \operatorname{ord}_\rho f \cdot \lim_{\varepsilon \to 0} \varphi(\varepsilon) = -\frac{\pi i}{3} \cdot \operatorname{ord}_\rho f$$

because of (3.8). Thus the first integral in (3.18) has the claimed value. For the second integral we proceed in the same way.

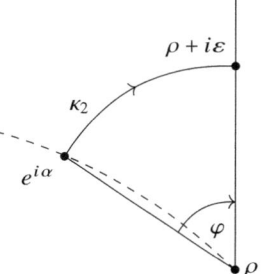

Figure 24: The path κ_2

Completely analogously, we ultimately get

$$\lim_{\varepsilon \to 0} \int_{\kappa_4} F(\tau) d\tau = -\frac{2\pi i}{2} \cdot \operatorname{ord}_i f \qquad (3.19)$$

and

$$\int_{\kappa_1'} F(\tau) d\tau + \int_{\kappa_1} F(\tau) d\tau = -2\pi i \cdot \operatorname{ord}_w f. \qquad (3.20)$$

For the *proof* of the weight formula we now start from (3.10) and add the integrals on the right-hand side using equations (3.12), (3.13), (3.16)-(3.20). □

3.3 Exercises.

1) For $0 \neq f \in \mathbb{V}_k$, we have

$6 \operatorname{ord}_i f + 4 \operatorname{ord}_\rho f \equiv k \pmod{12}$, $\operatorname{ord}_i f \equiv k/2 \pmod 2$ and $\operatorname{ord}_\rho f \equiv k \pmod 3$.

2) If $0 \neq f \in \mathbb{K}$, then $\operatorname{ord}_i f$ is even and $\operatorname{ord}_\rho f$ is divisible by 3.

3) A non-constant $f \in \mathbb{K}$ which is holomorphic on $\mathbb{H}$ has a pole at ∞

§ 4 Modular forms

As a first application of the weight formula in §3, we obtain the central results of §1 (about modular forms of negative weight), on the discriminant (in particular its non-vanishing on $\mathbb{H}$ according to Corollary I.3.18), and furthermore the dimension formula without using the theory of elliptic functions.

1. The dimension formula. As in §1, let $\mathbb{M}_k$ denote the $\mathbb{C}$–vector space of modular forms of weight k and $\mathbb{S}_k$ the subspace of all cusp forms. In Proposition 1.1, we showed that

$$\mathbb{M}_k = \{0\} \quad \text{for odd } k. \tag{4.1}$$

Therefore, if $0 \neq f \in \mathbb{M}_k$, then f is holomorphic in $\mathbb{H}$ and at ∞. In the Weight Formula 3.2,

$$\frac{1}{2}\text{ord}_i f + \frac{1}{3}\text{ord}_\rho f + \text{ord}_\infty f + \sum_{w \in \mathbb{F}\setminus\{i,\rho\}} \text{ord}_w f = \frac{k}{12}, \quad 0 \neq f \in \mathbb{M}_k, \tag{4.2}$$

there are only non-negative terms on the left-hand side.

For $k = 4, 6, 8, 10, 14$, therefore, (4.2) is only solvable in the cases

k	$k/12$	$\text{ord}_\infty f$	$\text{ord}_i f$	$\text{ord}_\rho f$	f has roots at
4	1/3	0	0	1	ρ
6	1/2	0	1	0	i
8	2/3	0	0	2	ρ
10	5/6	0	1	1	i and ρ
14	7/6	0	1	2	i and ρ

(4.3)

4.1 Proposition. *We have:*
a) $\mathbb{M}_k = \{0\}$ *for* $k < 0$.
b) $\mathbb{M}_0 = \mathbb{C}$ *and* $\mathbb{S}_0 = \{0\}$.
c) $\mathbb{M}_2 = \{0\}$.
d) $\mathbb{M}_k = \mathbb{C} \cdot E_k = \mathbb{C} \cdot E_k^*$ *and* $\mathbb{S}_k = \{0\}$ *for* $k = 4, 6, 8, 10$ *and* 14.

Proof a) If $0 \neq f \in \mathbb{M}_k$, the left side of (4.2) would be negative.
b) If f were not constant, the weight formula for $g = f - f(i) \in \mathbb{M}_0$ would give a contradiction.
c) $\alpha + \frac{1}{2}\beta + \frac{1}{3}\gamma = \frac{1}{6}$ is not solvable with non-negative integers α, β, γ.
d) We apply (4.3) to E_k and see, by the Roots Lemma 2.7, that E_k has roots of the appropriate order only at the given points. Then $f/E_k \in \mathbb{M}_0$ follows for $0 \neq f \in \mathbb{M}_k$, hence it is constant according to b). □

4.2 Corollary. a) E_4 *has a single root in* $\mathbb{F}$. *It is* ρ *and is of 1st order.*
b) E_6 *has a single root in* $\mathbb{F}$. *It is* i *and is of 1st order.*

Using the theory of the $\wp$-function shows that the discriminant Δ has no roots in $\mathbb{H}$. Without even using the theory of elliptic functions we can see the non-vanishing behavior directly from the Weight Formula 3.2. First, (2.16) yields

$$\text{ord}_\infty \Delta = 1. \tag{4.4}$$

Therefore, (4.2) leads to

4.3 Proposition. $\Delta(\tau) \neq 0$ *holds for all* $\tau \in \mathbb{H}$.

In (1.14), we showed that

§ 4 Modular forms

$$\mathbb{M}_k \cdot \mathbb{M}_\ell \subseteq \mathbb{M}_{k+\ell} \quad \text{and} \quad \mathbb{S}_k \cdot \mathbb{M}_\ell \subseteq \mathbb{S}_{k+\ell} \quad \text{for} \quad k, l \in \mathbb{Z} \tag{4.5}$$

The decisive tool now is the

4.4 Lemma. *For even $k \geq 0$, one has*

$$\mathbb{S}_k = \Delta \cdot \mathbb{M}_{k-12}.$$

Proof (4.5) yields $\Delta \cdot \mathbb{M}_{k-12} \subseteq \mathbb{S}_k$. By Proposition 4.3, Δ is zero-free in $\mathbb{H}$. For $f \in \mathbb{S}_k$, $g := f/\Delta$ is holomorphic in $\mathbb{H}$ and modular of weight $k - 12$. The FOURIER expansions

$$\Delta(\tau) = (2\pi)^{12} \cdot e^{2\pi i \tau} \cdot (1 + \ldots) \quad \text{and} \quad f(\tau) = e^{2\pi i \tau} \cdot (\alpha_f(1) + \ldots)$$

imply

$$g(\tau) = f(\tau)/\Delta(\tau) = (2\pi)^{-12} \cdot (\alpha_f(1) + \ldots).$$

Thus g has no pole at ∞ and is consequently a modular form. Therefore, it follows that $g \in \mathbb{M}_{k-12}$, hence $\mathbb{S}_k \subseteq \Delta \cdot \mathbb{M}_{k-12}$. □

In particular, Proposition 4.3 and Proposition 4.1 lead to

4.5 Corollary. $\mathbb{S}_{12} = \mathbb{C} \cdot \Delta = \mathbb{C} \cdot \Delta^*$.

This already gives the

4.6 Dimension Formula. *For even $k \geq 0$ the vector space $\mathbb{M}_k$ is finite–dimensional, and its dimension is*

$$\dim_\mathbb{C} \mathbb{M}_k = \begin{cases} \left[\frac{k}{12}\right], & \text{if } k \equiv 2 \pmod{12}, \\ \left[\frac{k}{12}\right] + 1, & \text{if } k \not\equiv 2 \pmod{12}. \end{cases}$$

Proof By Proposition 4.1 this dimension formula holds for $0 \leq k < 12$. If we allow that the dimension of $\mathbb{M}_k$ may be ∞, Proposition 2.5 yields

$$\dim \mathbb{M}_k = 1 + \dim \mathbb{S}_k, \quad k \geq 4,$$

and the Lemma 4.4 implies

$$\dim \mathbb{S}_k = \dim \mathbb{M}_{k-12}, \quad k \geq 12.$$

An induction on k shows that all $\mathbb{M}_k$ are finite–dimensional and that

$$\dim \mathbb{M}_k = 1 + \dim \mathbb{M}_{k-12} \quad \text{for} \quad k \geq 12$$

holds. Since the right-hand side of the dimension formula satisfies the same recursion formula as the left-hand side, the formula holds for all k. □

In analogy with (4.3), we use $6 \cdot \text{ord}_i f + 4 \cdot \text{ord}_\rho f \equiv k \pmod{12}$ for $0 \neq f \in \mathbb{M}_k$ due to (4.2) for the proof of

k (mod 12)	0	2	4	6	8	10
$\mathrm{ord}_i f$ (mod 2)	0	1	0	1	0	1
$\mathrm{ord}_\rho f$ (mod 3)	0	2	1	0	2	1

(4.6)

4.7 Corollary. *For $0 \neq f \in \mathbb{M}_k$, one has $\mathrm{ord}_\infty f < \dim \mathbb{M}_k$.*

In other words, Corollary 4.7 states that $\alpha_f(m) = 0$ for $0 \leq m < \dim \mathbb{M}_k$ always implies $f = 0$.

Proof The Weight Formula 3.2 and (4.6), in the case $f \neq 0$, yield

$$\frac{k}{12} \geq \begin{cases} \mathrm{ord}_\infty f + \frac{1}{2} + \frac{1}{3} + \frac{1}{3}, & \text{if } k \equiv 2 \pmod{12}, \\ \mathrm{ord}_\infty f, & \text{otherwise}, \end{cases}$$

hence $\dim \mathbb{M}_k - 1 \geq \mathrm{ord}_\infty f$, if we use the Dimension Formula 4.6. □

2. Simple conclusions. In Proposition 4.1, the vector spaces $\mathbb{M}_k$ were described for $k \leq 10$. The Dimension Formula 4.6 and Lemma 4.4 now immediately yield a description of the next spaces starting with

$$\left.\begin{aligned} \mathbb{M}_{12} &= \mathbb{C} \cdot E_{12} + \mathbb{C} \cdot \Delta, & \mathbb{S}_{12} &= \mathbb{C} \cdot \Delta, \\ \mathbb{M}_{14} &= \mathbb{C} \cdot E_{14}, & \mathbb{S}_{14} &= \{0\}, \\ \mathbb{M}_k &= \mathbb{C} \cdot E_k + \mathbb{C} \cdot \Delta \cdot E_{k-12}, & \mathbb{S}_k &= \mathbb{C} \cdot \Delta \cdot E_{k-12} \\ & & &\text{for } k = 16, 18, 20, 22, 26, \\ \mathbb{M}_{24} &= \mathbb{C} \cdot E_{24} + \mathbb{C} \cdot E_{12} \cdot \Delta + \mathbb{C} \cdot \Delta^2, & \mathbb{S}_{24} &= \mathbb{C} \cdot E_{12} \cdot \Delta + \mathbb{C} \cdot \Delta^2 \end{aligned}\right\} \quad (4.7)$$

etc. Of course, these relations remain valid if we use the normalized EISENSTEIN series E_k^* instead of E_k in each case.

This immediately gives rise to a plethora of identities: first, because E_4^{*2} belongs to $\mathbb{M}_8 = \mathbb{C} \cdot E_8^*$, there is a constant γ with $E_4^{*2} = \gamma \cdot E_8^*$. But since we are working with the normalized series, $\gamma = 1$ follows immediately, thus

$$E_4^{*2} = E_8^*, \quad \text{i.e. hence} \quad 3 \cdot E_4^2 = 7 \cdot E_8. \tag{4.8}$$

For this and the following relations for the series E_k, we refer to table (2.8). Analogously, we get

$$E_4^* \cdot E_6^* = E_{10}^*, \quad \text{hence} \quad 5 \cdot E_4 \cdot E_6 = 11 \cdot E_{10}, \tag{4.9}$$

$$E_4^{*2} \cdot E_6^* = E_{14}^*, \quad \text{hence} \quad 2 \cdot 3 \cdot 5 \cdot E_4^2 \cdot E_6 = 11 \cdot 13 \cdot E_{14}. \tag{4.10}$$

In the case $k = 12$, we also use the normalized discriminant Δ^* as in (2.17).

4.8 Lemma. *We have*

$$\Delta^* = \frac{1}{1\,728}(E_4^{*3} - E_6^{*2}) = \frac{691}{762\,048}(E_{12}^* - E_6^{*2}) = \frac{691}{432\,000}(E_4^{*3} - E_{12}^*).$$

§4 Modular forms

Proof Here, the first equation is just a repetition of (2.18). With the abbreviations

$$A := \sum_{m\geq 1} \sigma_{11}(m) \cdot e^{2\pi i m \tau} \quad \text{and} \quad B := \sum_{m\geq 1} \sigma_5(m) \cdot e^{2\pi i m \tau}$$

from (2.9) and (2.8) as well as (2.11), we have

$$E_{12}^*(\tau) = 1 + \frac{1008 \cdot 65}{691} A \quad \text{und} \quad E_6^*(\tau) = 1 - 504 B. \tag{4.11}$$

Thus,

$$E_{12}^* - E_6^{*2} = \frac{1008 \cdot (65 + 691)}{691} \cdot e^{2\pi i \tau} + \cdots = \frac{1008 \cdot 756}{691} \cdot e^{2\pi i \tau} + \cdots,$$

follows. Since the left-hand side is a cusp form in $\mathbb{M}_{12}$, i.e. a multiple of Δ^*, the second equation follows. We obtain the third equation analogously. $\square$

A direct consequence is the

4.9 RAMANUJAN Congruence. *We have*

$$\tau(m) \equiv \sigma_{11}(m) \pmod{691} \quad \text{for all} \quad m \in \mathbb{N}.$$

Proof According to the second equation in Lemma 4.8, together with (4.11), we get

$$756 \cdot \Delta^* = 65 \cdot A + 691 \cdot F,$$

where F is a FOURIER series with coefficients from $\mathbb{Z}$. A comparison of the coefficients yields the assertion. $\square$

The relations (4.8), (4.9) and (4.10) are already known from the theory of elliptic functions: they are special cases of the general recursion formula in Corollary I.3.11.

3. Bases of $\mathbb{M}_k$. The following considerations use the normalized series E_k^* and the normalized discriminant Δ^* as before. Completely analogous statements hold for the E_k and Δ. To simplify the formulas, we set

$$E_0 := E_0^* := 1. \tag{4.12}$$

Furthermore, let $k \geq 0$ always be even. Applying Lemma 4.4 repeatedly yields the

4.10 Theorem. *If* $k = 12l + r$, $l \in \mathbb{N}_0, r \in \{0, 4, 6, 8, 10, 14\}$ *then*

$$\mathbb{M}_k = \bigoplus_{0 \leq j \leq l} \mathbb{C} \cdot E_{k-12j}^* \cdot \Delta^{*j}.$$

The mapping

$$\mathbb{M}_k \to \mathbb{C}^{l+1}, \quad f \mapsto (\alpha_f(0), \ldots, \alpha_f(l))^{tr},$$

is an isomorphism of vector spaces.

Proof $\dim \mathbb{M}_k = l+1$ follows from the Dimension Formula 4.6. Moreover note that $E^*_{k-12j} \Delta^{*j}$ has a FOURIER series expansion of the form

$$\sum_{n=j}^{\infty} \alpha(n) e^{2\pi i n \tau}, \quad \alpha(j) = 1. \qquad \square$$

4.11 Proposition. *For each solution r, s of*

$$4r + 6s = k, \quad k \geq 4, \text{ even}, \tag{4.13}$$

in natural numbers $r, s \in \mathbb{N}_0$ one has

$$\mathbb{M}_k = \mathbb{C} \cdot E_4^{*r} \cdot E_6^{*s} \oplus \mathbb{S}_k \tag{4.14}$$

and there always exist solutions r, s.

Proof The solvability of (4.13) is clear for $k \equiv 0 \pmod 4$, in the case $k \equiv 2 \pmod 4$, i.e. $k = 4\ell + 2 \geq 6$, we choose, for example, $r = \ell - 1$ and $s = 1$. Of course, the right-hand side of (4.14) is contained in $\mathbb{M}_k$. However, if $f \in \mathbb{M}_k$, then $f - \alpha_f(0) \cdot E_4^{*r} \cdot E_6^{*s}$ is a cusp form. $\qquad \square$

4.12 Theorem. *If $k \geq 4$ is even,*

$$\mathbb{M}_k = \bigoplus_{r,s} \mathbb{C} \cdot E_4^{*r} E_6^{*s}, \tag{4.15}$$

holds, where the sum runs over all natural numbers $r, s \in \mathbb{N}_0$ with $4r + 6s = k$.

Proof Proposition 4.1 shows that (4.15) is correct for $4 \leq k < 12$. By Lemma 4.4 and (4.14) we have for $k \geq 12$

$$\mathbb{M}_k = \mathbb{C} \cdot E_4^{*r} \cdot E_6^{*s} \oplus \mathbb{M}_{k-12} \cdot \Delta^* \quad \text{with} \quad \Delta^* = \frac{1}{1728}(E_4^{*3} - E_6^{*2}).$$

An induction on k shows that the modular forms $E_4^{*r} \cdot E_6^{*s}$ with $4r + 6s = k$ span the vector space $\mathbb{M}_k$. With an induction on k, we can also easily show that the number of pairs $(r, s) \in \mathbb{N}_0 \times \mathbb{N}_0$ with $4r + 6s = k$ is $\left[\frac{k}{12}\right]$, if $k \equiv 2 \pmod{12}$ and $\left[\frac{k}{12}\right] + 1$ if $k \not\equiv 2 \pmod{12}$. Thus the given modular forms form a basis. $\qquad \square$

Consider the $\mathbb{C}$–vector space

$$\mathbb{M} := \bigoplus_{k \text{ even}} \mathbb{M}_k = \mathbb{C} \oplus \mathbb{M}_4 \oplus \mathbb{M}_6 \oplus \cdots. \tag{4.16}$$

From $\mathbb{M}_k \cdot \mathbb{M}_\ell \subseteq \mathbb{M}_{k+\ell}$, we conclude that $\mathbb{M}$ is a commutative $\mathbb{C}$-algebra with a unit element, which is graded by (4.5). The representation as a direct sum in Theorem 4.12 yields the

4.13 Corollary. $\mathbb{M} = \mathbb{C}[E_4^*, E_6^*]$ *holds and E_4^*, E_6^* are algebraically independent.*

Thus, especially, $E_k^* \in \mathbb{C}[E_4^*, E_6^*]$ holds for $k \geq 4$. Using the theory of the $\wp$-function, we have already seen a stronger version of this, namely $E_k \in \mathbb{Q}[E_4, E_6]$, in Corollary I.3.12.

4*. Modular forms with integer FOURIER coefficients. Let $\mathbb{M}_k^{\mathbb{Z}}$ denote the set of all $f \in \mathbb{M}_k$ whose FOURIER coefficients are are all integers,

$$f(\tau) = \sum_{m=0}^{\infty} \alpha_f(m) \cdot e^{2\pi i m \tau}, \quad \tau \in \mathbb{H}, \quad \alpha_f(m) \in \mathbb{Z} \text{ for all } m \geq 0. \quad (4.17)$$

Obviously, $\mathbb{M}_k^{\mathbb{Z}}$ is a $\mathbb{Z}$-module and we have

$$\mathbb{M}_k^{\mathbb{Z}} \cdot \mathbb{M}_\ell^{\mathbb{Z}} \subseteq \mathbb{M}_{k+\ell}^{\mathbb{Z}} \quad \text{for} \quad k, \ell \geq 0. \quad (4.18)$$

A modular form $f \in \mathbb{M}_k^{\mathbb{Z}}$ is called *normalized* if $\alpha_f(0) = 1$ holds. (2.10) and (2.11) imply $E_4^* \in \mathbb{M}_4^{\mathbb{Z}}$ and $E_6^* \in \mathbb{M}_6^{\mathbb{Z}}$. Both series are normalized. Thus also

$$E_4^{*r} \cdot E_6^{*s} \in \mathbb{M}_k^{\mathbb{Z}} \quad \text{for} \quad 4r + 6s = k.$$

We define a normalized modular form $g_k \in \mathbb{M}_k^{\mathbb{Z}}$ for $k = 4, 6, 8, 10, 12, 14$ by

k	4	6	8	10	12	14
g_k	E_4^*	E_6^*	E_4^{*2}	$E_4^* E_6^*$	E_4^{*3}	$E_4^{*2} E_6^*$

Thus we have

$$\mathbb{M}_k^{\mathbb{Z}} = \mathbb{Z} \cdot g_k \quad \text{for} \quad k = 4, 6, 8, 10, 14.$$

Using the normalized discriminant Δ^* we now get

4.14 Proposition. *If $k \geq 12$ one has*

$$\mathbb{M}_k^{\mathbb{Z}} = \mathbb{Z} \cdot g_k \oplus \mathbb{M}_{k-12}^{\mathbb{Z}} \cdot \Delta^* \quad \text{with} \quad \mathbb{M}_0^{\mathbb{Z}} = \mathbb{Z} \quad \text{and} \quad \mathbb{M}_2^{\mathbb{Z}} = \{0\} \quad (4.19)$$

for any normalized $g_k \in \mathbb{M}_k^{\mathbb{Z}}$.

In particular, then $\mathbb{M}_{12}^{\mathbb{Z}} = \mathbb{Z} \cdot g_{12} \oplus \mathbb{Z} \cdot \Delta^*$.

Proof First, the right-hand side of (4.19) is a direct sum of $\mathbb{Z}$-modules and is contained in the left-hand side. If $f \in \mathbb{M}_k^{\mathbb{Z}}$, then $f - \alpha_f(0) \cdot g_k$ is a cusp form in $\mathbb{M}_k^{\mathbb{Z}}$. But since $1/\Delta^*$ has FOURIER coefficients in $\mathbb{Z}$ due to (2.17), it follows that $f - \alpha_f(0) \cdot g_k = \Delta^* \cdot h$ with some $h \in \mathbb{M}_{k-12}^{\mathbb{Z}}$ from Proposition I.4.7. □

An induction on k yields

4.15 Corollary. $\mathbb{M}_k^{\mathbb{Z}}$ *is a free $\mathbb{Z}$-module whose rank is equal to the dimension of $\mathbb{M}_k$. Bases of $\mathbb{M}_k^{\mathbb{Z}}$ over $\mathbb{Z}$ are obtained in the form*

$$g_\nu \cdot \Delta^{*\nu}, \ 0 \leq \nu \leq \left[\frac{k}{12}\right] \quad \text{or} \quad 0 \leq \nu < \left[\frac{k}{12}\right] \quad \text{if } k \equiv 2 \pmod{12}, \quad (4.20)$$

where the $g_\nu \in \mathbb{M}^{\mathbb{Z}}_{k-12\nu}$ are normalized. The modular forms (4.20) are also a basis of the $\mathbb{C}$–vector space $\mathbb{M}_k$.

A $\mathbb{Z}$–basis of $\mathbb{M}^{\mathbb{Z}}_k$ which is also a $\mathbb{C}$–basis of $\mathbb{M}_k$ is called an *integral basis of* $\mathbb{M}_k$. So, in particular, (4.20) is an integral basis of $\mathbb{M}_k$.

Analogously, let $\mathbb{S}^{\mathbb{Z}}_k := \mathbb{M}^{\mathbb{Z}}_k \cap \mathbb{S}_k$ denote the $\mathbb{Z}$-module of cusp forms with integral FOURIER coefficients.

4.16 Proposition. *Let $k \geq 4$ be even. Then $\mathbb{S}^{\mathbb{Z}}_k$ is a free $\mathbb{Z}$-module.*
a) $\mathbb{M}^{\mathbb{Z}}_k = \mathbb{Z} \cdot E_4^{*r} \cdot E_6^{*s} \oplus \mathbb{S}^{\mathbb{Z}}_k$ *for any choice of $r, s \in \mathbb{N}_0$ with $4r + 6s = k$.*
b) $\mathbb{S}^{\mathbb{Z}}_k = \Delta^* \cdot \mathbb{M}^{\mathbb{Z}}_{k-12}.$

From a basis of $\mathbb{M}^{\mathbb{Z}}_{k-12}$ we obtain a basis of $\mathbb{S}^{\mathbb{Z}}_k$.

Proof a) The modular forms $E_4^{*r} \cdot E_6^{*s}$ are normalized.
b) Given $f \in \mathbb{S}^{\mathbb{Z}}_k$, there exists some $g \in \mathbb{M}_{k-12}$ with $f = g \cdot \Delta^*$. But since $(\Delta^*)^{-1}$ has a FOURIER expansion with integral coefficients by (2.17) and Proposition I.4.7, $g \in \mathbb{M}^{\mathbb{Z}}_{k-12}$ follows. $\square$

5*. Representation by theta series. The classical theta series

$$\vartheta(\tau) := \sum_{n \in \mathbb{Z}} e^{\pi i n^2 \tau}, \quad \tau \in \mathbb{H},$$

was considered in the Introduction. In (0.8) and (0.9), we derived

$$\vartheta(\tau + 2) = \vartheta(\tau) \quad \text{and} \quad \vartheta(-1/\tau) = \sqrt{\tau/i} \cdot \vartheta(\tau). \tag{4.21}$$

The definition directly yields

$$\vartheta(\tau) + \vartheta(\tau + 1) = \sum_{n \in \mathbb{Z}} e^{\pi i n^2 \tau} (1 + (-1)^n) = 2 \cdot \vartheta(4\tau).$$

Using (4.21), we get

$$\vartheta(1 - 1/\tau) = \sqrt{\tau/i} \cdot (\vartheta(\tau/4) - \vartheta(\tau)) = \sqrt{\tau/i} \cdot \sum_{n \in \mathbb{Z}} e^{\pi i (n+1/2)^2 \tau}. \tag{4.22}$$

4.17 Proposition. *We have for all $\tau \in \mathbb{H}$:*
a) $\vartheta^4(\tau) - \vartheta^4(\tau + 1) + \tau^{-2} \cdot \vartheta^4(1 - 1/\tau) = 0.$
b) $\vartheta^8(\tau) + \vartheta^8(\tau + 1) + \tau^{-4} \cdot \vartheta^8(1 - 1/\tau)$
$\quad = 2 \cdot \vartheta^8(\tau) + 2 \cdot \vartheta^8(\tau + 1) - 2 \cdot \vartheta^4(\tau)\vartheta^4(\tau + 1) = 2 \cdot E_4^*(\tau).$
c) $[\vartheta^4(\tau) + \vartheta^4(\tau + 1)] \cdot [\frac{5}{2} \cdot \vartheta^4(\tau)\vartheta^4(\tau + 1) - \vartheta^8(\tau) - \vartheta^8(\tau + 1)] = E_6^*(\tau).$
d) $\tau^{-4} \cdot \vartheta^8(\tau) \cdot \vartheta^8(\tau + 1) \cdot \vartheta^8(1 - 1/\tau) = 2^8 \cdot \Delta^*(\tau).$

Proof a) Denote the left-hand side by $f(\tau)$. Then we immediately verify that

$$f(\tau + 1) = -f(\tau) \text{ and } f(-1/\tau) = -\tau^2 \cdot f(\tau)$$

§ 4 Modular forms

by (4.21). Then (4.22) shows that $f^2 \in \mathbb{M}_4$ and its constant FOURIER coefficient is 0. So $f^2 = 0$ and thus $f = 0$ holds according to Proposition 4.1.

b), c) We proceed in the same way and use a).

d) Analogously we show that the left-hand side belongs to $\mathbb{M}_{12}$. Because of (4.21) the constant FOURIER coefficient is 0 and the coefficient of $e^{2\pi i \tau}$ is 2^8. Then we get the assertion from Corollary 4.5. □

Part d) and Proposition 4.3 immediately imply the

4.18 Corollary. $\vartheta(\tau) \neq 0$ *holds for all* $\tau \in \mathbb{H}$.

4.19 Remarks. a) Using Proposition 4.17, we can represent any $f \in \mathbb{M}_k$ as a homogeneous symmetric polynomial in $\vartheta^4(\tau)$ and $\vartheta^4(\tau + 1)$.

b) From (4.21) and Corollary II.3.9 $\vartheta^8|_4 M = \vartheta^8$ follows for all $M \in \Gamma_\vartheta$. Then we can express the parts b) and d) of Proposition 4.17 in the form

$$\sum_{M:\Gamma_\vartheta \backslash \Gamma} \vartheta^8|_4 M = 2 \cdot E_4^*, \qquad \prod_{M:\Gamma_\vartheta \backslash \Gamma} \vartheta^8|_4 M = 2^8 \cdot \Delta^*.$$

c) Using (4.22), we conclude that for $r = a/c \in \mathbb{Q}, a \in \mathbb{Z}, c \in \mathbb{N}, \gcd(a, c) = 1$,

$$\lim_{y \downarrow 0} |\vartheta(r + iy)| = \begin{cases} \infty, & \text{if } ac \text{ even,} \\ 0, & \text{if } ac \text{ odd.} \end{cases}$$

P. ULLRICH (*Res. Math.* **31**, 245-265 (1997)) has shown how this result implies that the continuous RIEMANN function

$$\sum_{n=1}^{\infty} \frac{\sin(n^2 \pi x)}{n^2}$$

is not differentiable at the points of the dense subset $\{a/c \ ; \ a, c \in \mathbb{Z} \text{ odd}\}$ of $\mathbb{R}$.

6*. The differential calculus. If $f : \mathbb{H} \to \mathbb{C}$ is meromorphic and $f \not\equiv 0$, then its logarithmic derivative $f'/f : \mathbb{H} \to \mathbb{C}$ is meromorphic again. Logarithmic differentiation of $f(M\tau) = (c\tau + d)^k \cdot f(\tau)$ for $M \in \Gamma$ yields

$$\frac{f'}{f}(M\tau) \cdot \frac{dM\tau}{d\tau} = \frac{kc}{c\tau + d} + \frac{f'}{f}(\tau) \quad \text{for} \quad M = \begin{pmatrix} a & b \\ c & d \end{pmatrix} \in \Gamma. \qquad (4.23)$$

In particular, this results in

$$f \in \mathbb{K} \quad \Rightarrow \quad f' \text{ and } f'/f \in \mathbb{V}_2. \qquad (4.24)$$

If f has a FOURIER expansion of the form

$$f(\tau) = \sum_{m \geq m_0} \alpha_f(m) \cdot e^{2\pi i m \tau}, \quad \alpha_f(m_0) \neq 0, \qquad (4.25)$$

for $\operatorname{Im}\tau > \gamma$ with $\gamma > 0$, then we conclude that f'/f has a FOURIER expansion of the form

$$\frac{f'}{f}(\tau) = 2\pi i \cdot m_0 + \sum_{m \geq 1} \beta(m) \cdot e^{2\pi i m \tau}, \quad \operatorname{Im}\tau > \gamma'. \tag{4.26}$$

In particular, f'/f is holomorphic at ∞.

If $f \in \mathbb{V}_k$ and $g \in \mathbb{V}_\ell$ are given, define a meromorphic function $[f, g]$ by

$$[f, g] := \ell \cdot f'g - k \cdot fg' = fg \cdot \left(\ell \frac{f'}{f} - k \frac{g'}{g} \right). \tag{4.27}$$

4.20 Proposition. a) *If $f \in \mathbb{V}_k$ and $g \in \mathbb{V}_\ell$, then $[f, g] \in \mathbb{V}_{k+\ell+2}$.*
b) *If $f \in \mathbb{M}_k$ and $g \in \mathbb{M}_\ell$, then $[f, g] \in \mathbb{S}_{k+\ell+2}$.*

Proof By (4.23) $[f, g]$ is modular of weight $k + \ell + 2$. Because of (4.26), $[f, g]$ has at most a pole at ∞. For holomorphic f, g, $[f, g]$ is also holomorphic in $\mathbb{H}$ and at ∞ and the FOURIER series of $[f, g]$ has no constant term. □

4.21 Corollary. *We have*

$$\Delta^* = \frac{1}{1728 \cdot 4\pi i} \cdot [E_4^*, E_6^*].$$

Proof The right-hand side belongs to $\mathbb{S}_{12} = \mathbb{C}\Delta^*$ due to Proposition 4.20. □

4.22 Corollary. *We have*

$$[E_4^*, \Delta^*] = -8\pi i \cdot E_6^* \cdot \Delta^*.$$

Proof The left-hand side belongs to $\mathbb{S}_{18} = \mathbb{C} \cdot E_6^* \cdot \Delta^*$ due to Proposition 4.20. □

If we insert the corresponding FOURIER series into Corollary 4.21, we obtain

4.23 Corollary. *If $m \in \mathbb{N}$, we have*

$$\tau(m) = \frac{m}{12}(5 \cdot \sigma_3(m) + 7 \cdot \sigma_5(m)) - 70 \cdot \sum_{\substack{r \geq 1, s \geq 1 \\ r+s=m}} (3r - 2s) \cdot \sigma_3(r) \cdot \sigma_5(s).$$

4.24 Corollary. *If $f \in \mathbb{M}_k$, we have*

$$\frac{1}{\Delta} \cdot [f, \Delta] = 12 \cdot f' - k \cdot f \cdot \frac{\Delta'}{\Delta} \in \mathbb{M}_{k+2}.$$

As in (4.16), we consider the commutative $\mathbb{C}$-algebra

$$\mathbb{M} := \bigoplus_k \mathbb{M}_k = \mathbb{C} \oplus \mathbb{M}_4 \oplus \mathbb{M}_6 \oplus \cdots = \mathbb{C}[E_4, E_6]. \tag{4.28}$$

4.25 Lemma. *Given $f, g, h \in \mathbb{M}$, the following assertions hold*:

§ 4 Modular forms

$$[f, g] + [g, f] \equiv 0, \tag{4.29}$$

$$[f, [g, h]] + [g, [h, f]] + [h, [f, g]] \equiv 0, \tag{4.30}$$

$$[f, g \cdot h] = [f, g] \cdot h + g \cdot [f, h]. \tag{4.31}$$

Proof For linearity reasons, we may assume $f \in \mathbb{M}_k$, $g \in \mathbb{M}_\ell$ and $h \in \mathbb{M}_m$. Then an elementary verification yields the assertion. □

The identities (4.29) and (4.30) state that $\mathbb{M}$ *with the product* $(f, g) \mapsto [f, g]$ *is a* LIE *algebra*. We denote this LIE algebra by LIE $\mathbb{M}$. The identity (4.31) states that the mapping $g \mapsto [f, g]$ with fixed $f \in \mathbb{M}$ is a *derivation* of the $\mathbb{C}$-algebra $\mathbb{M}$. Besides the algebra $\mathbb{M}$, we consider the ideal

$$\mathbb{S} := \bigoplus_k \mathbb{S}_k = \mathbb{C} \cdot \Delta \oplus \cdots \tag{4.32}$$

of $\mathbb{M}$. As $[\mathbb{M}, \mathbb{S}] \subseteq \mathbb{S}$, we conclude that $\mathbb{S}$ is an ideal of the LIE algebra LIE $\mathbb{M}$.

4.26 Remarks. a) The beginnings of this differential calculus can already be found in R. FRICKE [28], 313–318.
b) In a similar way, we can calculate $g := k \cdot f \cdot f'' - (k+1) \cdot f'^2 \in \mathbb{S}_{2k+4}$ for $f \in \mathbb{M}_k$ (cf. R.A. RANKIN [66], Theorem. 4.3.1).
c) Further results can be found in B. VAN DER POL (*Indagationes Math.* **13**, 261–271 and 272–284 (1951)) and in R.A. RANKIN (*J. Indian Math. Soc.* **20**, 103–116 (1956) and *Mich. Math. J.* **4**, 181–186 (1957)). A systematic consideration of the differential calculus was given by D. ZAGIER (*Proc. Indian Acad. Sci.* **104**, 57-75 (1994)).

4.27 Exercises.
1) Let $\tau \in \mathbb{H}$ and $k > 2$. There exists an $f \in \mathbb{M}_k$ with $f(\tau) \neq 0$ if and only if k is a multiple of the order of the stabilizer subgroup Γ_τ (cf. II.(2.13)).
2) Let $\tau \in \mathbb{H}$ and $k = 12$ or $k > 14$ such that k is a multiple of the order of the stabilizer subgroup Γ_τ. Then there is an $f \in \mathbb{S}_k$ with $f(\tau) \neq 0$.
3) Given $f \in \mathbb{M}_k$, we define

$$g(\tau) := f(2\tau) \cdot f(\tau/2) \cdot f((\tau + 1)/2)$$

Then $g \in \mathbb{M}_{3k}$ holds. If $f \in \mathbb{M}_k^{\mathbb{Z}}$ then $g \in \mathbb{M}_{3k}^{\mathbb{Z}}$. Also g is a cusp form if f is.
4) a) $\Delta^*(2\tau) \cdot \Delta^*(\tau/2) \cdot \Delta^*((\tau + 1)/2) = -\Delta^{*3}(\tau)$.
b) $E_4^*(2\tau) \cdot E_4^*(\tau/2) \cdot E_4^*((\tau + 1)/2) = E_4^{*3}(\tau) - 240 \cdot 225 \cdot \Delta^*(\tau)$.
5) Let $\wp(z; \tau, 1)$ be the $\wp$–function for the lattice $\mathbb{Z}\tau + \mathbb{Z}$ as in I(4.3). Then we have

$$\wp(1/2; \tau, 1) + \wp(\tau/2; \tau, 1) + \wp((\tau + 1)/2; \tau, 1) = 0,$$

$$\wp(1/2; \tau, 1)^2 + \wp(\tau/2; \tau, 1)^2 + \wp((\tau + 1)/2; \tau, 1)^2 = 30 \cdot E_4(\tau),$$

$$\wp(1/2; \tau, 1)^3 + \wp(\tau/2; \tau, 1)^3 + \wp((\tau + 1)/2; \tau, 1)^3 = 105 \cdot E_6(\tau).$$

Given $f \in \mathbb{M}_k$, k even, there is a homogeneous symmetric polynomial $P(X, Y, Z)$ of weight $k/2$ such that

$$P(\wp(1/2;\tau,1),\wp(\tau/2;\tau,1),\wp((\tau+1)/2;\tau,1)) = f(\tau).$$

The algebra $\mathbb{M}$ is isomorphic to the quotient algebra of the symmetric polynomials over $\mathbb{C}$ in X, Y, Z over to the ideal generated by $X + Y + Z$.

6) Let $\Omega = \mathbb{Z}\tau + \mathbb{Z}$ with $\tau \in \mathbb{H}$ and $n \in \mathbb{N}$. Let $\{v_0, \ldots, v_m\}$, $m = n^2 - 1$, be the system of representatives from Proposition I.7.1 with $v_0 = 0$. Given a modular form $f \in \mathbb{M}_k$, k even, there is a symmetric polynomial $P(X_1, \ldots, X_m)$ of weight $k/2$ with $P(\wp(v_1;\tau,1), \ldots, \wp(v_m;\tau,1)) = f(\tau)$ for all $\tau \in \mathbb{H}$.

7) Let $k \geq 12$ be even, $k \not\equiv 2 \pmod{12}$. Then there is exactly one $f \in \mathbb{M}_k$ whose FOURIER coefficients satisfy $\alpha_f(0) = 1$, $\alpha_f(m) = 0$ for $1 \leq m \leq \left[\frac{k}{12}\right]$.

8) If $f \in \mathbb{M}_k$ with $\alpha_f(m) \in \mathbb{Q}$ for all $m \geq 0$, then there exists some $\lambda \in \mathbb{N}$ satisfying $\lambda \cdot \alpha_f(m) \in \mathbb{Z}$ for all $m \geq 0$.

9) $E_{12}(i) \neq 0$ und $E_{12}(\rho) \neq 0$.

10) We have

$$\sum_{k=0}^{\infty} \dim \mathbb{M}_k \cdot x^k = \frac{1}{(1-x^4)(1-x^6)}, \quad \sum_{k=0}^{\infty} \dim \mathbb{S}_k \cdot x^k = \frac{x^{12}}{(1-x^4)(1-x^6)}$$

for $|x| < 1$.

11) There exists an $f \in \mathbb{M}_k$ with $f(\tau) \neq 0$ for all $\tau \in \mathbb{H}$ if and only if $k = 12l$ for an $l \in \mathbb{N}_0$. In this case, $f(\tau) = c \cdot \Delta(\tau)^l$ holds for some $0 \neq c \in \mathbb{C}$.

12) Determine the weight of a non-cusp form, which has only simple roots at i and $3i$ in $\mathbb{F}$.

13) Let $k \geq l \geq 4$ be even. Show that $E_k^* E_l^* = E_{k+l}$ holds if and only if we have $(k, l) = (4, 4), (6, 4), (8, 4), (10, 4)$.

14) Consider the JACOBI theta series $\vartheta_j(z, \tau)$, $0 \leq j \leq 3$ from sect. 2 in I, §8.

a) Show for all $\tau \in \mathbb{H}$ that

$$\vartheta_0(0;\tau) = 0, \qquad \vartheta_1(0,\tau) = \vartheta(\tau+1),$$

$$\vartheta_2(0;\tau) = \frac{1}{\sqrt{\tau/i}}\vartheta(1-1/\tau), \quad \vartheta_3(0;\tau) = \vartheta(\tau).$$

b) Any $f \in \mathbb{M}_k$ is a polynomial in the fourth powers of the JACOBI theta series $\vartheta_j(0;\tau)$, $j = 1, 3$.

15) Let $k > 12$ be even and

$$F(t) = e^{ikt/2} E_k(e^{it}), \quad \pi/3 < t < \pi/2.$$

Show that

a) $F(t) = 4\zeta(k)\cos(kt/2) + \dfrac{1}{2^{k-1}\cos^k(t/2)} + \dfrac{(-1)^{k/2}}{2^{k-1}\sin^k(t/2)} + R(t)$,

$$R(t) = \sum_{\substack{(c,d)\in\mathbb{Z}\times\mathbb{Z} \\ |cd|>1}} \left(ce^{it/2} + de^{-it/2}\right)^{-k}.$$

b) $|R(t)| \leq 4(\zeta(k/2)^2 - 1)$.
c) All roots of $E_k(\tau)$ in $\mathbb{F}$ satisfy $|\tau| = 1$ and they are of 1st order except for possibly i and ρ.

§ 5 Modular functions

1. The weight formula for modular functions. As a second application, we consider the Weight Formula 3.2 in the case $k = 0$. Any modular function $f \neq 0$, i.e. $0 \neq f \in \mathbb{K}$, satisfies the **weight formula**

$$\sum_{w \in \mathbb{F}^*} \frac{1}{\operatorname{ord} w} \cdot \operatorname{ord}_w f = 0. \tag{5.1}$$

5.1 Lemma. *If $f \in \mathbb{K}$ is holomorphic on $\mathbb{F}^*$, then f is constant.*

Proof By assumption, in the case $f \neq 0$, $\operatorname{ord}_w f \geq 0$ and $\operatorname{ord}_w f = 0$ always holds by (5.1). This implies $f(w) \neq 0$ for all $w \in \mathbb{F}^*$. Since with f also $f - f(i)$ also belongs to $\mathbb{K}$ with a root at $\tau = i$, f is constant. □

If we now apply (5.1) to $f - z$ instead of f, we obtain the following.

5.2 Proposition. *If f is not constant in $\mathbb{K}$, then f has poles in $\mathbb{F}^*$. If f is holomorphic in i and ρ, then for $z \in \mathbb{C}$ with $z \neq f(i)$ and $z \neq f(\rho)$ the number of roots of $f(\tau) - z$ in $\mathbb{F}^*$ is equal to the number of poles in $\mathbb{F}^*$ (counted with multiplicities).*

2. Application to the absolute invariant. According to (2.19), (2.21) as well as Proposition 4.3, we have

$$j = E_4^{*3}/\Delta^*, \quad \operatorname{ord}_\infty j = -1 \quad \text{as well as} \quad \operatorname{ord}_w j \geq 0 \quad \text{for all} \quad w \in \mathbb{H}. \tag{5.2}$$

For $z \in \mathbb{C}$, we apply the Weight Formula (5.1) to $f := j - z$ and get

$$\sum_{w \in \mathbb{F}} \frac{1}{\operatorname{ord} w} \cdot \operatorname{ord}_w(j - z) = 1. \tag{5.3}$$

5.3 Theorem. *The function $j : \mathbb{H} \to \mathbb{C}$ is holomorphic and attains every value of $\mathbb{C}$ in $\mathbb{F}$. More precisely*:
a) *Any complex number z different from 0 and $12^3 = 1728$ is attained exactly once in $\mathbb{F} \setminus \{i, \rho\}$ and the root of $j - z$ is of first order.*
b) *$j - 1728$ has a root of order 2 at $\tau = i$ and $j(\tau) \neq 1728$ holds for all $\tau \in \mathbb{F} \setminus \{i\}$.*
c) *j has a root of order 3 at $\tau = \rho$ and $j(\tau) \neq 0$ holds for all $\tau \in \mathbb{F} \setminus \{\rho\}$.*

Proof c) We choose $z = 0$ in (5.3). Because of (2.22), $j(\rho) = 0$, so $\operatorname{ord}_\rho j \geq 1$. Thus the left-hand side of (5.3) is greater than or equal to $1/3$ and it follows that $j(\tau) \neq 0$ for $\tau \in \mathbb{F} \setminus \{\rho\}$, because otherwise the left-hand side of (5.3) would be greater than 1 because of $j(i) \neq 0$.
b) We proceed analogously.

a) Because of $z \neq 0$ and $z \neq 1728$, $w = i$ and $w = \rho$ yield 0 in (5.3). □

5.4 Corollary. *The mapping $j : \mathbb{F} \to \mathbb{C}$ is a bijection.*

Every rational function of j is a modular function. The converse is also true.

5.5 Theorem. *We have $\mathbb{K} = \mathbb{C}(j)$.*

Proof Let $f \in \mathbb{K}$ be non-constant. For complex numbers $u \neq v$, consider the function

$$g(\tau) := \frac{f(\tau) - u}{f(\tau) - v}.$$

Since the poles of f cancel here, g (counted with multiplicities) has zeros in $\mathbb{F}$ at the roots p_ν of $f - u$ and poles at the roots q_μ of $f - v$. We can therefore choose $u \neq v$ such that g is holomorphic and non-zero at i and ρ. By Proposition 5.2, the numbers of roots and poles coincide. Proposition 5.3 shows that g and h,

$$h(\tau) = \prod_\nu \frac{j(\tau) - j(p_\nu)}{j(\tau) - j(q_\nu)},$$

have the same roots and poles in $\mathbb{F}^*$, such that g and h differ only by a constant factor due to Lemma 5.1. Then $g \in \mathbb{C}(j)$ follows and thus also

$$f = \frac{vg - u}{g - 1} \in \mathbb{C}(j).$$ □

5.6 Corollary. *$f \in \mathbb{K}$ is holomorphic on $\mathbb{H}$ if and only if f is a polynomial in j.*

If $k \geq 4$ is even and if $g, h \in \mathbb{M}_k$ are given, $h \neq 0$, then $g/h \in \mathbb{K}$ holds. Conversely, we have the

5.7 Proposition. *Every modular function is a quotient of two modular forms of the same weight.*

Proof If $f \in \mathbb{M}$ is not constant, we can write $f = P(j)/Q(j)$ with polynomials P and Q according to Theorem 5.5. If r denotes the maximum of the degrees of P and Q, then $P(j) \cdot \Delta^r$ and $Q(j) \cdot \Delta^r$ are modular forms of weight $12r$. □

To avoid the unusual number $12^3 = 1728$, we sometimes consider the function $J(\tau) := \frac{1}{1728} j(\tau)$ which does not attain the values 0 and 1 in $\mathbb{F} \setminus \{i, \rho\}$.

3. The conformal mapping given by j. We first study the mapping j on the boundary of $\mathbb{F}$:

5.8 Proposition. *The modular function j takes real values on the boundary of $\mathbb{F}$ and on the imaginary axis, but at no other point of $\mathbb{F}$. More precisely, j maps*
a) *the straight line from ∞ to ρ onto the interval $]-\infty, 0]$,*
b) *the arc from ρ to i onto the interval $[0, 1728]$,*
c) *the straight line from i to ∞ onto the interval $[1728, \infty[$.*

Proof Since the FOURIER coefficients of j are real, it immediately follows that

§ 5 Modular functions

$$j(-\bar{\tau}) = \overline{j(\tau)} \quad \text{for } \tau \in \mathbb{H}.$$

Hence, j is real on the imaginary axis. If $\tau := -\frac{1}{2} + iy$, $y > 0$, it further holds that

$$\overline{j(\tau)} = j(-\bar{\tau}) = j\left(\frac{1}{2} + iy\right) = j\left(-\frac{1}{2} + iy\right) = j(\tau).$$

Thus, j is also real on the two straight lines with $|\text{Re}(\tau)| = 1/2$. For $|\tau| = 1$ we have $-1/\tau = -\bar{\tau}$, hence also $\overline{j(\tau)} = j(\tau)$. The remaining statements result from continuity arguments and from the bijectivity of the mapping $j : \mathbb{F} \to \mathbb{C}$, because $\tau \in \overset{\circ}{\mathbb{F}}$ and $j(\tau) = \overline{j(-\bar{\tau})} \in \mathbb{R}$ show that $\tau = -\bar{\tau}$, i.e. $\tau \in i\mathbb{R}$. □

If we define, as in II(2.23), the "left half" of $\overline{\mathbb{F}}$ to be

$$\mathbb{L} := \left\{ \tau \in \mathbb{H} \,;\, -\frac{1}{2} \leq \text{Re } \tau \leq 0, \, |\tau| \geq 1 \right\},$$

then $\mathbb{L}$ is bijectively mapped onto $\overline{\mathbb{H}} := \mathbb{H} \cup \mathbb{R}$ by j.

4*. Representation. We represent each modular form f by E_4, E_6, Δ and j in terms of the roots of f.

5.9 Theorem *If $0 \neq f \in \mathbb{M}_k$, $k \geq 12$ even, there exists some $\alpha \in \mathbb{C}$ such that*

$$f = \alpha \cdot E_4^{\text{ord}_\rho f} \cdot E_6^{\text{ord}_i f} \cdot \Delta^{\text{ord}_\infty f + \gamma(f)} \cdot \prod_{w \in \mathbb{F} \setminus \{i, \rho\}} (j - j(w))^{\text{ord}_w f}, \quad (5.4)$$

where

$$\gamma(f) := \sum_{w \in \mathbb{F} \setminus \{i, \rho\}} \text{ord}_w f. \quad (5.5)$$

Proof Because $\Delta \cdot (j - j(w))$ belongs to $\mathbb{M}_{12}$, the right-hand side g of (5.4) is a modular form of weight

$$k' := 4 \cdot \text{ord}_\rho f + 6 \cdot \text{ord}_i f + 12 \cdot (\gamma(f) + \text{ord}_\infty f).$$

The Weight Formula 3.2 yields $k' = k$. Because of Corollary 4.2 and Theorem 5.3, f and g have the same orders everywhere. Therefore, the quotient f/g belongs to $\mathbb{M}_0$, hence is constant. □

5.10 Corollary. *We have*

$$E_4^* \cdot \frac{j'}{j} = -2\pi i \cdot E_6^*, \quad j' = -2\pi i \cdot \frac{E_{14}^*}{\Delta^*}.$$

Proof Because of (4.24) and (4.26), $E_4^* \cdot j'/j$ belongs to $\mathbb{M}_6$, as the roots of E_4^* and j at $\tau = \rho$ cancel. This gives the first equation. Then we substitute (2.19) for j and use (4.10). □

5*. Picard's Little Theorem can be directly derived from the mapping properties of j.

5.11 Picard's Little Theorem. *If f is a non-constant entire function, then f attains every value in $\mathbb{C}$ with at most one exception.*

Proof We assume that f is entire and does not attain the values a and b, $a \neq b$. Then we consider the entire function

$$g(z) := 1728 \cdot \frac{f(z) - a}{b - a},$$

which omits the values 0 and 1728. First, by Theorem 5.3, we determine some $\tau_0 \in \mathbb{F} \setminus \{i, \rho\}$ with $j(\tau_0) = g(0)$. Then $j'(\tau_0) \neq 0$ holds because of Theorem 5.3. Hence there is a function φ, holomorphic in a neighborhood of 0, with values in $\mathbb{H}$, such that

$$j(\varphi(z)) = g(z) \quad \text{and} \quad \varphi(0) = \tau_0.$$

Now $j'(\tau) \neq 0$ holds for all $\tau \in \mathbb{H}$ with $j(\tau) \in g(\mathbb{C}) \subseteq \mathbb{C} \setminus \{0, 1728\}$. Thus φ is analytically continuable along any path in $\mathbb{C}$. Since $\mathbb{C}$ is simply connected, by the monodromy theorem (cf. A. Hurwitz and R. Courant [40], 372), there exists an entire function

$$\varphi : \mathbb{C} \to \mathbb{H} \quad \text{with} \quad \varphi(0) = \tau_0 \quad \text{and} \quad j(\varphi(z)) = g(z) \quad \text{for all} \quad z \in \mathbb{C}.$$

Hence $e^{i\varphi(z)}$ is bounded. By Liouville's Theorem A.12, $e^{i\varphi(z)}$ and hence $\varphi(z)$ is constant. But then g and hence f is also constant. □

Picard's Little Theorem was first proved in 1879 by E. Picard [63], 1–21. Our proof follows the classical book of A. Hurwitz and R. Courant [40], 438–439. Alternatively, compare R. Remmert and G. Schumacher [67], 10.2.2, where the statement is derived from Bloch's Theorem.

6*. Weakly holomorphic modular forms can be viewed as generalizations of modular forms, where we allow poles at ∞. The vector space $\mathbb{M}_k^!$ of *weakly holomorphic modular forms* of weight $k \in \mathbb{Z}$ consists of all $f \in \mathbb{V}_k$ (cf. sect. 3 in §1) which are holomorphic on $\mathbb{H}$. Hence they posses a Fourier series expansion of the form

$$f(\tau) = \sum_{m=m_0}^{\infty} \alpha_f(m) e^{2\pi i m \tau}, \quad \tau \in \mathbb{H}, \quad m_0 \in \mathbb{Z}. \tag{5.6}$$

Then (1.9) and (1.10) imply for $k, l \in \mathbb{Z}$

$$\mathbb{M}_k^! \cdot \mathbb{M}_l^! \subseteq \mathbb{M}_{k+l}^!, \tag{5.7}$$

$$\mathbb{M}_k^! = \{0\} \quad \text{for odd} \quad k. \tag{5.8}$$

As $\Delta(\tau) \neq 0$ for all $\tau \in \mathbb{H}$ due to Proposition 4.3, we conclude

$$\frac{1}{\Delta} \in \mathbb{M}_{-12}^!. \tag{5.9}$$

5.12 Proposition. *a)* $\mathbb{M}_0^! = \mathbb{C}[j]$.
b) For all even $k \in \mathbb{Z}$ we have
$$\mathbb{M}_k^! = \Delta \cdot \mathbb{M}_{k-12}^!.$$

Proof a) Compare Corollary 5.6.
b) Apply $\Delta \in \mathbb{M}_{12} \subseteq \mathbb{M}_{12}^!$ as well as (5.7) and (5.9). □

Now we will show that the vector space $\mathbb{M}_k^!$ is large.

5.13 Theorem. *If $k \in \mathbb{Z}$ is even, $k = 12\nu + r$, $\nu \in \mathbb{Z}$, $r \in \{0, 4, 6, 8, 10, 14\}$ and $E_0^* := 1$, then we have*
$$\mathbb{M}_k^! = \mathbb{C}[j] \cdot \Delta^\nu \cdot E_r^*. \tag{5.10}$$
In particular, $\dim \mathbb{M}_k^! = \infty$.

Proof " $\subseteq$ " The claim follows from (5.7), (5.9) and Proposition 5.12.
" $\supseteq$ " Let $0 \neq f \in \mathbb{M}_k^!$ and therefore $g := f \cdot \Delta^{-\nu} \in \mathbb{M}_r^!$. Let $-m = \text{ord}_\infty g$, $m \in \mathbb{N}_0$. By adding suitable multiples of j^n, $0 \leq n \leq m$, we may successively reduce the pole order at ∞. Thus we obtain a polynomial $p(X) \in \mathbb{C}[X]$ such that
$$g - p(j)E_r^* \in \mathbb{M}_r^!$$
has a root at ∞ and thus belongs to $\mathbb{S}_r = \{0\}$ according to Proposition 4.1. The claim follows as $\dim \mathbb{C}[j] = \infty$. □

Just as in (4.16), consider the graded ring
$$\mathbb{M}^! = \bigoplus_{k \text{ even}} \mathbb{M}_k^!. \tag{5.11}$$

Using (5.2) and (2.14), Theorem 5.13 implies

5.14 Corollary. *The graded ring $\mathbb{M}^!$ consists of all polynomials in*
$$\frac{1}{\Delta}, \quad E_4^* \quad \text{and} \quad E_6^*. \tag{5.12}$$

Weakly holomorphic modular forms were applied by R. BORCHERDS (*Invent math.* **120**, 161–213 (1995)). He used them as input in order to construct higher dimensional automorphic forms as products.

5.15 Exercises.
1) For every $f \in \mathbb{V}_k$, there is some $\varphi \in \mathbb{C}[j]$ such that $\varphi \cdot f$ is holomorphic on $\mathbb{H}$.
2) Describe a biholomorphic mapping between $\overset{\circ}{\mathbb{F}}$ and $\mathbb{C} \setminus \{x \in \mathbb{R}\,;\, x \leq 0\}$.
3) $(2\pi i)^2 \cdot E_4^* = j'^2/j(j - 1728)$.
4) $(-2\pi i)^3 \cdot E_6^* = j'^3/j^2(j - 1728)$.
5) Let $w \in \mathbb{H}$ with $E_{12}^*(w) = 0$. Then $(2\pi i)^6 \cdot E_{12}^* = j'^6(j - j(w))/j^4(j - 1728)^3$.
6) $(2\pi i)^6 \cdot \Delta^* = j'^6/j^4(j - 1728)^3$.
7) There is exactly one $\tau \in \mathbb{H}$ with $E_4^{*3}(\tau) = \Delta^*(\tau)$ and this τ is of absolute value 1.

8) There is exactly one $\tau \in \mathbb{F}$ with $E_6^{*2}(\tau) = \Delta^*(\tau)$ and this τ lies on the imaginary axis.

9) If $\tau_1, \ldots, \tau_n \in \mathbb{F}$ are pairwise distinct and $\alpha_1, \ldots, \alpha_n \in \mathbb{C}$, then there exists a modular function $f \in \mathbb{K}$, holomorphic in $\mathbb{H}$, with $f(\tau_j) = \alpha_j$, $j = 1, \ldots, n$.

10) A non-constant meromorphic function on $\mathbb{C}$ attains every value with at most 2 exceptions.

11) If f and g are entire functions with $e^{f(z)} + e^{g(z)} = 1$, then f and g are constant.

12) If $\mathbb{M}_k^! = \mathbb{C}[j]g$, then g and $\Delta^\nu E_r^*$ in Theorem 5.13 are linearly dependent.

13) $\mathbb{M}_2^! = \mathbb{C}[j]j'$.

14) $\mathbb{K}$ is the quotient field of the graded ring $\mathbb{M}^!$, i.e. any modular function is the quotient of two weakly holomorphic modular forms of the same weight.

15) Let $k \in \mathbb{Z}$ be even. Construct a function f which is holomorphic on $\mathbb{H}$ and modular of weight k but does not belong to $\mathbb{M}_k^!$.

16) Derive a non-trivial algebraic relation among the weakly holomorphic modular forms in (5.12).

17) Given $f \in \mathbb{K}$ show that $\mathbb{K} = \mathbb{C}(f)$ holds if and only if $f = M\langle j \rangle$ for some matrix $M \in \mathrm{SL}(2; \mathbb{C})$.

§ 6 The DEDEKIND eta–function

The product formula of the discriminant is the only remaining content from the theory of elliptic functions. A direct proof of this is also given below. Because of the importance of this product formula, we also go into other possibilities for the proof.

1. The conditionally convergent EISENSTEIN series. Following G. EISENSTEIN (cf. I, §3), we define the conditionally convergent EISENSTEIN series

$$E_2(\tau) := \sum_{n \neq 0} n^{-2} + \sum_{m \neq 0} \left(\sum_{n \in \mathbb{Z}} (m\tau + n)^{-2} \right), \quad \tau \in \mathbb{H}. \tag{6.1}$$

By analogy with Proposition I.4.2 (but in contrast to (4.12)), we first have the

6.1 Proposition. *The function $E_2 : \mathbb{H} \to \mathbb{C}$ is holomorphic and satisfies*

$$E_2(\tau) = \frac{\pi^2}{3}\left(1 - 24 \cdot \sum_{n=1}^{\infty} \sigma_1(n) \cdot e^{2\pi i n \tau}\right) \text{ for all } \tau \in \mathbb{H}.$$

Proof Using Proposition I.4.2, we get

$$E_2(\tau) = 2\zeta(2) + 2 \cdot \sum_{m \geq 1}\left(\sum_{n \in \mathbb{Z}}(m\tau + n)^{-2}\right)$$
$$= 2\zeta(2) - 8\pi^2 \cdot \sum_{m \geq 1}\sum_{r \geq 1} r \cdot e^{2\pi i r m \tau}.$$

§ 6 The DEDEKIND eta–function

Because of the absolute convergence of the last series, we may collect all terms m and r with $rm = n$. We then use $\zeta(2) = \pi^2/6$. Obviously, the FOURIER series represents a holomorphic function on $\mathbb{H}$. □

Sometimes it is convenient to consider the normalized version

$$E_2^*(\tau) = \frac{1}{2\zeta(2)} \cdot E_2(\tau) = 1 - 24 \cdot \sum_{n=1}^{\infty} \sigma_1(n) \cdot e^{2\pi i n \tau}. \tag{6.2}$$

The function E_2 is not a modular form, but we can determine its transformation behavior under modular substitutions. We have $E_2(\tau + 1) = E_2(\tau)$ and

6.2 Proposition. *For $\tau \in \mathbb{H}$ we have*

$$E_2(-1/\tau) = \tau^2 \cdot E_2(\tau) - 2\pi i \tau.$$

Proof The equation (6.1) can be written as

$$E_2(\tau) = \frac{\pi^2}{3} \cdot \left(1 + \frac{1}{\tau^2}\right) + 2 \cdot \sum_{m \geq 1} \sum_{n \geq 1} \left((m\tau + n)^{-2} + (m\tau - n)^{-2}\right).$$

Thus we get

$$E_2(-1/\tau) = \frac{\pi^2}{3} \cdot (1 + \tau^2) + 2\tau^2 \cdot \sum_{m \geq 1} \sum_{n \geq 1} \left((n\tau - m)^{-2} + (n\tau + m)^{-2}\right).$$

This results in

$$F(\tau) := \frac{1}{2\tau^2} \cdot \left(\tau^2 E_2(\tau) - E_2(-1/\tau)\right) = \sum_{m \geq 1} \sum_{n \geq 1} A_{mn} - \sum_{n \geq 1} \sum_{m \geq 1} A_{mn}, \tag{$*$}$$

where

$$A_{mn} := (m\tau + n)^{-2} + (m\tau - n)^{-2}.$$

The series on the right-hand side of $(*)$ are not absolutely convergent, so the summation may not be interchanged. We proceed as follows: let

$$B_{mn} := \frac{1}{m\tau + n - 1} - \frac{1}{m\tau + n} + \frac{1}{m\tau - n} - \frac{1}{m\tau - n + 1}, \quad m \geq 1 \text{ and } n \geq 1,$$

$$= \frac{1}{(m\tau + n)(m\tau + n - 1)} + \frac{1}{(m\tau - n)(m\tau - n + 1)}.$$

Then, according to Proposition 2.1,

$$A_{mn} - B_{mn} = O\left((m^2 + n^2)^{-3/2}\right) = O\left(m^{-3/2} \cdot n^{-3/2}\right),$$

if we observe $m^2 + n^2 > mn$. Because of the absolute convergence, we obtain

$$\sum_{m\geq 1}\sum_{n\geq 1}(A_{mn} - B_{mn}) = \sum_{n\geq 1}\sum_{m\geq 1}(A_{mn} - B_{mn}),$$

and (∗) yields

$$F(\tau) = \sum_{m\geq 1}\sum_{n\geq 1}B_{mn} - \sum_{n\geq 1}\sum_{m\geq 1}B_{mn}. \tag{**}$$

We immediately obtain

$$\sum_{n\geq 1}B_{mn} = 0$$

from the definition. On the other hand, we get

$$\tau \cdot \sum_{m\geq 1}B_{mn} = \sum_{m\geq 1}\left(\frac{1}{m+(n-1)/\tau} - \frac{1}{m+n/\tau} + \frac{1}{m-n/\tau} - \frac{1}{m-(n-1)/\tau}\right)$$
$$= \sum_{m\geq 1}\left(\frac{2(n-1)/\tau}{[(n-1)/\tau]^2 - m^2} - \frac{2n/\tau}{[n/\tau]^2 - m^2}\right) = \varphi(n-1) - \varphi(n),$$

where, due to the partial fractional expansion of the cotangent (cf. B.2),

$$\varphi(\xi) := \begin{cases} \pi \cot(\pi\xi/\tau) - \dfrac{1}{\xi/\tau} & \text{for } \xi \neq 0, \\ 0 & \text{for } \xi = 0 \end{cases}$$

holds. Therefore, (∗∗) leads to

$$\tau \cdot F(\tau) = -\tau \cdot \sum_{n\geq 1}\sum_{m\geq 1}B_{mn} = -\sum_{n\geq 1}(\varphi(n-1) - \varphi(n)) = -\varphi(0) + \lim_{n\to\infty}\varphi(n).$$

If $z = x + iy$, then

$$\cot z = i \cdot \frac{e^{ix-y} + e^{-ix+y}}{e^{ix-y} - e^{-ix+y}}, \quad \text{hence} \quad \lim_{y\to-\infty}\cot z = i.$$

Because of Im $(1/\tau) < 0$, it follows that $\tau \cdot F(\tau) = \pi i$, thus the assertion. □

6.3 Remark. The idea of this proof can already be found in G. EISENSTEIN [20], vol. I, 357–478; a precise implementation of the idea of the proof is probably given for the first time by A. HURWITZ in his thesis [39], vol. I, 23–26. Compare R. FUETER [30], 21–23, J.–P. SERRE [73], 95–96, and N. KOBLITZ [47], Proposition III.2.7.

2. The transformation behavior of η. Following R. DEDEKIND [16], vol. I, 159–173, we define a holomorphic function $\eta : \mathbb{H} \to \mathbb{C}$ by

$$\eta(\tau) := e^{\pi i \tau/12} \cdot \prod_{m=1}^{\infty}\left(1 - e^{2\pi i m \tau}\right). \tag{6.3}$$

This DEDEKIND *eta function* must not be mixed up with the η–function introduced in I(6.4)! Obviously,

§ 6 The DEDEKIND eta–function

$$\eta(\tau + 1) = e^{\pi i/12} \cdot \eta(\tau) \tag{6.4}$$

holds. Since the product converges absolutely, because of the absolute convergence of the geometric series $\sum_{m=1}^{\infty} e^{2\pi i m\tau}$, $\tau \in \mathbb{H}$, we also have

$$\eta(\tau) \neq 0 \quad \text{for all} \quad \tau \in \mathbb{H}. \tag{6.5}$$

6.4 Eta Transformation Formula. *We have*

$$\eta(-1/\tau) = \sqrt{\tau/i} \cdot \eta(\tau) \quad \text{for all} \quad \tau \in \mathbb{H}.$$

The branch of the root to be chosen is the one that, for positive arguments, itself becomes positive.

Proof For $\tau \in \mathbb{H}$, consider the function $f(\tau) := \eta'(\tau)/\eta(\tau)$. From (6.3) we directly conclude

$$f(\tau) = \frac{\pi i}{12} \cdot \left(1 - 24 \cdot \sum_{m \geq 1} m \cdot \frac{e^{2\pi i m\tau}}{1 - e^{2\pi i m\tau}}\right)$$

$$= \frac{\pi i}{12} \cdot \left(1 - 24 \cdot \sum_{m \geq 1} \sum_{r \geq 1} m \cdot e^{2\pi i r m\tau}\right)$$

$$= \frac{\pi i}{12} \cdot \left(1 - 24 \cdot \sum_{n \geq 1} \sigma_1(n) \cdot e^{2\pi i n\tau}\right)$$

$$= \frac{i}{4\pi} \cdot E_2(\tau),$$

if we use Proposition 6.1. Proposition 6.2 thus translates into

$$f\left(-\frac{1}{\tau}\right) \cdot \frac{1}{\tau^2} - f(\tau) - \frac{1}{2\tau} = 0. \tag{*}$$

Let

$$g(y) := \frac{\eta(i/y)}{\eta(iy)\sqrt{y}}, \quad y > 0.$$

Then

$$\frac{g'(y)}{g(y)} = f\left(\frac{i}{y}\right) \cdot \frac{-i}{y^2} - i \cdot f(iy) - \frac{1}{2y} = 0$$

holds due to (∗). So there is a constant γ with $\eta(i/y) = \gamma \cdot \sqrt{y} \cdot \eta(iy)$. For $y = 1$ we get $\gamma = 1$, so the assertion follows with the Identity Theorem A.1. □

6.5 Proposition. *We have* $\eta^{24} = \Delta^*$.

Proof $f := \eta^{24}$ is also holomorphic on $\mathbb{H}$. (6.3) leads to

$$f(\tau) = e^{2\pi i \tau} \cdot \prod_{m=1}^{\infty} \left(1 - e^{2\pi i m\tau}\right)^{24} = e^{2\pi i \tau} + \cdots, \tag{*}$$

hence f possesses a FOURIER expansion with $\alpha_f(0) = 0$ and $\alpha_f(1) = 1$. (6.4) and Proposition 6.4 show that
$$f|_{12} M = f \tag{**}$$
is satisfied for $M = T$ and $M = J$. Since T and J generate the modular group Γ by Theorem II.2.2, (**) holds for all $M \in \Gamma$. Thus $f \in \mathbb{S}_{12}$ follows, hence $f = \Delta^*$ with Corollary 4.5. □

Now (*) yields a new proof of the product expansion of Δ.

6.6 Corollary. *We have*
$$\Delta^*(\tau) = e^{2\pi i \tau} \cdot \prod_{m=1}^{\infty} \left(1 - e^{2\pi i m \tau}\right)^{24} \quad \text{for all } \tau \in \mathbb{H}.$$

6.7 Remarks. a) A comparison of Theorem 6.4 with the Theta Transformation Formula 0.1 shows that for
$$\psi(\tau) := \vartheta(\tau)/\eta(\tau), \quad \tau \in \mathbb{H},$$
the transformation formula $\psi(-1/\tau) = \psi(\tau)$ applies. According to (6.5) ψ is holomorphic on $\mathbb{H}$. But because of (6.4) and (0.8), $\psi(\tau + 2) = e^{\pi i/6} \cdot \psi(\tau)$ also holds. Thus there exists a 12th root of unity $v(M)$ such that
$$\psi(M\tau) = v(M) \cdot \psi(\tau) \quad \text{for all} \quad M \in \Gamma_\vartheta.$$
For the definition of the theta group Γ_ϑ compare II, §3.

b) L. EULER considered the η-product already in 1747 ([22], posthuma I, 76–84) in connection with the *partition function $p(n)$*
$$\prod_{m=1}^{\infty} (1 - x^m) = \sum_{n=0}^{\infty} p(n) \cdot x^n.$$

c) The product representation of Δ^* in Corollary 6.6 is due to C.G.J.JACOBI [42], vol. I, 154.

3*. The general transformation behavior of η. R. DEDEKIND, in his *Erläuterungen zu zwei Fragmenten von RIEMANN*, already determined the transformation behavior of $\log \eta(\tau)$ under arbitrary modular substitutions. We find a modern exposition e.g. in J. LEHNER [55], 338–344. A central role is played by the so-called DEDEKIND *sum $s(h, k)$*, which for coprime integers h, k with $k > 0$ is defined by
$$s(h, k) := \sum_{r=1}^{k-1} \left(\frac{r}{k} - \frac{1}{2}\right) \cdot \left(\frac{rh}{k} - \left[\frac{rh}{k}\right] - \frac{1}{2}\right). \tag{6.6}$$

We formulate the general transformation behavior of η as

6.8 DEDEKIND's Theorem. *For* $M = \begin{pmatrix} a & b \\ c & d \end{pmatrix} \in \Gamma$ *with* $c > 0$, *one has*

§ 6 The DEDEKIND eta–function

$$\eta(M\tau) = v(M) \cdot \sqrt{\frac{c\tau + d}{i}} \cdot \eta(\tau) \quad \text{with} \quad v(M) := e^{\pi i \left(\frac{a+d}{12c} + s(-d,c) - \frac{1}{4}\right)}.$$

A *proof* can be found, for example, in J. LEHNER [55], 338–344, or T.M. APOSTOL [5], Theorem 3.4.

4*. Various proofs of the eta transformation formula.

a) C.L. SIEGEL ([78], vol. III, 188) gave a one-page proof in 1954, which he himself explained in much more detail in his *Lectures on advanced analytic number theory* (Tata Institute of Fundamental Research, Bombay 1961) and *Analytic Number Theory II* (Göttingen 1963/64, 11–17). Compare K. CHANDRASEKHARAN [10], 126–131.
A short proof using the WEIERSTRASS ζ-function is due to H. PETERSSON and was extended in 2006 by J. ELSTRODT (*Manuscripta Math.* **121**, 457-459).

b) For $\tau \in \mathbb{H}$ and $s \in \mathbb{C}$ with Re $s > 0$, the series

$$E(\tau;s) := {\sum_{m,n}}' (m\tau + n)^{-2} \cdot |m\tau + n|^{-s}$$

is absolutely convergent by the Convergence Lemma 2.2. The series $E(\tau;s)$ is certainly not holomorphic in τ, but for fixed $\tau \in \mathbb{H}$ is holomorphic in s for Re $s > 0$. But $E(\tau;s)$ has a clear behavior under all modular substitutions $\tau \mapsto M\tau$, $M \in \Gamma$. We now try to continue $E(\tau;s)$ holomorphically to $s = 0$:

6.9 Proposition. a) *For fixed $\tau \in \mathbb{H}$, $E(\tau;s)$ possesses a continuation as an entire function into the s-plane.*
b) *We have*

$$E(\tau;0) = \frac{\pi^2}{3} - \frac{\pi}{\operatorname{Im}\tau} - 8\pi^2 \cdot \sum_{m \geq 1} \sigma_1(m) \cdot e^{2\pi i m\tau}.$$

c) *For any $M \in \Gamma$,*

$$E(\tau;0)|_2 M = E(\tau;0)$$

holds.

For a *proof* compare B. SCHOENEBERG [71], 63–68, or T. MIYAKE [59], §7.2. This summation procedure is often named after E. HECKE, who already used it in 1925 [36], 412.
A comparison with Proposition 6.1 shows that $E(\tau;0) + \pi/\operatorname{Im}\tau = E_2(\tau)$ holds. Part c) of Proposition 6.9 can be obtained from Proposition 6.2.

c) B. SCHOENEBERG (*Mitt. Math. Ges. Hamburg* **9**, Heft 4, 4–11) found a proof based on the functional equation of a simple L-series in 1968. We also refer to the references cited there.
d) Another proof will be given in IV, §4.

5*. Extremal modular forms.
In this section we describe the modular form $f \in \mathbb{M}_k$ which attains the value 1 with the highest possible order at ∞. This modular form will be used in chapter V to characterize extremal lattices. The results are due to C.L. SIEGEL [78], vol. IV, 82–97.

In this section, let $k \geq 4$ always be even and $t := \dim \mathbb{M}_k$. If we set

$$E_0^* := 1, \tag{6.7}$$

then Proposition 4.1 and the Dimension Formula 4.6 lead to

$$\frac{E_{14}^*}{E_{12t-k+2}^*} = E_{k-12t+12}^*. \tag{6.8}$$

Now we consider the weakly holomorphic modular form

$$g_k := E_{12t-k+2}^* \cdot \Delta^{*-t} \in \mathbb{M}_{2-k}^!. \tag{6.9}$$

Because of $j' = -2\pi i \cdot E_{14}^* \cdot \Delta^{*-1}$ by Corollary 5.10, we immediately obtain

$$g_k = -\frac{1}{2\pi i} \cdot \frac{1}{E_{k-12t+12}^* \cdot \Delta^{*t-1}} \cdot j'. \tag{6.10}$$

from (6.8). Corollary 6.6 leads to

$$\Delta^{*-t}(\tau) = e^{-2\pi i t \tau} \cdot \prod_{m=1}^{\infty} \left(\sum_{n=0}^{\infty} e^{2\pi i m n \tau} \right)^{24t}. \tag{6.11}$$

Consequently, g_k has a FOURIER expansion of the form

$$g_k(\tau) = \sum_{m \geq -t} \beta_k(m) \cdot e^{2\pi i m \tau}, \quad \beta_k(-t) = 1. \tag{6.12}$$

6.10 Lemma. *Let $f \in \mathbb{M}_k$ with* FOURIER *coefficients $\alpha_f(m)$. Then*

$$\sum_{m=0}^{t} \alpha_f(m) \cdot \beta_k(-m) = 0. \tag{6.13}$$

Proof We consider the modular form $h := E_{k-12t+12}^* \cdot \Delta^{*t-1} \in \mathbb{M}_k$. The roots of h in $\mathbb{H}$ can be calculated from Corollary 4.2 as well as (4.8), (4.9). From (4.10) it immediately follows that $f/h \in \mathbb{K}$ is holomorphic on $\mathbb{H}$ and thus by Corollary 5.6 is a polynomial in j. (6.10) thus implies

$$f \cdot g_k \in \mathbb{C}[j] \cdot j'. \tag{6.14}$$

Because of (6.11) the left-hand side of (6.13) is its constant FOURIER coefficient. However, for $\ell \geq 0$, we have

$$j^\ell \cdot \frac{dj}{d\tau} = \frac{1}{\ell+1} \frac{dj^{\ell+1}}{d\tau}.$$

§ 6 The DEDEKIND eta–function

Hence the constant FOURIER coefficient of a weakly holomorphic modular form in $\mathbb{C}[j] \cdot j' = \mathbb{M}_2^!$ is 0. □

It is now essential to show that $\beta_k(0) \neq 0$ holds.

6.11 Proposition. *For even $k \geq 4$, one has*

$$(-1)^{k/2} \beta_k(0) < 0.$$

Proof (i) Let $k \equiv 2 \pmod 4$. Then we have

$$E^*_{12t-k+2} = E_4^{*\nu}, \quad \nu \in \{0, 1, 2\},$$

if we observe (4.8) and (2.10). Then (6.9) and (6.11) show that

$$\beta_k(m) > 0 \quad \text{for all} \quad m \geq -t.$$

(ii) Let $k \equiv 0 \pmod 4$, $k \equiv 4\nu \pmod{12}$ with $\nu \in \{0, 1, 2\}$. Then

$$E^*_{k-12t+12} = E_4^{*\nu}.$$

If we use (6.10) and (2.19), a simple calculation yields

$$2\pi i \cdot g_k = -E_4^{*-\nu} \cdot \Delta^{*1-t} \cdot \frac{dj}{d\tau} = \frac{3}{\nu - 3} \Delta^{*1-t-\nu/3} \cdot \frac{dj^{1-\nu/3}}{d\tau}$$

$$= \frac{3}{\nu - 3} \cdot \frac{d(E_4^{*3-\nu} \cdot \Delta^{*-t})}{d\tau} + \frac{3t + \nu - 3}{(3-\nu)t} E_4^{*3-\nu} \cdot \frac{d\Delta^{*-t}}{d\tau}.$$

From the last expression, we conclude that $\beta_k(0)$ is also the constant coefficient in the FOURIER series expansion of

$$\frac{1}{2\pi i} \cdot \frac{3t + \nu - 3}{(3-\nu)t} \cdot E_4^{*3-\nu} \cdot \frac{d}{d\tau} \Delta^{*-t}.$$

All FOURIER coefficients $\alpha(m)$, $m \geq 0$, of $E_4^{*3-\nu}$ are positive. From (6.11) we conclude that the FOURIER coefficients $\gamma(m)$ of $\frac{1}{2\pi i} \cdot \frac{d\Delta^{*-t}}{d\tau}$ satisfy

$$\gamma(m) < 0 \quad \text{for} \quad -t \leq m < 0 \quad \text{and} \quad \gamma(0) = 0.$$

Thus we get

$$\beta_k(0) < 0. \quad \square$$

We now describe the "extremal" modular form.

6.12 Corollary. *Let $k \geq 4$ be even. There is a unique modular form $f \in \mathbb{M}_k$ with the properties*

$$\alpha_f(0) = 1 \quad \text{and} \quad \alpha_f(m) = 0 \quad \text{for} \quad 0 < m < t. \tag{6.15}$$

In this case

$$(-1)^{k/2}\alpha_f(t) > 0.$$

We call f the *extremal modular form of weight k*.

Proof The claim follows from Theorem 4.10 and Lemma 6.10. □

6.13 Remarks. a) We have $\beta_{12}(0) = -196560$ and $\beta_{12}(-1) = 24$. So in the case $k \equiv 0 \pmod 4$ the numbers $\beta_k(m)$, $-t < m \leq 0$, do not necessarily have the same sign. Further statements about the arithmetic character of $\beta_k(m)$ can be found in C.L. SIEGEL [78], vol. IV, 87.

b) $f(\tau) = E_{12}^*(\tau) - \frac{65520}{691} \cdot \Delta^*(\tau)$ is the extremal modular form of the weight 12. In Corollary V.2.20 it will turn out that $f(\tau)$ is the theta series of the LEECH lattice.

6.14 Exercises.

1) If $f \in \mathbb{M}_k$, then $g(\tau) := k \cdot f(\tau) \cdot E_2(\tau) + 2\pi i \cdot f'(\tau)$ belongs to $\mathbb{M}_{k+2}$. Here g is a cusp form if and only if f is a cusp form.

2) $\frac{k\pi^2}{3} \cdot E_{k+2}^* = k \cdot E_k^* \cdot E_2 + 2\pi i \cdot E_k^{*'}$ for $k = 4, 6, 8, 12$.

3) $E_2(i) = \pi$ and $E_2(\rho) = 2\pi/\sqrt{3}$ for $\rho = \left(1 + i\sqrt{3}\right)/2$.

4) $E_2|_2 M(\tau) = E_2(\tau) - 2\pi i c/(c\tau + d)$ for all $M \in \Gamma$.

5) If $n \in \mathbb{N}$ and $g(\tau) := E_2(\tau) - n \cdot E_2(n\tau)$, then $g|_2 M = g$ holds for all $M \in \Gamma_0[n]$.

6) $\chi(M) := \eta^2|_1 M(\tau)/\eta^2(\tau)$, $M \in \Gamma$, is independent of $\tau \in \mathbb{H}$. χ is an abelian character of Γ and $\chi^{12}(M) = 1$.

7) The abelian characters of Γ are exactly the mappings $\chi^j, 0 \leq j < 12$.

8) $\beta_k(m) \in \mathbb{Z}$ holds for all $m \geq -t$ in sect. 5.

9) Describe the extremal modular form for $k = 12, 16, 18, 20, 22, 26$ concretely and calculate the FOURIER coefficient of 2 explicitly.

10) If $f \in \mathbb{M}_k$, then $g := f/E_{k-12t+12}^*$ belongs to $\mathbb{M}_{12(t-1)}$. If f is a cusp form, then g is a cusp form. If $f \in \mathbb{M}_k^\mathbb{Z}$ then g has integral FOURIER coefficients.

11) Let $f \in \mathbb{M}_k$ have rational FOURIER coefficients and $\tau_0 \in \mathbb{H}$ with $f(\tau_0) = 0$. Then $j(\tau_0)$ is algebraic over $\mathbb{Q}$ of degree $\leq \dim \mathbb{M}_k$. In particular, $j(\tau_0)$ is algebraic if $E_k^*(\tau_0) = E_k(\tau_0) = 0$.

12) Let f be holomorphic on $\mathbb{H}$ and at ∞. Then $f|_k M = \pm f$ holds for all $M \in \Gamma$ if and only if $f \in \mathbb{M}_k$ or $f \in \eta^{12} \cdot \mathbb{M}_{k-6}$.

13) For $g(\tau)$ from (0.11), we have $g = -4i \cdot \eta(\tau)^6$.

14) Apply the Partial Fractions Development of the Cotangent B.2 in order to reprove that the EISENSTEIN series $E_2(\tau)$ in (6.1) does not converge absolutely.

15) Apply JACOBI's Triple Product Identity I.6.21 in order to demonstrate that the theta series (0.7) can be expressed as

$$\vartheta(\tau) = \eta((\tau+1)/2)^2/\eta(\tau+1), \tau \in \mathbb{H}.$$

16) Show that $v(M)$ in Remark 6.7 a) is an abelian character of Γ_ϑ. Can v be extended to an abelian character of Γ?

§ 7 Modular forms for congruence subgroups

In this section, the central finiteness statements for modular forms with respect to congruence groups will be derived from the results for the full modular group Γ.

1. The notion of a modular form for a congruence subgroup. If $n \in \mathbb{N}$ is given, we repeat the definition of the *prinicipal congruence group* (mod n) from II(3.4):

$$\Gamma[n] := \{M \in \Gamma \,;\, M \equiv I \pmod{n}\}.$$

A subgroup Λ of Γ is called a *congruence subgroup* if there is an $n \in \mathbb{N}$ satisfying $\Gamma[n] \subseteq \Lambda$. Each congruence subgroup has finite index in Γ. A group homomorphism

$$\chi : \Lambda \longrightarrow \{z \in \mathbb{C} \,;\, |z| = 1\}$$

is called an *abelian character of* Λ. The character, which maps every $M \in \Lambda$ to 1, is called the *trivial character* and is by abuse of notation denoted by 1. An abelian character χ is called *finite* if there is a positive $m \in \mathbb{Z}$ with $\chi^m \equiv 1$. We say that χ is a *character* mod n *of* Λ if $\Gamma[n] \subseteq \Lambda$ and $\chi(M) = 1$ for all $M \in \Gamma[n]$ holds. Clearly, the abelian characters of Λ form a group under pointwise multiplication.

As important examples, we note

$$\Lambda = \Gamma_0[p], \ p > 2 \ \text{prime number}, \ \chi(M) = \left(\frac{d}{p}\right) \quad \text{Legendre symbol} \tag{7.1}$$

(cf. II, §3 and V, §2) and with II(3.14)

$$\Lambda = \Gamma_\vartheta \,,\ \chi_\vartheta(M) = \begin{cases} 1, & \text{if } M \in \Gamma[2] \,, \\ -1, & \text{if } M \notin \Gamma[2] \,. \end{cases} \tag{7.2}$$

7.1 Lemma. *Let Λ be a congruence group. Then every abelian character* mod n *of Λ is a finite character.*

Proof $\Lambda/\Gamma[n]$ is a finite group whose order we denote by m. According to Lagrange's Theorem for the factor group, we have $L^m \in \Gamma[n]$ for all $L \in \Lambda$, hence $\chi^m(L) = \chi(L^m) = 1$. □

Now let $k \in \mathbb{Z}$, Λ be a congruence group and χ be an abelian character of Λ. A function $f : \mathbb{H} \to \mathbb{C}$ is called a *modular form of weight k for the congruence group Λ and the character χ* if

(MC.1) f is holomorphic on $\mathbb{H}$.
(MC.2) $f|_k L = \chi(L) \cdot f$ for all $L \in \Lambda$.
(MC.3) $f|_k M$ is holomorphic at ∞ for every $M \in \Gamma$.

The set $\mathbb{M}_k(\Lambda, \chi)$ of all modular forms of weight k for Λ and χ is obviously a vector space over $\mathbb{C}$. For the trivial character we use the abbreviation $\mathbb{M}_k(\Lambda) := \mathbb{M}_k(\Lambda, 1)$. Obviously, we have

$$\mathbb{M}_k = \mathbb{M}_k(\Gamma).$$

As a first new example, (7.2), Corollary II(3.9), the Theta Transformation Formula 0.1 and (4.22) imply

$$\vartheta^4 \in \mathbb{M}_2(\Gamma_\vartheta, \chi_\vartheta). \tag{7.3}$$

Of course, we can multiply modular forms for a congruence group Λ and abelian characters χ, χ':

$$\mathbb{M}_k(\Lambda, \chi) \cdot \mathbb{M}_\ell(\Lambda, \chi') \subseteq \mathbb{M}_{k+\ell}(\Lambda, \chi \cdot \chi'). \tag{7.4}$$

Using (MC.2) for $L = -I$, we get a first trivial existence condition.

7.2 Proposition. *Let $k \in \mathbb{Z}$, Λ be a congruence group with $-I \in \Lambda$ and χ an abelian character of Λ. If $\chi(-I) \neq (-1)^k$ holds, then $\mathbb{M}_k(\Lambda, \chi) = \{0\}$.*

As a direct conclusion from the definition, we note

7.3 Proposition. *Let $k \in \mathbb{Z}$, Λ be a congruence group, χ an abelian character of Λ, and $M \in \Gamma$. Then*

$$\chi_M(K) := \chi(MKM^{-1}) \quad \text{for all} \quad K \in M^{-1}\Lambda M$$

is an abelian character of $M^{-1}\Lambda M$ and the map

$$\mathbb{M}_k(\Lambda, \chi) \longrightarrow \mathbb{M}_k(M^{-1}\Lambda M, \chi_M), \ f \longmapsto f|_k M,$$

is an isomorphism of vector spaces.

2. The FOURIER expansions. In view of (MC.3), we are interested in the FOURIER expansions at arbitrary cusps of Λ.

7.4 Proposition. *Let $k \in \mathbb{Z}$, $\Lambda \supseteq \Gamma[n]$ be a congruence group and χ an abelian character mod n of Λ. If $f \in \mathbb{M}_k(\Lambda, \chi)$ and $M \in \Gamma$, then $f|_k M$ has a FOURIER series expansion of the form*

$$f|_k M(\tau) = \sum_{m=0}^{\infty} \alpha_f(m; M) \cdot e^{2\pi i m \tau/n}, \ \tau \in \mathbb{H}, \tag{7.5}$$

which for each $\varepsilon > 0$ converges absolutely uniformly on the set $\{\tau \in \mathbb{H} \ ; \ \text{Im}\,\tau \geq \varepsilon\}$. The FOURIER coefficients $\alpha_f(m; M)$ are uniquely determined and satisfy

$$\alpha_f(m; LM) = \chi(L) \cdot \alpha_f(m; M) \text{ for all } m \in \mathbb{N}_0, \ L \in \Lambda \text{ and } M \in \Gamma. \tag{7.6}$$

Proof Since $\Gamma[n]$ is a normal subgroup of Γ, we have $M\Gamma[n]M^{-1} = \Gamma[n]$. So $f|_k M \in \mathbb{M}_k(\Gamma[n])$ follows from Proposition 7.3. If we consider

$$g(\tau) := f|_k M(n\tau), \ \tau \in \mathbb{H},$$

then g is holomorphic in $\mathbb{H}$ and at ∞ as well as periodic with period 1. Applying Lemma 1.2 to g, we obtain the existence and uniqueness of the FOURIER series expansion (7.5) for $f|_k M$. If $L \in \Lambda$, then we use the FOURIER series expansion for

§ 7 Modular forms for congruence subgroups

$f|_k LM$. If we apply (MC.2), then (7.6) follows from the uniqueness of the FOURIER coefficients. □

3. The transition to the full modular group. In this section we describe two ways of constructing modular forms for the full modular group from modular forms for congruence groups. This principle will be used again for the HECKE operators in Chapter IV. The basis is the following purely algebraic consideration.

7.5 Lemma. *Let S be a subgroup of a group G with finite index m. If $g \in G$ and $g_1, \ldots, g_m$ is a system of representatives of the right cosets of G for S, i.e.*

$$G = \bigcup_{j=1}^{m} Sg_j. \tag{7.7}$$

Then $g_1 g, \ldots, g_m g$ is also a system of representatives of the right cosets.

Proof By assumption, G has exactly m right cosets with respect to the subgroup S. However, as the right cosets $Sg_1, \ldots, Sg_m$ are disjoint, so are the right cosets $Sg_1 g, \ldots, Sg_m g$. □

We now apply these considerations to modular forms.

7.6 Corollary. *Let $k \in \mathbb{Z}$ and Λ be a congruence group of index m in Γ. If $M_1, \ldots, M_m$ form a system of representatives of the right cosets in Γ with respect to Λ and if $f \in \mathbb{M}_k(\Lambda)$, then*

a) $\operatorname{tr}(f) := \displaystyle\sum_{j=1}^{m} f|_k M_j \in \mathbb{M}_k,$

b) $\pi(f) := \displaystyle\prod_{j=1}^{m} f|_k M_j \in \mathbb{M}_{km}.$

Proof Because of $f|_k L = f$ for $L \in \Lambda$, the definitions of $\operatorname{tr}(f)$ and $\pi(f)$ do not depend on the choice of representatives of the right cosets. Then Lemma 7.5 and (7.4) lead to

$$\operatorname{tr}(f)|_k M = \sum_{j=1}^{m} f|_k (M_j M) = \operatorname{tr}(f)$$

and

$$\pi(f)|_{km} M = \prod_{j=1}^{m} f|_k (M_j M) = \pi(f) \quad \text{for all } M \in \Gamma. \qquad \square$$

We call $\operatorname{tr}(f)$ the *trace of f*.

4. Negative weight. If the weight k is non-positive, the results are similar to the results for $\mathbb{M}_k$.

7.7 Proposition. *Let $k \in \mathbb{Z}$, Λ be a congruence group and χ be a finite abelian character of Λ. Then*
a) $\mathbb{M}_k(\Lambda, \chi) = \{0\}$ if $k < 0$.
b) $\mathbb{M}_0(\Lambda) = \mathbb{C}$ and $\mathbb{M}_0(\Lambda, \chi) = \{0\}$ if $\chi \neq 1$.

Proof Let $m \in \mathbb{N}$ with $\chi^m = 1$ and $k \leq 0$. If $f \in \mathbb{M}_k(\Lambda, \chi)$, then $g := f^m$ belongs to $\mathbb{M}_{km}(\Lambda)$ by (7.4). With $\ell := [\Gamma : \Lambda]$, we consider $\pi(g) \in \mathbb{M}_{\ell km}$ according to Corollary 7.6.

a) As $\ell km < 0$, it follows that $\pi(g) = 0$ from Proposition 1.6. The Identity Theorem A.1 implies $g = 0$ and hence $f = 0$.

b) We first show that g is constant. Because the constants belong to $\mathbb{M}_0(\Lambda)$, we may assume $\alpha_g(0; I) = 0$ in the FOURIER expansion of g by Proposition 7.4. From this the claim follows, due to

$$\lim_{y \to \infty} \pi(g)(iy) = 0, \text{ hence } \pi(g) = 0$$

with Proposition 4.1. Then we get $g = 0$ and $f = 0$. Consequently, an arbitrary modular form $f \in \mathbb{M}_0(\Lambda, \chi)$ is constant. For $\chi \neq 1$ we get $f = 0$ from (MC.2). □

5. Positive weight. For positive weights, we give a dimension bound.

7.8 Proposition. *Let $k \in \mathbb{N}$, Λ be a congruence group, χ an abelian character mod n of Λ. Let $\Lambda^* := \{L \in \Lambda ; \chi(L) = 1\}$ and $\ell := [\Gamma : \Lambda^*]$. If $f \in \mathbb{M}_k(\Lambda, \chi)$ and $M \in \Gamma$ with the property*

$$\alpha_f(m; M) = 0 \quad \text{for} \quad 0 \leq m \leq \frac{\ell k n}{12}, \tag{7.8}$$

then $f = 0$.

Proof We consider f as an element of $\mathbb{M}_k(\Lambda^*)$ and $g = \pi(f)$ in $\mathbb{M}_{\ell k}$ according to Corollary 7.6. Multiplying the FOURIER series in Proposition 7.4, we obtain

$$\alpha_g(m) = 0 \quad \text{for} \quad 0 \leq m \leq \frac{\ell k}{12}$$

from (7.8). Then $g = 0$ follows from Corollary 4.7, hence $f = 0$ due to the Identity Theorem A.1. □

7.9 Corollary. *Under the assumptions of Proposition 7.8, we have*

$$\dim \mathbb{M}_k(\Lambda, \chi) \leq \left\lfloor \frac{\ell k n}{12} \right\rfloor + 1.$$

An exact determination of the dimension seems to be impossible with the methods used so far. Applying the RIEMANN–ROCH Theorem, we find explicit formulas, e.g. by B. SCHOENEBERG [71], G. SHIMURA [74] or T. MIYAKE [59].

6. Cusp forms. Let $k \in \mathbb{Z}$, Λ be a congruence subgroup and χ an abelian character of Λ. An $f \in \mathbb{M}_k(\Lambda, \chi)$ is called a *cusp form* if $f|_k M$ has a root at ∞ for every $M \in \Gamma$. We denote the subspace of cusp forms by $\mathbb{S}_k(\Lambda, \chi)$. From Proposition 7.3, we immediately conclude

$$f \in \mathbb{S}_k(\Lambda, \chi), \ M \in \Gamma \implies f|_k M \in \mathbb{S}_k(M^{-1}\Lambda M, \chi_M). \tag{7.9}$$

If $f : \mathbb{H} \to \mathbb{C}$, then we define, just as in (1.15),
$$\tilde{f} : \mathbb{H} \longrightarrow \mathbb{R}, \quad \tau \longmapsto y^{k/2} \cdot |f(\tau)|. \tag{7.10}$$

7.10 Proposition. *Let $k \in \mathbb{N}$, Λ a congruence subgroup, χ an abelian character mod n of Λ and $f \in \mathbb{M}_k(\Lambda, \chi)$. Then:*
a) $\tilde{f}$ *is Λ–invariant, i.e. $\tilde{f}(L\tau) = \tilde{f}(\tau)$ for all $L \in \Lambda$.*
b) $\tilde{f}$ *is bounded on $\mathbb{H}$ if and only if f is a cusp form.*
c) *If $f \in \mathbb{S}_k(\Lambda, \chi)$, then $\alpha_f(m; M) = O\left(m^{k/2}\right)$ for all $m \in \mathbb{N}$ and $M \in \Gamma$.*

Proof a) We use II(1.14) and (MC.2) in sect. 1.
b) As in the proof of Proposition 7.7, we conclude that $\tilde{f}$ is bounded in $\mathbb{F}$ if and only if $\alpha_f(0; I) = 0$. Now let $\ell := [\Gamma : \Lambda']$, $\Lambda' = \{\pm L; L \in \Lambda\}$ and $M_1, \ldots, M_\ell$ be a system of representatives of the right cosets. By Theorem II.3.1,
$$\mathcal{F}(\Lambda) := \bigcup_{1 \le \nu \le \ell} M_\nu \overline{\mathbb{F}}$$
is a fundamental domain with respect to Λ. Thus, because of (7.9), $\tilde{f}$ is bounded in $\mathcal{F}(\Lambda)$, if and only if $\alpha_f(0; M_\nu) = 0$ for $1 \le \nu \le \ell$, hence $\alpha_f(0; M) = 0$ for all $M \in \Gamma$ because of (MC.2) and (7.6). Then the assertion follows from a).
c) According to (7.9) and part b), there exists a constant C with the property
$$\widetilde{f|M}(\tau) \le C \quad \text{for all} \quad \tau \in \mathbb{H} \quad \text{and} \quad M \in \Gamma. \tag{7.11}$$

In the explicit formula
$$\alpha_f(m; M) = \frac{1}{n} \cdot e^{2\pi m y/n} \cdot \int_0^n f|M(x + iy) \cdot e^{-2\pi i m x/n} dx,$$
we use (7.11) and then substitute $y = 1/m$. Thus we get
$$|\alpha_f(m; M)| \le \frac{1}{n} \cdot y^{-k/2} \cdot e^{2\pi m y/n} \cdot \int_0^n \widetilde{f|M}(x + iy)\, dx \le C \cdot e^{2\pi/n} \cdot m^{k/2}. \qquad \square$$

7.11 Corollary. *We have*
$$\mathbb{S}_2(\Gamma_0[2]) = \{0\}.$$

Proof We have for $f \in \mathbb{S}_2(\Gamma_0[2])$ in the notation of Corollary 7.6
$$\pi(f) \in \mathbb{S}_6 = \{0\}. \qquad \square$$

7.12 Remark. A. WILES (*Ann. Math.* **141**, 443-551 (1995)) was able to show that every semistable elliptic curve is modular, i.e. comes from a non-trivial cusp form of weight 2 for $\Gamma_0[r]$. G. FREY described a construction which assigns an elliptic curve arising from a non-trivial cusp form of weight 2 with respect to $\Gamma_0[2]$ to each non-trivial solution $x^n + y^n = z^n$, $n \geq 3$, of the FERMAT equation. Such a cusp form cannot exist according to Corollary 7.11.

7*. Sums of squares. As a number theoretic application of the previous results, we describe the representation numbers of a natural number as the sum of 4 or 8 squares. For $k \in \mathbb{N}$ and $m \in \mathbb{N}_0$ let

$$\delta_k(m) := \#\{(g_1, \ldots, g_k)^{tr} \in \mathbb{Z}^k;\ g_1^2 + \ldots + g_k^2 = m\}, \tag{7.12}$$

so in particular $\delta_k(0) = 1$. Then, by the theta series $\vartheta(\tau)$ in (0.7), it follows that

$$\vartheta^k(\tau) = \sum_{g_1, \ldots, g_k \in \mathbb{Z}} e^{\pi i \tau (g_1^2 + \ldots + g_k^2)} = \sum_{m=0}^{\infty} \delta_k(m) \cdot e^{\pi i m \tau}. \tag{7.13}$$

We adopt the definition of the normalized EISENSTEIN series $E_k^*(\tau)$ from (2.4) and (6.2), respectively.

7.13 Proposition. a) *For all $\tau \in \mathbb{H}$, we have*

$$\vartheta^4(\tau) = \frac{1}{3}\left(4 \cdot E_2^*(2\tau) - E_2^*\left(\frac{\tau}{2}\right)\right), \tag{7.14}$$

$$\vartheta^8(\tau) = \frac{1}{15}\left(16 \cdot E_4^*(\tau) - E_4^*\left(\frac{\tau+1}{2}\right)\right). \tag{7.15}$$

b) *For all $m \in \mathbb{N}$, we have*

$$\delta_4(m) = 8 \cdot \sum_{d \mid m,\, 4 \nmid d} d, \tag{7.16}$$

$$\delta_8(m) = 16 \cdot \sum_{d \mid m} (-1)^{m-d} d^3. \tag{7.17}$$

Proof From Corollary II(3.9) and Proposition 4.17, we conclude $\vartheta^8 \in \mathbb{M}_4(\Gamma_\vartheta)$. On the other hand, let

$$f(\tau) = \frac{1}{15}\left(16 \cdot E_4^*(\tau) - E_4^*\left(\frac{\tau+1}{2}\right)\right) = \sum_{m=0}^{\infty} \alpha_f(m) \cdot e^{\pi i m \tau}.$$

Then, by (2.10), $\alpha_f(0) = 1$ holds and for $m \in \mathbb{N}$, $\alpha_f(m)$ is equal to

$$\begin{cases} 16^2 \sigma_3(m/2) - 16 \sigma_3(m) \\ 16 \sigma_3(m) \end{cases} = \begin{cases} 16\left(2 \sum_{d \mid m,\, 2 \mid d} d^3 - \sum_{d \mid m} d^3\right), & \text{if } 2 \mid m, \\ 16 \sigma_3(m), & \text{if } 2 \nmid m \end{cases} \tag{*}$$

$$= 16 \cdot \sum_{d \mid m} (-1)^{m-d} d^3.$$

$E_4^* \in \mathbb{M}_4$ leads to $f|_4 J = f|_4 T^2 = f$. From

§ 7 Modular forms for congruence subgroups 209

$$f(\tau + 1) = \frac{1}{15}(16 E_4^*(\tau) - E_4^*(\tau/2)), \quad \tau^{-4} \cdot f(1 - 1/\tau) = \frac{16}{15}(E_4^*(\tau) - E_4^*(2\tau))$$

we also get $f \in \mathbb{M}_4(\Gamma_\vartheta)$. Now we verify directly that

$$\alpha_f(0) = \delta_8(0) = 1, \quad \alpha_f(1) = \delta_8(1) = 16, \quad \alpha_f(2) = \delta_8(2) = 112.$$

If we now apply Proposition 7.8 to $\vartheta^8 - f$, $\Lambda = \Lambda^* = \Gamma_\vartheta$, $k = 4$, $l = 3$, $n = 2$, then (7.15) follows . A comparison of the FOURIER coefficients in (7.13) and (∗) yields (7.17).

Corollary II.3.9, (7.2) and Proposition 4.17 lead to $\vartheta^4 \in \mathbb{M}_2(\Gamma_\vartheta, \chi_\vartheta)$. On the other hand, let

$$g(\tau) = \frac{1}{3}(4 \cdot E_2^*(2\tau) - E_2^*(\tau/2)) = \sum_{m=0}^{\infty} \alpha_g(m) \cdot e^{\pi i m \tau}.$$

Then $\alpha_g(0) = 1$ holds by (6.2), and for $m \in \mathbb{N}$,

$$\alpha_g(m) = \begin{cases} 8\sigma_1(m) - 32\sigma_1(m/4), & \text{if } 4 \mid m \\ 8\sigma_1(m), & \text{if } 4 \nmid m \end{cases} = \sum_{d \mid m, 4 \nmid d} d. \quad (**)$$

We have $g(\tau + 2) = g(\tau)$ and by Proposition 6.2 also $g|_2 J = -g$. From

$$g(\tau + 1) = \frac{1}{3}\left(4 \cdot E_2^*(2\tau) - E_2^*\left(\frac{\tau+1}{2}\right)\right),$$

$$\tau^{-2} \cdot g\left(1 - \frac{1}{\tau}\right) = \frac{1}{3}\left(E_2^*\left(\frac{\tau}{2}\right) - E_2^*\left(\frac{\tau+1}{2}\right)\right)$$

and Corollary II.3.9 we also get $g \in \mathbb{M}_2(\Gamma_\vartheta, \chi_\vartheta)$. Now we verify directly that

$$\alpha_g(0) = \delta_4(0) = 1, \quad \alpha_g(1) = \delta_4(1) = 8, \quad \alpha_g(2) = \delta_4(2) = 24.$$

Applying Proposition 7.8 to $\vartheta^4 - g$ with $\Lambda = \Gamma_\vartheta$, $\Lambda^* = \Gamma[2]$, $k = 2$, $l = 6$, $n = 2$, (7.14) follows. A comparison of the FOURIER coefficients yields (7.16). □

From (7.16) we get $\delta_4(m) \geq 8$ for all $m \in \mathbb{N}$. So the consequence is

7.14 Corollary. *Every positive integer is the sum of 4 squares of integers.*

7.15 Remarks. a) The *Four-Squares-Theorem* in Corollary 7.14 was already conjectured in 1659 by P. DE FERMAT. L. EULER was able to reduce the assertion to prime numbers in 1748. The first complete proof was given by J. LAGRANGE in 1770 ([51], vol. III, 189-201). We therefore also call Corollary 7.14 the LAGRANGE *Theorem*.
b) Explicit formulas for $\delta_k(m)$ are known for even $2 \leq k \leq 18$. Compare J. GLAISHER (*Proc. London Math. Soc., Ser. 2*, **5**, 479–490 (1907)).

8. Modular functions. Let Λ be a subgroup of Γ of finite index. A function $f : \mathbb{H} \to \mathbb{C} \cup \{\infty\}$ is called a *modular function for* Λ if

(MF.1) f is meromorphic on $\mathbb{H}$.

(MF.2) $f \mid_0 L = f$ for all $L \in \Lambda$.
(MF.3) $f \mid_0 M$ has at most a pole at ∞ for all $M \in \Gamma$.

The set $\mathbb{K}(\Lambda)$ of modular functions for Λ is obviously a field containing $\mathbb{K}$. As $-I$ acts trivially on $\mathbb{H}$, we conclude

$$\mathbb{K}(\Lambda) = \mathbb{K}(\Lambda'), \quad \Lambda' = \{\pm M; \ M \in \Lambda\}.$$

7.16 Theorem. *If Λ is a subgroup of Γ and $\Lambda' = \{\pm M; \ M \in \Lambda\}$ with finite index $[\Gamma : \Lambda'] = l \in \mathbb{N}$, then the following holds:*
a) *Any $f \in \mathbb{K}(\Lambda)$ is a root of the polynomial*

$$P(X) = \prod_{M : \Lambda' \backslash \Gamma} (X - f \mid_0 M) \in \mathbb{K}[X]. \tag{7.18}$$

b) *$\mathbb{K}(\Lambda)/\mathbb{K}$ is a finite field extension of degree $\leq l$ and there exists some $\varphi \in \mathbb{K}(\Lambda)$ satisfying*

$$\mathbb{K}(\Lambda) = \mathbb{C}(j)[\varphi]. \tag{7.19}$$

c) *Given $f, g \in \mathbb{K}(\Lambda)$, there exists a non-trivial polynomial $Q(X, Y) \in \mathbb{C}[X, Y]$ satisfying*

$$Q(f, g) = 0.$$

Proof a) Due to (MF.2) the definition of the polynomial $P(X)$ does not depend on the choice of the system of representatives. Let $M_1, \ldots, M_l$ be such a system. The coefficients of the polynomial $P(X)$ are exactly the elementary symmetric polynomials in $f \mid_0 M_1, \ldots, f \mid_0 M_l$ up to a sign. In view of Lemma 7.5, they are invariant under the action of any $M \in \Gamma$. Thus they are modular functions for Γ.
b) We obtain φ from the Primitive Element Theorem (cf. S. LANG [53], Theorem V.4.6). The degree of the field extension is the degree of the minimal polynomial of φ and therefore $\leq l$ due to a).
c) The claim follows as $\mathbb{K}(\Lambda)$ has transcendence degree 1 over $\mathbb{C}$ due to b). □

We consider a particular case.

7.17 Theorem. *If Λ is a normal subgroup of Γ and $\Lambda' = \{\pm M; \ M \in \Lambda\}$ with finite index in Γ, then the factor group Γ/Λ' acts on $\mathbb{K}(\Lambda)$ via*

$$f \mapsto f^M := f \mid_0 M, \quad M \in \Gamma. \tag{7.20}$$

The fixed group of this action is equal to $\mathbb{K}$ and $\mathbb{K}(\Lambda)/\mathbb{K}$ is a GALOIS extension whose GALOIS group is a subgroup of Γ/Λ'.

Proof f^M belongs to $\mathbb{K}(\Lambda)$ due to Proposition 7.3. Now apply Theorem 7.16. Consider the polynomial $P(X)$ in (7.18) for φ. Thus $\mathbb{K}(\Lambda)$ is the splitting field of $P(X) \in \mathbb{K}[X]$. This yields the assertion. □

More results on the field of modular functions for congruence subgroups can be found in B. SCHOENEBERG [71].

9*. The commutator subgroup of Γ. If G is a group, the *commutator subgroup*

§ 7 Modular forms for congruence subgroups

CG is generated by all commutators $aba^{-1}b^{-1}$, $a, b \in G$. It is a normal subgroup of G and the factor group G/CG is abelian. If the index $[G : CG]$ is finite, the factor group G/CG is isomorphic to the group of abelian characters of G.
We define the particular mapping

$$\chi_\Gamma : \Gamma \to \mathbb{C}^*, \quad M \mapsto \frac{\eta^2 |_1 M}{\eta^2}, \tag{7.21}$$

in view of (6.5). The value does not depend on τ due to (6.4) and Theorem 6.4. Then χ_Γ is an abelian character of Γ by (1.2) and it satisfies

$$\chi_\Gamma(J) = -i, \quad \chi_\Gamma(T) = e^{\pi i/6} \tag{7.22}$$

due to Theorem 6.4 and (6.4).

7.18 Theorem. *The group of abelian characters of Γ is cyclic of order 12 and is generated by χ_Γ. We have $C\Gamma = \ker \chi_\Gamma$ and*

$$\Gamma/C\Gamma \cong \mathbb{Z}/12\mathbb{Z} \tag{7.23}$$

as well as

$$\Gamma = \bigcup_{j=0,\ldots,11} C\Gamma T^j. \tag{7.24}$$

Proof First of all, χ_Γ has the order 12 due to (7.22), because J and T generate Γ. As $\Gamma/C\Gamma$ is abelian, we conclude that the set

$$\{U^\rho J^\mu; \ \rho = 0, 1, 2, \ \mu = 1, 2, 3, 4\}$$

contains representatives of the cosets of $C\Gamma$ in Γ due to Corollary II.2.4. Thus

$$C\Gamma \subseteq \ker \chi_\Gamma =: \Lambda, \quad [\Gamma : C\Gamma] \leq 12, \quad [\Gamma : \Lambda] = 12$$

follows because of

$$\Gamma/\Lambda \cong \chi_\Gamma(\Gamma) = \{e^{2\pi i j/12}; \ 0 \leq j < 12\} \cong \mathbb{Z}/12\mathbb{Z}$$

by the homomorphism theorem for groups. The cosets ΛT^j, $0 \leq j < 12$, are mutually distinct as $\chi_\Gamma(T^j) = e^{\pi i j/6}$. Thus $\Lambda = C\Gamma$ and therefore the claim follows. □

It is easy to describe the corresponding spaces of modular forms.

7.19 Corollary. *Let $0 < j < 12$ and $k \in \mathbb{Z}$. Then*

$$\mathbb{M}_k(\Gamma, \chi_\Gamma^j) = \mathbb{S}_k(\Gamma, \chi_\Gamma^j) = \eta^{2j} \mathbb{M}_{k-j}.$$

In particular,

$$\mathbb{M}_j(\Gamma, \chi_\Gamma^j) = \mathbb{S}_j(\Gamma, \chi_\Gamma^j) = \mathbb{C}\eta^{2j}.$$

Proof Any $f \in M_k(\Gamma, \chi_\Gamma^j)$ satisfies $f(\tau + 1) = e^{\pi i j/6} f(\tau)$ and therefore possesses a FOURIER series expansion of the form

$$f(\tau) = \sum_{n \in \mathbb{N}, n \equiv j \pmod{12}} \alpha_f(n) e^{2\pi i n \tau/12}.$$

Thus, f is a cusp form and f/η^{2j} is modular of weight $k - j$, holomorphic on $\mathbb{H}$ and at ∞, thus an element of M_{k-j}. Finally, note that $M_0 = \mathbb{C}$. □

7.20 Exercises.
1) $\kappa(M) := (-1)^{ac+bc+bd}$ defines an abelian character of Γ. It is the only non-trivial real character and it satisfies $\kappa = \chi_\Gamma^6$ as well as $\chi_\vartheta = \kappa|_{\Gamma_\vartheta}$.
2) $\eta^{12} \in S_6(\Gamma, \kappa)$.
3) If $n \in \mathbb{Z}$, $n > 1$ and $f \in M_k$, then $g(\tau) := f(n\tau)$ and $h(\tau) := \sum_{j=0}^{n-1} f((\tau+j)/n)$ belong to $M_k(\Gamma_0[n])$.
4) Let $n \in \mathbb{Z}$, $n > 1$. The function $f(\tau) := \Delta(n\tau)/\Delta(\tau)$ is holomorphic on $\mathbb{H}$ and satisfies $f(M\tau) = f(\tau)$ for all $M \in \Gamma_0[n]$. Does this contradict Proposition 7.7?
5) Let Λ be a congruence subgroup of Γ and $k \in \mathbb{Z}$ be even. Then tr : $M_k(\Lambda) \longrightarrow M_k$ (cf. Corollary 7.6) is a surjective homomorphism of vector spaces.
6) Let $n \in \mathbb{Z}$, $n > 1$. The mapping

$$\Phi_n : M_k(\Gamma_0[n]) \longrightarrow M_k(\Gamma_0[n]), \quad f(\tau) \longmapsto (\sqrt{n}\tau)^{-k} \cdot f(-1/n\tau),$$

is an isomorphism of vector spaces with $\Phi_n \circ \Phi_n = \text{id}$. If $\varepsilon = \pm 1$, we define

$$M_k^\varepsilon(\Gamma_0[n]) := \{f \in M_k(\Gamma_0[n]) \,;\, \Phi_n(f) = \varepsilon f\}.$$

Then

$$M_k(\Gamma_0[n]) = M_k^+(\Gamma_0[n]) \oplus M_k^-(\Gamma_0[n]).$$

7) Let $\mathbb{F}_\vartheta$ denote the fundamental domain of Γ_ϑ from Exercise II(3.18) 4). Consider $0 \neq f \in M_k(\Gamma_\vartheta)$. For $q \in \mathbb{Q}$, let

$$\text{ord}_q f := \text{ord}_\infty f|M, \quad \text{if } M \in \Gamma \text{ mit } M\infty = q.$$

Then the weight formula says that

$$2\text{ord}_\infty f + \text{ord}_1 f + \frac{1}{2}\text{ord}_i f + \sum_{w \in \mathbb{F}_\vartheta, w \neq i} \text{ord}_w f = \frac{k}{4}.$$

8) Let p be a prime number and $f \in M_k$. Then

$$f(\tau) - \frac{1}{p}\sum_{j=0}^{p-1} f(\tau + j/p) = \sum_{m \geq 0, p \nmid m} \alpha_f(m) e^{2\pi i m \tau}$$

§ 7 Modular forms for congruence subgroups

belongs to $\mathbb{M}_k(\Gamma_0[p^2])$.

9) Let $G(\sqrt{p})$, $p = 2, 3$, be the HECKE group from Exercise II.4.7 1). We define the corresponding modular forms $\mathbb{M}_k(G(\sqrt{p}))$ when replacing Γ by $G(\sqrt{p})$ in sect. 4 of §1. Then $\mathbb{M}_k(G(\sqrt{p}))$ is isomorphic to $\mathbb{M}_k^+(\Gamma_0[p])$ (cf. Exercise 6).

10) Let $f \in \mathbb{M}_k$, $N \in \mathrm{SL}(2; \mathbb{Q})$ and $n \in \mathbb{N}$ be such that nN is integral. Then $f|_k N$ belongs to $\mathbb{M}_k(\Gamma[n^2])$.

11) Let $f \in \mathbb{M}_k$ and $N \in \mathrm{Mat}(2; \mathbb{Z})$ with $\det N = n > 0$. Then $f|_k N \in \mathbb{M}_k(\Gamma[n])$.

12) Let $n \in \mathbb{N}$ and $k = \frac{24}{n+1} \in \mathbb{N}$. Then $(\eta(\tau)\eta(n\tau))^k$ belongs to $\mathbb{S}_k(\Gamma_0[n])$.

13) Let $n \in \mathbb{N}$ be a divisor of 12 and $k = 12/n$. Then η^{2k} belongs to $\mathbb{S}_k(\Gamma[n])$.

14)* Let $n > 1$, $f \in \mathbb{M}_k(\Gamma_0[n])$, χ be a DIRICHLET character mod n and let $\chi(M) := \chi(d)$ as well as

$$f_\chi(\tau) := \sum_{m=0}^{\infty} \chi(m)\alpha_f(m)e^{2\pi im\tau}.$$

a) $f_\chi(\tau) = \frac{1}{n}\sum_{l,r \bmod n} \chi(l) e^{-2\pi i l r/n} f(\tau + r/n)$.
b) $f_\chi \in \mathbb{M}_k(\Gamma_0[n^2], \chi^2)$.
c) If χ is primitive, then $f_\chi(\tau) = c \cdot \sum_{r \bmod n} \overline{\chi}(r) f(\tau + r/n)$ and $\Phi_{n^2} f_\chi = \tilde{c} \cdot f_{\overline{\chi}}$ (cf. Exercise 6).

15) Let χ be a DIRICHLET character mod N. For even $k > 2$,

$$E_k(\tau; \chi) := \sideset{}{'}\sum_{m,n \in M} \chi(m)(m\tau + n)^{-k}$$

belongs to $\mathbb{M}_k(\Gamma_0[N], \chi)$, $\chi(M) := \chi(d)$, $M \in \Gamma_0[N]$.

16) For a congruence subgroup Λ, we have $\dim \mathbb{M}_k(\Lambda) - \dim \mathbb{S}_k(\Lambda) \leq [\Gamma : \Lambda]$.

17) Given $k \in \mathbb{Z}$, show that

$$\mathbb{M}_k(C\Gamma) = \bigoplus_{j=0}^{11} \mathbb{M}_k\left(\Gamma, \chi_\Gamma^j\right).$$

18) Show that the graded ring $\mathbb{M}(C\Gamma) = \bigoplus_{k \in \mathbb{Z}} \mathbb{M}_k(C\Gamma)$ is generated by

$$\eta^2, \; E_4^*, \; E_6^*.$$

19) Show that for any $d \in \mathbb{N}$, $d|12$,

$$\chi_\Gamma(M)^{12/d} = 1 \quad \text{for all} \quad M \in \Gamma[d]$$

and that $C\Gamma$ is a congruence subgroup of level 12 (Hint: Theorem II.3.8, Exercise II.3.18 13) and 14)).

20) The commutator subgroup $C\Gamma$ consists of all words in J and U where the number of J is a multiple of 4 and the number of U is a multiple of 3. One has

$$J \in C\Gamma T^9 \quad \text{and} \quad -I \in C\Gamma T^6.$$

21) Describe a fundamental domain with respect to the commutator subgroup $C\Gamma$.
22) Let $(u, v) \in \mathbb{Q} \times \mathbb{Q}$, $(u, v) \notin \mathbb{Z} \times \mathbb{Z}$ and $r \in \mathbb{N}$ such that $(ru, rv) \in \mathbb{Z} \times \mathbb{Z}$. Show that the map
$$\tau \mapsto \wp(u\tau + v, \mathbb{Z}\tau + \mathbb{Z})$$
belongs to $\mathbb{M}_2(\Gamma[r])$.

23) Let $n \in \mathbb{N}, k > 2$ be even and
$$E_{k,n}^*(\tau) := \sum_{M : \Gamma_\infty \backslash \Gamma_0[n]} 1 \mid_k M(\tau), \quad \tau \in \mathbb{H}.$$

a) $E_{k,n}^* \in \mathbb{M}_k(\Gamma_0[n])$,

b) $\lim_{y \to \infty} E_{k,n}^* \mid_k L(iy) = \begin{cases} 1, & \text{if } L \in \Gamma_0[n], \\ 0, & \text{if } L \in \Gamma \backslash \Gamma_0[n], \end{cases}$

c) $\mathrm{tr}(E_{k,n}^*) = E_k^*$.

24) Let $n \in \mathbb{N}, k > 2$ be even and
$$E_{k,n}(\tau) := \sideset{}{'}\sum_{(c,d) \in \mathbb{Z} \times \mathbb{Z}} (nc\tau + d)^k, \quad \tau \in \mathbb{H}.$$

Then $E_{k,n} \in \mathbb{M}_k(\Gamma_0[n])$. Are $E_{k,n}$ and $E_{k,n}^*$ from Exercise 23) linearly dependent?

25) Let Λ be a normal subgroup with finite index in Γ and $\phi \in \mathbb{K}(\Lambda)$ such that $\Lambda = \{M \in \Gamma; \phi \mid_0 M = \phi\}$. Then
$$\mathbb{K}(\Lambda) = \mathbb{K}[\phi].$$

26) Let $\Gamma_N[2]$ be the normal subgroup of Γ of index 2 (cf. II(3.16)) and let Λ be the normal subgroup of Γ of index 3 from Exercise II.3.18 17). Show that
a) $\mathbb{K}(\Gamma_N[2]) = \mathbb{C}(\varphi)$, $\varphi = E_6/\eta^{12}$.
b) $\mathbb{K}(\Lambda) = \mathbb{C}(\psi)$, $\psi = E_4/\eta^8$.

27) Show that
$$\mathbb{K}(C\Gamma) = \mathbb{K}[f], \quad f = \eta^4 E_4/E_6 = \psi/\varphi.$$

$\mathbb{K}(C\Gamma)/\mathbb{K}$ is a GALOIS extension of degree 6 and the GALOIS group is cyclic generated by
$$\mathbb{K}(C\Gamma) \to \mathbb{K}(C\Gamma), \quad g(\tau) \mapsto g(\tau + 1).$$

28) Recall
$$\lambda(\tau) = \frac{\wp(1/2; \tau, 1) - \wp((\tau + 1)/2; \tau, 1)}{\wp(\tau/2; \tau, 1) - \wp((\tau + 1)/2; \tau, 1)}$$

from Exercise I.4.18 7). Show that
a) $\mathbb{K}(\Gamma[2]) = \mathbb{C}(\lambda)$.
b) $f(\tau) = \eta(\tau/2)^8 \eta(2\tau)^{16}/\eta(\tau)^{24}$ belongs to $\mathbb{K}(\Gamma[2])$ and represent f as a rational function of λ.

Chapter IV
The HECKE–PETERSSON theory

Introduction

1. The historical approach. In [36], E. HECKE (1887-1947) started to develop a systematic theory of certain arithmetically defined endomorphisms

$$T_n : \mathbb{M}_k \to \mathbb{M}_k, \quad n \in \mathbb{N},$$

of the vector space of modular forms of weight k. If p is a prime, the image of $f \in \mathbb{M}_k$ under T_p can be described as

$$(T_p f)(\tau) = p^{k-1} f(p\tau) + \frac{1}{p} \sum_{j=1}^{p} f((\tau + j)/p), \quad \tau \in \mathbb{H}. \tag{0.1}$$

These HECKE *operators* commute mutually and generate a commutative algebra $\mathcal{H}_k$ of endomorphisms of $\mathbb{M}_k$. Moreover, the subspace $\mathbb{S}_k$ of cusp forms is mapped into itself. Their importance for modular forms with respect to congruence subgroups cannot be overestimated: especially the discriminant Δ is an eigenfunction under all T_n (cf. Theorem 1.10):

$$T_n \Delta = \tau(n) \Delta \quad \text{for all} \quad n \in \mathbb{N}.$$

This result was the origin of the construction (0.1), which is due to L.J. MORDELL (*Proc. Camb. Philos. Soc.* **19**, 117-124) from 1917. His work used the investigations by S. RAMANUJAN (*Trans. Camb. Phil. Soc.* **22**, 159-184 (1916)) and J.W.L. GLAISHER (*Quart. J. Math.* **36**, 305-358 (1905)). The second important step was done by H. PETERSSON (1902–1984). He introduced a canonical scalar product on $\mathbb{S}_k$ in 1939 (*Jahresber. Deutsch. Math. Verein.* **49**, 49-75). The HECKE operators turn out to be self-adjoint with respect to this scalar product. Simple results from linear algebra then show that the HECKE operators can be transformed onto diagonal form simultaneously. Thus, each vector space $\mathbb{M}_k$ possesses a basis which consists of simultaneous eigenforms under all T_n, $n \in \mathbb{N}$. In what follows, we will describe 3 different approaches which give rise to (0.1).

2. The trace operator.

In Corollary III.7.6, we introduced the trace operator in order to lift a modular form for a congruence subgroup to a modular form for the full modular group. This construction can be viewed as a HECKE operator. Let us start with $f \in \mathbb{M}_k$ and a prime p. Then

$$g(\tau) := f(p\tau) \in \mathbb{M}_k(\Gamma_0[p]) \tag{0.2}$$

holds due to

$$pM\tau = \tilde{M} <p\tau>, \quad \tilde{M} = \begin{pmatrix} a & bp \\ c/p & d \end{pmatrix}.$$

If $M = \begin{pmatrix} a & b \\ c & d \end{pmatrix} \in \Gamma$ and $p \nmid c$, then there exists a unique $j \in \{1, \ldots, p\}$ such that

$$-cj + d \equiv 0 \pmod{p}.$$

This leads to the disjoint coset decomposition

$$\Gamma = \Gamma_0[p] \cup \bigcup_{j=1}^{p} \Gamma_0[p] \begin{pmatrix} 0 & 1 \\ -1 & -j \end{pmatrix}.$$

If $P = \begin{pmatrix} p & 0 \\ 0 & 1 \end{pmatrix}$, then Corollary III.7.6 yields

$$\text{tr}(g) = g + \sum_{j=1}^{p} g \mid \begin{pmatrix} 0 & 1 \\ -1 & -j \end{pmatrix}$$

$$= f|P + \sum_{j=1}^{p} f|JP \begin{pmatrix} 0 & 1 \\ -1 & -j \end{pmatrix}$$

$$= f(p\tau) + \sum_{j=1}^{p} p^{-k} f((\tau + j)/p).$$

A comparison with (0.1) yields the

0.1 Lemma. *If $f \in \mathbb{M}_k$ and p is a prime, then $g(\tau) = f(p\tau)$ satisfies*

$$T_p f = p^{k-1} \text{tr}(g).$$

3. HECKE operators on lattices.

We start as in Chapter I with a function F on all lattices Ω in $\mathbb{C}$ satisfying

$$F(\lambda\Omega) = \lambda^{-k} F(\Omega) \quad \text{for all} \quad 0 \neq \lambda \in \mathbb{C}. \tag{0.3}$$

Let us consider an averaging operator where we sum over all sublattices Ω' of Ω with index p, p prime:

IV The Hecke–Petersson theory

$$G(\Omega) := \sum_{\Omega':[\Omega:\Omega']=p} F(\Omega') \tag{0.4}$$

If $\Omega = \mathbb{Z}\tau + \mathbb{Z}$, $\tau \in \mathbb{H}$, without loss of generality, we have

$$\Omega' = \mathbb{Z}(a\tau + b) + \mathbb{Z}(c\tau + d), \quad M = \begin{pmatrix} a & b \\ c & d \end{pmatrix} \in \mathrm{Mat}\,(2;\mathbb{Z}),\ \det M = p.$$

Changing the basis means replacing M by $KM, K \in \Gamma$. We obviously know that $\gcd(a, c) \in \{1, p\}$. Distinguishing these two cases, we find some $K \in \Gamma$ and a unique representative

$$KM = \begin{pmatrix} p & 0 \\ 0 & 1 \end{pmatrix} \quad \text{or} \quad \begin{pmatrix} 1 & j \\ 0 & p \end{pmatrix},\ j = 1, \ldots, p.$$

Setting $f(\tau) = F(\mathbb{Z}\tau + \mathbb{Z})$, then (0.3) says that

$$f\left(\frac{\alpha\tau + \beta}{\gamma\tau + \delta}\right) = (\gamma\tau + \delta)^k f(\tau) \quad \text{for all} \quad \begin{pmatrix} \alpha & \beta \\ \gamma & \delta \end{pmatrix} \in \Gamma.$$

In this case, (0.4) has the form

$$g(\tau) = G(\mathbb{Z}\tau + \mathbb{Z}) = F(\mathbb{Z}p\tau + \mathbb{Z}) + \sum_{j=1}^{p} F(\mathbb{Z}(\tau + j) + \mathbb{Z}p)$$

$$= f(p\tau) + p^{-k} f((\tau + j)/p)$$

A comparison with (0.1) yields the

0.3 Lemma. *Under the above assumptions one has*

$$g = p^{1-k} T_p f.$$

This approach was pointed out by J.-P. Serre [73].

4. The abstract Hecke algebra. In this section, we describe an algebraic approach which in a much more general setting is due to G. Shimura [74]. The Hecke *algebra* consists of all formal linear combinations of double cosets

$$\Gamma M \Gamma = \{KML;\ K, L \in \Gamma\},\ M \in \mathrm{GL}\,(2; \mathbb{Q}),\ \det M > 0.$$

Each double coset decomposes into finitely many right cosets

$$\Gamma M \Gamma = \bigcup_{i=1,\ldots,m}^{\bullet} \Gamma M_i.$$

If $\Gamma N \Gamma = \bigcup_{j=1,\ldots,n}^{\bullet} \Gamma N_j$, we define the product by linear extension of

$$\Gamma M \Gamma \cdot \Gamma N \Gamma = \sum \alpha_L \Gamma L \Gamma,$$

where
$$\alpha_L = \#\{(i,j); \ \Gamma M_i N_j = \Gamma L\}.$$

This leads to a commutative $\mathbb{C}$-algebra $\mathcal{H}$ with identity element $\Gamma = \Gamma I \Gamma$, called the HECKE *algebra associated with* Γ. HECKE operators are endomorphisms of the vector space $\mathbb{M}_k$; more precisely, they consist of the following ring homomorphisms $T : \mathcal{H} \to \text{End } \mathbb{M}_k$, given by linear extension of

$$T_{\Gamma M \Gamma} f = \sum_{j=1}^{m} f | M_j. \qquad (0.5)$$

If $M = \begin{pmatrix} 1 & 0 \\ 0 & p \end{pmatrix}$ with a prime p then the results of sect. 3 show that

$$T_{\Gamma M \Gamma} f = p^{k-1} T_p f \qquad (0.6)$$

if we use (0.1). These HECKE operators concide with the HECKE operators described in this chapter. This algebraic approach is outlined in the exercises and is described in detail in A. KRIEG [49], chap. V.

5. The importance of HECKE's construction. HECKE operators provide a tool to obtain arithmetical information about the FOURIER coefficients from the vector space of modular forms. First of all the action of a HECKE operator can completely be described by identities among the FOURIER coefficients. We introduce a scalar product on the space of cusp forms such that it becomes a unitary vector space. HECKE operators turn out to be self-adjoint with respect to this scalar product and commute among each other. A result from linear algebra says that there exists a basis of simultaneous HECKE eigenforms. If such a simultaneous HECKE eigenform is normalized the FOURIER coefficients are mulitplicative, i.e.

$$\alpha_f(mn) = \alpha_f(m) \cdot \alpha_f(n) \quad \text{if} \quad \gcd(m,n) = 1. \qquad (0.7)$$

The DIRICHLET series associated with the FOURIER coefficients therefore possesses an EULER product expansion. Moreover there is a recursion formula among the FOURIER coefficients of the powers of a prime p (cf. Corollary 2.8). Hence all the FOURIER coefficients can easily be described from the values $\alpha_f(p)$ for all primes p. Finally in § 4 we will give a precise description which arithmetical functions appear as FOURIER coefficients of modular forms.

§ 1 HECKE operators

1. HECKE operators on the vector space $V(\mathbb{H})$. Let $V(\mathbb{H})$ denote the $\mathbb{C}$–vector space of functions f with the following properties:

(MP.1) f is meromorphic on $\mathbb{H}$.

§ 1 HECKE operators

(MP.2) f is periodic with period 1.
(MP.3) f has at most a pole at ∞.

For any $f \in V(\mathbb{H})$, by Lemma III.1.2, there exists some $\gamma > 0$ such that f can be represented by a FOURIER series of the form

$$f(\tau) = \sum_{m \geq m_0} \alpha_f(m) \cdot e^{2\pi i m \tau}, \quad m_0 \in \mathbb{Z}, \tag{1.1}$$

which is absolutely and uniformly convergent for $\operatorname{Im} \tau > \gamma$. Now, for positive integers a, d and $f \in V(\mathbb{H})$ define a meromorphic function $T_{a,d} f$ by

$$(T_{a,d} f)(\tau) := \sum_{b (\bmod d)} f((a\tau + b)/d) . \tag{1.2}$$

Here, b runs through a full coset system $(\bmod\ d)$. Since the representatives of any two full coset systems $(\bmod\ d)$ differ only by integer multiples of d up to order, (1.2) does not depend on the choice of the coset system $(\bmod\ d)$ because of (MP.2). For example, we can sum from 1 to d. Obviously, $T_{a,d}$ is linear in f, i.e.

$$T_{a,d}(\alpha f + \beta g) = \alpha \cdot T_{a,d} f + \beta \cdot T_{a,d} g \text{ for all } f, g \in V(\mathbb{H}) \text{ and } \alpha, \beta \in \mathbb{C}. \tag{1.3}$$

1.1 Proposition. *For $f \in V(\mathbb{H})$ one has $T_{a,d} f \in V(\mathbb{H})$ and the FOURIER series of $T_{a,d} f$ is given by*

$$(T_{a,d} f)(\tau) = d \cdot \sum_{m \geq m_0/d} \alpha_f(md) \cdot e^{2\pi i m a \tau} . \tag{1.4}$$

Proof The FOURIER series of $T_{a,d} f$ is given by

$$(T_{a,d} f)(\tau) = \sum_{m \geq m_0} \alpha_f(m) \cdot e^{2\pi i m a \tau / d} \cdot \sum_{b (\bmod d)} e^{2\pi i m b / d} .$$

We apply the summation formula for the finite geometric series and get

$$\sum_{b (\bmod d)} \left(e^{2\pi i m / d} \right)^b = \begin{cases} d, & \text{if } d | m, \\ 0, & \text{otherwise.} \end{cases}$$

Insert this and replace m by md to get (1.4). Thus, (MP.2) and (MP.3) also hold for $T_{a,d} f$. □

For $n \in \mathbb{N}$ and $k \in \mathbb{Z}$, the HECKE *operator* $T_n^{(k)}$ on $V(\mathbb{H})$ is defined by

$$T_n^{(k)} f := n^{k-1} \cdot \sum_{ad=n, d>0} d^{-k} \cdot T_{a,d} f . \tag{1.5}$$

Explicitly, this means

$$\left(T_n^{(k)}f\right)(\tau) = n^{k-1} \cdot \sum_{ad=n, d>0} d^{-k} \cdot \sum_{b \pmod{d}} f((a\tau+b)/d). \tag{1.6}$$

The factor n^{k-1} is more or less arbitrary. This choice yields "nicer" formulas in the sequel. Therefore, for a prime number p, the following holds:

$$\left(T_p^{(k)}f\right)(\tau) = p^{k-1} \cdot f(p\tau) + \frac{1}{p} \cdot \sum_{b \pmod{p}} f((\tau+b)/p). \tag{1.7}$$

To prove the differentiated equation

$$n\left(T_n^{(k)}f\right)' = T_n^{(k+2)}f' \quad \text{for} \quad f \in V(\mathbb{H}), \tag{1.8}$$

we only have to note $a = n/d$ on the right-hand side of (1.6).

Because of (1.3) and (1.4), all $T_n^{(k)}$ are endomorphisms of the vector space $V(\mathbb{H})$.

Proposition 1.1 leads to

1.2 Lemma. *For $f \in V(\mathbb{H})$ the* FOURIER *coefficients of $g = T_n^{(k)}f$ are given by*

$$\alpha_g(m) = \sum_{d \mid \gcd(m,n)} d^{k-1} \cdot \alpha_f\left(mn/d^2\right) \text{ with } m \geq \begin{cases} 0, & \text{if } m_0 = 0, \\ 1, & \text{if } m_0 > 0, \\ nm_0, & \text{if } m_0 < 0. \end{cases} \tag{1.9}$$

In particular,

$$\alpha_g(0) = \sigma_{k-1}(n) \cdot \alpha_f(0) \quad \text{and} \quad \alpha_g(1) = \alpha_f(n).$$

Here and in the following, we only sum over positive divisors d and we always let (m, n) denote the greatest common divisor of m and n.

Proof Because of (1.5) and (1.4), we have

$$\begin{aligned}
\left(T_n^{(k)}f\right)(\tau) &= n^{k-1} \cdot \sum_{ad=n} d^{-k} \cdot (T_{a,d}f)(\tau) \\
&= n^{k-1} \cdot \sum_{ad=n} d^{1-k} \cdot \sum_{m \geq m_0/d} \alpha_f(md) \cdot e^{2\pi i m a \tau} \\
&= \sum_{ad=n} a^{k-1} \cdot \sum_{m \geq am_0/n} \alpha_f(mn/a) \cdot e^{2\pi i m a \tau}.
\end{aligned}$$

Now we collect the terms with $ma = r$ and obtain

$$\left(T_n^{(k)}f\right)(\tau) = \sum_{r,a} a^{k-1} \cdot \alpha_f\left(nr/a^2\right) \cdot e^{2\pi i r \tau},$$

where the sum over all integers r and a is given by

$$a > 0, \ a|n, \ a|r \quad \text{and} \quad r \geq a^2 m_0/n.$$

Since a runs exactly through the positive divisors of (n, r), we obtain (1.9) if we note that

$$a^2 m_0/n \geq \begin{cases} 0, & \text{if } m_0 = 0, \\ m_0/n > 0, & \text{if } m_0 > 0, \\ nm_0, & \text{if } m_0 < 0. \end{cases} \qquad \square$$

We formulate the special case of a prime number as

1.3 Corollary. *If $f \in V(\mathbb{H})$ and p is a prime number, then*

$$\alpha_{T_p^{(k)} f}(m) = \begin{cases} \alpha_f(mp), & \text{if } p \nmid m, \\ \alpha_f(mp) + p^{k-1} \cdot \alpha_f(m/p), & \text{if } p | m, \end{cases}$$

holds.

Instead of $T_n^{(k)} f$, we sometimes write $f|T_n^{(k)}$ or $f|_k T_n$.

2. Transformations of order n. The original definition of HECKE operators T_n is based on the integer 2×2 matrices of determinant n, the — as we used to say — *transformations of order n*

$$\Gamma_n := \{M \in \text{Mat}(2; \mathbb{Z}) \, ; \, \det M = n\}, \quad n \in \mathbb{N}. \tag{1.10}$$

Obviously, $\Gamma_1 = \Gamma$ is the modular group. Furthermore, Γ acts on Γ_n by multiplication from the left and from the right,

$$\Gamma \cdot \Gamma_n = \Gamma_n = \Gamma_n \cdot \Gamma.$$

A subset $\mathcal{R} \subseteq \Gamma_n$ is called a *system of representatives for the right cosets of Γ_n modulo Γ* if:

(RS.1) For every $M \in \Gamma_n$, there is some $L \in \Gamma$ with $LM \in \mathcal{R}$.
(RS.2) If $M_1, M_2 \in \mathcal{R}$ with $M_1 = LM_2$ for some $L \in \Gamma$, then $L = I$ holds.

These conditions are equivalent to

$$\Gamma_n = \bigcup_{M \in \mathcal{R}} \Gamma M \quad \text{disjoint}. \tag{RS}$$

Analogously a system of representatives for the left cosets of Γ_n modulo Γ can be defined.

If we call M_1 and M_2 of Γ_n *equivalent*, in symbols $M_1 \sim M_2$, if there exists some $L \in \Gamma$ with $M_1 = LM_2$, then any system of representatives for the right cosets of Γ_n modulo Γ consists of inequivalent matrices. Thus because of (RS.1), we can also speak of a *complete system of inequivalent matrices*.

Systems of representatives for the right cosets are, of course, not uniquely determined.

If the choice of such a system is not important, it shall be denoted by

$$\mathcal{R} = \Gamma : \Gamma_n.$$

However, in the present case, there is a *standard system of representatives for the right cosets*:

1.4 Proposition. *The set*

$$\Gamma : \Gamma_n = \left\{ M = \begin{pmatrix} a & b \\ 0 & d \end{pmatrix} \in \mathrm{Mat}\,(2;\mathbb{Z})\,;\ ad = n,\ d > 0,\ b\ (\mathrm{mod}\ d) \right\} \quad (1.11)$$

is a system of representatives for the right cosets of Γ_n modulo Γ.

Proof (RS.1) Let $M = \begin{pmatrix} a & b \\ c & d \end{pmatrix} \in \Gamma_n$. We choose coprime $\gamma, \delta \in \mathbb{Z}$ with $\gamma a + \delta c = 0$. Then there are $\alpha, \beta \in \mathbb{Z}$ with $\alpha\delta - \beta\gamma = 1$. Thus we calculate

$$LM = \begin{pmatrix} \alpha & \beta \\ \gamma & \delta \end{pmatrix} \begin{pmatrix} a & b \\ c & d \end{pmatrix} = \begin{pmatrix} a' & b' \\ 0 & d' \end{pmatrix} \quad \text{with} \quad L := \begin{pmatrix} \alpha & \beta \\ \gamma & \delta \end{pmatrix} \in \Gamma.$$

Because $\det(LM) = \det M = n$, we get $a'd' = n$. So we may assume $d' > 0$, after replacing L by $-L$ if necessary. Because

$$T^m LM = \begin{pmatrix} 1 & m \\ 0 & 1 \end{pmatrix} \begin{pmatrix} a' & b' \\ 0 & d' \end{pmatrix} = \begin{pmatrix} a' & b' + md' \\ 0 & d' \end{pmatrix}$$

we can reduce b' modulo d'. Thus $T^m LM$ is contained in (1.11).
(RS.2) Given M and M' from (1.11) and $L \in \Gamma$ with $M = LM'$, thus

$$\begin{pmatrix} a & b \\ 0 & d \end{pmatrix} = \begin{pmatrix} \alpha & \beta \\ \gamma & \delta \end{pmatrix} \begin{pmatrix} a' & b' \\ 0 & d' \end{pmatrix}, \quad \alpha\delta - \beta\gamma = 1.$$

This implies $\gamma a' = 0$, hence $\gamma = 0$ and – since d and d' are positive – also $\alpha = \delta = 1$. Now $a' = a$, $d' = d$ and $b = b' + \beta d$ hold. But since b and b' belong to a coset system (mod d), it follows that $b' = b$ and $\beta = 0$, hence $L = I$. □

1.5 Corollary. *The order of each system of representatives for the right cosets of Γ_n modulo Γ is*

$$\sigma_1(n) = \sum_{d \mid n} d.$$

Proof The elements of two such systems of representatives for the right cosets differ only by a permutation and by left-hand side factors from Γ. Hence all the systems of representatives for the right cosets have the same number of elements. But the number of (1.11) is equal to

$$\sum_{d \mid n} \left(\sum_{b (\mathrm{mod}\ d)} 1 \right) = \sum_{d \mid n} d = \sigma_1(n). \qquad \square$$

1.6 Remarks. a) Since Γ acts on the set Γ_n by multiplication from the left, the quotient space is

$$\Gamma \backslash \Gamma_n := \{\Gamma \cdot M \, ; \, M \in \Gamma_n\}, \quad \Gamma \cdot M := \{LM \, ; \, L \in \Gamma\},$$

with the canonical surjective mapping

$$\pi : \Gamma_n \longrightarrow \Gamma \backslash \Gamma_n, \quad \pi(M) = \Gamma \cdot M.$$

By the definition of a system of representatives for the right cosets $\mathcal{R}$ of Γ_n modulo Γ, the restriction $\pi|_\mathcal{R} : \mathcal{R} \to \Gamma \backslash \Gamma_n$ is a bijection.
b) If $\mathcal{R}$ is a system of representatives for the right cosets of Γ_n modulo Γ, then

$$\mathcal{R}^{tr} := \{M^{tr} \, ; \, M \in \mathcal{R}\}, \quad \mathcal{R}^{\#} = \{M^{\#}; \, M \in \mathcal{R}\},$$

are clearly systems of representatives for the left cosets of Γ_n modulo Γ, where M^{tr} denotes the transpose and $M^{\#} = (\det M) \cdot M^{-1}$ the adjugate of M. Thus, any systems of representatives for the right and left cosets have the same number of elements.

3. Hecke operators for modular forms. In analogy to the definition of $f|M$ for $M \in \mathrm{SL}(2; \mathbb{R})$ in III(1.1), now $f|_k M = f|M$ is also defined for $M \in \mathrm{GL}(2; \mathbb{R})$ with $\det M > 0$ and a function f meromorphic on $\mathbb{H}$ by

$$(f|M)(\tau) := (f|_k M)(\tau) := (c\tau + d)^{-k} \cdot f(M\tau) . \tag{1.12}$$

We verify again that

$$(\alpha f + \beta g)|M = \alpha \cdot f|M + \beta \cdot g|M \quad \text{and} \quad (f|M)|N = f|(MN) \tag{1.13}$$

for $M, N \in \mathrm{GL}(2; \mathbb{R})$ with $\det M > 0$ and $\det N > 0$.

According to III,§1, meromorphic modular forms of weight k belong to $V(\mathbb{H})$, hence $\mathbb{V}_k \subseteq V(\mathbb{H})$.

1.7 Proposition. *If* $\Gamma : \Gamma_n$ *is an arbitrary system of representatives for the right cosets of* Γ_n *modulo* Γ *and* $f \in \mathbb{V}_k$, *then*

$$T_n f := T_n^{(k)} f = n^{k-1} \cdot \sum_{M \in \Gamma : \Gamma_n} f|_k M \tag{1.14}$$

and $T_n f$ *again belongs to* $\mathbb{V}_k$.

Note that by Corollary 1.5, this is a finite sum. The weight k is usually omitted here. We call T_n a Hecke *operator*.

Proof Denote the right-hand side of (1.14) by f^*. Since the elements of two systems of representatives for the right cosets of Γ_n modulo Γ, only differ by left-hand factors from Γ up to order, $f|L = f$ for $L \in \Gamma$ shows that f^* does not depend on the choice of the system of representatives for the right cosets $\Gamma : \Gamma_n$. The use of the symbol $\Gamma : \Gamma_n$ is therefore justified. Thus we may take the standard system of representatives for the right cosets from Proposition 1.4 on the right-hand side of (1.14). Then (1.6) yields

$$f^*(\tau) = n^{k-1} \cdot \sum_{ad=n} d^{-k} \cdot \sum_{b \pmod n} f((a\tau+b)/d) = \left(T_n^{(k)} f\right)(\tau) \in V(\mathbb{H}).$$

If $N \in \Gamma$ is fixed and $\Gamma : \Gamma_n$, is a system of representatives for the right cosets, $\{MN \; ; \; M \in \Gamma : \Gamma_n\}$ is also a system of representatives for the right cosets of Γ_n modulo Γ. Therefore, (1.13) leads to

$$T_n f = n^{k-1} \cdot \sum_{M \in \Gamma : \Gamma_n} f|_k(MN) = n^{k-1} \cdot \sum_{M \in \Gamma : \Gamma_n} (f|M)|N = (T_n f)|N$$

such that $T_n f$ is also modular of weight k. □

Together with Lemma 1.2 we obtain the

1.8 Corollary. *All* Hecke *operators* $T_n^{(k)} : \mathbb{M}_k \to \mathbb{M}_k$, $n \geq 1$, *are endomorphisms which map cusp forms into cusp forms.*

Finally, we write down the special cases $m = 0, 1$ of Lemma 1.2 as

$$\alpha_{T_n f}(0) = \sigma_{k-1}(n) \cdot \alpha_f(0), \quad \alpha_{T_n f}(1) = \alpha_f(n). \tag{1.15}$$

Obviously, Proposition 1.7 is proved in a completely analogous way as Corollary III.7.6. Indeed, we can define Hecke operators as the trace of a suitable modular form with respect to the congruence group $\Gamma_0[n]$ (cf. Exercise 2).

4. Simultaneous eigenforms. A function $0 \neq f \in V(\mathbb{H})$ is called an *eigenform of the* Hecke *operator* $T_n^{(k)}$, $n \geq 1$ *with the eigenvalue* $\lambda_f(n) \in \mathbb{C}$ if

$$T_n^{(k)} f = \lambda_f(n) \cdot f. \tag{1.16}$$

If f is an eigenform of all Hecke operators $T_n^{(k)}$, $n \geq 1$, then f is called a *simultaneous eigenform*. By Lemma 1.2, $f \neq 0$ is a simultaneous eigenform if and only if its Fourier coefficients satisfy the conditions

$$\lambda_f(n) \cdot \alpha_f(m) = \sum_{d|(m,n)} d^{k-1} \cdot \alpha_f\left(mn/d^2\right) \tag{1.17}$$

for all $m, n \in \mathbb{N}$. If we set $m = 1$, we get in particular

$$\lambda_f(n) \cdot \alpha_f(1) = \alpha_f(n), \tag{1.18}$$

and obtain the surprising

1.9 Lemma. *For a non–constant* $f \in \mathbb{M}_k$, *the following assertions are equivalent*:

(i) f *is a simultaneous eigenform.*
(ii) $\alpha_f(1) \neq 0$, *and all* $m \in \mathbb{N}_0$, $n \in \mathbb{N}$ *satisfy*

§ 1 Hecke operators

$$\alpha_f(m) \cdot \alpha_f(n) = \alpha_f(1) \cdot \sum_{d \mid (m,n)} d^{k-1} \cdot \alpha_f(mn/d^2).$$

In this case, the eigenvalues are $\lambda_f(n) = \alpha_f(n)/\alpha_f(1)$, $n \in \mathbb{N}$, *and*

$$\alpha_f(m) \cdot \alpha_f(n) = \alpha_f(1) \cdot \alpha_f(mn) \tag{1.19}$$

holds for all coprime m and n.

As a first (and perhaps most important) application, we consider $k = 12$ and the discriminant Δ or the normalized discriminant Δ^*. According to III(2.17), the following holds:

$$\Delta^*(\tau) = (2\pi)^{-12} \cdot \Delta(\tau) = \sum_{m=1}^{\infty} \tau(m) \cdot e^{2\pi i m \tau} \quad \text{with} \quad \tau(1) = 1,$$

and all $\tau(m)$ are integers. By Corollary III.4.5, Δ is the only cusp form of weight 12 up to a constant factor. Since the HECKE operators $T_n = T_n^{(k)}$ map cusp forms into cusp forms according to Corollary 1.8, we get – if we still keep Lemma 1.9 in mind – the fundamental

1.10 Theorem. *The discriminant Δ is a simultaneous eigenform:*

$$T_n \Delta = \tau(n) \cdot \Delta \quad \text{for all} \quad n \in \mathbb{N}$$

and all $m, n \in \mathbb{N}$ *satisfy*

$$\tau(m) \cdot \tau(n) = \sum_{d \mid (m,n)} d^{11} \cdot \tau(mn/d^2). \tag{1.20}$$

In particular, we have

$$\tau(mn) = \tau(m) \cdot \tau(n), \quad \text{if } (m, n) = 1,$$

and for prime numbers p,

$$\tau(p^{r+1}) = \tau(p^r) \cdot \tau(p) - p^{11} \cdot \tau(p^{r-1}), \quad r \geq 1. \tag{1.21}$$

The vector spaces $\mathbb{S}_k$ are also one-dimensional for $k = 16, 18, 20, 22, 26$. The cusp forms $E_{k-12} \cdot \Delta$ are therefore simultaneous eigenforms for these k as well.

The eigenvalues cannot become arbitrarily large for cusp forms. As a simple bound we obtain the

1.11 Proposition. *Let $n > 1$ and $0 \neq f \in \mathbb{S}_k$. If $T_n f = \lambda_f(n) \cdot f$ with a $\lambda_f(n) \in \mathbb{C}$, it follows that*

$$|\lambda_f(n)| \leq n^{k/2} \cdot \sigma_{-1}(n).$$

Proof According to Proposition III.1.7, there exists some $w = u + iv \in \mathbb{H}$ with the property

$$\tilde{f}(\tau) := y^{k/2} \cdot |f(\tau)| \leq \tilde{f}(w) \quad \text{for all} \quad \tau \in \mathbb{H}. \tag{1.22}$$

Then (1.6) yields

$$|\lambda_f(n) \cdot \tilde{f}(w)| = |\widetilde{T_n f}(w)| = \left| n^{k-1} v^{k/2} \cdot \sum_{a,b,d} d^{-k} \cdot f((aw+b)/d) \right|$$

$$\leq n^{-1+k/2} \cdot \sum_{a,b,d} \tilde{f}((aw+b)/d) \leq n^{-1+k/2} \cdot \sigma_1(n) \cdot \tilde{f}(w),$$

if we observe (1.22). In view of $\tilde{f}(w) \neq 0$ and $\sigma_1(n) = n \cdot \sigma_{-1}(n)$, the assertion follows. □

Proposition 1.11 was proved in 1917 by L.J. MORDELL (*Proc. Cambridge. Phil. Soc.* **19**, 117–124). His proof essentially uses what we now call HECKE operators, i.e. the endomorphisms $T_n^{(12)}$. But only E. HECKE recognized 20 years later ([36], 644–707) the wider meaning of this construction.

5*. Application to the absolute invariant. According to III(2.19)

$$j := (720 E_4)^3 / \Delta \tag{1.23}$$

is a modular function which is holomorphic on $\mathbb{H}$ and which has a 1st order pole at ∞. Moreover,

$$j(\tau) = e^{-2\pi i \tau} + \sum_{m=0}^{\infty} j_m \cdot e^{2\pi i m \tau}, \quad \tau \in \mathbb{H}, \tag{1.24}$$

with positive integers j_m. Then, for $T_n := T_n^{(0)}$, the function $T_n j$ is again a modular function by Proposition 1.7, which is also holomorphic on $\mathbb{H}$ by (1.6). Then Corollary III.5.6 implies

1.12 Theorem. *For every $n \geq 1$, $T_n j$ is a polynomial of degree n in j.*

The "principal part" of the FOURIER series of $T_n j$ is as simple as it could be:

1.13 Proposition. *For $n \geq 1$,*

$$n \cdot (T_n j)(\tau) = e^{-2\pi i n \tau} + 744 \cdot \sigma_1(n) + n j_n \cdot e^{2\pi i \tau} + \cdots$$

holds.

Proof Due to (1.9) we have $j_{-1} = 1$ and

$$\alpha_{T_n j}(m) = \sum_{d \mid (m,n)} d^{-1} \cdot j_{mn/d^2} \quad \text{for} \quad m \geq -n$$

because of $m_0 = -1$. We get a non-zero coefficient for negative m only if the index is $mn = -d^2$, i.e. if $d = n$ and $m = -n$. Now we determine the missing coefficients according to (1.9) and observe $n \cdot \sigma_{-1}(n) = \sigma_1(n)$ as well as $j_0 = 744$. □

§ 1 Hecke operators

1.14 Corollary. *For every $m \in \mathbb{N}$, there are γ_m and γ_{mn} in $\mathbb{Z}$ such that*

$$j^m = \gamma_m + \sum_{n=1}^{m} n\gamma_{mn} \cdot T_n j.$$

Proof By taking the m-th power in (1.24), we get a Fourier series for j^m with coefficients in $\mathbb{Z}$. Therefore,

$$j^m - \sum_{n=1}^{m} n\gamma_{mn} \cdot T_n j$$

is a modular function and, for suitable $\gamma_{mn} \in \mathbb{Z}$, it is holomorphic on $\mathbb{H}$ and at ∞, i.e. a constant γ_m. Since j^m and all $n \cdot T_n j$ have integer constant Fourier coefficients, γ_m is also an integer. □

As in the proof of Proposition 1.13, we obtain for powers of j the

1.15 Corollary. *For $n \geq 1$,*
$$T_n \mathbb{Q}[j] \subseteq \mathbb{Q}[j].$$

6*. The modular equation is to be regarded as an analogon of the n-partition equation of the $\wp$–function in I, §7 for the j–function. The Hecke theory will be applied here.

1.16 Proposition. *For every $n \in \mathbb{N}$, there exists a uniquely determined polynomial $F_n(X, Y) \in \mathbb{Q}[X, Y]$ with the property*

$$F_n(X, j(\tau)) = \prod_{M \in \Gamma : \Gamma_n} (X - j(M\tau)) \quad \text{for all } \tau \in \mathbb{H}. \tag{1.25}$$

$F_n(X, Y)$ has degree $\sigma_1(n)$ as a polynomial in X and Y respectively.

Proof Define

$$F_n(X) := \prod_{M \in \Gamma : \Gamma_n} (X - j(M\tau)). \tag{1.26}$$

Because j is already Γ–invariant, (1.26) does not depend on the choice of the system of representatives. Let $r = \sigma_1(n)$ and $M_1, \ldots, M_r$ be the standard system of representatives for the right cosets (1.11). Then it follows that

$$F_n(X) = \sum_{k=0}^{r} (-1)^k P_k(j(M_1\tau), \ldots, j(M_r\tau)) \cdot X^{r-k},$$

where P_k is the k–th elementary symmetric polynomial in r indeterminates. The power sums satisfy

$$\sum_{\nu=1}^{r} j(M_\nu \tau)^k = (nT_n(j^k))(\tau) \in \mathbb{Q}[j(\tau)] \tag{1.27}$$

according to Corollary 1.15. As is well-known the elementary symmetric polynomials can be expressed rationally in terms of the power sums, i.e.

$$P_k(j(M_1\tau), \ldots, j(M_r\tau)) \in \mathbb{Q}[j(\tau)]. \tag{1.28}$$

Because $j(\tau)$ is transcendental, there exists a unique polynomial $F_n(X, Y) \in \mathbb{Q}[X, Y]$ with the property (1.25). Obviously, $F_n(X, Y)$ has degree r in X.

The highest occurring pole at ∞ in the functions (1.28) has order $r = \sigma_1(n)$ and this occurs only at P_r, because in this case the FOURIER expansion begins with the term

$$\prod_{d|n} \prod_{b \pmod d} e^{-2\pi i(a\tau+b)/d} = \pm e^{-2\pi i r \tau}.$$

Consequently, all of the $P_k(j(M_1\tau), \ldots, j(M_r\tau))$ are polynomials in $j(\tau)$ of degree $< r$ for $0 \leq k < r$. For $k = r$, it is a polynomial in $j(\tau)$ of degree r. Accordingly, $F_n(X, Y)$ has also degree r in Y. □

Following I, §7, we call $F_n(X, Y) = 0$ the *modular equation of degree n*. Trivially $F_1(X, Y) = X - Y$ holds. A longer calculation with the FOURIER expansion of $j(\tau)$ yields

$$F_2(X, Y) = X^3 + Y^3 - X^2 Y^2 + 1\,488(X^2 Y + XY^2) - 162\,000(X^2 + Y^2)$$
$$+ 40\,773\,375\, XY + 8\,748\,000\,000(X + Y) - 157\,464\,000\,000\,000.$$

To proceed further, we need the following tool, which describes the SMITH *normal form* for 2×2 matrices.

1.17 Lemma. *If $n \in \mathbb{N}$ and $M \in \Gamma_n$, there are matrices $K, L \in \Gamma$ satisfying*

$$KML = \begin{pmatrix} r & 0 \\ 0 & rs \end{pmatrix},$$

where $r \in \mathbb{N}$ is the gcd of the entries of M and $r^2 s = n$.

Proof Consider $\frac{1}{r}M$, if necessary, and assume that the entries of M are coprime. Because of (1.11), we can assume the form $M = \begin{pmatrix} a & b \\ 0 & d \end{pmatrix}$ with $ad = n$. Then a, b, d are coprime. By Lemma II.3.3, there exists some $x \in \mathbb{Z}$ such that $xa + b$ and d are already coprime. We now choose $\alpha, \beta \in \mathbb{Z}$ with $\alpha(xa + b) + \beta d = 1$. Then the assertion follows with

$$K = \begin{pmatrix} \alpha & \beta \\ -d & xa+b \end{pmatrix}, \quad L = \begin{pmatrix} x & -1 \\ 1 & 0 \end{pmatrix} \cdot \begin{pmatrix} 1 & \alpha a \\ 0 & 1 \end{pmatrix} \in \Gamma. \quad \square$$

Let $\mathbb{K} = \mathbb{V}_0$ again be the field of modular functions. Then we obtain the central

1.18 Theorem. *Let $n \in \mathbb{N}$, $n > 1$ be square-free. Then*

$$F_n(X, j) \in \mathbb{K}[X]$$

is irreducible. Moreover,

$$F_n(X,Y) = F_n(Y,X) \quad \text{and} \quad F_n(X,X) \in \mathbb{Q}[X]\setminus\{0\}$$

hold.

Proof For fixed $L \in \Gamma$, ML runs through a system of representatives for the right cosets of Γ_n modulo Γ as M does. Thus $f(\tau) \mapsto f(L\tau)$ is an automorphism of the splitting field $\mathbb{K}' = \mathbb{K}[j(M\tau), M \in \Gamma_n]$ of $F_n(X,j)$ over $\mathbb{K}$. If $M \in \Gamma_n$, we choose $K, L \in \Gamma$ by Lemma 1.17 and get

$$j(M\tau) = j((L^{-1}\tau)/n).$$

Consequently, all roots $X = j(M\tau)$, $M \in \Gamma_n$, of the polynomial $F_n(X,j)$ can be obtained from the root $X = j(\tau/n)$ by applying automorphisms of $\mathbb{K}'$ over $\mathbb{K}$. Therefore, $F_n(X,j)$ is irreducible over $\mathbb{K}$.

From $\begin{pmatrix} n & 0 \\ 0 & 1 \end{pmatrix} \in \Gamma_n$, we conclude $F_n(j(n\tau), j(\tau)) = 0$ for all $\tau \in \mathbb{H}$. If we replace τ by τ/n, then $F_n(j(\tau), j(\tau/n)) = 0$ holds for all $\tau \in \mathbb{H}$. So $F_n(j(\tau), Y) \in \mathbb{K}[Y]$ is divisible by $Y - j(\tau/n)$ and because of irreducibility also by $F_n(Y, j(\tau))$. Since both polynomials have degree $\sigma_1(n)$ by Proposition 1.16, there is some $c \in \mathbb{Q}$ with

$$F_n(X,Y) = c\, F_n(Y,X).$$

For reasons of symmetry, $c = \pm 1$ follows. If $c = -1$, we have $F_n(X,X) = 0$. Then $F_n(X,j)$ is divisible by $X - j$, which because of $\sigma_1(n) > 1$ contradicts irreducibility. From this we get $c = 1$ and $F_n(X,X) \neq 0$. □

7*. The singular values. As an application of Theorem 1.18, we show the

1.19 Proposition. *If $\tau \in \mathbb{H}$ belongs to an imaginary–quadratic number field, then $j(\tau)$ is an algebraic number.*

Proof If $\tau \in \mathbb{H}$ belongs to an imaginary–quadratic number field, then τ has the form

$$\tau = \frac{1}{d}(b + ia\sqrt{D}), \quad b \in \mathbb{Z}, \quad a, d, D \in \mathbb{N}, \quad D \text{ square-free}.$$

First, $\begin{pmatrix} 1 & 0 \\ 0 & D \end{pmatrix} \langle i\sqrt{D} \rangle = i/\sqrt{D} = J\langle i\sqrt{D} \rangle$, thus

$$j(i\sqrt{D}) = j(i\sqrt{D}/D).$$

We have $j(i) = 1728 \in \mathbb{Q}$, and $j(i\sqrt{D})$ is a root of $F_D(X, j(i\sqrt{D}))$ for $D > 1$ by Proposition 1.16 and Theorem 1.18. These values are roots of the polynomial $F_D(X,X) \in \mathbb{Q}[X]\setminus\{0\}$. Consequently, $j(i\sqrt{D})$ is an algebraic number. Let $ad = n$, so $M = \begin{pmatrix} a & b \\ 0 & d \end{pmatrix} \in \Gamma_n$ with $M\langle i\sqrt{D} \rangle = \tau$. By Proposition 1.16, $j(\tau)$ is the root of the monic polynomial

$$F_n(X, j(i\sqrt{D})) \in \mathbb{Q}(j(i\sqrt{D}))[X],$$

thus algebraic over $\mathbb{Q}(j(i\sqrt{D}))$ and thus also algebraic over $\mathbb{Q}$. □

If Ω is a lattice with complex multiplication, we call $j(\Omega)$ a *singular value* of j, following L. KRONECKER. According to Proposition I.7.16, Ω then has the form $\Omega = \lambda(\mathbb{Z}\tau + \mathbb{Z})$ with some $0 \neq \lambda \in \mathbb{C}$ and some $\tau \in \mathbb{H}$ belonging to an imaginary–quadratic number field. Because of $j(\Omega) = j(\tau)$, Proposition 1.19 yields

1.20 Corollary. *The singular values of j are algebraic numbers.*

1.21 Corollary. *Let $n \in \mathbb{N}$, $n > 1$ be square–free and $c \in \mathbb{C}$ with $F_n(c,c) = 0$. Then there exists some $\tau \in \mathbb{H}$ with $j(\tau) = c$, and each such τ belongs to an imaginary–quadratic number field.*

Proof The existence of τ follows from Theorem III.5.3 or I.4.10. According to Proposition 1.16, $c = j(\tau)$ is a root of the polynomial $F_n(X, j(\tau))$, thus we get $j(\tau) = j(M\tau)$ for an $M \in \Gamma_n$. According to Theorem III.5.3 or I.4.10, there exists some $K \in \Gamma$ with $\tau = KM\tau$. Because n is square–free, we conclude $KM \notin \mathbb{Z}I$. Then sect. 5 in I, §7 implies that τ belongs to an imaginary–quadratic number field. □

1.22 Remark (KRONECKER's Jugendtraum (youthful dream)). According to a classical result of L. KRONECKER, every abelian extension of $\mathbb{Q}$, i.e. every GALOIS extension of $\mathbb{Q}$ with an abelian GALOIS group, is contained in a field generated by roots of unity. Thus the roots of unity, i.e. the values of the exponential function $e^{2\pi i \alpha}$, $\alpha \in \mathbb{Q}$, generate the maximal abelian extensions of $\mathbb{Q}$. This result is known as *Theorem of* KRONECKER–WEBER. A first (incomplete) proof is due to L. KRONECKER ([50], vol. IV, 3–11), a second (somewhat incomplete) to H. WEBER (1842–1913), *Acta Math.* **8**, 193–263 (1886), and finally a complete proof to D. HILBERT ([37], vol. I, 53–62). A modern exposition in classical language can be found in L.C. WASHINGTON [86], chap. 14.

An analogous construction is also possible for imaginary-quadratic number fields K, if we replace the exponential function by the j-function. For $\tau \in K \cap \mathbb{H}$, Proposition 1.19 states that $K(j(\tau))$ is a finite extension of K. With more effort (cf. H. WEBER [87], § 121) we can show that this field extension is also abelian. An answer to the question of which abelian extensions of K arise in this way can be found in S. LANG [52], Chap. 10.

The 12th HILBERT problem ([37], vol. III, 290–329) poses the problem of describing all abelian extensions of an arbitrary algebraic number field.

1.23 Exercises.
1) Let $\mathcal{G} = \{M \in \mathrm{GL}(2; \mathbb{Q}); \det M > 0\}$.
a) Show that $\Gamma \cap M^{-1}\Gamma M$, $M \in \mathcal{G}$, has finite index in Γ and conclude that $\Gamma M \Gamma$ decomposes into finitely many right cosets modulo Γ.
b) Show that each double coset $\Gamma M \Gamma$, $M \in \mathcal{G}$, contains a unique representative of the form $q \begin{pmatrix} 1 & 0 \\ 0 & n \end{pmatrix}$, $q \in \mathbb{Q}$, $q > 0$, $n \in \mathbb{N}$.
c) Show that each HECKE operator $T_{\Gamma M \Gamma}$, $M \in \mathcal{G}$, in (0.5) belongs to $\mathcal{H}_k$.
d) Show that each HECKE operator $\mathbb{T}_n^{(k)}$ is a linear combination of HECKE operators $T_{\Gamma M \Gamma}$, $M \in \mathcal{G}$.
2) Let $n \in \mathbb{N}$, $f \in \mathbb{M}_k$, $g(\tau) = f(n\tau) \in \mathbb{M}_k(\Gamma_0[n])$. Determine a HECKE operator

§ 1 Hecke operators

$T \in \mathcal{H}_k$ such that $\text{tr}(g) = Tf$.

3) $|\tau(n)| \leq n^5 \cdot \sigma_1(n)$ holds for all $n \in \mathbb{N}$.

4) For every $n \geq 1$, $T_n^{(2)} j' = p(j) \cdot j'$ holds, where $p(j)$ is a polynomial in j of degree $n-1$.

5) Let $n > 1$ and $\mathcal{R} = \{M_1, \ldots, M_r\}$, $r = \sigma_1(n)$, be a system of representatives for the right cosets of Γ_n modulo Γ. Let $P(X_1, \ldots, X_r)$ be a homogeneous symmetric polynomial of degree m. Then

$$P_n^{(k)} : \mathbb{M}_k \longrightarrow \mathbb{M}_{mk}, \quad f \longmapsto P(f|M_1, \ldots, f|M_r),$$

is a homomorphism of vector spaces. By the standard system of representatives for the right cosets (1.11),

$$\alpha_{P_n^{(k)}(f)}(0) = n^{k-1} \cdot P(d_1, \ldots, d_r) \cdot \alpha_f(0).$$

If Q and R are homogeneous symmetric polynomials with $P = Q \cdot R$, then

$$P_n^{(k)}(f) = Q_n^{(k)}(f) \cdot R_n^{(k)}(f) \text{ for all } f \in \mathbb{M}_k.$$

6) $\{P_2^{(4)}(E_4) \, ; \, P \in \mathbb{C}[X_1, X_2, X_3] \text{ symmetric}\} = \bigoplus_{k \geq 0} \mathbb{M}_{4k}$.

7) Let $f \in \mathbb{M}_k$, p be a prime number, $g(\tau) = f(p\tau) \in \mathbb{M}_k(\Gamma_0[p])$, $r = p+1$ and $P(X_1, \ldots, X_r) = X_1 \cdot \ldots \cdot X_r$. Then, in the notation of Corollary III.7.6, $\pi(g) = P_p^{(k)}(f)$.

8) Let p be a prime number and $f \in \mathbb{M}_k$. Then

$$f(\tau) - (T_p f)(p\tau) + p^{k-1} \cdot f(p^2 \tau) = \sum_{m \geq 1, p \nmid m} \alpha_f(m) \cdot e^{2\pi i m \tau} =: f_p(\tau).$$

Show that $f_p \in \mathbb{M}_k(\Gamma_0[p^2])$.

9) Let $f \in \mathbb{M}_k$ and p be a prime such that $\alpha_f(pm) = 0$ for all $m \geq 0$, then $f = 0$.

10) If $f \in \mathbb{M}_k$ and p is a prime with $\alpha_f(m) = 0$ for all $m \in \mathbb{N}_0$ with $p \nmid m$, then f is constant.

11) Which $f \in \mathbb{K} = V_0$ are simultaneous eigenforms under all Hecke operators?

12) The Hecke operators $T_n^{(k)}$ map the subspaces $A(\mathbb{H})$ and $B(\mathbb{H})$ (cf. Exercise III.1.8) of $V(\mathbb{H})$ into themselves.

13) a) Calculate an explicit representation of $2T_2 j$ and $3T_3 j$ as a polynomial in j.
b) Express j^2 and j^3 explicitly in the form of Corollary 1.14.

14) $-F_p(X, X)$ is a monic polynomial in $\mathbb{Q}[X]$ of degree $2p$ for any prime p.

15) $-F_2(X, X) = (X - 12^3) \cdot (X - 20^3) \cdot (X + 15^3)^2$ holds. Conclude $j(i\sqrt{2}) = 20^3$ and furthermore $j(\frac{1}{2}(1 + i\sqrt{7})) = -15^3$.

16) If $k \neq 0$ is even, then there is no non-trivial simultaneous eigenform in $\mathbb{M}_k^!$.

17) Show that the numbers of left and right cosets for Γ in Γ_n coincide for every $n \in \mathbb{N}$. Construct a set of simultaneous representatives for the left and right cosets.

18) Let $n \in \mathbb{N}$ be square–free. Show that a system of representatives for the right

cosets $\Gamma : \Gamma_n$ can be given in the form $\begin{pmatrix} n & 0 \\ 0 & 1 \end{pmatrix} M$, where M runs through a system of representatives for the right cosets $\Gamma_0[n]\backslash\Gamma$.

19) Let p be a prime and $f \in \mathbb{M}_k$. Then $T_p f = 0$ holds if and only if $f \in \mathbb{S}_k$ and $\alpha_f(p^{2r+1}m) = 0$, $\alpha_f(p^{2r}m) = (-1)^r p^{r(k-1)} \alpha_f(m)$ for all $r \in \mathbb{N}_0$ and $m \in \mathbb{N}$ with $p \nmid m$.

20) Let $f \in \mathbb{M}_k$, $k > 0$ such that $\alpha_f(m) \neq 0$ implies $m = n^2$ for some $n \in \mathbb{N}_0$. Show that $T_p f = 0$ for every prime p.

21) Consider $f(\tau) = \left(\frac{\pi}{\sin(\pi\tau)}\right)^2$ in $V(\mathbb{H})$ and compute $T_{a,d} f$ for $a, d \in \mathbb{N}$.

§ 2 The algebra of HECKE operators

1. The multiplicativity of HECKE operators.

2.1 Theorem *For coprime numbers $m, n \in \mathbb{N}$ and $f \in V(\mathbb{H})$*

$$T_{mn}^{(k)} f = T_m^{(k)} T_n^{(k)} f = T_n^{(k)} T_m^{(k)} f \qquad (2.1)$$

holds.

Here, the successive application of the mappings T_m and T_n is denoted by $T_n T_m$. The claimed commutativity of T_m and T_n for coprime m, n simply follows from the fact that the left-hand side of (2.1) is symmetric in m and n.

The proof of multiplicativity is reduced to a corresponding statement about systems of representatives for the right cosets.

2.2 Lemma. *Let $m, n \in \mathbb{N}$ be coprime, $a_1 d_1 = m$, $a_2 d_2 = n$ with integers a_1, a_2, d_1, d_2, and let b_1 (mod d_1) and b_2 (mod d_2) each run through a full residue system. Then*

$$b_{12} := a_2 b_1 + b_2 d_1 \qquad (2.2)$$

is a full residue system (mod $d_1 d_2$).

Proof The numbers of the form (2.2) are distinct (mod $d_1 d_2$): in obvious notation,

$$a_2 \tilde{b}_1 + \tilde{b}_2 d_1 \equiv a_2 b_1 + b_2 d_1 \pmod{d_1 d_2}$$

immediately yields $a_2 \tilde{b}_1 \equiv a_2 b_1$ (mod d_1), hence – since a_2 and d_1 are coprime – $\tilde{b}_1 \equiv b_1$ (mod d_1) follows. Since $\tilde{b}_1$ and b_1 belong to a residue system (mod d_1), it follows that $\tilde{b}_1 = b_1$ and $\tilde{b}_2 d_1 \equiv b_2 d_1$ (mod $d_1 d_2$). This yields $\tilde{b}_2 \equiv b_2$ (mod d_2) and thus $\tilde{b}_2 = b_2$. □

Now for the *proof* of (2.1), we have, according to (1.6),

§ 2 The algebra of Hecke operators

$$\left[\left(T_m^{(k)} T_n^{(k)}\right) f\right](\tau)$$

$$= (mn)^{k-1} \cdot \sum_{a_1 d_1 = m} \sum_{a_2 d_2 = n} (d_1 d_2)^{-k} \cdot \sum_{b_1 (\mathrm{mod}\ d_1)} \sum_{b_2 (\mathrm{mod}\ d_2)} f((a_2 a_1 \tau + b_{12})/d_1 d_2),$$

where b_{12} is described by (2.2). Since we get the divisors of mn exactly once in the form $d_1 d_2$ with $d_1 | m$ and $d_2 | n$, we can apply (1.6) again and get (2.1). □

2. A recursion formula for T_{p^r}. We obtain

2.3 Theorem *For all primes p, all $r \in \mathbb{N}$ and all $f \in V(\mathbb{H})$ one has*

$$T_{p^r} T_p f = T_{p^{r+1}} f + p^{k-1} \cdot T_{p^{r-1}} f. \tag{2.3}$$

Here $T_{p^0} = T_1$ is, of course, the identity map. The proof of (2.3) is again reduced to an appropriate statement about residue systems:

2.4 Lemma. *If b_ν runs through a residue system (mod p^ν) and a runs through a residue system (mod p), then*

$$c_\nu := b_\nu + a p^\nu \tag{2.4}$$

runs through a residue system (mod $p^{\nu+1}$) exactly once.

Proof In obvious notation, the congruence

$$\tilde{b}_\nu + \tilde{a} p^\nu \equiv b_\nu + a p^\nu \pmod{p^{\nu+1}}$$

yields $\tilde{b}_\nu \equiv b_\nu \pmod{p^\nu}$, hence $\tilde{b}_\nu = b_\nu$ and consequently $\tilde{a} \equiv a \pmod{p}$. □

2.5 Lemma. *If $\nu \geq 1$, we reduce for a residue system b_ν (mod p^ν) modulo $p^{\nu-1}$, we obtain p copies of a residue system c_ν (mod $p^{\nu-1}$).*

Proof Analogously. □

For the *proof* of (2.3), we use (1.6), insert (1.7), and treat the case $\nu = 0$ separately:

$$p^{-r(k-1)} \left(T_{p^r} T_p f\right)(\tau)$$

$$= \sum_{\nu=0}^{r} p^{-\nu k} \cdot \sum_{b_\nu (\mathrm{mod}\ p^\nu)} T_p f\left((p^{r-\nu} \tau + b_\nu)/p^\nu\right)$$

$$= p^{k-1} \cdot f(p^{r+1} \tau) + \frac{1}{p} \cdot \sum_{a (\mathrm{mod}\ p)} f((p^r \tau + a)/p) +$$

$$\sum_{\nu=1}^{r} p^{-\nu k} \sum_{b_\nu (\mathrm{mod}\ p^\nu)} \left\{ p^{k-1} f\left(\frac{p^{r-\nu} \tau + b_\nu}{p^{\nu-1}}\right) + \frac{1}{p} \sum_{a (\mathrm{mod}\ p)} f\left(\frac{p^{r-\nu} \tau + c_\nu}{p^{\nu+1}}\right) \right\},$$

where c_ν are defined as in (2.4). Lemma 2.5 and Lemma 2.4, respectively, lead to

$$(T_{p^r} T_p f)(\tau)$$
$$= p^{(r+1)(k-1)} \cdot f(p^{r+1}\tau)$$
$$+ p^{r(k-1)} \cdot \sum_{\nu=1}^{r} p^{-(\nu-1)k} \cdot \sum_{b_\nu \pmod{p^{\nu-1}}} f\left((p^{(r-1)-(\nu-1)}\tau + b_\nu)/p^{\nu-1}\right)$$
$$+ p^{r(k-1)} \cdot \sum_{\nu=0}^{r} p^{-\nu k-1} \cdot \sum_{c_\nu \pmod{p^{\nu+1}}} f\left((p^{(r+1)-(\nu+1)}\tau + c_\nu)/p^{\nu+1}\right)$$
$$= p^{(r+1)(k-1)} \cdot f(p^{r+1}\tau) + p^{r(k-1)-(r-1)(k-1)} \cdot T_{p^{r-1}} f(\tau)$$
$$+ p^{(r+1)(k-1)} \cdot \sum_{\nu=0}^{r} p^{-(\nu+1)k} \cdot \sum_{c_\nu \pmod{p^{\nu+1}}} f\left((p^{(r+1)-(\nu+1)}\tau + c_\nu)/p^{\nu+1}\right)$$
$$= p^{k-1} \cdot T_{p^{r-1}} f(\tau) + T_{p^{r+1}} f(\tau). \qquad \square$$

2.6 Corollary. *For a prime p and $r, s \in \mathbb{N}_0$*

$$T_{p^r} T_{p^s} = \sum_{\nu=0}^{\min(r,s)} p^{\nu(k-1)} \cdot T_{p^{r+s-2\nu}}$$

holds. In particular, T_{p^r} and T_{p^s} commute.

Proof By induction on s: The case $s = 0$ is trivial and $s = 1$ was proven in (2.3). Then (2.3) and the induction hypothesis yield

$$T_{p^r} T_{p^{s+1}} = \sum_{\nu=0}^{\min(r,s)} p^{\nu(k-1)} \cdot T_{p^{r+s-2\nu}} T_p - p^{k-1} \cdot \sum_{\nu=0}^{\min(r,s-1)} p^{\nu(k-1)} \cdot T_{p^{r+s-1-2\nu}}.$$

Thus (2.3) yields

$$T_{p^r} T_{p^{s+1}} = \sum_{\nu=0}^{\min(r,s)} p^{\nu(k-1)} \cdot T_{p^{r+s-2\nu+1}} + \sum_{\substack{\nu=0 \\ 2\nu+1 \leq r+s}}^{\min(r,s)} p^{(\nu+1)(k-1)} \cdot T_{p^{r+s-2\nu-1}}$$
$$- \sum_{\nu=0}^{\min(r,s-1)} p^{(\nu+1)(k-1)} \cdot T_{p^{r+s-2\nu-1}}$$
$$= \sum_{\nu=0}^{\min(r,s)} p^{\nu(k-1)} \cdot T_{p^{r+s+1-2\nu}} + \sum_{\nu} p^{(\nu+1)(k-1)} \cdot T_{p^{r+s-2\nu-1}},$$

where the last sum runs over the ν with

$$\min(r, s-1) < \nu \leq \min(r, s) \quad \text{and} \quad 2\nu + 1 \leq r + s.$$

§ 2 The algebra of HECKE operators   235

But this second sum is not empty if and only if $r > s$, i.e. if $\min(r, s + 1) = s + 1$ holds. In this case, it consists of the single term with $v = s$. This completes the induction. □

3. The algebra of HECKE operators. According to Proposition 1.7 or Corollary 1.8, the HECKE operators $T_n = T_n^{(k)}$, $n \geq 1$, are endomorphisms of the vector space $\mathbb{V}_k$ which map the subspace $\mathbb{M}_k$ of modular forms into itself. Let $\mathcal{H}_k$ denote the $\mathbb{C}$-vector space of endomorphisms of $\mathbb{M}_k$ spanned by the $T_n = T_n^{(k)}$, $n \geq 1$. By definition, $\mathcal{H}_k$ consists of all finite linear combinations of the form

$$\sum_{n \geq 1} \alpha_n T_n \quad \text{with} \quad \alpha_n \in \mathbb{C}. \tag{2.5}$$

As usual, let End $\mathbb{M}_k$ denote the $\mathbb{C}$-algebra of all endomorphisms of $\mathbb{M}_k$.

2.7 Theorem. *$\mathcal{H}_k$ is a commutative subalgebra of* End $\mathbb{M}_k$ *which, as a $\mathbb{C}$-algebra with unit, is generated by T_p for primes p. For $m, n \geq 1$ the following composition rules hold*

$$T_m T_n = \sum_{d | (m,n)} d^{k-1} \cdot T_{mn/d^2}. \tag{2.6}$$

In particular, each HECKE operator T_n, $n \in \mathbb{N}$, is a rational polynomial in the T_p, p prime.

The sum in (2.6) again runs over all positive common divisors of m and n.

We call $\mathcal{H}_k$ the HECKE *algebra (of weight k)*.

Proof (i) If m and n are coprime, then (2.6) expresses the multiplicativity of T_n proved in Theorem 2.1. Each T_n can therefore be written as a product of mappings of the form T_{p^r} with primes p and $r \in \mathbb{N}$.
(ii) For a prime p, T_{p^r}, $r \in \mathbb{N}$, is a polynomial in T_p. For the proof, we use Theorem 2.3 and an induction on r.
(iii) $\mathcal{H}_k$ is commutative, because by (i) and (ii) every T_n is a polynomial in T_p's for p prime. The HECKE operators T_p and T_q commute for different primes p and q according to Theorem 2.1.
(iv) The proof of (2.6) is now given by an induction on the number of different prime divisors of mn: if mn contains only one prime divisor p, then $m = p^r$ and $n = p^s$ holds with $r + s \geq 0$. Corollary 2.6 yields

$$T_{p^r} T_{p^s} = \sum_{v=0}^{\min(r,s)} p^{v(k-1)} \cdot T_{p^{r+s-2v}} = \sum_{d | (p^r, p^s)} d^{k-1} \cdot T_{p^r p^s / d^2}, \tag{*}$$

i.e. assertion (2.6).

If m and n are not coprime (otherwise (2.6) has already been proven), then we choose a prime number p with

$$m = m' p^r, \; n = n' p^s \quad \text{and} \quad (m', p) = (n', p) = 1, \; r \geq 1, s \geq 1.$$

So by the induction hypothesis, (2.6) holds for m' and n' instead of m and n. It now follows with (i), (iii) and ($*$) that

$$\begin{aligned} T_m T_n &= T_{m'} T_{n'} T_{p^r} T_{p^s} \\ &= \sum_{t \mid (m',n')} t^{k-1} \cdot T_{m'n'/t^2} \cdot \sum_{d \mid (p^r, p^s)} d^{k-1} \cdot T_{p^r p^s / d^2}, \\ &= \sum_{t \mid (m',n')} \sum_{d \mid (p^r, p^s)} (td)^{k-1} \cdot T_{mn/(td)^2}. \end{aligned}$$

But since td runs exactly through the positive common divisors of m and n, equation (2.6) follows. □

2.8 Corollary. *If $f \in \mathbb{M}_k$ is not constant, the following assertions are equivalent:*
(i) *f is a simultaneous eigenform of all* HECKE *operators.*
(ii) *For every prime p, there is some $\lambda_f(p) \in \mathbb{C}$ with $T_p f = \lambda_f(p) \cdot f$.*
(iii) *For each prime p and all $m \in \mathbb{N}_0$,*

$$\alpha_f(p) \cdot \alpha_f(m) = \alpha_f(1) \cdot (\alpha_f(mp) + p^{k-1} \cdot \alpha_f(m/p)) \tag{2.7}$$

holds, where $\alpha_f(m/p) = 0$ if $p \nmid m$.
In this case, $\alpha_f(1) \neq 0$ holds, as well as

$$T_n f = \frac{\alpha_f(n)}{\alpha_f(1)} \cdot f \quad \text{for all} \quad n \in \mathbb{N}.$$

Proof (i) $\iff$ (ii): Every T_n is a polynomial in certain T_p's.
(ii) $\iff$ (iii): We apply Corollary 1.3 as well as (1.18). □

2.9 Remarks. a) The power n^{k-1} in definition (1.14) of the operator T_n is more or less arbitrary. If we change it, the composition rule (2.6) changes. Thus we obtain, e.g. for T_n^* defined by

$$T_n^* := n^{(1-k)/2} \cdot T_n$$

the simpler rule

$$T_m^* T_n^* = \sum_{d \mid (m,n)} T_{mn/d^2}^*.$$

b) By Theorem 2.7, every HECKE operator is a polynomial in T_p, p prime. If we write Corollary 2.6 in the form introduced in a) as T_n^*, then

$$T_{p^{r+1}}^* + T_{p^{r-1}}^* = T_p^* \cdot T_{p^r}^* \quad \text{for} \quad r \geq 1.$$

Thus we can express $T_{p^r}^*$ by the CHEBYSHEV *polynomials* P_r and Q_r,

$$\left(X + \sqrt{X^2 - 1}\right)^r = P_r(X) + Q_r(X) \cdot \sqrt{X^2 - 1},$$

§ 2 The algebra of HECKE operators

(cf. G. ANDREWS, R. ASKEY and R. ROY [4]): we obtain

$$T^*_{p^r} = Q_{r+1}(X) \quad \text{with} \quad X := \frac{1}{2}T^*_p.$$

In view of $Q_0(X) = 0$, $Q_1(X) = 1$, we easily verify the recursion formula

$$Q_{r+1}(X) = 2XQ_r(X) - Q_{r-1}(X) \quad \text{for } r \geq 1.$$

c) According to Theorem 1.10, the FOURIER coefficients $\tau(m)$ of Δ satisfy

$$\tau(p^{r+1}) + p^{11} \cdot \tau(p^{r-1}) = \tau(p^r) \cdot \tau(p)$$

for each prime p and $r \geq 1$. Thus, if we define $\tau^*(m) := m^{-11/2} \cdot \tau(m)$, then we get a perfect analogon of b):

$$\tau^*(p^r) = Q_{r+1}(x) \quad \text{with} \quad x := \frac{1}{2}\tau^*(p).$$

Obviously, the RAMANUJAN Conjecture III.2.9 is equivalent to $|x| \leq 1$, i.e. to the existence of a *real t* satisfying $x = \cos t$.

d) If we define T_{-1} on $\mathbb{M}_k$ by $(T_{-1}f)(\tau) := \overline{f(-\bar{\tau})}$, let $\mathcal{H}'_k$ be the commutative algebra $\mathcal{H}_k + \mathcal{H}_k T_{-1}$. If $f \neq 0$ is a simultaneous eigenform of $\mathcal{H}'_k$, then there exists some $0 \neq \alpha \in \mathbb{C}$ such that αf has real FOURIER coefficients (cf. Lemma 3.7).

e) A generalization of the theory of HECKE operators to congruence subgroups of the modular group can be found in G. SHIMURA [74], chap. 3, and T. MIYAKE [59], chap. 2, 4.

4. The EISENSTEIN series as simultaneous eigenforms. Using Corollary 2.8, it shall be shown that the EISENSTEIN series are simultaneous eigenforms.

2.10 Proposition. *If $k > 1$ and $\alpha(m) := \sigma_{k-1}(m)$, then for every prime p and all $m \geq 1$,*

$$\alpha(p) \cdot \alpha(m) = \alpha(mp) + p^{k-1} \cdot \alpha(m/p)$$

holds.

Proof We first show the assertion for $m = p^r$, $r \geq 1$:

$$\alpha(p) \cdot \alpha(p^r) = \alpha(p^{r+1}) + p^{k-1} \cdot \alpha(p^{r-1}). \tag{$*$}$$

Because

$$\alpha(p^r) = \sigma_{k-1}(p^r) = (q^{r+1} - 1)/(q - 1) \quad \text{for} \quad q := p^{k-1},$$

this is correct. If p is not a divisor of m, then the assertion is certainly correct because of the multiplicativity of α. If p is a divisor of m, we write $m = m'p^r$ with $(m', p) = 1$ and get

$$\alpha(p) \cdot \alpha(m) = \alpha(m') \cdot (\alpha(p) \cdot \alpha(p^r))$$
$$= \alpha(m') \cdot (\alpha(p^{r+1}) + p^{k-1} \cdot \alpha(p^{r-1}))$$
$$= \alpha(pm) + p^{k-1} \cdot \alpha(m/p). \qquad \square$$

2.11 Corollary. *For even $k \geq 4$, the* EISENSTEIN *series E_k is a simultaneous eigenform; more precisely*
$$T_n E_k = \sigma_{k-1}(n) \cdot E_k \quad \text{for all } n \geq 1.$$

Proof Because of III(2.3) and (1.15), we only need to prove (2.7) for the sequence $\alpha(m) := \sigma_{k-1}(m), m \in \mathbb{N}$. But this is the assertion of Proposition 2.10. $\square$

We will indicate another proof in Exercise 2.20 1).

The EISENSTEIN series are the only simultaneous eigenforms whose constant FOURIER coefficients do not vanish. More generally, we get

2.12 Theorem. *Let $k \geq 4$ be even and $f \in \mathbb{M}_k$ with $\alpha_f(0) = 1$. If f is an eigenform of T_n for an $n > 1$, then*
$$f = E_k^*.$$

Proof We get the eigenvalue $\lambda_f(n) = \sigma_{k-1}(n)$ from (1.15). If $f \neq E_k^*$, then
$$g := f - E_k^* \in \mathbb{S}_k, \quad g \neq 0,$$
as well as
$$T_n g = \sigma_{k-1}(n) \cdot g.$$
Proposition 1.11 leads to
$$\sigma_{k-1}(n) - n^{k/2} \cdot \sigma_{-1}(n) \leq 0. \qquad (*)$$
But we have
$$2\sigma_{k-1}(n) - 2n^{k/2} \cdot \sigma_{-1}(n) = \sum_{d|n} \left(d^{k-1} + \left(\frac{n}{d}\right)^{k-1} - n^{k/2}\left(\frac{1}{d} + \frac{d}{n}\right) \right)$$
$$= \sum_{d|n} \frac{n^{k/2}}{d} \left(1 - \left(\frac{\sqrt{n}}{d}\right)^{k-2}\right) \left(\left(\frac{d}{\sqrt{n}}\right)^k - 1\right) > 0,$$
since in the last sum every summand except for possibly $d = \sqrt{n}$ is positive due to $k \geq 4$. Thus we have a contradiction and $f = E_k^*$ follows. $\square$

2.13 Remark. Consider the conditionally convergent EISENSTEIN series E_2 in III(6.1). By Proposition III.6.1, E_2 belongs to $V(\mathbb{H})$. Proposition 2.10 and an analogon of Lemma 1.9 now show that E_2 is a simultaneous eigenform:
$$T_n E_2 = \sigma_1(n) \cdot E_2 \quad \text{for all} \quad n \in \mathbb{N}.$$

§2 The algebra of HECKE operators

5*. An integral representation of the HECKE operators. In III, §4 we saw that the $\mathbb{Z}$-modules $\mathbb{M}_k^{\mathbb{Z}}$ and $\mathbb{S}_k^{\mathbb{Z}}$ respectively are free for $k \geq 4$ and that

$$\mathbb{M}_k^{\mathbb{Z}} = \mathbb{Z} \cdot g \oplus \mathbb{S}_k^{\mathbb{Z}} \quad \text{for each} \quad g = E_4^{*r} \cdot E_6^{*s} \quad \text{with} \quad 4r + 6s = k \tag{2.8}$$

holds. For $k > 0$, Lemma 1.2 immediately yields

$$T_n f \in \mathbb{M}_k^{\mathbb{Z}} \quad \text{for all} \quad f \in \mathbb{M}_k^{\mathbb{Z}}, \tag{2.9}$$

as well as

$$T_n f \in \mathbb{S}_k^{\mathbb{Z}} \quad \text{for all} \quad f \in \mathbb{S}_k^{\mathbb{Z}}. \tag{2.10}$$

We choose an integral basis $g_1, g_2, \ldots, g_t$, $t := t_k := \dim_{\mathbb{C}} \mathbb{S}_k$ of $\mathbb{S}_k$ and denote by g the column vector with components $g_1, g_2, \ldots, g_t$. Let $T_n g$ be the column vector obtained by applying T_n to each component. Then we get

2.14 Proposition. *For $k > 0$ one has:*
a) *For each $n \in \mathbb{N}$ there is a unique matrix $A(n) \in \text{Mat}\,(t; \mathbb{Z})$ with the property*

$$T_n g = A(n) g. \tag{2.11}$$

b) *The matrices $A(n), n \in \mathbb{N}$, commute and satisfy*

$$A(m) \cdot A(n) = \sum_{d \mid (m,n)} d^{k-1} \cdot A(mn/d^2) \quad \text{for all } m, n \in \mathbb{N}. \tag{2.12}$$

c) *The $\mathbb{Z}$-module $\mathcal{H}_k^{\mathbb{Z}}$ spanned by the matrices $A(n)$, $n \in \mathbb{N}$, in $\text{Mat}\,(t; \mathbb{Z})$ is a commutative subring generated by $A(p)$, p prime and the identity matrix.*

Proof According to (2.10), there are unique $a_{\nu\mu}(n) \in \mathbb{Z}$ with

$$T_n g_\nu = \sum_{\mu=1}^{t} a_{\mu\nu}(n) \cdot g_\mu \quad \text{for} \quad \nu = 1, 2, \ldots, t.$$

Now we set $A(n) := (a_{\mu\nu}(n))$ and get part a). The remaining assertions follow from Theorem 2.7. □

2.15 Corollary. *The eigenvalues of the matrices $A(n)$, $n \in \mathbb{N}$, are algebraic integers of a degree $\leq t$.*

Proof The eigenvalues are the roots of the characteristic polynomial

$$\det(XI - A(n)) \in \mathbb{Z}[X]. \qquad \square$$

The eigenvalues of $A(n)$ are the eigenvalues of T_n on $\mathbb{S}_k$. By Corollary 2.11, it follows that

2.16 Corollary. *The eigenvalues of T_n, $n \in \mathbb{N}$, on $\mathbb{M}_k$, $k > 0$, are algebraic integers of degree $\leq t$.*

Thus Lemma 1.9 implies

2.17 Corollary. *Let $f \in \mathbb{M}_k$ be a simultaneous eigenform with $\alpha_f(1) = 1$. Then all FOURIER coefficients $\alpha_f(n)$, $n \in \mathbb{N}$, are algebraic integers of degree $\leq t$.*

6*. The first new case. If $k = 24, 28, 30, 32, 34$ and 38, then $\mathbb{S}_k$ has dimension $t = 2$ by the Dimension Formula III.4.6. We briefly treat the first case $k = 24$:

Let $q := e^{2\pi i \tau}$. According to Proposition III. 4.16,

$$g_1 := E_6^{*2} \cdot \Delta^* = \sum_{m \geq 1} \alpha_1(m) q^m, \quad g_2 := \Delta^{*2} = \sum_{m \geq 2} \alpha_2(m) q^m, \quad q = e^{2\pi i \tau}$$

are an integral basis of $\mathbb{S}_{24}^{\mathbb{Z}}$. Using III(2.11) and I(4.24), we calculate

$$g_1 = q - 2^3 \cdot 3 \cdot 43 \cdot q^2 + 2^2 \cdot 3^2 \cdot 7^2 \cdot 139 \cdot q^3 + 2^6 \cdot 31 \cdot 5527 \cdot q^4 + \cdots,$$
$$g_2 = q^2 - 2^4 \cdot 3 \cdot q^3 + 2^3 \cdot 3^3 \cdot 5 \cdot q^4 + \cdots.$$

According to Lemma 1.2, the FOURIER coefficients $\alpha_f(m)$ of a modular form f of weight 24 satisfy

$$\alpha_{T_p f}(m) = \begin{cases} \alpha_f(pm) + p^{23} \cdot \alpha_f(m/p), & \text{if } p \mid m, \\ \alpha_f(pm), & \text{if } p \nmid m, \end{cases}$$

for any prime p. With this, we calculate

$$T_2 g_1 = -2^3 \cdot 3 \cdot 43 \cdot g_1 + 2^9 \cdot 3^6 \cdot 7^2 \cdot g_2 \quad \text{and} \quad T_2 g_2 = g_1 + 2^6 \cdot 3 \cdot 11 \cdot g_2,$$

thus

$$A(2) = -2^3 \cdot 3 \cdot 43 \cdot I + A \quad \text{with} \quad A = \begin{pmatrix} 0 & 2^9 \cdot 3^6 \cdot 7^2 \\ 1 & 2^3 \cdot 3 \cdot 131 \end{pmatrix}. \tag{2.13}$$

Accordingly, with the abbreviations

$$\xi(p) := \alpha_1(2p) + 2^3 \cdot 3 \cdot 43 \alpha_1(p), \quad \eta(p) := \alpha_2(2p) + 2^3 \cdot 3 \cdot 43 \alpha_2(p), \tag{2.14}$$

we get for each prime $p > 2$:

$$A(p) = \begin{pmatrix} \alpha_1(p) & \xi(p) \\ \alpha_2(p) & \eta(p) \end{pmatrix} = \alpha_1(p) \cdot I + \begin{pmatrix} 0 & \xi(p) \\ \alpha_2(p) & \eta(p) - \alpha_1(p) \end{pmatrix}. \tag{2.15}$$

This is, of course, only a formula. The non-trivial result arises because $A(2)$ and $A(p)$ commute:

2.18 Proposition. *For every prime $p \geq 2$, one has*

$$A(p) = \alpha_1(p) \cdot I + \alpha_2(p) \cdot A.$$

Proof We use (2.13) and (2.15). As $A(2)$ and $A(p)$ commute, we immediately obtain

§ 2 The algebra of HECKE operators 241

$$\xi(p) = 2^9 \cdot 3^6 \cdot 7^2 \cdot \alpha_2(p) \quad \text{and} \quad \eta(p) - \alpha_1(p) = 2^3 \cdot 3 \cdot 131 \cdot \alpha_2(p),$$

hence the assertion. □

2.19 Corollary. *The eigenvalues of all matrices A(p), p prime, belong to the field* $\mathbb{Q}\left[\sqrt{144\,169}\right]$.

Proof The eigenvalues of an integral 2×2 matrix B lie in the field

$$\mathbb{Q}\left[\sqrt{(\operatorname{tr} B)^2 - 4 \det B}\right].$$

□

2.20 Exercises.
1) Let $n \in \mathbb{N}$ and $\mathcal{R}$ be the standard system of representatives for the right cosets from (1.11). Two matrices M_1 and M_2 of Γ_n are called Γ_∞-*equivalent* if there exists some $K \in \Gamma_\infty$ with $KM_1 = M_2$ (cf. III(2.5)). If M runs through the set $\mathcal{R}$ and L through a system of representatives for the right cosets of Γ for Γ_∞, then both LM and ML run through a complete system of Γ_∞–inequivalent matrices in Γ_n. Conclude that the EISENSTEIN series E_k, $k > 4$ even, are simultaneous eigenforms.
2) For even $k \geq 4$ and $m \in \mathbb{N}_0$, recall the definition of the POINCARÉ series $Q_{k,m}$ from Exercise III.2.13 3). For $n \in \mathbb{N}$,

$$T_n Q_{k,m} = \sum_{d \mid (m,n)} \left(\frac{n}{d}\right)^{k-1} \cdot Q_{k, mn/d^2},$$

and in particular,

$$T_n Q_{k,1} = n^{k-1} \cdot Q_{k,n} \quad \text{and} \quad T_n E_k^* = \sigma_{k-1}(n) \cdot E_k^*.$$

3) Recall the definition of the POINCARÉ series $P_k(\cdot, w)$ from Exercise III.2.13 5) for even $k \geq 4$ and $w \in \mathbb{H}$. For $n \in \mathbb{N}$, let $\mathcal{R}$ denote the standard system of representatives for the right cosets of Γ_n modulo Γ from (1.11). Then again

$$T_n P_k(\tau, w) = n^{k-1} \cdot \sum_{M \in \mathcal{R}} d^{-k} \cdot P_k(\tau, Mw) = n^{k-1} \cdot \sum_{M \in \Gamma_n} (M\{\tau, w\})^{-k}.$$

4) Let $\varphi : \mathbb{H} \to \mathbb{C}$ be holomorphic, bounded and periodic modulo 1, in particular $\varphi \in V(\mathbb{H})$. Recall the definition of $P_k(\tau, \varphi)$ from Exercise III.2.13 9). Given $n \in \mathbb{N}$ show that

$$P_k(\cdot, \varphi) \mid_k T_n = P_k(\cdot, T_n^{(k)} \varphi).$$

5) Derive the analogon of sect. 6 for $k = 28, 30, 32, 34$ and 38.
6) Let $\mathcal{H}_0$ be the $\mathbb{C}$–subalgebra of End $\mathbb{K}$ generated by T_n, $n \geq 1$. Then

$$\{Tj\,;\,T \in \mathcal{H}_0\} = \mathbb{C}[j].$$

7) Show that $\mathbb{K}$ does not contain a basis of simultaneous eigenforms.
8) Let $T = \sum_{n=0}^m \alpha_n T_n$, $\alpha_n \in \mathbb{C}$, $\alpha_m \neq 0$. Then there is some $k \in \mathbb{N}$ and some $f \in M_k$

with $Tf \neq 0$.

9) Under which conditions on $a, a', d, d' \in \mathbb{N}$ does

$$T_{a,d}T_{a',d'} = T_{aa',dd'} \quad \text{hold on} \quad V(\mathbb{H})?$$

§ 3 Petersson scalar product

1. The invariant volume element. For $\tau = x + iy \in \mathbb{H}$, denote by $dxdy$ the two-dimensional Lebesgue measure on $\mathbb{C} \cong \mathbb{R}^2$. Consider the volume element

$$dv := dv(\tau) := y^{-2}dxdy. \tag{3.1}$$

We write integrals over a Lebesgue measurable subset Ω of $\mathbb{H}$ and a continuous function $\varphi : \mathbb{H} \to \mathbb{C}$ in the form

$$\int_\Omega \varphi \, dv = \int_\Omega \varphi(\tau)dv(\tau).$$

3.1 Proposition. *The volume element* (3.1) *is invariant under all mappings*

$$\mathbb{H} \to \mathbb{H}, \quad \tau \mapsto M\tau, \quad M \in \mathrm{GL}(2;\mathbb{R}), \quad \det M > 0.$$

We call (3.1) the $\mathbb{H}$-*volume element* or *hyperbolic volume element*.

Proof If f is holomorphic on a domain, then the absolute value of the functional determinant of the mapping defined by f is equal to $|f'|^2$. We thus obtain

$$dv(M\tau) = (\mathrm{Im}\, M\tau)^{-2} \left|\frac{dM\tau}{d\tau}\right|^2 dxdy = dv(\tau),$$

using II(1.14) and II(1.7). □

For a Lebesgue measurable subset Ω of $\mathbb{H}$, we call $v(\Omega) := \int_\Omega dv$ the $\mathbb{H}$-*area* or *hyperbolic area* of Ω. The fundamental domain $\mathbb{F}$ of the modular group Γ has finite $\mathbb{H}$-area; more precisely

3.2 Lemma. $v(\mathbb{F}) = v(\overline{\mathbb{F}}) = v(\overset{\circ}{\mathbb{F}}) = \pi/3.$

Proof According to II(2.7), we have

$$v(\mathbb{F}) = \int_{-\frac{1}{2}}^{\frac{1}{2}} \int_{\sqrt{1-x^2}}^{\infty} y^{-2}dydx = \int_{-\frac{1}{2}}^{\frac{1}{2}} \frac{1}{\sqrt{1-x^2}}dx = 2\arcsin\tfrac{1}{2} = \frac{\pi}{3}.$$

Clearly, the boundary of $\mathbb{F}$ has Lebesgue measure 0. □

3.3 Corollary. *If* $\varphi : \mathbb{H} \to \mathbb{C}$ *is continuous and bounded, then* $\int_\mathbb{F} \varphi dv$ *is absolutely* *convergent.*

3.4 Remarks. a) The volume element (3.1) is the volume element belonging to the invariant metric in $\mathbb{H}$ introduced in sect. 6 of II, §1.
b) If D is a hyperbolic triangle in $\mathbb{H}$ formed by orthogonal circles (cf. II, §1), then $v(D) = \pi - (\alpha + \beta + \gamma)$ if α, β, γ denote the interior angles of the triangle D. For the proof, we represent D as a union of triangles with a corner at ∞. In this sense $\overline{\mathbb{F}}$ is a hyperbolic triangle.

2. The PETERSSON scalar product. In analogy with the function $\tilde{f}$ defined in III(1.15), we now consider a corresponding construction for $f, g \in \mathbb{M}_k$. More generally, we define for $k \in \mathbb{Z}$ and arbitrary functions $f, g : \mathbb{H} \to \mathbb{C}$

$$\varphi_{f,g}(\tau) := f(\tau) \cdot \overline{g(\tau)} \cdot (\operatorname{Im} \tau)^k. \tag{3.2}$$

From II(1.14) and (1.12), we conclude, for all $M \in \operatorname{GL}(2; \mathbb{R})$, $\det M > 0$,

$$(\det M)^k \cdot \varphi_{f|M, g|M}(\tau) = \varphi_{f,g}(M\tau). \tag{3.3}$$

3.5 Proposition. *Let $f, g \in \mathbb{M}_k$, $k > 0$.*
a) $\varphi_{f,g}(M\tau) = \varphi_{f,g}(\tau)$ *holds for all $M \in \Gamma$, $\tau \in \mathbb{H}$.*
b) $\varphi_{f,g}$ *is bounded on $\mathbb{H}$ if and only if f or g is a cusp form.*

Proof a) We use (3.3) and (M.2') in sect. 4 in III, §1.
b) Because $|\varphi| = \widetilde{(fg)}$ in III(1.15) and $fg \in \mathbb{M}_{2k}$, the assertion follows from Proposition III.1.7. □

Thus we obtain a mapping $\mathbb{M}_k \times \mathbb{S}_k \to \mathbb{C}$.

3.6 Theorem. *If $f, g \in \mathbb{M}_k$ and either f or g is a cusp form, then the integral*

$$\langle f, g \rangle := \int_{\mathbb{F}} f(\tau) \cdot \overline{g(\tau)} \cdot (\operatorname{Im} \tau)^k \, dv(\tau) \tag{3.4}$$

is absolutely convergent and

$$\langle g, f \rangle = \overline{\langle f, g \rangle}. \tag{3.5}$$

$$\langle f, g \rangle \quad \text{is} \quad \mathbb{C}\text{-linear in} \quad f. \tag{3.6}$$

$$\langle f, f \rangle \geq 0 \quad \text{for} \quad f \in \mathbb{S}_k \quad \text{and} \quad \langle f, f \rangle = 0 \quad \text{only if} \quad f = 0. \tag{3.7}$$

We call $\langle f, g \rangle$ the PETERSSON *scalar product of f and g.*

Proof The absolute convergence of (3.4) follows from Corollary 3.3 and Proposition 3.5. The assertions (3.5) to (3.7) are simple consequences. □

Note that the PETERSSON scalar product does not exist whenever $k > 0$ and neither f nor g is a cusp form because of $|E_k^*(\tau)^2 - 1| = O(e^{-2\pi \operatorname{Im} \tau})$ on $\mathbb{F}$.

As in Remark 2.9, we define, for each map $f : \mathbb{H} \to \mathbb{C}$, another map

$$\overline{f} = T_{-1}f : \mathbb{H} \to \mathbb{C}, \quad \overline{f}(\tau) := \overline{f(-\overline{\tau})}.$$

If f is holomorphic then $\overline{f}$ is also holomorphic in $\mathbb{H}$.

3.7 Lemma. *Let $f \in \mathbb{M}_k$. Then $\overline{f}$ also belongs to $\mathbb{M}_k$. The FOURIER coefficients of f are all real if and only if $\overline{f} = f$ and all purely imaginary if and only if $\overline{f} = -f$.*

Proof If f is bounded then $\overline{f}$ is also bounded on $\mathbb{F}$. To prove $\overline{f}|M = \overline{f}$ for $M \in \Gamma$, we may restrict ourselves to $M = T$ and $M = J$ due to Theorem II.2.2. In both cases we easily verify the assertion. The explicit description of the Fourier expansion of $\overline{f}$ completes the proof. □

Since $-\overline{\tau}$ runs through the closure $\overline{\mathbb{F}}$ as τ does and the boundary of $\mathbb{F}$ has LEBESGUE measure 0, we obtain

3.8 Corollary. *If $f, g \in \mathbb{M}_k$ and f or g is a cusp form, then*

$$\langle \overline{f}, g \rangle = \langle \overline{g}, f \rangle. \tag{3.8}$$

We note that the mapping $\mathbb{M}_k \to \mathbb{M}_k$, $f \mapsto \overline{f}$, is $\mathbb{C}$-antilinear.

3. Integration over invariant functions. Let $\varphi : \mathbb{H} \to \mathbb{C}$ be continuous. We consider the *invariance group* $\Gamma(\varphi)$ of φ,

$$\Gamma(\varphi) := \{M \in \Gamma \; ; \; \varphi(M\tau) = \varphi(\tau) \text{ for all } \tau \in \mathbb{H}\}. \tag{3.9}$$

In general, of course, $\Gamma(\varphi) = \{\pm I\}$. The subgroups Λ of Γ considered below are always assumed to contain $-I$.

3.9 Proposition. *If Ω is a measurable subset of $\mathbb{H}$ then*

$$\int_\Omega \varphi \, dv = \int_{M\Omega} \varphi \, dv$$

for any $M \in \Gamma(\varphi)$, provided that either of the two integrals exists.

Proof Because $\varphi(M\tau)dv(M\tau) = \varphi(\tau)dv(\tau)$ according to Proposition 3.1, the assertion follows from the transformation formula for area integrals. □

3.10 Lemma. *Let Λ be a subgroup of $\Gamma(\varphi)$. If $\mathcal{F}_1$ and $\mathcal{F}_2$ are two fundamental domains for Λ, then*

$$\int_{\mathcal{F}_1} \varphi \, dv = \int_{\mathcal{F}_2} \varphi \, dv,$$

provided one of the two integrals exists.

Proof $\mathcal{F}_1$ and $\mathcal{F}_2$ are LEBESGUE measurable as they are relatively closed sets. By Proposition 3.9, we get

§ 3 Petersson scalar product

$$\int_{\mathcal{F}_1} \varphi \, dv = \frac{1}{2} \sum_{M \in \Lambda} \int_{M\mathcal{F}_2 \cap \mathcal{F}_1} \varphi \, dv = \frac{1}{2} \sum_{M \in \Lambda} \int_{\mathcal{F}_2 \cap M^{-1}\mathcal{F}_1} \varphi \, dv$$

$$= \frac{1}{2} \sum_{M \in \Lambda} \int_{M\mathcal{F}_1 \cap \mathcal{F}_2} \varphi \, dv = \int_{\mathcal{F}_2} \varphi \, dv,$$

because Λ is countable and M^{-1} runs through Λ as M does. □

With this result, we can now define

$$\int_{\Lambda \backslash \mathbb{H}} \varphi \, dv := \int_{\mathcal{F}} \varphi \, dv, \qquad (3.10)$$

provided the integral exists whenever $\mathcal{F}$ is an arbitrary fundamental domain for $\Lambda \subseteq \Gamma(\varphi)$.

3.11 Proposition. *If Λ is a subgroup of $\Gamma(\varphi)$ with finite index, then*

$$\frac{1}{[\Gamma(\varphi) : \Lambda]} \cdot \int_{\Lambda \backslash \mathbb{H}} \varphi \, dv = \int_{\Gamma(\varphi) \backslash \mathbb{H}} \varphi \, dv,$$

provided one of the two integrals exists.

Proof Let $\mathcal{F}$ be a fundamental domain for $\Gamma(\varphi)$ and let

$$\Gamma(\varphi) = \bigcup_\nu \Lambda M_\nu$$

be a decomposition into distinct right cosets. In analogy with Theorem II.3.1, we conclude that

$$\mathcal{G} := \bigcup_\nu M_\nu \mathcal{F}$$

is a fundamental domain for Λ. Proposition 3.9 and Lemma 3.10 lead to

$$\int_{\Lambda \backslash \mathbb{H}} \varphi \, dv = \int_\mathcal{G} \varphi \, dv = \sum_\nu \int_{M_\nu \mathcal{F}} \varphi \, dv = \sum_\nu \int_\mathcal{F} \varphi \, dv = [\Gamma(\varphi) : \Lambda] \cdot \int_{\Gamma(\varphi) \backslash \mathbb{H}} \varphi \, dv. \quad \square$$

4. Cusp forms as the orthogonal complement of the Eisenstein series. In Proposition III.2.5,

$$\mathbb{M}_k = \mathbb{C} E_k^* \oplus \mathbb{S}_k \qquad (3.11)$$

was shown for even $k \geq 4$. In this section, we show that the direct sum in (3.11) is orthogonal with respect to the Petersson scalar product.

3.12 Theorem. *Let $k \geq 4$ be even and $f \in \mathbb{S}_k$. Then*

$$\langle E_k, f \rangle = \langle E_k^*, f \rangle = 0 \tag{3.12}$$

holds.

Proof We use the representation III(2.6) for E_k^*. For this we fix a system of representatives for right cosets $\mathcal{R}$ of the right cosets of Γ for Γ_∞. As $f|M = f$ holds for all $M \in \Gamma$, we conclude that

$$\langle E_k^*, f \rangle = \int_{\mathbb{F}} \sum_{M \in \mathcal{R}} 1|M(\tau) \cdot \overline{f(\tau)} \cdot (\operatorname{Im}\tau)^k \, dv(\tau)$$

$$= \int_{\mathbb{F}} \sum_{M \in \mathcal{R}} \varphi_{1|M,f|M} \, dv = \sum_{M \in \mathcal{R}} \int_{M\overline{\mathbb{F}}} \varphi_{1,f} \, dv ,$$

where we used (3.3) and Proposition 3.1 in the last step. $\bigcup_{M \in \mathcal{R}} M\overline{\mathbb{F}}$ is a fundamental domain for Γ_∞ by Theorem II.3.1. Because of $\Gamma_\infty \subseteq \Gamma(\varphi_{1,f})$, we can choose an arbitrary fundamental domain for Γ_∞ by Lemma 3.10, e.g.

$$\mathbb{F}_\infty := \{\tau \in \mathbb{H} \,;\, 0 \leq x \leq 1\}.$$

Accordingly, we have

$$\langle E_k^*, f \rangle = \int_{\mathbb{F}_\infty} \overline{f(x+iy)} \cdot y^{k-2} \, dx dy.$$

Since f is a cusp form,

$$\int_0^1 \overline{f(x+iy)} dx = \overline{\alpha_f(0)} = 0$$

holds for any $y > 0$. □

5. Self-adjointness of the HECKE operators. As a tool, we need the

3.13 Lemma. *If p is a prime, there exists a simultaneous system of representatives for the left and right cosets of Γ_p modulo Γ.*

Proof We use

$$\begin{pmatrix} p & 0 \\ 0 & 1 \end{pmatrix}, \quad \begin{pmatrix} 1 & 0 \\ k & 1 \end{pmatrix} \begin{pmatrix} 1 & k \\ 0 & p \end{pmatrix} = \begin{pmatrix} 1 & k \\ k & p+k^2 \end{pmatrix}, \quad k = 0, \ldots, p-1.$$

By Proposition 1.4, we obtain a system of representatives for the right cosets of Γ_p modulo Γ, which consists of symmetric matrices. But by transposing we get a system of representatives for the left cosets. □

§ 3 PETERSSON scalar product   247

This brings us to the fact that T_n is *self-adjoint* with respect to the PETERSSON scalar product formulated in the

3.14 Main Theorem. *If* $f, g \in \mathbb{M}_k$ *and* f *or* g *is a cusp form, then*

$$\langle T_n f, g \rangle = \langle f, T_n g \rangle \tag{3.13}$$

holds for all $n \in \mathbb{N}$.

The **proof** is divided into several steps: since T_n is a rational polynomial in the T_p by Theorem 2.7, it suffices to prove the assertion for any prime $n = p$. First let

$$M^\sharp := \begin{pmatrix} d & -b \\ -c & a \end{pmatrix} \quad \text{for} \quad M = \begin{pmatrix} a & b \\ c & d \end{pmatrix}. \tag{3.14}$$

It obviously follows that

$$M^\sharp M = M M^\sharp = (\det M) \cdot I. \tag{3.15}$$

Analogously to II(3.4), let $\Gamma(p)$ denote the extended principal congruence subgroup

$$\Gamma(p) := \{ M \in \Gamma \,;\, M \equiv \pm I \pmod{p} \} \tag{3.16}$$

of level p. Because of $f|(\lambda I) = \lambda^{-k} \cdot f$, (3.2), (3,3) and (3.4) lead to the

Assertion A. $\varphi_{f|M,g}(M^{-1}\tau) = \varphi_{f,g|M^\sharp}(\tau)$ *holds for all* $M \in \mathrm{GL}(2;\mathbb{R})$ *with* $\det M > 0$.

Assertion B. Let $f, g \in \mathbb{M}_k$ *and* $M \in \Gamma_p$.
a) $\Gamma(p)$ *and* $M^{-1}\Gamma(p)M$ *are contained in the invariance subgroup* $\Gamma(\varphi_{f|M,g})$ *with finite index.*
b) $\Gamma(p)$ *and* $M\Gamma(p)M^{-1}$ *are contained in the invariance subgroup* $\Gamma(\varphi_{f,g|M^\sharp})$ *with finite index.*

Proof a) For $L \in \Gamma(p)$, $L = \pm I + pA$, $A \in \mathrm{Mat}(2;\mathbb{Z})$ we get

$$L_0 := MLM^{-1} = \pm I + MAM^\sharp \in \mathrm{Mat}(2;\mathbb{Z})$$

because of (3.15). We obtain $L_0 \in \Gamma$ and $ML = L_0 M$, thus

$$f|M|L = f|(ML) = (f|L_0)|M = f|M \quad \text{and} \quad g|L = g.$$

Hence, (3.3) implies

$$\varphi_{f|M,g}(L\tau) = \varphi_{f|M|L,g|L}(\tau) = \varphi_{f|M,g}(\tau),$$

thus $\Gamma(p) \subseteq \Gamma(\varphi_{f|M,g})$. The index is finite because of $[\Gamma : \Gamma(p)] < \infty$. For $M^{-1}\Gamma(p)M \supseteq \Gamma(p^2)$ the assertion is immediately clear.
b) We swap f and g and apply a) to $M^\sharp$ instead of M. $\square$

Assertion C. For $M \in \Gamma_p$ *one has*

$$[\Gamma : \Gamma(p)] = [\Gamma : M^{-1}\Gamma(p)M].$$

Proof Let $\mathcal{F}$ be a fundamental domain with respect to $\Gamma(p)$. Then $M^{-1}\mathcal{F}$ is a fundamental domain with respect to $M^{-1}\Gamma(p)M$. Lemma 3.2, Proposition 3.11 and Proposition 3.1 lead to

$$[\Gamma : M^{-1}\Gamma(p)M] \cdot \frac{\pi}{3} = \int_{M^{-1}\mathcal{F}} dv = \int_{\mathcal{F}} dv = [\Gamma : \Gamma(p)] \cdot \frac{\pi}{3}. \qquad \square$$

Assertion D. If $M \in \Gamma_p$ and $\mathcal{F}$ is a fundamental domain with respect to $\Gamma(p)$, then

$$\int_{\mathcal{F}} \varphi_{f|M,g}\, dv = \int_{M^{-1}\mathcal{F}} \varphi_{f|M,g}\, dv.$$

Proof $M^{-1}\Gamma(p)M$ and $\Gamma(p)$, have the same index in Γ by Assertion C, thus by Assertion B also in $\Gamma(\varphi_{f|M,g})$. Then the statement follows immediately from Proposition 3.11. $\qquad \square$

Proof of the Main Theorem. Because of Assertion B, for any system of representatives for the right cosets $\mathcal{R}$ of Γ_p modulo Γ, we have by Proposition 3.11

$$\langle T_p f, g \rangle = \frac{p^{k-1}}{[\Gamma : \Gamma(p)]} \cdot \int_{\Gamma(p)\backslash \mathbb{H}} \sum_{M \in \mathcal{R}} \varphi_{f|M,g}\, dv.$$

If we use Assertion D and A, it follows that

$$\langle T_p f, g \rangle = \frac{p^{k-1}}{[\Gamma : \Gamma(p)]} \cdot \sum_{M \in \mathcal{R}} \int_{\Gamma(p)\backslash \mathbb{H}} \varphi_{f,g|M^\sharp}\, dv.$$

We now choose $\mathcal{R}$ to be a two-sided system of representatives for the cosets using Lemma 3.13. Since $M^\sharp = pM^{-1}$ also runs through a system of representatives for the right cosets, as M does, we get the assertion (3.13). $\qquad \square$

Using the SMITH normal form in Lemma 1.17, we can easily show that, for every $n \in \mathbb{N}$, there exists a simultaneous system of representatives for the left and right cosets of Γ_n modulo Γ consisting of symmetric matrices.

6. A distinguished basis of $\mathbb{S}_k$ is described in the

3.15 Theorem. *There exists a basis $f_1, \ldots, f_t$, $t = \dim \mathbb{S}_k$, of the vector space $\mathbb{S}_k$ of cusp forms of weight k with the following properties:*
a) $\langle f_\nu, f_\mu \rangle = \delta_{\nu\mu}$ *for* $\nu, \mu = 1, \ldots, t$.
b) *Each f_ν is a simultaneous eigenform with respect to all* HECKE *operators.*
c) *The* FOURIER *coefficients of all f_ν, $\nu = 1, \ldots, t$, are real.*

Proof Because of (3.5) to (3.7), the PETERSSON scalar product $\langle , \rangle$ is a positive definite Hermitian form on $\mathbb{S}_k$. Thus $(\mathbb{S}_k; \langle , \rangle)$ is a unitary vector space. According

§ 3 Petersson scalar product

to the Main Theorem 3.14,

$$\langle T_n f, g \rangle = \langle f, T_n g \rangle \quad \text{holds for all} \quad f, g \in \mathbb{S}_k, \ n \in \mathbb{N}. \tag{3.17}$$

Moreover, the Hecke operators commute

$$T_n T_m = T_m T_n \quad \text{for all} \quad n, m \in \mathbb{N}. \tag{3.18}$$

Given an arbitrary but fixed orthonormal basis $g_1, \ldots, g_t$, $t = \dim \mathbb{S}_k$ and $n \in \mathbb{N}$, there is a matrix

$$H_n := (h_{\nu\mu}^{(n)}) \in \mathrm{Mat}\,(t; \mathbb{C})$$

with the property

$$T_n g_\mu = \sum_{\nu=1}^{t} h_{\nu\mu}^{(n)} g_\nu, \quad \mu = 1, \ldots, t. \tag{3.19}$$

$$\overline{H}_n^{tr} = H_n \quad \text{and} \quad H_n H_m = H_m H_n \quad \text{for} \quad m, n \in \mathbb{N}. \tag{3.20}$$

holds because of (3.17) and (3.18). □

We need a result from linear algebra.

3.16 Lemma. *Let $\mathcal{M} \subseteq \mathrm{Mat}\,(q; \mathbb{C})$ be a set of Hermitian matrices which mutually commute. Then there exists a unitary $q \times q$ matrix W such that all matrices $\overline{W}^{tr} HW$, $H \in \mathcal{M}$, are in diagonal form.*

Proof We use induction on q, where $q = 1$ is trivial. In the case $q > 1$, we assume that there is a matrix S in $\mathcal{M}$ which is not in diagonal form. There exists a unitary matrix W with

$$\overline{W}^{tr} SW = \begin{pmatrix} \lambda I & 0 \\ 0 & T \end{pmatrix}, \quad I = I^{(r)}, \ 1 \leq r < q,$$

due to the Spectral Theorem (cf. S. Lang [53], XV §6), where $\lambda \in \mathbb{R}$ is not an eigenvalue of T. For any matrix $H \in \mathcal{M}$, we now write

$$\overline{W}^{tr} HW = \begin{pmatrix} A & B \\ C & D \end{pmatrix} \quad \text{with} \quad A \in \mathrm{Mat}\,(r, \mathbb{C}), \ B = \overline{C}^{tr}.$$

As S and H commute, we have

$$\begin{pmatrix} \lambda E & 0 \\ 0 & T \end{pmatrix} \begin{pmatrix} A & B \\ C & D \end{pmatrix} = \begin{pmatrix} A & B \\ C & D \end{pmatrix} \begin{pmatrix} \lambda E & 0 \\ 0 & T \end{pmatrix}, \tag{$*$}$$

so $TC = \lambda C$. Thus, for each column vector v of C, we have $Tv = \lambda v$. But since λ is not an eigenvalue of T, it follows that $v = 0$, i.e. $C = 0$ and $B = 0$. From $(*)$ we now get

$$\overline{W}^{tr} HW = \begin{pmatrix} A_H & 0 \\ 0 & D_H \end{pmatrix} \quad \text{for all} \quad H \in \mathcal{M}.$$

Hence
$$\mathcal{M}_1 := \{A_H \,;\, H \in \mathcal{M}\} \quad \text{and} \quad \mathcal{M}_2 := \{D_H \,;\, H \in \mathcal{M}\}$$
are sets of Hermitian matrices of smaller size which mutually commute. By induction, $\mathcal{M}_1$ and $\mathcal{M}_2$ can be simultaneously transformed to diagonal forms. □

Now Lemma 3.16 is applied to the set $\mathcal{M} := \{H_n \,;\, n \in \mathbb{N}\}$ as in (3.19). Due to (3.20), $\mathcal{M}$ consists of Hermitian matrices which mutually commute. So there is a unitary $t \times t$ matrix $W = (w_{\nu\mu})$ such that

$$H_n = \overline{W}^{tr} L_n W, \quad L_n = \begin{pmatrix} \lambda_1(n) & & 0 \\ & \ddots & \\ 0 & & \lambda_t(n) \end{pmatrix} \tag{3.21}$$

holds for all $n \in \mathbb{N}$. We now set

$$f_\mu := \sum_{\nu=1}^t \overline{w}_{\mu\nu} g_\nu, \quad \mu = 1, \ldots, t,$$

and obtain

$$\langle f_\nu, f_\mu \rangle = \delta_{\nu\mu} \quad \text{for} \quad \nu, \mu = 1, \ldots, t. \tag{3.22}$$

Then (3.22), (3.19) and (3.21) imply

$$T_n f_\mu = \sum_\nu \overline{w}_{\mu\nu} \cdot T_n g_\nu = \sum_{\nu,\sigma} \overline{w}_{\mu\nu} \cdot h_{\sigma\nu}^{(n)} \cdot g_\sigma = \sum_\sigma (H_n \overline{W}^t)_{\sigma\mu} \cdot g_\sigma$$

$$= \sum_\sigma (\overline{W}^t L_n)_{\sigma\mu} \cdot g_\sigma = \sum_\sigma \overline{w}_{\mu\sigma} \cdot \lambda_\mu(n) \cdot g_\sigma = \lambda_\mu(n) \cdot f_\mu,$$

for $n \in \mathbb{N}$, thus

$$T_n f_\mu = \lambda_\mu(n) \cdot f_\mu \quad \text{for all} \quad n \in \mathbb{N} \quad \text{and} \quad \mu = 1, \ldots, t. \tag{3.24}$$

By Corollary 2.8, $\alpha_{f_\mu}(1) \neq 0$. After multiplying f_μ by a complex number of absolute value 1, we may assume $\alpha_{f_\mu}(1) \in \mathbb{R}$. The $\lambda_\mu(n)$ are real, being eigenvalues of Hermitian matrices. Then, by Corollary 2.8, all FOURIER coefficients $\alpha_{f_\mu}(n) = \lambda_\mu(n) \cdot \alpha_{f_\mu}(1)$, $n \in \mathbb{N}$, are also real. This proves the assertion. □

Thus $f_1, \ldots, f_t$ is an orthonormal basis of $\mathbb{S}_k$ consisting of simultaneous eigenforms with respect to all HECKE operators.

The deeper meaning of the PETERSSON scalar product is outlined by J. ELSTRODT and F. GRUNEWALD (*Jahresber. Dtsch. Math. Verein.* **100**, 253-283 (1998)).

7*. Algebraic properties of the eigenvalues are described in

3.17 Corollary. *The eigenvalues of the* HECKE *operators T_n, $n \in \mathbb{N}$, on $\mathbb{M}_k$, $k > 0$,*

§ 3 Petersson scalar product

are real; more precisely they are algebraic integers and are totally real over $\mathbb{Q}$ of a degree $\leq \dim \mathbb{S}_k$.

Proof Let $0 \neq f \in \mathbb{M}_k$ with $T_n f = \lambda_f(n) \cdot f$. If $f \notin \mathbb{S}_k$, i.e. $\alpha_f(0) \neq 0$, then $\lambda_f(n) = \sigma_{k-1}(n) \in \mathbb{Z}$ follows from (1.15). We get the assertion for $f \in \mathbb{S}_k$ from Corollary 2.16. □

Using Lemma 1.9, this yields

3.18 Corollary. *If $0 \neq f \in \mathbb{M}_k$, $k > 0$, is a simultaneous eigenform with respect to all Hecke operators, then $\alpha_f(1) \neq 0$ holds and the quotients $\alpha_f(n)/\alpha_f(1)$, $n \in \mathbb{N}$, are algebraic integers and totally real over $\mathbb{Q}$ of degree $\leq \dim \mathbb{S}_k$.*

In particular, Corollary 3.18 holds for the basis $f_1, \ldots, f_t$ of Theorem 3.15, and thus strengthens its part c).

$\mathbb{M}_0$ consists only of the constant functions, which are eigenforms of the Hecke operator T_n, $n \in \mathbb{N}$, with eigenvalue $\sigma_{-1}(n) \in \mathbb{Q}$.

3.19 Exercises.

In all exercises, let $k \geq 4$ be even. Recall the definition of Poincaré series from Exercises III.2.18 3) and 5).

1) The $\mathbb{H}$–area of a vertical strip (cf. Exercise III.2.13) of height ε in $\mathbb{H}$ is finite: $v(\mathcal{V}_\varepsilon) = 2/\varepsilon^2$ for $\varepsilon > 0$.
2) For $m \in \mathbb{N}$ and $f \in \mathbb{S}_k$, $\langle f, Q_{k,m} \rangle = \frac{(k-2)!}{(4\pi m)^{k-1}} \cdot \alpha_f(m)$ holds.
3) If $m \geq 1$ and the m–th Fourier coefficient of $Q_{k,m}$ is equal to 0, then $Q_{k,m} = 0$.
4) For $m, n \geq 1$,
$$n^{k-1} \cdot \alpha_{Q_{k,n}}(m) = m^{k-1} \cdot \alpha_{Q_{k,m}}(n).$$
5) The Poincaré series $Q_{k,m}$, $m \geq 1$, span the space $\mathbb{S}_k$ of cusp forms. For $k = 12$ or even $k \geq 16$, $Q_{k,1}, \ldots, Q_{k,t}$, $t = \dim \mathbb{S}_k$, form a basis of $\mathbb{S}_k$.
6) For $w \in \mathbb{Q}$ and $f \in \mathbb{S}_k$ we have $\langle f, P_k(\cdot, w) \rangle = i^k \frac{\pi}{2^{k-3}(k-1)} \cdot f(-\overline{w})$.
7) Let $k = 12$ or $k \geq 16$. For $w \in \mathbb{H}$, the following assertions are equivalent:
(i) $P_k(\cdot, w) \neq 0$.
(ii) $P_k(-\overline{w}, w) \neq 0$.
(iii) k is a multiple of the order of the stabilizer subgroup Γ_w.
8) For all $\tau, w \in \mathbb{H}$ we have $P_{12}(\tau, w) = P_{12}(-\overline{w}, w) \cdot \Delta(\tau)/\Delta(w)$ and moreover $P_{12}(-\overline{w}, w) \neq 0$.
9) The Poincaré series $P_k(\cdot, w)$, $w \in \mathbb{H}$, span the vector space $\mathbb{S}_k$. For $k = 12$ or even $k \geq 16$ and any mutually distinct $w_1, \ldots, w_t \in \mathbb{F} \setminus \{i, \rho\}$, $t = \dim \mathbb{S}_k$, Poincaré series $P_k(\cdot, w_1), \ldots, P_k(\cdot, w_t)$ form a basis of $\mathbb{S}_k$.
10) Derive a new proof for the self-adjointness of the Hecke operators T_n from Exercises 9), 5) and 2.20 2).
11) There exists an $f \in \mathbb{S}_k$ with the property $\{Tf \; ; \; T \in \mathcal{H}_k\} = \mathbb{S}_k$.
12) Let k, ℓ be even with $k, \ell, k - \ell \geq 4$. For $f \in \mathbb{S}_k$ and $g \in \mathbb{S}_\ell$, we have

$$\langle f, E^*_{k-\ell} \cdot g \rangle = (k-1)! \cdot \sum_{n=1}^{\infty} \alpha_f(n) \overline{\alpha_g(n)} (4\pi n)^{1-k}.$$

§ 4 DIRICHLET series with functional equation

In this paragraph, we discuss the HECKE correspondence between modular forms and DIRICHLET series with a functional equation.

1. DIRICHLET series. Let $(\alpha_m)_{m \geq 1}$ be a complex sequence. Then we call

$$D(s) := \sum_{m=1}^{\infty} \alpha_m \cdot m^{-s}, \quad s \in \mathbb{C}, \tag{4.1}$$

the *associated* DIRICHLET *series*. The standard example is the RIEMANN *zeta function*

$$\zeta(s) := \sum_{m=1}^{\infty} m^{-s}, \tag{4.2}$$

associated with the sequence $\alpha_m = 1$, $m \in \mathbb{N}$. We write $\alpha_m = O(m^\chi)$ with $\chi \in \mathbb{R}$ if there exists some $C > 0$ satisfying

$$|\alpha_m| \leq C \cdot m^\chi \quad \text{for all } m \geq 1.$$

The convergence behavior of DIRICHLET series is described in the

4.1 Proposition. *If $(\alpha_m)_{m \geq 1}$ is a complex sequence with associated DIRICHLET series $D(s)$, then there exists some $\sigma_0 \in \mathbb{R} \cup \{\pm\infty\}$ with the following properties:*
(i) *$D(s)$ converges absolutely for $\sigma = \operatorname{Re} s > \sigma_0$.*
(ii) *$D(s)$ does not converge absolutely for $\sigma = \operatorname{Re} s < \sigma_0$.*
$D(s)$ is a holomorphic function on the half-plane $\{s \in \mathbb{C}; \operatorname{Re} s > \sigma_0\}$. For $\rho \in \mathbb{R}$ with $\rho > \sigma_0$, $D(s)$ is absolutely and uniformly convergent and bounded on the half-plane $\{s \in \mathbb{C}; \operatorname{Re} s \geq \rho\}$.

We call σ_0 the *absolute convergence abscissa* of the DIRICHLET *series $D(s)$*.

Proof For $s = \sigma + it$, we have $|\alpha_m \cdot m^{-s}| = |\alpha_m| \cdot m^{-\sigma}$. The assertion follows with

$$\sigma_0 := \inf \left\{ \sigma \in \mathbb{R}; \sum_{m=1}^{\infty} |\alpha_m| \cdot m^{-\sigma} \text{ converges} \right\}.$$

For $\rho \in \mathbb{R}$ with $\rho > \sigma_0$,

$$\sum_{m=1}^{\infty} |\alpha_m \cdot m^{-s}| \leq \sum_{m=1}^{\infty} |\alpha_m| \cdot m^\rho < \infty \quad \text{for } \operatorname{Re} s \geq \rho.$$

Since each summand is holomorphic, $D(s)$ is also holomorphic for $\operatorname{Re} s > \sigma_0$ according to the WEIERSTRASS M-Test A.7. □

Clearly, $\sigma_0 = 1$ is the absolute convergence abscissa of the RIEMANN zeta function.

§4 Dirichlet series with functional equation

4.2 Lemma. *If $(\alpha_m)_{m\geq 1}$ is a sequence with $\alpha_m = O(m^\chi)$ for some $\chi \in \mathbb{R}$, then the absolute convergence abscissa σ_0 of the associated Dirichlet series $D(s)$ satisfies*

$$\sigma_0 \leq \chi + 1.$$

Proof There exists some $C > 0$ with $|\alpha_m| \leq C \cdot m^\chi$ for all $m \geq 1$. Thus it follows that

$$\sum_{m=1}^{\infty} |\alpha_m| \cdot m^{-\sigma} \leq C \cdot \zeta(\sigma - \chi) < \infty \quad \text{for} \quad \sigma > \chi + 1. \quad \square$$

Further statements about the convergence behavior of Dirichlet series can be found e.g. in D.B. Zagier [90], §§1, 2, or T.M. Apostol [6], chap. 11.

2. Mellin transform. For real $\chi \geq 0$, let $\mathcal{A}_\chi$ denote the set of continuous functions $g :]0, \infty[\to \mathbb{C}$ with the property $g(y) = O(y^{-\sigma})$ for all $\sigma > \chi$. This means that for $\sigma > \chi$, there exists some $C_\sigma > 0$ such that

$$|g(y)| \leq C_\sigma \cdot y^{-\sigma} \quad \text{for all} \quad y > 0. \tag{4.3}$$

Clearly, $\mathcal{A}_\chi$ is a $\mathbb{C}$–vector space. The function $y \mapsto e^{-y}$ belongs to $\mathcal{A}_0$.

4.3 Proposition. *Given $g \in \mathcal{A}_\chi$,*

$$\{s \in \mathbb{C} \,;\, \sigma = \operatorname{Re} s > \chi\} \to \mathbb{C}, \, s \mapsto M_g(s) := \int_0^\infty g(y) y^{s-1} \, dy, \tag{4.4}$$

is a holomorphic function.

We call $M_g(s)$ the Mellin *transform of* g.

Proof For $\chi < \alpha < \sigma < \beta$ we have, according to (4.4),

$$\int_0^1 |g(y) \cdot y^{s-1}| \, dy \leq C_\alpha \cdot \int_0^1 y^{\sigma-\alpha-1} \, dy = C_\alpha \cdot \frac{1}{\sigma - \alpha},$$

$$\int_1^\infty |g(y) \cdot y^{s-1}| \, dy \leq C_\beta \cdot \int_1^\infty y^{\sigma-\beta-1} \, dy = C_\beta \cdot \frac{1}{\beta - \sigma}.$$

In each vertical strip $\{s \in \mathbb{C} \,;\, \alpha + \varepsilon \leq \sigma = \operatorname{Re} s \leq \beta - \varepsilon\}$, $\varepsilon > 0$, the integral converges absolutely uniformly. The Weierstrass M-Test A.7 shows that $M_g(s)$ is holomorphic on the half-plane $\{s \in \mathbb{C} \,;\, \sigma > \chi\}$. $\square$

Now we shall show that we can reverse the transformation (4.4). Let $\mathcal{B}_\chi$ denote the set of holomorphic functions f on the half-plane $\{s \in \mathbb{C} \,;\, \sigma = \operatorname{Re} s > \chi\}$ such that for all α, β with $\chi < \alpha < \beta$, there are $\gamma > 1$ and $C > 0$ with the property

$$|f(s)| \leq C \cdot (1 + |t|)^{-\gamma} \quad \text{for} \quad s = \sigma + it, \, \alpha \leq \sigma \leq \beta. \tag{4.5}$$

Obviously, $\mathcal{B}_\chi$ is a $\mathbb{C}$-vector space, even a $\mathbb{C}$-algebra. We formulate the MELLIN *inversion formula* as

4.4 Theorem. *For $f \in \mathcal{B}_\chi$, $\chi \geq 0$, $y > 0$ and $\sigma > \chi$, the integral along the line $(\sigma) := \sigma + i\mathbb{R}$,*

$$f^*(y) := \frac{1}{2\pi i} \cdot \int_{(\sigma)} f(s) \cdot y^{-s} \, ds := \frac{1}{2\pi} \cdot \int_{-\infty}^{\infty} f(\sigma + it) \cdot y^{-\sigma - it} \, dt, \qquad (4.6)$$

converges absolutely and

a) *f^* is independent of σ, $\sigma > \chi$.*

b) *$f^* \in \mathcal{A}_\chi$.*

c) *For $s \in \mathbb{C}$ with $\sigma = \operatorname{Re} s > \chi$*

$$f(s) = M_{f^*}(s) = \int_0^\infty f^*(y) \cdot y^{s-1} \, dy \qquad (4.7)$$

holds.

Proof Because of (4.5), we have for $\sigma > \chi$ and some $\gamma > 1$

$$\int_{(\sigma)} |f(s) \cdot y^{-s}| \cdot |ds| \leq 2C \cdot y^{-\sigma} \cdot \int_0^\infty (1+t)^{-\gamma} \, dt = \frac{2C}{\gamma - 1} \cdot y^{-\sigma}. \qquad (*)$$

Hence the integral is absolutely convergent and represents a continuous function on the interval $]0, \infty[$.

a) According to the CAUCHY's Integral Theorem A.6,

$$\int_{\partial R} f(s) \cdot y^{-s} \, ds = 0$$

holds for each axis-parallel rectangle R in the half-plane $\{s \in \mathbb{C}; \sigma > \chi\}$.
Due to (4.5), the integrals over the sides parallel to the real axis vanish as $a \to \infty$, $b \to \infty$. It follows that

$$\int_{(\alpha)} f(s) \cdot y^{-s} \, ds = \int_{(\beta)} f(s) \cdot y^{-s} \, ds.$$

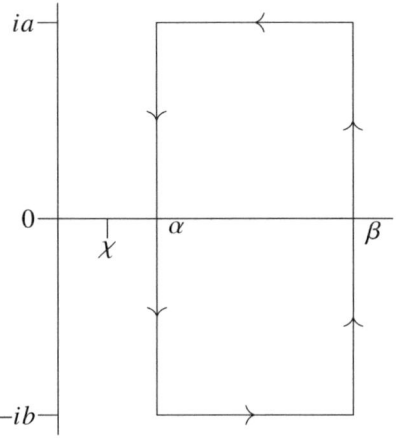

Figure 25: Integration path

b) We use $(*)$.

c) Due to b) and Proposition 4.3, $M_{f^*}(s)$ exists. Following a), we have

§ 4 Dirichlet series with functional equation

$$\int_0^\infty f^*(y) y^{s_0-1} \, dy$$

$$= \frac{1}{2\pi i} \int_0^1 \left(\int_{(\alpha)} f(s) y^{-s} \, ds \right) y^{s_0-1} \, dy + \frac{1}{2\pi i} \int_1^\infty \left(\int_{(\beta)} f(s) y^{-s} \, ds \right) y^{s_0-1} \, dy$$

for $\chi < \alpha < \sigma_0 = \operatorname{Re} s_0 < \beta$. Because of (∗), we can interchange the integrals in both cases and get

$$\frac{1}{2\pi} \int_{-\infty}^\infty \left(\int_0^1 y^{s_0-\alpha-it-1} \, dy \right) f(\alpha+it) \, dt + \frac{1}{2\pi} \int_{-\infty}^\infty \left(\int_1^\infty y^{s_0-\beta-it-1} \, dy \right) f(\beta+it) \, dt$$

$$= \frac{1}{2\pi} \left(\int_{-\infty}^\infty \frac{f(\alpha+it)}{s_0 - (\alpha+it)} \, dt + \int_{-\infty}^\infty \frac{f(\beta+it)}{\beta+it-s_0} \, dt \right).$$

According to Cauchy's Integral Theorem A.6 and (4.5), this is equal to

$$\frac{1}{2\pi i} \int_{\partial R} \frac{f(z)}{z - s_0} \, dz$$

for any axis-parallel positively oriented rectangle R containing the point s_0 and lying in the strip $\alpha \leq \operatorname{Re} s \leq \beta$. Using the Cauchy's Integral Formula, we get $f(s_0)$ as its value. Hence (4.7) holds. □

4.5 Remark. The transformation (4.4), introduced by H. Mellin is closely related to the Fourier transformation. If we substitute $y = e^x$ in (4.4) and moreover write $G(x) := g(e^x)$, we have

$$M_g(s) = \int_{-\infty}^\infty G(x) e^{sx} \, dx.$$

The inverse formula described in Theorem 4.4 is due to H. Mellin (*Math. Ann.* **68**, 305–337 (1910)). For a systematic investigation of the Mellin transformation, see P.L. Butzer and S. Jansche, (*J. Fourier Anal. Appl.* **3**, 325-376 (1997)).

3. Application to the gamma function. The function $g(y) := e^{-y}$ belongs to $\mathcal{A}_0$. So we can define the *gamma function* by Proposition 4.3 for $\operatorname{Re} s > 0$ as the Mellin transform of g, i.e.

$$\Gamma(s) := \int_0^\infty e^{-y} y^{s-1} \, dy, \quad s \in \mathbb{C}, \quad \operatorname{Re} s > 0. \tag{4.8}$$

The gamma function can be continued into the entire s–plane as a meromorphic function. It is holomorphic except for 1st order poles at the points $-n, n \in \mathbb{N}_0$, with residues

$$\operatorname{res}_{s=-n} \Gamma(s) = (-1)^n/n! \,. \tag{4.9}$$

It satisfies the functional equation

$$\Gamma(s+1) = s \cdot \Gamma(s) \tag{4.10}$$

and has no roots such that

$$1/\Gamma(s) \quad \text{is an entire function.} \tag{4.11}$$

Also of importance are the *doubling formula*

$$\Gamma(s) = \frac{1}{\sqrt{\pi}} \cdot 2^{s-1} \cdot \Gamma\left(\frac{s}{2}\right) \cdot \Gamma\left(\frac{s+1}{2}\right), \quad \text{in particular} \quad \Gamma\left(\frac{1}{2}\right) = \sqrt{\pi}, \tag{4.12}$$

and the *complex* STIRLING *formula*, according to which

$$\log \Gamma(s) - \left(s - \frac{1}{2}\right) \log s + s \quad \text{for} \quad \operatorname{Re} s \geq 0, \ |s| \geq \varepsilon > 0 \tag{4.13}$$

is bounded. Compare A.13.

4.6 Lemma. $\Gamma \in \mathcal{B}_0$.

Proof Let $0 < \alpha < \beta$. By (4.13), there exists some $C > 0$ such that

$$|\Gamma(s)| \leq C \cdot |e^{(s-1/2)\log s - s}| \quad \text{for} \quad \operatorname{Re} s \geq \alpha.$$

For $s = \sigma + it$, $\alpha \leq \sigma \leq \beta$, we get

$$\operatorname{Re}\left(\left(s - \tfrac{1}{2}\right)\log s - s\right) = \left(\sigma - \tfrac{1}{2}\right) \log \sqrt{\sigma^2 + t^2} - \sigma - t \arg s \leq \delta - \tfrac{\pi}{4}|t|$$

for some $\delta > 0$ depending only on α and β because L'HOSPITAL's theorem and the growth of the logarithm yield

$$\tfrac{\pi}{2}|t| - t \arg s \leq \tfrac{\delta}{2} \quad \text{and} \quad (\sigma - \tfrac{1}{2}) \log |s| - \sigma \leq \tfrac{\delta}{2} + \tfrac{\pi}{4}|t|.$$

Since

$$|\Gamma(s)| \leq C \cdot e^{\delta - \pi |t|/4} = O\left((1 + |t|)^{-2}\right)$$

holds, Γ belongs to $\mathcal{B}_0$ due to (4.5). □

Now Theorem 4.4 can be applied to Γ. In the following, let

$$z^s := e^{s(\log |z| + i \arg(z))}$$

for numbers $z, s \in \mathbb{C}$, $z \neq 0$, always denote the *principal value*.

4.7 Proposition. *For $\sigma > 0$ and $z \in \mathbb{C}$ with $\operatorname{Re} z > 0$,*

$$\frac{1}{2\pi i} \int_{(\sigma)} \Gamma(s) \cdot z^{-s} \, ds = e^{-z} \tag{4.14}$$

holds.

Proof By Lemma 4.6 and Theorem 4.4, $\Gamma^*(y)$ is defined for $y > 0$. By differentiation under the integral sign, we get

$$\frac{d\Gamma^*(y)}{dy} = \frac{d}{dy} \frac{1}{2\pi i} \int_{(\sigma)} \Gamma(s) \cdot y^{-s} \, ds = \frac{1}{2\pi i} \int_{(\sigma)} \Gamma(s)(-s) \cdot y^{-s-1} \, ds$$

$$= \frac{-1}{2\pi i} \int_{(\sigma)} \Gamma(s+1) \cdot y^{-s-1} \, ds = \frac{-1}{2\pi i} \int_{(\sigma+1)} \Gamma(s) \cdot y^{-s} \, ds = -\Gamma^*(y),$$

if we apply (4.10) and Theorem 4.4. It follows that $\Gamma^*(y) = c \cdot e^{-y}$ with a constant c. Then (4.7) implies

$$\Gamma(s) = c \cdot \int_0^\infty e^{-y} y^{s-1} \, dy,$$

thus $c = 1$. Accordingly, (4.14) holds for all $z = y > 0$. The integral is again holomorphic in z due to the WEIERSTRASS M-Test A.7. Hence the assertion follows by analytic continuation. □

4. DIRICHLET series associated with modular forms. For even $k \geq 4$, consider the vector space $\mathbb{M}_k$ of modular forms of weight k as defined in III, §1. Each $f \in \mathbb{M}_k$ has a FOURIER series expansion of the form

$$f(\tau) = \sum_{m=0}^\infty \alpha_f(m) \cdot e^{2\pi i m \tau}, \quad \tau \in \mathbb{H}. \tag{4.15}$$

From III(2.12), it follows that

$$\alpha_f(m) = O(m^{k-1}).$$

By Lemma 4.2, the DIRICHLET *series associated with f*

$$D_f(s) := \sum_{m=1}^\infty \alpha_f(m) \cdot m^{-s} \tag{4.16}$$

converges absolutely for $\operatorname{Re} s > k$ and represents a holomorphic function in the half-plane $\{s \in \mathbb{C}; \operatorname{Re} s > k\}$. Now we set

$$\mathbb{D}_f(s) := (2\pi)^{-s} \cdot \Gamma(s) \cdot D_f(s), \quad \operatorname{Re} s > k. \tag{4.17}$$

Moreover, we call

$$\mathcal{V}_{\alpha,\beta} := \{s \in \mathbb{C}; \ \alpha \leq \operatorname{Re} s \leq \beta\}, \quad \alpha, \beta \in \mathbb{R}, \ \alpha < \beta,$$

a *vertical strip* in $\mathbb{C}$.

4.8 Theorem. *Let* $f \in \mathbb{M}_k$, $k \geq 4$ *even.*
a) *The* DIRICHLET *series* $D_f(s)$ *associated with* f *possesses a meromorphic continuation into the whole s-plane and is holomorphic except for a possible simple pole at* $s = k$ *with the residue*

$$\operatorname{res}_{s=k} D_f(s) = \frac{(2\pi i)^k}{(k-1)!} \cdot \alpha_f(0). \tag{4.18}$$

Moreover, one has

$$D_f(0) = -\alpha_f(0) \quad \text{and} \quad D_f(-n) = 0 \quad \text{for} \quad n = 1, 2, 3, \ldots. \tag{4.19}$$

b) *The function*

$$\mathbb{D}_f^*(s) = \mathbb{D}_f(s) - \alpha_f(0) \cdot \left(\frac{i^k}{s-k} - \frac{1}{s} \right) \tag{4.20}$$

$$= \int_1^\infty \left[f(iy) - \alpha_f(0) \right] \cdot \left[y^s + i^k y^{k-s} \right] \frac{dy}{y}$$

is entire and $\mathbb{D}_f(s)$ *satisfies the functional equation*

$$\mathbb{D}_f(k - s) = i^k \cdot \mathbb{D}_f(s). \tag{4.21}$$

For each vertical strip $\mathcal{V}_{\alpha,\beta}$, *there is some* $c > 0$ *such that*

$$|\mathbb{D}_f^*(s)| \leq \frac{c}{1 + |t|} \quad \text{for all} \quad s = \sigma + it \in \mathcal{V}_{\alpha,\beta}. \tag{4.22}$$

Proof In the integral representation (4.8)

$$\Gamma(s) = \int_0^\infty e^{-r} r^{s-1} \, dr, \quad \operatorname{Re} s > 0,$$

we substitute $r = 2\pi m y$ and get

$$\Gamma(s) \cdot (2\pi m)^{-s} = \int_0^\infty e^{-2\pi m y} y^{s-1} \, dy,$$

hence

§ 4 Dirichlet series with functional equation

$$\mathbb{D}_f(s) = \sum_{m=1}^{\infty} \alpha_f(m)\Gamma(s)(2\pi m)^{-s} = \int_0^{\infty} \sum_{m=1}^{\infty} \alpha_f(m) e^{-2\pi m y} y^{s-1} \, dy \qquad (4.23)$$

$$= \int_0^{\infty} \left[f(iy) - \alpha_f(0) \right] \cdot y^{s-1} \, dy.$$

The interchange of summation and integration here is justified by the Dominated Convergence Theorem, since we have

$$\sum_{m=1}^{\infty} |\alpha_f(m)| \cdot \int_0^{\infty} e^{-2\pi m y} y^{\sigma-1} \, dy < \infty$$

for $\sigma > k$. Now we decompose the integral in (4.23). The substitution $y \mapsto 1/y$ and the identity $f(i/y) = (iy)^k \cdot f(iy)$ for $\sigma > k$ lead to

$$\mathbb{D}_f(s) = \int_1^{\infty} \left[f(iy) - \alpha_f(0) \right] \cdot y^{s-1} \, dy + \int_0^1 \left[f(iy) - \alpha_f(0) \right] \cdot y^{s-1} \, dy$$

$$= \int_1^{\infty} \left[f(iy) - \alpha_f(0) \right] \cdot y^{s-1} \, dy + \int_1^{\infty} \left[f(i/y) - \alpha_f(0) \right] \cdot y^{-s-1} \, dy$$

$$= \int_1^{\infty} \left[f(iy) - \alpha_f(0) \right] \cdot y^{s-1} \, dy + \int_1^{\infty} \left[f(iy) - \alpha_f(0) \right] \cdot i^k y^{k-s-1} \, dy$$

$$+ \alpha_f(0) \cdot \int_1^{\infty} \left[i^k y^{k-s-1} - y^{-s-1} \right] \, dy$$

$$= \int_1^{\infty} \left[f(iy) - \alpha_f(0) \right] \cdot \left[y^s + i^k y^{k-s} \right] \frac{dy}{y} + \alpha_f(0) \cdot \left(\frac{i^k}{s-k} - \frac{1}{s} \right),$$

hence (4.20). We have $f(iy) - \alpha_f(0) = O\left(e^{-2\pi y}\right)$ on $[1, \infty[$ by Lemma III.1.4. Thus, according to the dominated convergence, the right-hand side of (4.20) is an entire function.

From (4.20) and $i^{2k} = 1$, the functional equation (4.21) follows immediately. $\mathbb{D}_f(s)$ is holomorphic except for possible simple poles at $s = k$ and $s = 0$. Because of (4.11), this is also true for

$$D_f(s) = (2\pi)^s \cdot \frac{1}{\Gamma(s)} \cdot \mathbb{D}_f(s).$$

Because $\Gamma(s)$ has a simple pole with residue 1 at the point $s = 0$, $D_f(s)$ has a removable singularity at $s = 0$, with $D_f(0) = -\alpha_f(0)$ because of (4.20). The poles of $\Gamma(s)$ in $s = -n$, $n \in \mathbb{N}$, according to (4.9), provide the roots of $D_f(s)$ in (4.19). Moreover we get

$$\mathrm{res}_{s=k} D_f(s) = (2\pi)^k \cdot \frac{1}{\Gamma(k)} \cdot \mathrm{res}_{s=k} \mathbb{D}_f(s) = \frac{(2\pi i)^k}{(k-1)!} \cdot \alpha_f(0),$$

hence (4.18). Finally, let $\alpha < \beta$, $s = \sigma + it \in \mathcal{V}_{\alpha,\beta}$, $|t| \geq 1$. Then integration by parts in (4.20) yields

$$\mathbb{D}_f^*(s) = -\left(f(i) - \alpha_f(0)\right)\left(\frac{1}{s} + \frac{i^k}{k-s}\right)$$
$$- \int_1^\infty if'(iy)\left[\frac{y^s}{s} + i^k \frac{y^{k-s}}{k-s}\right] dy,$$

because $f(iy) - \alpha_f(0)$ is exponentially decreasing on $[1, \infty[$ due to Lemma III.1.4. Thus we obtain

$$|\mathbb{D}_f^*(s)| \leq \left[2\left(|f(i)| + |\alpha_f(0)|\right) + \int_1^\infty |f'(iy)|\left[y^{\beta+1} + y^k\right] dy\right] \frac{1}{|t|}$$
$$= c/|t|.$$

As $\mathbb{D}_f^*(s)$ is bounded on the compact set $\{s \in \mathcal{V}_{\alpha,\beta}; |t| \leq 1\}$, the claim follows. □

Using Proposition III.1.7 and $\alpha_f(0) = 0$ for $f \in \mathbb{S}_k$, we immediately obtain the

4.9 Corollary. *If $f \in \mathbb{S}_k$ is a cusp form, $D_f(s)$ converges absolutely for $\mathrm{Re}\, s > \frac{k}{2} + 1$ and represents an entire function.*

5. Riemann zeta function. Using the same method as in sect. 4, we will now consider the Riemann zeta function (cf. B.1)

$$\zeta(s) := \sum_{m=1}^\infty m^{-s}, \quad s \in \mathbb{C}, \ \mathrm{Re}\, s > 1.$$

For this purpose, we use the standard notation

$$\xi(s) := \pi^{-s/2} \cdot \Gamma(s/2) \cdot \zeta(s), \quad \mathrm{Re}\, s > 1, \tag{4.24}$$

and consider the theta series from III(0.7)

$$\vartheta(\tau) = \sum_{n \in \mathbb{Z}} e^{\pi i n^2 \tau} = 1 + 2 \cdot \sum_{n=1}^\infty e^{\pi i n^2 \tau}, \quad \tau \in \mathbb{H}. \tag{4.25}$$

§4 Dirichlet series with functional equation

4.10 Theorem. a) *The Riemann zeta function $\zeta(s)$ possesses a meromorphic continuation into the whole s-plane. It is holomorphic except for a simple pole at $s = 1$ with the residue 1. Also,*

$$\zeta(0) = -\frac{1}{2} \quad \text{and} \quad \zeta(-2n) = 0 \quad \text{for} \quad n = 1, 2, 3, \ldots .$$

hold.
b) *The function*

$$\xi(s) - \left(\frac{1}{s-1} - \frac{1}{s} \right)$$

is an entire function and satisfies the functional equation

$$\xi(1-s) = \xi(s) \quad \text{for} \quad s \in \mathbb{C}.$$

Proof Because of $\vartheta(i/y) = \sqrt{y} \cdot \vartheta(iy)$ according to the Theta Transformation Formula III.0.1, we can proceed as in the proof of Theorem 4.8:

$$\xi(2s) = \sum_{n=1}^{\infty} \Gamma(s) \cdot (\pi n^2)^{-s} = \int_0^{\infty} \sum_{n=1}^{\infty} e^{-\pi n^2 y} y^{s-1} \, dy$$

$$= \frac{1}{2} \int_0^{\infty} [\vartheta(iy) - 1] \cdot y^{s-1} \, dy$$

$$= \frac{1}{2} \int_1^{\infty} [\vartheta(iy) - 1] \cdot y^{s-1} \, dy + \frac{1}{2} \int_1^{\infty} [\vartheta(i/y) - 1] \cdot y^{-s-1} \, dy$$

$$= \frac{1}{2} \int_1^{\infty} [\vartheta(iy) - 1] \cdot y^{s-1} \, dy + \frac{1}{2} \int_1^{\infty} [\vartheta(iy) - 1] \cdot y^{-s-1/2} \, dy$$

$$+ \frac{1}{2} \int_1^{\infty} \left[y^{-s-1/2} - y^{-s-1} \right] dy$$

$$= \frac{1}{2} \int_1^{\infty} [\vartheta(iy) - 1] \cdot \left[y^s + y^{-s+1/2} \right] \frac{dy}{y} + \left(\frac{1}{2s-1} - \frac{1}{2s} \right).$$

From this representation we get the assertions as in the proof of Theorem 4.8. □

The proof of the analytic continuation and functional equation given here goes back to B. Riemann ([68], 177–187). Of course we can prove Theorem 4.8 and 4.10 simultaneously. Compare E. Hecke [36], 591–626.

6. Hecke's Converse Theorem. In this section, we shall construct a modular form from a Dirichlet series with functional equation. This leads to a characterization which sequences can occur as Fourier coefficients of modular forms.

4.11 Theorem. *Consider an even number $k \geq 4$ and a* DIRICHLET *series*

$$D(s) := \sum_{m=1}^{\infty} \alpha_m \cdot m^{-s} \tag{4.26}$$

with absolute convergence abscissa $\chi < \infty$. Assume that the function

$$\mathbb{D}(s) := (2\pi)^{-s} \cdot \Gamma(s) \cdot D(s) \tag{4.27}$$

possesses a meromorphic continuation into the whole s-plane and satisfies the functional equation

$$\mathbb{D}(k-s) = i^k \cdot \mathbb{D}(s). \tag{4.28}$$

If there is some $\alpha_0 \in \mathbb{C}$ such that the function

$$\mathbb{D}^*(s) := \mathbb{D}(s) - \alpha_0 \cdot \left(\frac{i^k}{s-k} - \frac{1}{s} \right) \tag{4.29}$$

is entire and bounded in each vertical strip $\mathcal{V}_{\alpha,\beta}$ in $\mathbb{C}$, then

$$f(\tau) := \sum_{m=0}^{\infty} \alpha_m \cdot e^{2\pi i m \tau} \tag{4.30}$$

is a modular form of weight k with DIRICHLET *series $D_f(s) = D(s)$.*

Proof From the convergence of $D(\chi + \varepsilon)$, $\varepsilon > 0$, it follows that the summands are bounded, i.e. $\alpha_m = O(m^{\chi+\varepsilon})$. Consequently, $f : \mathbb{H} \to \mathbb{C}$ in (4.30) is holomorphic. Because the sequence of summands in (4.26) is bounded for $\sigma \geq \chi + \varepsilon$, $\varepsilon > 0$, $\mathbb{D} \in \mathcal{B}_\chi$ follows by Lemma 4.6 and (4.27). Finally, Proposition 4.7 yields for $\sigma > \chi$

$$\frac{1}{2\pi i} \int_{(\sigma)} \mathbb{D}(s) y^{-s} \, ds = \frac{1}{2\pi i} \cdot \sum_{m=1}^{\infty} \alpha_m \int_{(\sigma)} \Gamma(s)(2\pi m y)^{-s} \, ds \tag{$*$}$$
$$= f(iy) - \alpha_0,$$

where the interchange of summation and integration is justified according to the dominated convergence.

Now we consider the integral

$$I := \frac{1}{2\pi i} \int_{\partial R} \mathbb{D}(s) y^{-s} \, ds$$

§ 4 Dirichlet series with functional equation

over a positively oriented axis-parallel rectangle R containing the points 0, k and χ. The Residue Theorem A.5 and (4.30) imply

$$I = \alpha_0 \cdot \left((iy)^{-k} - 1\right).$$

$\mathbb{D}^*(s)$ is bounded on $\mathcal{V}_{-a,b}$ by assumption. $|\mathbb{D}^*(s)| \leq c/(1+|t|)$ holds for $\sigma = b$ in view of $\mathbb{D} \in \mathcal{B}_\chi$. Then (4.28) shows for $k + a > \chi$ that this bound also holds for $\sigma = -a$. Thus the Phragmen-Lindelöf Theorem A.16 implies that it holds for all $s \in \mathcal{V}_{-a,b}$ and moreover

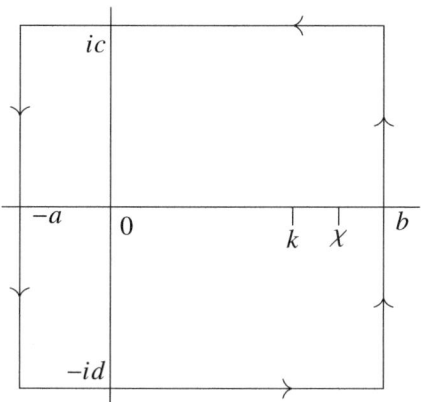

Figure 26: Integration path

$$|\mathbb{D}(s)| \leq \tilde{c}/(1+|t|) \quad \text{for} \quad s \in \mathcal{V}_{-a,b}, \ |t| \geq 1.$$

Hence the integrals over the parallels to the real axis vanish as $c \to \infty$ and $d \to \infty$. Thus

$$\alpha_0 + \frac{1}{2\pi i} \int_{(b)} \mathbb{D}(s) y^{-s}\, ds = \alpha_0 (iy)^{-k} + \frac{1}{2\pi i} \int_{(-a)} \mathbb{D}(s) y^{-s}\, ds. \tag{4.31}$$

By $(*)$, the left-hand side of (4.31) is $f(iy)$. On the right-hand side of (4.31) we use (4.28) and again $(*)$:

$$\alpha_0 (iy)^{-k} + \frac{1}{2\pi i} \int_{(-a)} i^{-k} \mathbb{D}(k-s) y^{-s}\, ds$$

$$= \alpha_0 (iy)^{-k} + \frac{1}{2\pi i} \int_{(k+a)} i^{-k} \mathbb{D}(s) y^{s-k}\, ds$$

$$= (iy)^{-k} \cdot f(i/y).$$

Using the Identity Theorem A.1, we conclude $f(\tau) = \tau^{-k} \cdot f(-1/\tau)$ for all $\tau \in \mathbb{H}$. Because of the definition as a Fourier series (4.30), we have $f(\tau + 1) = f(\tau)$, hence $f \in \mathbb{M}_k$ with $D_f = \mathbb{D}$. □

4.12 Remark. The Correspondence Theorem 4.11 goes back to E. Hecke's classical work *Über die Bestimmung Dirichletscher Reihen durch ihre Funktionalgleichung* from 1936 (cf. [36], 591–626). If an additional parameter is introduced, we can prove the results on $D_f(s)$ and $\zeta(s)$ simultaneously. The analog of Theorem 4.11 is known as Hamburger's *Theorem (Math. Z.* **10**, 240–254 (1921), **11**, 224–245 (1921), **13**, 240–254 (1922)). Compare E. Freitag and R. Busam [27], VII, §3 as well as B.C. Berndt and M. Knopp [7] for a more general treatment.

7. Products of Dirichlet series. A product of two power series with equal development point is known to be such a power series again. An analogous statement holds for Dirichlet series:

4.13 Proposition. *If*

$$A(s) := \sum_{m=1}^{\infty} \alpha_m \cdot m^{-s} \quad \text{and} \quad B(s) := \sum_{m=1}^{\infty} \beta_m \cdot m^{-s}$$

are absolutely convergent Dirichlet series for $\operatorname{Re} s > \kappa$, *then*

$$A(s) \cdot B(s) = \sum_{m=1}^{\infty} \gamma_m \cdot m^{-s} \quad \text{with} \quad \gamma_m := \sum_{d \mid m} \alpha_d \beta_{m/d} \tag{4.32}$$

holds for $\operatorname{Re} s > \kappa$.

Here, of course, we again sum only over the positive divisors d of m.

Proof As the product of two absolutely convergent series is absolutely convergent again, we can rearrange the series

$$A(s) \cdot B(s) = \sum_{\nu,\mu=1}^{\infty} \alpha_\nu \beta_\mu \cdot (\nu\mu)^{-s}, \quad \operatorname{Re} s > \kappa,$$

arbitrarily. We combine the terms with fixed $\nu\mu = m$ and obtain the representation (4.32) with

$$\gamma_m = \sum_{\nu\mu=m} \alpha_\nu \beta_\mu.$$

But this is the assertion. □

4.14 Corollary. *For* $r \in \mathbb{R}$ *and* $\operatorname{Re} s > \max\{1, r+1\}$ *one has*

$$\zeta(s) \cdot \zeta(s-r) = \sum_{m=1}^{\infty} \sigma_r(m) \cdot m^{-s}.$$

Proof We choose $\beta_m = 1$ and $\alpha_m = m^r$ in Proposition 4.13. Hence $\gamma_m = \sigma_r(m)$ holds. □

4.15 Corollary. *For even* $k \geq 4$, *the Dirichlet series associated to the Eisenstein series* E_k *is equal to*

$$2\frac{(2\pi i)^k}{(k-1)!} \cdot \zeta(s) \cdot \zeta(s+1-k) \quad \text{for} \quad \operatorname{Re} s > k.$$

Proof We use III(2.3) and Corollary 4.14. □

As usual, we call a function $\alpha : \mathbb{N} \to \mathbb{C}$ *multiplicative* if $\alpha(mn) = \alpha(m) \cdot \alpha(n)$ holds for all coprime $m, n \in \mathbb{N}$.

§ 4 DIRICHLET series with functional equation 265

4.16 Lemma. *If $\alpha : \mathbb{N} \to \mathbb{C}$ is multiplicative and $\sum_{n=1}^{\infty} \alpha(n) \neq 0$ is absolutely convergent, then*

$$\sum_{n=1}^{\infty} \alpha(n) = \prod_p \left(\sum_{r=0}^{\infty} \alpha(p^r) \right),$$

holds, where the product is extended over all primes p.

Such a product is called an EULER *product.*

Proof The multiplicativity yields $\alpha(1) = 1$. For $N > 1$,

$$\prod_N := \prod_{p \leq N} \left(\sum_{r=0}^{\infty} \alpha(p^r) \right) = \sum_{r_1 \geq 0} \sum_{r_2 \geq 0} \cdots \sum_{r_q \geq 0} \alpha\left(p_1^{r_1}\right) \alpha\left(p_2^{r_2}\right) \cdot \ldots \cdot \alpha\left(p_q^{r_q}\right),$$

where the primes $\leq N$ are denoted by $p_1, \ldots, p_q$. Because of the multiplicativity of α, we get

$$\prod_N = \sum_{r_1 \geq 0} \sum_{r_2 \geq 0} \cdots \sum_{r_q \geq 0} \alpha\left(p_1^{r_1} p_2^{r_2} \cdot \ldots \cdot p_q^{r_q}\right) = \sum_{n=1}^{N} \alpha(n) + \sum_{n \in E_N} \alpha(n),$$

with

$$E_N := \{n \in \mathbb{N}; n > N, n \text{ has at most } p_1, \ldots, p_q \text{ as prime divisors}\}.$$

Thus,

$$\left| \prod_N - \sum_{n=1}^{N} \alpha(n) \right| \leq \sum_{n > N} |\alpha(n)|$$

follows and the claim has been proved. □

4.17 Corollary. *One has*

$$\zeta(s) = \prod_p (1 - p^{-s})^{-1} \quad \text{for} \quad \text{Re } s > 1.$$

Proof Taking $\alpha(n) = n^{-s}$ in Lemma 4.16 we immediately obtain

$$\zeta(s) = \prod_p \left(\sum_{r=0}^{\infty} p^{-rs} \right).$$

Now we apply the sum formula for the geometric series in order to obtain the assertion. □

8. Simultaneous eigenforms under all HECKE operators. Let $0 \neq f \in \mathbb{S}_k$ be a simultaneous eigenform with respect to all HECKE operators T_n as in (1.16) with the FOURIER expansion

$$f(\tau) = \sum_{m=1}^{\infty} \alpha_f(m) \cdot e^{2\pi i m \tau}, \quad \tau \in \mathbb{H}.$$

By Lemma 1.9 we may assume $\alpha_f(1) = 1$ without loss of generality and then we have

$$\alpha_f(m) \cdot \alpha_f(n) = \sum_{d \mid (m,n)} d^{k-1} \cdot \alpha_f\left(mn/d^2\right) \quad \text{for all } m, n \geq 1.$$

The DIRICHLET series associated with f is denoted by D_f, i.e.

$$D_f(s) = \sum_{m=1}^{\infty} \alpha_f(m) \cdot m^{-s} \quad \text{for} \quad \operatorname{Re} s > 1 + k/2.$$

According to Corollary 4.9, $D_f(s)$ possesses a continuation to an entire function in the complex s-plane that satisfies the functional equation $\mathbb{D}_f(k-s) = i^k \cdot \mathbb{D}_f(s)$ for $s \in \mathbb{C}$. Now, as a crucial point, $D_f(s)$ has a representation as an EULER product:

4.18 Theorem. *Let $f \in \mathbb{S}_k$, $k \geq 12$ be even, $k \neq 14$, with $\alpha_f(1) = 1$. Then the following assertions are equivalent:*
(i) *f is a simultaneous eigenform with respect to all HECKE operators.*
(ii) $D_f(s) = \prod_p \left(1 - \alpha_f(p) \cdot p^{-s} + p^{k-1-2s}\right)^{-1}$ *for* $\operatorname{Re} s > 1 + k/2$.

Proof (i) $\Longrightarrow$ (ii): Since α_f is multiplicative by Lemma 1.9, Lemma 4.16 yields

$$D_f(s) = \prod_p F_p \quad \text{with} \quad F_p := \sum_{r=0}^{\infty} \alpha_f(p^r) p^{-rs}.$$

As Corollary 2.8 yields

$$\alpha_f(p^{r-1}) \cdot \alpha_f(p) = \alpha_f(p^r) + p^{k-1} \cdot \alpha_f(p^{r-2}) \quad \text{for } r \geq 2, \tag{4.33}$$

we obtain

$$F_p \cdot (1 - \alpha_f(p) p^{-s} + p^{k-1-2s})$$
$$= \sum_{r=0}^{\infty} \alpha_f(p^r) \cdot p^{-rs} - \sum_{r=1}^{\infty} \alpha_f(p^{r-1}) \cdot \alpha_f(p) p^{-rs} + \sum_{r=2}^{\infty} \alpha_f(p^{r-2}) \cdot p^{k-1-rs}$$
$$= 1 + \sum_{r=2}^{\infty} \left[\alpha_f(p^r) - \alpha_f(p^{r-1}) \cdot \alpha_f(p) + p^{k-1} \cdot \alpha_f(p^{r-2})\right] \cdot p^{-rs} = 1.$$

But this is the assertion (ii).
(ii) $\Longrightarrow$ (i): Since D_f does not vanish identically, the multiplicativity of α follows from the existence of the EULER product and the Identity Theorem A.1. Furthermore the same calculations as above show that (4.33) is also correct. Thus, we easily obtain

§ 4 Dirichlet series with functional equation

$$\alpha_f(p) \cdot \alpha_f(m) = \alpha_f(mp) + p^{k-1} \cdot \alpha_f(m/p) \quad \text{for all } m \geq 1.$$

Corollary 2.8 and $\alpha_f(0) = 0$ imply (i). □

4.19 Remark. If $f \in \mathbb{M}_k$, $f \notin \mathbb{S}_k$, $k > 0$, is a simultaneous eigenform with respect to all Hecke operators with $\alpha_f(1) = 1$, then $f = c \cdot E_k$ follows from Theorem 2.12 as well as $c = \frac{1}{2}(2\pi i)^{-k}(k-1)!$ from III(2.3). Using Corollaries 4.15 and 4.17, we now conclude

$$D_f(s) = c \cdot D_{E_k}(s) = \zeta(s) \cdot \zeta(s+1-k) = \prod_p \left(1 - p^{-s}\right)^{-1} \cdot \left(1 - p^{k-1-s}\right)^{-1}$$

$$= \prod_p \left(1 - \sigma_{k-1}(p) \cdot p^{-s} + p^{k-1-2s}\right)^{-1}$$

for Re $s > k$ as an analogon of Theorem 4.18.

9*. The eta transformation formula. In this section we give a new proof of the Eta Transformation Formula III.6.4 which uses the Hecke theory. Recall that $\eta : \mathbb{H} \to \mathbb{C}$ is a holomorphic function given by

$$\eta(\tau) = e^{\pi i \tau/12} \cdot \prod_{m=1}^{\infty}(1 - e^{2\pi i m \tau}), \quad \tau \in \mathbb{H} \tag{4.34}$$

4.20 Theorem. *For all $\tau \in \mathbb{H}$, we have*

$$\eta(-1/\tau) = \sqrt{\tau/i} \cdot \eta(\tau). \tag{4.35}$$

Proof Because of holomorphy, it is sufficient to prove (4.35) for $\tau = iy$, $y > 0$. Then all factors in (4.35) are positive real numbers. Thus, the assertion (4.35) is equivalent to

$$\log \eta(i/y) = \log \eta(iy) + \frac{1}{2} \log y. \tag{4.36}$$

We get

$$\varphi(y) := -\frac{\pi y}{12} - \log \eta(iy) = -\sum_{m=1}^{\infty} \log(1 - e^{-2\pi m y})$$

$$= \sum_{m=1}^{\infty} \sum_{n=1}^{\infty} \frac{1}{n} \cdot e^{-2\pi m n y} = \sum_{m=1}^{\infty} \sigma_{-1}(m) \cdot e^{-2\pi m y} \tag{4.37}$$

from (4.34). Corollary 4.14 now yields

$$D(s) = \sum_{m=1}^{\infty} \sigma_{-1}(m) \cdot m^{-s} = \zeta(s) \cdot \zeta(s+1) \quad \text{for } \text{Re } s > 1.$$

From (4.12) and Theorem 4.10 we get

$$\mathbb{D}(s) := (2\pi)^{-s} \cdot \Gamma(s) \cdot D(s) = \tfrac{1}{2}\xi(s) \cdot \xi(s+1), \quad \mathbb{D}(-s) = \mathbb{D}(s), \quad s \in \mathbb{C}.$$

Due to Theorem 4.10, $\xi(-1) = \xi(2) = \pi/6$ holds. The fact that $\mathbb{D}(s)$ is an even function then shows that

$$\mathbb{D}(s) - \left(\frac{\pi}{12(s-1)} - \frac{\pi}{12(s+1)} - \frac{1}{2s^2} \right)$$

possesses a continuation as an entire function and it decays in each vertical strip in $\mathbb{C}$. Now we proceed as in the proof of Theorem 4.11 and integrate over a rectangle formed analogously. The function $y^{-s} \cdot \mathbb{D}(s)$ only has poles at $s = 1, 0 - 1$ with residues $\frac{\pi}{12y}$, $\frac{1}{2}\log y$, $-\frac{\pi y}{12}$. Proposition 4.7 leads to

$$\varphi(y) = \sum_{m=1}^{\infty} \sigma_{-1}(m) \cdot e^{-2\pi m y} = \frac{1}{2\pi i} \cdot \sum_{m=1}^{\infty} \sigma_{-1}(m) \int\limits_{(2)} \Gamma(s)(2\pi m y)^{-s} \, ds$$

$$= \frac{1}{2\pi i} \int\limits_{(2)} \mathbb{D}(s) y^{-s} \, ds = \frac{1}{2\pi i} \int\limits_{(-2)} \mathbb{D}(s) y^{-s} \, ds + \frac{\pi}{12y} - \frac{\pi y}{12} + \frac{1}{2} \log y$$

$$= \frac{1}{2\pi i} \int\limits_{(2)} \mathbb{D}(s) y^{s} \, ds + \frac{\pi}{12y} - \frac{\pi y}{12} + \frac{1}{2} \log y$$

$$= \varphi(1/y) + \frac{\pi}{12y} - \frac{\pi y}{12} + \frac{1}{2} \log y,$$

where we take into account that $\mathbb{D}(s) = \mathbb{D}(-s)$ holds. Because of (4.37), we finally obtain (4.36). □

4.21 Exercises. Let $f \in \mathbb{M}_k$ be a simultaneous eigenform of all HECKE operators with $\alpha_f(1) = 1$.

1) For $\operatorname{Re} s > k+1$, we have

$$\zeta(2s+2-2k) \cdot \sum_{m=1}^{\infty} \alpha_f(m^2) \cdot m^{-s}$$

$$= \prod_p \left(1 - \alpha_f(p^2) p^{-s} + \alpha_f(p^2) p^{k-1-2s} - p^{3k-3-3s} \right)^{-1}.$$

2) For $\operatorname{Re} s > k+1$ we have

$$\zeta(s+1-k) \cdot \sum_{m=1}^{\infty} \alpha_f(m^2) \cdot m^{-s} = \sum_{m=1}^{\infty} \alpha_f(m)^2 \cdot m^{-s}.$$

3) If μ denotes the MÖBIUS function, then we have for all $m, n \geq 1$

§ 4 DIRICHLET series with functional equation

$$\alpha_f(mn) = \sum_{d|(m,n)} \mu(d) \cdot \alpha_f(m/d) \cdot \alpha_f(n/d).$$

4) For each $m \geq 1$, the following holds

$$D_f^{(m)}(s) := \sum_{n=1}^{\infty} \alpha_f(mn) \cdot n^{-s} = \left(\sum_{d|m} \mu(d) \alpha_f(m/d) \, d^{k-1-s} \right) \cdot D_f(s).$$

5) For every $m \geq 1$, $D_f^{(m)}(s)$ in Exercise 4) has a meromorphic continuation into the entire complex s-plane with at most one 1st order pole at $s = k$ and residue

$$i^k \frac{(2\pi)^k}{(k-1)!} \alpha_f(0) \cdot \sum_{d|m} \frac{\mu(d)}{d} \cdot \alpha_f\left(\frac{m}{d}\right).$$

6) Let $k \in \mathbb{N}$, $\delta_k(n) := \#\{(g_1, \ldots, g_k)^{tr} \in \mathbb{Z}^k; \, g_1^2 + \ldots + g_k^2 = n\}$ and moreover $\zeta_k(s) := \sum_{n=1}^{\infty} \delta_k(n) \cdot n^{-s}$. Then $\zeta_1(s) = 2\zeta(2s)$ holds. The DIRICHLET series $\zeta_k(s)$ is absolutely convergent for $\operatorname{Re} s > k/2$. $\zeta_k(s)$ has a meromorphic continuation into the s-plane and is holomorphic except for one simple pole at $s = k/2$ with the residue 1. It satisfies $\zeta_k(0) = -1$ and $\zeta_k(-n) = 0$ for $n = 1, 2, \ldots$. The function $\xi_k(s) := \pi^{-s} \Gamma(s) \zeta_k(s)$ has a meromorphic continuation into the whole s-plane, and $\xi_k(s) - \left(\frac{1}{s-k/2} - \frac{1}{s} \right)$ is entire and satisfies the functional equation $\xi_k(\frac{k}{2} - s) = \xi_k(s)$.
7) In Exercise 6) one has $\zeta_2(s) = 4\zeta(s) \cdot \sum_{n=1}^{\infty} \chi(n) \cdot n^{-s}$, where χ is the non-trivial DIRICHLET character mod 4.
8) Consider the conditionally convergent EISENSTEIN series E_2 from III, §6 and show that the associated DIRICHLET series is equal to $8\pi^2 \zeta(s)\zeta(s-1)$. Derive from this the transformation formula in Proposition III.6.2.

$$E_2(-1/\tau) = \tau^2 E_2(\tau) - 2\pi i \tau.$$

9) Let sequences $(\alpha_m)_{m \geq 0}$ and $(\beta_m)_{m \geq 0}$ be given which grow at most polynomially and define functions

$$g(\tau) := \sum_{m=0}^{\infty} \alpha_m \cdot e^{2\pi i m \tau} \quad \text{and} \quad h(\tau) := \sum_{m=0}^{\infty} \beta_m \cdot e^{2\pi i m \tau}.$$

Assume for some $n \in \mathbb{N}$ that

$$g(\tau) = (-i\sqrt{n}\tau)^{-k} \cdot h(-1/n\tau).$$

Then the function

$$\mathbb{D}_g(s) := \left(\frac{2\pi}{\sqrt{n}} \right)^{-s} \cdot \Gamma(s) \cdot \sum_{m=1}^{\infty} \alpha_m \cdot m^{-s}$$

has a meromorphic continuation into the s-plane,

$$\mathbb{D}_g(s) - \left(\frac{\alpha_0}{s-k} - \frac{\beta_0}{s} \right)$$

is entire and the functional equation

$$\mathbb{D}_g(k-s) = \mathbb{D}_h(s).$$

holds with the analogously defined function $\mathbb{D}_h(s)$

10) Let $f \in M_k$, k even, satisfy $\overline{f(-\bar{\tau})} = f(\tau)$ for all $\tau \in \mathbb{H}$. Then $i^{-k/2}\mathbb{D}_f\left(\frac{k}{2}+it\right)$ is real for all $t \in \mathbb{R}$.

11) $s(s-1)\zeta(s)$ is an entire function which takes real values only for $s \in \mathbb{R}$ and $s \in \frac{1}{2}+i\mathbb{R}$.

12) Show that the RIEMANN zeta function satisfies the functional equation

$$\zeta(1-s) = 2(2\pi)^{-s}\Gamma(s)\cos(\pi s/2)\zeta(s).$$

Moreover $\zeta(1-k) = (-1)^{k+1}B_k/k$ holds for all $k \in \mathbb{N}$.

13) Show that for $f \in S_k$ the so called periods of f satisfy

$$\int_0^\infty f(iy)y^n dy = n!(2\pi)^{-n-1}D_f(n+1), \quad 0 \le n \le k-2.$$

14) Let χ be a primitive DIRICHLET character mod n, $n \in \mathbb{N}$, $f \in M_k(\Gamma_0[n])$,

$$D_f(s,\chi) := \sum_{m=1}^\infty \chi(m) \cdot \alpha_f(m) \cdot m^{-s} \quad \text{and} \quad \mathbb{D}_f(s,\chi) := \left(\frac{2\pi}{n}\right)^{-s} \cdot \Gamma(s) \cdot D_f(s).$$

Then $D_f(s,\chi)$ and $\mathbb{D}_f(s,\chi)$ possess meromorphic continuations to the whole s-pole. There exists some $\varepsilon \in \mathbb{C}$ with $|\varepsilon|=1$ such that

$$\mathbb{D}_f(k-s,\chi) = \varepsilon \cdot D_f(s,\bar{\chi}).$$

15) Show that $f \in M_k$, $k>0$, in Exercise 1.23 20) vanishes identically.

§ 5* Maximal discrete groups

In this paragraph we describe an application of §3 to discrete subgroups of $\mathrm{SL}(2;\mathbb{R})$, which were already considered in Chapter II, §4. We call a subgroup Δ of $\mathrm{SL}(2;\mathbb{R})$ **maximal discrete** if for any discrete subgroup Δ' of $\mathrm{SL}(2;\mathbb{R})$ which contains Δ, we already have $\Delta' = \Delta$.

5.1 Theorem. *Let Λ be a subgroup of Γ of finite index and Δ a discrete subgroup of $\mathrm{SL}(2;\mathbb{R})$ containing Λ. Then Δ is countable and*

§ 5* Maximal discrete groups 271

$$[\Delta : \Lambda] < \infty.$$

Proof Let $-I \in \Lambda$ without loss of generality. In view of

$$\Delta = \bigcup_{n=1}^{\infty} \{M \in \Delta;\ \mathrm{tr}\,(M^{tr}M) \leq n\}$$

and the fact that every set on the right-hand side is finite, we conclude that Δ is countable. If $[\Delta : \Lambda] = \infty$, there exists a decomposition

$$\Delta = \bigcup_{j \in \mathbb{N}}^{\cdot} \Lambda M_j.$$

If $\mathcal{F}$ is a fundamental domain for Δ as in Theorem II.4.5, then

$$\mathcal{G} := \bigcup_{j \in \mathbb{N}} M_j \mathcal{F}$$

is a fundamental domain for Λ due to an analog of Theorem II.3.1. Thus, the hyperbolic area $v(\mathcal{G})$ of $\mathcal{G}$ is equal to

$$v(\mathcal{G}) = \begin{cases} \infty, & \text{if } v(\mathcal{F}) > 0, \\ 0, & \text{if } v(\mathcal{F}) = 0. \end{cases}$$

The same procedure applied to Γ instead of Δ as well as Lemma 3.2 and Lemma 3.10, imply

$$v(\mathcal{G}) = [\Gamma : \Lambda] \cdot \pi/3.$$

This is a contradiction and the claim follows. □

We need two purely algebraic and number theoretic auxiliary results.

5.2 Lemma *If Λ is a subgroup of a group Δ with finite index $r = [\Delta : \Lambda]$, then*

$$M^{r!} \in \Lambda \quad \text{for all} \quad M \in \Delta.$$

Proof Let $M \in \Delta$. Among the $r + 1$ right cosets ΛM^j, $0 \leq j \leq r$, at least two coincide. Thus, there exist $0 \leq k < l \leq r$ such that

$$\Lambda M^k = \Lambda M^l, \quad \text{i.e.} \quad M^{l-k} \in \Lambda.$$

As $l - k$ divides $r!$, the claim follows. □

If Λ is a subgroup of a group Δ, the *normalizer* of Λ in Δ is defined by

$$\mathcal{N}_\Delta(\Lambda) = \{M \in \Delta;\ M\Lambda M^{-1} = \Lambda\}.$$

Clearly, it is the biggest subgroup of Δ which contains Λ as a normal subgroup.

5.3 Lemma. Let $x, y \in \mathbb{R}\setminus\{0\}$ satisfy $x^2, xy, y^2 \in \mathbb{Q}$. Then there exists an $n \in \mathbb{N}$ such that
$$x, y \in \frac{1}{\sqrt{n}}\mathbb{Z}.$$

Proof Choose $m \in \mathbb{N}$ such that $mx^2, mxy, my^2 \in \mathbb{Z}$. Let
$$mx^2 = r^2 s, \quad my^2 = t^2 u, \quad r, s, t, u \in \mathbb{N}, \; s, u \text{ square-free}.$$

We obtain
$$x = \pm\frac{rs}{\sqrt{ms}}, \quad y = \pm\frac{tu}{\sqrt{mu}}, \quad (mxy)^2 = (rt)^2 su.$$

As the last quantity is a square in $\mathbb{Z}$, we conclude $s = u$. Thus the claim follows with $n = ms$. □

This leads to a technical result.

5.4 Lemma. Let Λ be a subgroup of Γ of finite index. If Δ is a subgroup of $\mathrm{SL}(2;\mathbb{R})$ containing Λ and satisfying
(i) $[\Delta : \Lambda] < \infty$ or
(ii) $\Delta \subseteq \mathcal{N}_{\mathrm{SL}(2;\mathbb{R})}(\Lambda)$,
then for any $M \in \Delta$ there exists an $n \in \mathbb{N}$ such that
$$\sqrt{n}M \in \Gamma_n.$$

Proof Let $M = \begin{pmatrix} a & b \\ c & d \end{pmatrix} \in \Delta$. Multiplying by suitable powers of T from the left and from the right, we may assume $a, b, c, d \in \mathbb{R}\setminus\{0\}$. As $[\Gamma : \Lambda] < \infty$, there is some $l \in \mathbb{N}$ such that
$$\begin{pmatrix} 1 & l \\ 0 & 1 \end{pmatrix}, \begin{pmatrix} 1 & 0 \\ l & 1 \end{pmatrix} \in \Lambda. \tag{5.1}$$

There exists some $s \in \mathbb{N}$ such that
$$\left(M^{-1}LM\right)^s, \left(MLM^{-1}\right)^s \in \Lambda \subseteq \Gamma \quad \text{for all} \quad L \in \Lambda, \tag{5.2}$$

namely $s = r!$ due to Lemma 5.2 in the case (i) and $s = 1$ in the case (ii). We apply (5.2) to the matrices (5.1) and obtain
$$lsa^2, \; lsb^2, \; lsc^2, \; lsd^2, \; lsab, \; lsac, \; lsbd, \; lscd \in \mathbb{Z}.$$

Thus we can apply Lemma 5.3 and obtain an $n \in \mathbb{N}$ such that
$$M \in \frac{1}{\sqrt{n}}\mathrm{Mat}(2;\mathbb{Z}).$$

Finally, $\det M = 1$ yields $\sqrt{n}M \in \Gamma_n$. □

The first application is

§ 5* Maximal discrete groups

5.5 Theorem. *Γ is a maximal discrete subgroup of* $SL(2;\mathbb{R})$.

Proof Let Δ be a discrete subgroup of $SL(2;\mathbb{R})$ containing Γ and $M \in \Delta$. We choose $n \in \mathbb{N}$ minimal such that $\sqrt{n}M \in \Gamma_n$ using Lemma 5.4. Applying Lemma 1.17, we obtain $U, V \in \Gamma$ such that

$$U(\sqrt{n}M)V = \begin{pmatrix} 1 & 0 \\ 0 & n \end{pmatrix}, \quad i.e. \quad UMV = \begin{pmatrix} 1/\sqrt{n} & 0 \\ 0 & \sqrt{n} \end{pmatrix}.$$

If $n > 1$, the cosets $\Gamma(UMV)^j \subseteq \Delta$, $j \in \mathbb{N}$, are mutually distinct. This contradicts Theorem 5.1. Thus we have

$$UMV = I, \quad M = U^{-1}V^{-1} \in \Gamma. \qquad \square$$

The same procedure yields

5.6 Theorem. *Γ coincides with its normalizer in* $SL(2;\mathbb{R})$.

Proof Let $M \in \mathcal{N}_{SL(2;\mathbb{R})}(\Gamma)$. The same arguments as in the proof of Theorem 5.5 show that there exist $n \in \mathbb{N}$ and $U, V \in \Gamma$ such that

$$UMV = \begin{pmatrix} 1/\sqrt{n} & 0 \\ 0 & \sqrt{n} \end{pmatrix}.$$

Then

$$(UMV)T(UMV)^{-1} = \begin{pmatrix} 1 & 1/n \\ 0 & 1 \end{pmatrix} \in \Gamma$$

yields $n = 1$ and $M = U^{-1}V^{-1} \in \Gamma$. $\qquad \square$

We give an application to principal congruence subgroups.

5.7 Corollary. *Let $n \in \mathbb{N}$. Then Γ is the normalizer of the principal congruence group $\Gamma[n]$ in* $SL(2;\mathbb{R})$.

Proof As $\Gamma[n]$ is a normal subgroup of Γ, we have $\Gamma \subseteq \Delta := \mathcal{N}_{SL(2;\mathbb{R})}(\Gamma[n])$. The same procedure as before shows that for $M \in \Delta$ there exist $m \in \mathbb{N}$ and $U, V \in \Gamma$ such that

$$UMV = \begin{pmatrix} 1/\sqrt{m} & 0 \\ 0 & \sqrt{m} \end{pmatrix} \in \Delta.$$

Then

$$\begin{pmatrix} 1/\sqrt{m} & 0 \\ 0 & \sqrt{m} \end{pmatrix} T^n \begin{pmatrix} 1/\sqrt{m} & 0 \\ 0 & \sqrt{m} \end{pmatrix}^{-1} = \begin{pmatrix} 1 & n/m \\ 0 & 1 \end{pmatrix} \in \Gamma[n]$$

yields $m = 1$ and $M \in \Gamma$. $\qquad \square$

5.8 Remark. The result of Theorem 5.5 is due to E. HECKE in an unpublished paper. A more general case was dealt with by H. PETERSSON (*Hamb. Abh.* **12**, 180-199 (1938)). In this context we also refer to the ATKIN-LEHNER theory (*Math. Ann.* **185**, 134-160 (1970)). Confer also T. MIYAKE [59].

5.9 Exercises.

1) Show that
$$\mathcal{N}_{SL(2;\mathbb{R})}(\Gamma_\infty) = \left\{\pm \begin{pmatrix} 1 & \beta \\ 0 & 1 \end{pmatrix};\ \beta \in \mathbb{R}\right\}.$$

2) If $\Lambda = \{\pm I, \pm J\}$ then
$$\mathcal{N}_{SL(2;\mathbb{R})}(\Lambda) = SO(2;\mathbb{R}).$$

3) Let $n \in \mathbb{N}$. Then
$$\mathcal{N}_\Gamma(\Gamma_1[n]) = \Gamma_0[n].$$

4) For $n \in \mathbb{N}$, we have
$$\mathcal{N}_\Gamma(\Gamma_0[n]) = \Gamma_0[n].$$

5) Let $n \in \mathbb{N}$ and $M = \begin{pmatrix} a & b \\ c & d \end{pmatrix} \in SL(2;\mathbb{R})$. Then $M \in \mathcal{N}_{SL(2;\mathbb{R})}(\Gamma_0[n])$ is equivalent to
$$a^2, d^2, ab, bd, nb^2 \in \mathbb{Z} \quad \text{and} \quad c^2, ac, cd \in n\mathbb{Z}.$$

6) Let $n \in \mathbb{N}$ be square-free.
a) $M \in SL(2;\mathbb{R})$ belongs to $\mathcal{N}_{SL(2;\mathbb{R})}(\Gamma_0[n])$ if and only if
$$M = M_r = \frac{1}{\sqrt{r}} \begin{pmatrix} \alpha r & \beta \\ \gamma n & \delta r \end{pmatrix},\ \alpha, \beta, \gamma, \delta \in \mathbb{Z},\ r|n,\ \alpha\delta r - \beta\gamma n/r = 1.$$

The coset $\Gamma_0[n]M = M\Gamma_0[n]$ only depends on r.
b) We have $M_r^2 \in \Gamma_0[n]$ in a).
c) If $s \in \mathbb{N}$, $s|n$, $\gcd(r, s) = 1$, then
$$M_r M_s \in M_{rs}\Gamma_0[n] = \Gamma_0[n]M_{rs}.$$

d) $\mathcal{N}_{SL(2;\mathbb{R})}(\Gamma_0[n])/\Gamma_0[n]$ is isomorphic to $(\mathbb{Z}/2\mathbb{Z})^r$, where r is the number of prime divisors of n.
The matrices M_r are called ATKIN-LEHNER *involutions*.

7) Let $n \in \mathbb{N}$. Show that $\mathcal{N}_{SL(2;\mathbb{R})}(\Gamma_0[n]) = \Gamma_0[n]$ holds if and only if $\gcd(n, 36)$ is square-free.

8) Let $M \in \text{Mat}(2;\mathbb{Z})$ with $\det M = m \neq 0$. Then $M^{-1}\Gamma M$ is a maximal discrete subgroup of $SL(2;\mathbb{R})$ containing $\Gamma[m]$.

9) Let $C = \begin{pmatrix} 1 & -i \\ 1 & i \end{pmatrix}$ be the matrix of the CAYLEY transformation. Consider
$$\Gamma^* := SL(2;\mathbb{R}) \cap C^{-1} SL(2;\mathbb{Z}[i]) C.$$

Then Γ^* is a maximal discrete subgroup of $SL(2;\mathbb{R})$ satisfying
$$\Gamma^* \cap \Gamma = \Gamma_\vartheta \quad \text{and} \quad [\Gamma^* : \Gamma_\vartheta] = 3.$$

Are Γ and Γ^* conjugate in $SL(2;\mathbb{R})$?

Chapter V
Theta series

Introduction

The classical theta series

$$\vartheta(\tau) = \sum_{m \in \mathbb{Z}} e^{\pi i m^2 \tau}, \quad \tau \in \mathbb{H}, \tag{0.1}$$

appears – as mentioned in the Introduction of Chapter III – in the works of L. EULER in 1748 first. Already in the 19th century multiple theta series were investigated. The general form of the multiple theta series with characteristic is given by

$$\vartheta(Z, p, q) := \sum_{g \in \mathbb{Z}^n} e^{\pi i (g+p)^{tr} Z (g+p) + 2\pi i (g+p)^{tr} q}, \tag{0.2}$$

where $p, q \in \mathbb{C}^n$, and Z is a complex symmetric $n \times n$ matrix whose imaginary part is positive definite. Much effort has been devoted to the study of their convergence behavior. A good survey of results known in the 19th century on theta series can be found in the textbook by A. KRAZER [48]. However, KRAZER's work is not easily readable, since he does not use the matrix calculus.

The idea of viewing the theta series as a function of two variables, $\tau \in \mathbb{H}$ and a positive definite quadratic form S, was probably stated for the first time by L. KRONECKER. He generalized RIEMANN's proof of the analytic continuation and functional equation of the RIEMANN zeta function (cf. IV, §4) to the EPSTEIN zeta function associated with a positive definite binary quadratic form. Instead of (0.1), KRONECKER considered in the notations of (0.2), both the Nullwert (zero value) $\vartheta(\tau S, 0, 0)$ ([50], vol. IV, 363) as well as more generally $\vartheta(\tau S, p, q)$ ([50], vol. IV, 483), and he derived the special case of the general theta transformation formula

$$\vartheta(-Z^{-1}, -q, p) = \left(\det \tfrac{1}{i} Z\right)^{1/2} \cdot e^{-2\pi i p^{tr} q} \cdot \vartheta(Z, p, q) \tag{0.3}$$

which is proved, e.g. by A. KRAZER [48], III, §5.

The theta series $\vartheta(\tau S, 0, 0)$ have a central place in the theory of quadratic forms. We can consider them as generating functions of the representation numbers of quadratic forms. Arithmetic statements can thus sometimes be proved by analytic methods (cf. e.g. Corollary 2.17, Proposition 2.23, Corollary 3.18).

To obtain modular forms for the full modular group, we need even unimodular positive definite quadratic forms. We can construct the associated lattices using, for example, certain codes (cf. J.H. CONWAY and N.J.A. SLOANE [14]). Thus, theta series also create a link between the theory of modular forms and coding theory.

§ 1 Integral and positive definite matrices

In this paragraph, the necessary tools are provided to study theta series.

1. The unimodular group. In this section, we describe the analogon of sect. 4 of I, §1 for $n \times n$ matrices. The set Mat $(n; \mathbb{Z})$ of $n \times n$ matrices with coefficients in $\mathbb{Z}$ is known to form a ring with the identity element

$$I := I^{(n)} := [1, \ldots, 1].$$

For square matrices $D_1, \ldots, D_r$, we use the more general abbreviation

$$[D_1, \ldots, D_r] := \begin{pmatrix} D_1 & & 0 \\ & \ddots & \\ 0 & & D_r \end{pmatrix}.$$

We denote the group of units of this ring by

$$\mathcal{U}_n := \{U \in \text{Mat}(n; \mathbb{Z}) \,;\, \text{there is some } V \in \text{Mat}(n; \mathbb{Z}) \text{ with } UV = VU = I\}.$$

$\mathcal{U}_n = \text{GL}(n; \mathbb{Z})$ is the *unimodular group of degree* n; its elements are called *unimodular matrices*.

1.1 Equivalence Theorem for Unimodular Matrices. *For $U \in \text{Mat}(n; \mathbb{Z})$, the following assertions are equivalent*:

  (i) $U \in \mathcal{U}_n$.

  (ii) $U^{tr} \in \mathcal{U}_n$.

  (iii) $\det U = \pm 1$.

  (iv) *U is invertible (over $\mathbb{Q}$) and $U^{-1} \in \text{Mat}(n; \mathbb{Z})$.*

  (v) *The mapping $U : \mathbb{Z}^n \to \mathbb{Z}^n$, $g \mapsto Ug$, is bijective.*

  (vi) *The mapping $U : \mathbb{Z}^n \to \mathbb{Z}^n$, $g \mapsto Ug$, is surjective.*

Proof Among the implications

§ 1 Integral and positive definite matrices

$$(\text{i}) \iff (\text{ii}) \iff (\text{iv})$$
$$\Downarrow \qquad\qquad \Uparrow$$
$$(\text{v}) \implies (\text{vi}) \implies (\text{iii})$$

only the following two need to be explained.

(vi) $\implies$ (iii): If $e_1, \ldots, e_n$ is the standard basis of $\mathbb{Z}^n$, then there exist vectors $v_1, \ldots, v_n \in \mathbb{Z}^n$ with $Uv_j = e_j$, $j = 1, \ldots, n$, hence $V = (v_1, \ldots, v_n) \in \text{Mat}(n; \mathbb{Z})$ with $UV = I$. Calculating determinants, $\det U = \pm 1$ follows.

(iii) $\implies$ (iv): We use the representation

$$U^{-1} = \frac{1}{\det U} U^{\sharp},$$

where the adjugate or complementary matrix $U^{\sharp}$ consisting of all $(n-1) \times (n-1)$ minors again has integral coefficients. □

The *special unimodular group of degree n*

$$\text{SL}(n; \mathbb{Z}) := \{U \in \text{Mat}(n; \mathbb{Z})\,;\, \det U = 1\}$$

is a normal subgroup of $\mathcal{U}_n = \text{GL}(n; \mathbb{Z})$ of index 2:

$$\mathcal{U}_n = \text{SL}(n; \mathbb{Z}) \cup F \cdot \text{SL}(n; \mathbb{Z}) \quad \text{with} \quad F = [-1, 1, \ldots, 1].$$

Examples of unimodular matrices include the so-called *permutation matrices*, which have exactly one 1 in each row and column and zeros otherwise. The permutation matrices form a group isomorphic to the symmetric group S_n of permutations of the set $\{1, \ldots, n\}$. Multiplication with a permutation matrix from the left (or right) induces the corresponding permutation of the rows (or columns), respectively.

More generally, in the case $n \geq m$, we call a matrix $P \in \text{Mat}(n, m; \mathbb{Z})$ *primitive* if there exists a unimodular matrix $U = (P, *) \in \mathcal{U}_n$. Thus, in the case $n = m$, the primitive matrices are exactly the unimodular matrices. A characterization is given in

1.2 Lemma. *For a matrix $P \in \text{Mat}(n, m; \mathbb{Z})$, $n \geq m$, the following assertions are equivalent*:
 (i) *P is primitive.*
 (ii) *UPV is primitive for all $U \in \mathcal{U}_n$ and $V \in \mathcal{U}_m$.*
 (iii) *There exists a matrix $W \in \mathcal{U}_n$ satisfying $WP = \binom{I}{0}$, $I = I^{(m)}$.*

Proof Note that $WP = \binom{I}{0}$ is equivalent to $W^{-1} = (P, *)$. Thus, we can easily verify the equivalences. □

2. Systems of representatives of left cosets. In this section, we adapt the results of IV, §1 to $n \times n$ matrices.

1.3 Lemma. *For $0 \neq g = (g_1, \ldots, g_n)^{tr} \in \mathbb{Z}^n$, there is a matrix $U \in \mathcal{U}_n$ satisfying*

$$Ug = (\delta, 0, \ldots, 0)^{tr}, \quad \delta = \gcd(g_1, \ldots, g_n).$$

Proof We use induction on n, where $n = 1$ is trivial. In the case $n = 2$, there are $\alpha, \beta \in \mathbb{Z}$ satisfying $\alpha g_1 + \beta g_2 = \gcd(g_1, g_2) = \delta$. The assertion is then valid with

$$U := \begin{pmatrix} \alpha & \beta \\ -g_2/\delta & g_1/\delta \end{pmatrix} \in \mathrm{SL}(2; \mathbb{Z}).$$

In the case $n > 2$ we first multiply by a matrix $\begin{pmatrix} I & 0 \\ 0 & V \end{pmatrix}$, $V \in \mathcal{U}_2$, to obtain $g_n = 0$. Then using the induction hypothesis, we find a matrix $W \in \mathcal{U}_{n-1}$ such that that the multiplication by $\begin{pmatrix} W & 0 \\ 0 & 1 \end{pmatrix}$ yields the desired form.

Since the coefficients of Ug are integer linear combinations of the coefficients of g, we know that $\gcd(g)$ is a divisor of $\gcd(Ug)$. The same conclusion for U^{-1} instead of U and Ug instead of g yields $\gcd(g) = \gcd(Ug)$. □

1.4 Corollary. *A vector $0 \neq g \in \mathbb{Z}^n$ is primitive, i.e. a column of a unimodular matrix if and only if $\gcd(g) = 1$.*

Proof If $\gcd(g) = 1$, then g is primitive by Lemma 1.3 and Lemma 1.2. Conversely, if $U = (g, *) \in \mathcal{U}_n$, then it follows from the LAPLACE Expansion Theorem for determinants that $\gcd(g)$ is a divisor of $\det U = \pm 1$, thus $\gcd(g) = 1$. □

Lemma 1.3 leads to

1.5 Corollary. a) *Given $G \in \mathrm{Mat}(n, m; \mathbb{Z})$, there exist $U, V \in \mathcal{U}_n$ such that UG has upper triangular form and VG has lower triangular form.*
b) *Given $G \in \mathrm{Mat}(n, m; \mathbb{Z})$, there exist $U, V \in \mathcal{U}_m$ such that GU has upper and GV has lower triangular form.*

Here, *upper* (or *lower*) *triangular* form for a matrix $H = (h_{\nu\mu}) \in \mathrm{Mat}(n, m; \mathbb{Z})$ means $h_{\nu\mu} = 0$ for $\nu > \mu$ (or for $n - \nu > m - \mu$, respectively).

Proof a) We use an induction where $n = 1$ is trivial. Applying Lemma 1.3 to the first column of G, we may assume that G has the form $\begin{pmatrix} \delta & * \\ 0 & * \end{pmatrix}$. Then the claim follows with a matrix $U = \begin{pmatrix} 1 & 0 \\ 0 & U_1 \end{pmatrix}$, $U_1 \in \mathcal{U}_{n-1}$. If we apply Lemma 1.3 to the last column of G, we may also assume that G hat the form $\begin{pmatrix} * & 0 \\ * & \gamma \end{pmatrix}$. Then the claim follows with a matrix $V = \begin{pmatrix} V_1 & 0 \\ 0 & 1 \end{pmatrix}$, $V_1 \in \mathcal{U}_{n-1}$.
b) We apply a) to G^{tr} and transpose. □

Matrices $A, B \in \mathrm{Mat}(n; \mathbb{Z})$ are called *equivalent*, in symbols $A \sim B$, if there is a matrix $U \in \mathcal{U}_n$ with $AU = B$. Thus, the equivalence classes with respect to $\sim$ are exactly the *left cosets modulo* $\mathcal{U}_n$

§ 1 Integral and positive definite matrices

$$A\,\mathcal{U}_n = \{AU \; ; \; U \in \mathcal{U}_n\}.$$

If $\det A \neq 0$, each equivalence class contains a canonical representative.

1.6 Theorem. *Let $A \in \mathrm{Mat}\,(n; \mathbb{Z})$ with $\det A \neq 0$. Then the left coset $A\,\mathcal{U}_n$ contains a unique representative of the form*:

$$B = \begin{pmatrix} b_1 & & b_{\nu\mu} \\ & \ddots & \\ 0 & & b_n \end{pmatrix}, \quad \begin{matrix} b_1, \ldots, b_n \text{ are positive integers}, \\ 0 \leq b_{\nu\mu} < b_\nu \text{ for } 1 \leq \nu < \mu \leq n. \end{matrix} \quad (1.1)$$

Proof We use induction on n, where $n = 1$ is trivial. In the case $n > 1$, we can assume by Corollary 1.5 that A has upper triangular form, i.e.

$$A = \begin{pmatrix} \alpha & a^{tr} \\ 0 & A' \end{pmatrix}, \quad A' \in \mathrm{Mat}\,(n-1; \mathbb{Z}), \quad a \in \mathbb{Z}^{n-1}, \quad \alpha \in \mathbb{Z}, \quad \alpha \det A' \neq 0. \quad (1.2)$$

Now we choose U in the form

$$U = \begin{pmatrix} \varepsilon & u^{tr} \\ 0 & V \end{pmatrix}, \quad V \in \mathcal{U}_{n-1}, \quad u \in \mathbb{Z}^{n-1}, \quad \varepsilon = \pm 1. \quad (1.3)$$

Since

$$AU = \begin{pmatrix} \alpha\varepsilon & \alpha u^{tr} + a^{tr}V \\ 0 & A'V \end{pmatrix}, \quad (1.4)$$

we should choose V and ε by the induction hypothesis and then u such that the coefficients of $\alpha u^{tr} + a^{tr}V$ are between 0 and $|\alpha| - 1$.

Now, if A and B are given in the form (1.1) and $U \in \mathcal{U}_n$ satisfies $AU = B$, it follows from $U = A^{-1}B$ that U also has upper triangular form. Therefore, we assume that U is given by (1.3), A by (1.2) and B analogously. Then (1.4) implies

$$A'V = B' \quad \text{and} \quad \varepsilon\alpha = \beta,$$

hence $V = I$ and $\varepsilon = 1$ by the induction hypothesis. Now $b = a + \alpha u$ also yields $u = 0$, thus $A = B$ and $U = I$. □

We sometimes say that the matrices in (1.1) are in HERMITE *normal form* (cf. M. NEWMAN [60], 15).

A finiteness statement is derived in

1.7 Corollary. *If q is a positive integer, then the set*

$$\{A \in \mathrm{Mat}\,(n; \mathbb{Z}) \; ; \; |\det A| = q\}$$

decomposes into finitely many left cosets modulo $\mathcal{U}_n$.

Proof The statement is equivalent to saying that there are only finitely many integral matrices B of the form (1.1) with $\det B = q$. Because the diagonal elements are divisors of q, there are only finitely many possibilities for the diagonal. But since there are only b_ν possibilities for $b_{\nu\mu}$, there exist only finitely many B. □

As the mapping $\mathbb{Z}^n \to \mathbb{Z}^n$, $x \mapsto Ux$, for $U \in \mathcal{U}_n$ is a bijection according to the Equivalence Theorem 1.1, we immediately get the

1.8 Corollary. *For $A \in \mathrm{Mat}\,(n;\mathbb{Z})$ with $\det A \neq 0$,*

$$\#(\mathbb{Z}^n/A\mathbb{Z}^n) = |\det A|.$$

3. Positive definite matrices. We use the notation
$$\mathrm{Sym}\,(n;\mathbb{R}) := \{S \in \mathrm{Mat}\,(n;\mathbb{R})\,;\, S^{tr} = S\} \tag{1.5}$$

for the real vector space of symmetric $n \times n$ matrices of dimension $n(n+1)/2$. For $S \in \mathrm{Sym}\,(n;\mathbb{R})$ and a real $n \times m$ matrix A, we define

$$S[A] := A^{tr} S A \in \mathrm{Sym}\,(m;\mathbb{R}). \tag{1.6}$$

If B is a real $m \times p$ matrix, we obtain

$$(S[A])[B] = S[(AB)]. \tag{1.7}$$

A straightforward verification yields the

1.9 Theorem on Completing the Square. *Given a symmetric matrix of the form* $S = \begin{pmatrix} S_1 & S_2 \\ S_2^{tr} & S_3 \end{pmatrix} \in \mathrm{Sym}\,(n;\mathbb{R})$ *with* $S_1 \in \mathrm{Sym}\,(m;\mathbb{R})$, $\det S_1 \neq 0$, $1 \leq m < n$, *then*

$$\begin{pmatrix} S_1 & S_2 \\ S_2^{tr} & S_3 \end{pmatrix} \left[\begin{pmatrix} I & -S_1^{-1} S_2 \\ 0 & I \end{pmatrix} \right] = \begin{pmatrix} S_1 & 0 \\ 0 & S_3 - S_1^{-1}[S_2] \end{pmatrix}. \tag{1.8}$$

If $S \in \mathrm{Sym}\,(n;\mathbb{R})$, then the mapping

$$\mathbb{R}^n \longrightarrow \mathbb{R},\ g \mapsto g^{tr} S g,$$

is called the *quadratic form* associated to S. The matrix S or its associated quadratic form is called *positive definite* if

$$S[g] = g^{tr} S g > 0 \quad \text{for all}\quad 0 \neq g \in \mathbb{R}^n. \tag{1.9}$$

$\mathrm{Pos}\,(n;\mathbb{R})$ stands for the set of positive definite matrices in $\mathrm{Sym}\,(n;\mathbb{R})$. If $1 \leq m \leq n$, let

$$S_{(m)} := S\left[\begin{pmatrix} I \\ 0 \end{pmatrix}\right] \in \mathrm{Sym}\,(m;\mathbb{R}),\quad I = I^{(m)}, \tag{1.10}$$

be the upper left $m \times m$ block of S. Then $\det S_{(m)}$ is called the *m-th principal minor of S*. The definition directly yields

$$S \in \mathrm{Pos}\,(n;\mathbb{R}) \Longrightarrow S_{(m)} \in \mathrm{Pos}\,(m;\mathbb{R}) \quad \text{for } 1 \leq m \leq n. \tag{1.11}$$

§ 1 Integral and positive definite matrices

From linear algebra, we adopt the following characterization.

1.10 Equivalence Theorem for Positive Definite Matrices. *For $S \in \mathrm{Sym}\,(n; \mathbb{R})$, the following assertions are equivalent*:

(i) $S \in \mathrm{Pos}\,(n; \mathbb{R})$.

(ii) *There exists a matrix $W \in \mathrm{GL}\,(n; \mathbb{R})$ with $S = W^{tr}W$.*

(iii) *The inverse matrix S^{-1} exists and is positive definite.*

(iv) *For $A \in \mathrm{GL}\,(n; \mathbb{R})$ one has $S[A] \in \mathrm{Pos}\,(n; \mathbb{R})$.*

(v) *All principal minors $\det S_{(m)}$, $1 \leq m \leq n$, are positive.*

(vi) *There is a matrix $V \in \mathrm{GL}\,(n; \mathbb{R})$ with $V^{tr}V = I$ and a diagonal matrix D with positive diagonal elements such that $S[V] = D$.*

(vii) *All eigenvalues of S are positive.*

Using part (vi) or (vii), we immediately obtain

$$\det S > 0 \quad \text{and} \quad \mathrm{tr}\, S > 0 \quad \text{for all} \quad S \in \mathrm{Pos}\,(n; \mathbb{R}). \tag{1.12}$$

It follows easily from characterization (v) that $\mathrm{Pos}\,(n; \mathbb{R})$ is an open subset of $\mathrm{Sym}\,(n; \mathbb{R})$ in the natural topology from the identification of $\mathrm{Sym}\,(n; \mathbb{R})$ with $\mathbb{R}^{n(n+1)/2}$.

We also need the

1.11 Square Root Theorem. *For every $S \in \mathrm{Pos}(n; \mathbb{R})$, there is a uniquely determined $P \in \mathrm{Pos}(n; \mathbb{R})$ with $S = P^2$.*

Proof We may assume that $S = [d_1, \ldots, d_n]$ is a diagonal matrix due to the Equivalence Theorem 1.10. Then $P = [\sqrt{d_1}, \ldots, \sqrt{d_n}]$ is positive definite and it satisfies $P^2 = S$. If Q is an arbitrary positive definite solution of $Q^2 = S$, we obtain

$$Q(Q - P)e_j = Se_j - \sqrt{d_j}Qe_j = d_je_j - \sqrt{d_j}Qe_j = -\sqrt{d_j}(Q - P)e_j,$$

hence $(Q - P)e_j = 0$ for $j = 1, \ldots, n$, as Q is positive definite. This yields $P = Q$, hence also the uniqueness. □

We call P the *square root of S* and use the notation

$$S^{1/2} := P. \tag{1.13}$$

4. A partial ordering on $\mathrm{Sym}\,(n; \mathbb{R})$. For $S, T \in \mathrm{Sym}\,(n; \mathbb{R})$, we define

$$S > T \iff S - T \in \mathrm{Pos}\,(n; \mathbb{R}).$$

Thus in particular

$$\mathrm{Pos}\,(n; \mathbb{R}) = \{S \in \mathrm{Sym}\,(n; \mathbb{R})\,;\, S > 0\}.$$

As a simple conclusion we get, for matrices $S, T, R \in \mathrm{Sym}\,(n; \mathbb{R})$, $\lambda \in \mathbb{R}$, $\lambda > 0$ and $A \in \mathrm{GL}\,(n; \mathbb{R})$,

$$S > T \iff \lambda S + R > \lambda T + R \iff S[A] > T[A]. \qquad (1.14)$$

1.12 Lemma. *For $S, T \in \mathrm{Sym}\,(n; \mathbb{R})$ with $S > T$ and $0 \neq G \in \mathrm{Mat}\,(n, m; \mathbb{R})$,*

$$\mathrm{tr}(S[G]) > \mathrm{tr}(T[G]t)$$

holds. In particular, $S[g] > T[g]$ for all $0 \neq g \in \mathbb{R}^n$.

Proof Let $G = (g_1, \ldots, g_m)$ and $g_r \neq 0$. Then (1.9) leads to

$$\mathrm{tr}(S[G]) - \mathrm{tr}(T[G]) = \mathrm{tr}((S - T)[G]) = \sum_{j=1}^{m} (S - T)[g_j] \geq (S - T)[g_r] > 0. \qquad \square$$

We can compare any positive definite matrix in this partial ordering with positive multiples of the identity matrix.

1.13 Proposition. *Given $S \in \mathrm{Pos}\,(n; \mathbb{R})$, there are positive $\alpha, \beta \in \mathbb{R}$ with the property*

$$\alpha I > S > \beta I.$$

Proof We apply part (vi) of the Equivalence Theorem 1.10 and obtain an orthogonal matrix V and a positive definite diagonal matrix D with the eigenvalues λ_j of S as diagonal elements such that $S[V] = D$. Now we choose positive $\alpha, \beta \in \mathbb{R}$ with $\alpha > \lambda_j > \beta$ for $j = 1, \ldots, n$. Then the claim follows from

$$\alpha I > D > \beta I,$$

hence the assertion due to (1.14) and $VV^{tr} = I$. $\qquad \square$

An important finiteness statement is contained in the following

1.14 Proposition. *Let $S \in \mathrm{Pos}\,(n; \mathbb{R})$, $t \in \mathbb{R}$ and $m \geq 1$. Then the set*

$$\{G \in \mathrm{Mat}\,(n, m; \mathbb{Z})\,;\, \mathrm{tr}(S[G]) \leq t\}$$

is finite.

Proof Given S, we choose $\beta > 0$ with $S > \beta I$ according to Proposition 1.13. Due to Lemma 1.12, the set we are looking for is contained in

$$\{G \in \mathrm{Mat}\,(n, m; \mathbb{Z})\,;\, \mathrm{tr}(G^{tr}G) \leq t/\beta\}.$$

If we write $G = (g_{\nu\mu})$, the assertion follows from

$$\mathrm{tr}(G^{tr}G) = \sum_{\nu=1}^{n} \sum_{\mu=1}^{m} g_{\nu\mu}^2. \qquad \square$$

§ 1 Integral and positive definite matrices

As a conclusion, we get the finiteness of certain representation sets.

1.15 Corollary. *Let* $S \in \text{Pos}(n; \mathbb{R})$ *and* $T \in \text{Sym}(m; \mathbb{R})$. *Then the set*

$$\mathcal{D}(S, T) := \{G \in \text{Mat}(n, m; \mathbb{Z}) \; ; \; S[G] = T\} \tag{1.15}$$

is finite and

$$\text{Aut } S := \mathcal{D}(S, S) = \{U \in \text{Mat}(n; \mathbb{Z}) \; ; \; S[U] = S\}$$

is a finite subgroup of $\mathcal{U}_n$.

Proof The finiteness follows from Proposition 1.14. (1.12) yields $\det S > 0$. By comparing determinants we get $\text{Aut } S \subseteq \mathcal{U}_n$ from the Equivalence Theorem 1.1. We can easily verify the subgroup property using (1.7). □

The order of $\sharp \mathcal{D}(S, T) =: \sharp (S, T)$ is called the *representation number of T by S*. Aut S is called *automorphism group of S*.

We define an equivalence relation on $\text{Pos}(n; \mathbb{R})$, called *integral equivalence*, by

$$S \sim T \iff \text{there is some } U \in \mathcal{U}_n \text{ satisfying } S[U] = T. \tag{1.16}$$

The *class* of S,

$$\langle S \rangle := \{T \in \text{Pos}(n; \mathbb{R}) \; ; \; T \sim S\} = \{S[U] \; ; \; U \in \mathcal{U}_n\} \tag{1.17}$$

is its equivalence class with respect to $\sim$. Therefore, two classes are either disjoint or equal. A mapping $\varphi : \text{Pos}(n; \mathbb{R}) \to \mathbb{C}$ is called a *class invariant* if the values of φ depend only on the class, i.e.

$$\varphi(S[U]) = \varphi(S) \quad \text{for all} \quad S \in \text{Pos}(n; \mathbb{R}) \text{ and all } U \in \mathcal{U}_n. \tag{1.18}$$

By the Equivalence Theorem 1.1, the determinant is obviously a class invariant. We obtain

$$\mathcal{D}(S[U], T[V]) = U^{-1} \mathcal{D}(S, T) V$$

for matrices $U \in \mathcal{U}_n$ and $V \in \mathcal{U}_m$ in the notation of Corollary 1.15. Thus the representation numbers $\sharp (S, T)$ are also class invariants.

1.16 Remarks. In the *reduction theory of quadratic forms* the objective is to determine a system of representatives of the classes. The so-called MINKOWSKI reduction theory is described, e.g. in H. MAASS, *Siegel's modular forms and Dirichlet series*, Lect. Notes Math. **216**, Springer–Verlag, Berlin–Heidelberg–New York 1971, in §9. Another method can be found in A. TERRAS [82].

5. Finiteness of the class numbers. If $S \in \text{Pos}(n; \mathbb{R})$, we define

$$\mu(S) := \inf\{S[g] \; ; \; 0 \neq g \in \mathbb{Z}^n\}, \tag{1.19}$$

and for $1 \leq m \leq n$,

$$\mu_m(S) := \inf \left\{ t \in \mathbb{R}; \; \begin{array}{l} \text{there is } G = (g_1, \ldots, g_m) \in \text{Mat}(n, m; \mathbb{Z}), \\ G \text{ primitive}, \; S[g_j] \leq t \text{ for } j = 1, \ldots, m \end{array} \right\}. \tag{1.20}$$

By Proposition 1.14, the set $\{g \in \mathbb{Z}^n \;;\; S[g] \leq t\}$ is finite. Hence, the infimum in (1.19) and (1.20) is attained. $\mu(S)$ is called the *minimum* of S and $\mu_m(S)$ is called the *m–th (primitive) minimum* of S. If $g \in \mathbb{Z}^n$ satisfies $\mu(S) = S[g]$, then the coefficients of g are coprime, i.e. g is primitive by Corollary 1.4. It follows that

$$\mu(S) = \mu_1(S) \leq \mu_2(S) \leq \ldots \leq \mu_n(S) \tag{1.21}$$

and

$$\mu_m(S) > \mu_m(T) \quad \text{for} \quad S > T \quad \text{and} \quad 1 \leq m \leq n. \tag{1.22}$$

Because of Lemma 1.2, the values $\mu_m(S)$ are class invariants, i.e.

$$\mu_m(S[U]) = \mu_m(S) \quad \text{for } U \in \mathcal{U}_n \quad \text{and} \quad 1 \leq m \leq n. \tag{1.23}$$

Since the infimum in (1.20) is attained and the primitive matrix G can be completed to a unimodular matrix, there is a matrix $U = U_m \in \mathcal{U}_n$ for $1 \leq m \leq n$ satisfying

$$S[U] = T = (t_{ij}), \quad t_{11} \leq t_{22} \leq \ldots \leq t_{mm} = \mu_m(S). \tag{1.24}$$

An inequality between class invariants is now provided by the

1.17 Proposition. *For $S \in \operatorname{Pos}(n; \mathbb{R})$,*

$$\mu_1(S) \cdot \ldots \cdot \mu_n(S) \leq \left(\frac{4}{3}\right)^{n(n-1)/2} \cdot \det S$$

holds.

Proof The proof proceeds by induction on n, where $n = 1$ is trivial. Now let $S = (s_{ij}) \in \operatorname{Pos}(n+1; \mathbb{R})$, $n \geq 1$. Because of (1.24), we can assume without loss of generality that

$$s_{11} = \mu_1(S) =: \sigma > 0.$$

We determine a representation of S by completing the square as in Theorem 1.9

$$S = \begin{pmatrix} \sigma & 0 \\ 0 & T \end{pmatrix} \left[\begin{pmatrix} 1 & s^{tr} \\ 0 & I \end{pmatrix} \right] = \begin{pmatrix} \sigma & \sigma s^{tr} \\ \sigma s & T + \sigma s s^{tr} \end{pmatrix}, \quad T \in \operatorname{Pos}(n, \mathbb{R}), \; s \in \mathbb{R}^n.$$

For $1 \leq m \leq n$, (1.20) leads to a primitive matrix $G = (g_1, \ldots, g_m) \in \operatorname{Mat}(n, m; \mathbb{Z})$ with the property

$$T[g_j] \leq \mu_m(T), \quad 1 \leq j \leq m.$$

We now choose a vector $h = (\eta_1, \ldots, \eta_m)^{tr} \in \mathbb{Z}^m$ such that the coefficients of $h + G^{tr} s$ belong to the interval $\left[-\frac{1}{2}, \frac{1}{2}\right]$. Along with G, the matrix

$$H := \begin{pmatrix} 1 & h^{tr} \\ 0 & G \end{pmatrix} \in \operatorname{Mat}(n+1, m+1; \mathbb{Z})$$

§ 1 Integral and positive definite matrices

is also primitive. Due to the choice of η_j we get, for $1 \le j \le m$,

$$S\left[\begin{pmatrix}\eta_j\\g_j\end{pmatrix}\right] = \begin{pmatrix}\sigma & 0\\0 & T\end{pmatrix}\left[\begin{pmatrix}\eta_j + s^{tr}g_j\\g_j\end{pmatrix}\right]$$
$$= \sigma(\eta_j + s^{tr}g_j)^2 + T[g_j] \le \tfrac{1}{4}\mu_1(S) + \mu_m(T)$$

and

$$s_{11} = \mu_1(S) \le S\left[\begin{pmatrix}\eta_j + s^{tr}g_j\\g_j\end{pmatrix}\right] \le \tfrac{1}{4}\mu_1(S) + \mu_m(T).$$

Since H is primitive, it follows from (1.21) that

$$\mu_{m+1}(S) \le \frac{1}{4}\mu_1(S) + \mu_m(T), \quad \text{hence} \quad \mu_{m+1}(S) \le \frac{4}{3}\mu_m(T).$$

As $\mu_1(S) \cdot \det T = \det S$, we get

$$\mu_1(S) \cdot \ldots \cdot \mu_{n+1}(S) \le \left(\tfrac{4}{3}\right)^n \cdot \mu_1(S) \cdot \mu_1(T) \cdot \ldots \cdot \mu_n(T)$$
$$\le \left(\tfrac{4}{3}\right)^{n(n+1)/2} \cdot \mu_1(S) \cdot \det T = \left(\tfrac{4}{3}\right)^{n(n+1)/2} \cdot \det S$$

from the induction hypothesis applied to T. □

If we use (1.21), we immediately obtain

1.18 Corollary. (HERMITE's inequality) *The constant*

$$\gamma_n := \inf\left\{\frac{\det S}{\mu(S)^n} \; ; \; S \in \mathrm{Pos}\,(n;\mathbb{R})\right\}$$

satisfies

$$\gamma_n \ge \left(\frac{3}{4}\right)^{n(n-1)/2}. \tag{1.25}$$

The inequality (1.25) is only sharp for $n = 1, 2$. The values γ_n are known for $n \le 10$ (cf. B.L. VAN DER WAERDEN and H. GROSS, *Studien zur Theorie der quadratischen Formen*, Birkhäuser, Basel–Stuttgart 1968, 28).

For positive $q \in \mathbb{Z}$, let $k_n(q)$ denote the number of classes $\langle S \rangle$ of all matrices $S \in \mathrm{Pos}\,(n;\mathbb{Z}) := \mathrm{Pos}\,(n;\mathbb{R}) \cap \mathrm{Mat}\,(n;\mathbb{Z})$ with $\det S = q$. The numbers $k_n(q)$, $q \ge 1$, are called *class numbers*.

1.19 Corollary. *The class numbers $k_n(q)$, $q \ge 1$, are finite and satisfy*

$$k_n(q) = O\left(q^{n(n+1)/2}\right) \quad \text{for} \quad q \to \infty.$$

Proof Because of (1.24), it is sufficient to calculate the number of all the matrices $S = (s_{ij}) \in \mathrm{Pos}\,(n;\mathbb{Z})$ with $s_{jj} \le \mu_n(S)$ for $1 \le j \le n$ and $\det S = q$. As S is an integral matrix, all minima are positive integers and, by Proposition 1.17, they satisfy

$$1 \leq s_{jj} \leq \mu_n(S) \leq \mu_1(S) \cdot \ldots \cdot \mu_n(S) \leq \left(\frac{4}{3}\right)^{n(n-1)/2} \cdot q, \quad 1 \leq j \leq n.$$

Since S is positive definite, the matrices $\begin{pmatrix} s_{ii} & s_{ij} \\ s_{ij} & s_{jj} \end{pmatrix}$, $1 \leq i < j \leq n$ are positive definite. Thus (1.12) yields

$$s_{ii}s_{jj} - s_{ij}^2 > 0, \quad \text{hence} \quad |s_{ij}| < \left(\frac{4}{3}\right)^{n(n-1)/2} \cdot q. \qquad \square$$

1.20 Exercises.
1) Let $n \geq m$. A matrix $G \in \text{Mat}(n, m; \mathbb{Z})$ is primitive if and only if there exists a matrix $H \in \text{Mat}(m, n; \mathbb{Z})$ with $HG = I^{(m)}$. In this case, H^{tr} is also primitive. Is H uniquely determined by G?
2) For $q \geq 1$, the set $\{G \in \text{Mat}(n; \mathbb{Z}) \,;\, \det G \neq 0, \, qG^{-1} \in \text{Mat}(n; \mathbb{Z})\}$ decomposes into finitely many left cosets modulo $\mathcal{U}_n$.
3) What are representatives of the right cosets $\mathcal{U}_n A$ for $A \in \text{Mat}(n; \mathbb{Z})$ satisfying $\det A \neq 0$ in analogy with (1.1)?
4) (Inequality of HADARMARD) For $S = (s_{ij}) \in \text{Pos}(n; \mathbb{R})$ one has

$$\det S \leq s_{11} \cdot s_{22} \cdot \ldots \cdot s_{nn}$$

and equality holds if and only if S is a diagonal matrix.
5) For $S, T \in \text{Pos}(n; \mathbb{R})$ one has $\text{tr}(ST) \geq n \cdot (\det S)^{1/n} \cdot (\det T)^{1/n}$.
6) (JACOBI *decomposition*) Each matrix $S \in \text{Pos}(n; \mathbb{R})$ has a unique representation of the form $S = D[B]$, where D is a positive definite diagonal matrix and $B \in \text{GL}(n; \mathbb{R})$ is an upper triangular matrix with 1's on the diagonal.
7) For $S \in \text{Pos}(n; \mathbb{R})$ and $1 \leq m \leq n$, we define the *successive minima* by

$$v_m(S) := \inf \left\{ t \in \mathbb{R} \,;\, \begin{array}{l} \text{there is } G = (g_1, \ldots, g_m) \in \text{Mat}(n, m; \mathbb{Z}) \text{ with} \\ \text{rank } G = m \text{ and } S[g_j] \leq t \text{ for } j = 1, \ldots, m \end{array} \right\}.$$

There is a matrix $G \in \text{Mat}(n; \mathbb{Z})$ such that $\det G \neq 0$ and

$$S[G] = T = (t_{ij}), \quad t_{mm} = v_m(S), \, 1 \leq m \leq n.$$

The $v_m(S)$ are class invariants and satisfy $v_m(S) \leq \mu_m(S)$ as well as

$$\det S \leq v_1(S) \cdot \ldots \cdot v_n(S) \leq \mu_1(S) \cdot \ldots \cdot \mu_n(S).$$

8) The matrix $S = \begin{pmatrix} 2I & e \\ e^{tr} & 5/2 \end{pmatrix}$, $I = I^{(4)}$, $e = (1, 1, 1, 1)^{tr}$, in $\text{Pos}(5; \mathbb{R})$ satisfies

$$\mu_m(S) = v_m(S) = v_5(S) = 2 \quad \text{for} \quad 1 \leq m \leq 4 \quad \text{and} \quad \mu_5(S) = 5/2.$$

9) Let $S = (s_{ij}) \in \text{Pos}(2; \mathbb{R})$ with $s_{11} < s_{22} = \mu_2(S)$. Show that $2|s_{12}| \leq s_{11}$. Is the statement also true for $s_{11} = s_{22}$?
10) For $S \in \text{Pos}(n; \mathbb{Z})$ there is a matrix $U \in \mathcal{U}_n$ such that $S[U] = (t_{ij})$ is tridiagonal,

i.e. $t_{ij} = 0$ for $|i - j| \geq 2$.
11) If $S \in \mathrm{Pos}\,(n;\mathbb{R})$, then

$$\mathrm{Aut}\,S^{-1} = \{U^{tr};\ U \in \mathrm{Aut}\,S\}.$$

If $S \in \mathrm{Pos}\,(n;\mathbb{R}) \cap \mathrm{SL}\,(n;\mathbb{Z})$, then

$$\{S^{-1}U^{tr}S;\ U \in \mathrm{Aut}\,S\} = \mathrm{Aut}\,S.$$

12) For $S \in \mathrm{Pos}\,(n;\mathbb{R})$ and an orthogonal $n \times n$ matrix U one has $(S[U])^{1/2} = S^{1/2}[U]$.

§ 2 Theta series as modular forms

In this paragraph, we first introduce a class of theta series that generalize the classical theta series from the Introduction of Chapter III or (0.1). The general theta transformation formula is proved. Thus, we show that all theta series associated with even, unimodular, positive definite quadratic forms are modular forms. In particular, we will see that every modular form is a linear combination of theta series if and only if the weight is a multiple of 4.

1. Theta series with characteristics. For $\tau \in \mathbb{H}$, $S \in \mathrm{Pos}\,(n;\mathbb{R})$ and $p, q \in \mathbb{C}^n$, we formally define the *theta series associated with S and the characteristic* (p, q) by

$$\Theta_{p,q}(\tau; S) := \sum_{g \in \mathbb{Z}^n} e^{\pi i \tau S[g+p] + 2\pi i (g+p)^{tr} q}. \tag{2.1}$$

In the special case $p = q = 0$ we speak of the *Theta Nullwert* (*theta zero value*) and use the abbreviation

$$\Theta(\tau; S) := \Theta_{0,0}(\tau; S) = \sum_{g \in \mathbb{Z}^n} e^{\pi i \tau S[g]}. \tag{2.2}$$

We obtain the classical theta series $\vartheta(\tau)$ from (0.1) by setting $n = 1$, $S = (1)$ in (2.2). The convergence properties follow from the

2.1 Lemma. *The theta series* $\Theta_{p,q}(\tau; S)$ *converges absolutely and uniformly on each compact subset of*

$$(p, q, \tau, S) \in \mathbb{C}^n \times \mathbb{C}^n \times \mathbb{H} \times \mathrm{Pos}\,(n;\mathbb{R}).$$

For a fixed $S \in \mathrm{Pos}\,(n;\mathbb{R})$, $\Theta_{p,q}(\tau; S)$ *is holomorphic in* $(p, q, \tau) \in \mathbb{C}^n \times \mathbb{C}^n \times \mathbb{H}$.

Here, a function $\varphi : \mathcal{D} \to \mathbb{C}$, $\mathcal{D} \subseteq \mathbb{C}^m$ open, is called holomorphic if φ is holomorphic in every variable.

Proof Let $C \subseteq \mathbb{C}^n \times \mathbb{C}^n \times \mathbb{H} \times \mathrm{Pos}\,(n;\mathbb{R})$ be compact. Then all $(p, q, \tau, S) \in C$, $p = p_1 + ip_2$, $q = q_1 + iq_2$, $\tau = x + iy$ satisfy

$$|\Theta_{p,q}(\tau;S)|$$
$$\leq \sum_{g\in\mathbb{Z}^n} |e^{\pi i\tau S[g+p]+2\pi i(g+p)^{tr}q}|$$
$$= \sum_{g\in\mathbb{Z}^n} e^{-\pi y S[g]-2\pi g^{tr}(ySp_1+xSp_2+q_2)-\pi y S[p_1]+\pi y S[p_2]-2\pi x p_1^{tr} Sp_2 - 2\pi p_2^{tr} q_1 - 2\pi p_1^{tr} q_2}.$$

Using the CAUCHY SCHWARZ inequality, we conclude that there are positive constants $\alpha, \beta, \gamma, \delta$, which only depend on C such that

$$|\Theta_{p,q}(\tau;S)| \leq \gamma \sum_{g\in\mathbb{Z}^n} e^{-2\pi\alpha g^{tr}g+\beta\sqrt{g^{tr}g}}$$
$$\leq \delta + \gamma \sum_{g\in\mathbb{Z}^n} e^{-\pi\alpha g^{tr}g} = \delta + \gamma \cdot \vartheta(i\alpha)^n.$$

The holomorphy of the summands implies the holomorphy of $\Theta_{p,q}(\tau;S)$. □

The following elementary statement describes the behavior of Θ under certain changes of parameters.

2.2 Proposition. *Let $\tau \in \mathbb{H}$, $S \in \operatorname{Pos}(n;\mathbb{R})$, $p, q \in \mathbb{C}^n$. Then all $h \in \mathbb{Z}^n$ and all $U \in \mathcal{U}_n$ satisfy:*

a) $\Theta_{p+h,q}(\tau;S) = \Theta_{p,q}(\tau;S)$,

b) $\Theta_{p,q+h}(\tau;S) = e^{2\pi i p^{tr}h} \cdot \Theta_{p,q}(\tau;S)$,

c) $\Theta_{p,q}(\tau;S[U]) = \Theta_{Up,U^{tr-1}q}(\tau;S)$,

in particular,

$$\Theta(\tau;S[U]) = \Theta(\tau;S).$$

Proof Because of Lemma 2.1, we may rearrange the series. In c), we additionally use part (v) of the Equivalence Theorem 1.1. □

Thus, the theta zero values $\Theta(\tau;S)$ are class invariants in the sense of (1.18).

For $S \in \operatorname{Pos}(n;\mathbb{R})$ and $T \in \operatorname{Pos}(m;\mathbb{R})$ we define the *direct sum* by

$$S \oplus T := \begin{pmatrix} S & 0 \\ 0 & T \end{pmatrix} \in \operatorname{Pos}(n+m;\mathbb{R}). \tag{2.3}$$

If we write $g = \binom{a}{b} \in \mathbb{Z}^{n+m}$ with $a \in \mathbb{Z}^n$, $b \in \mathbb{Z}^m$, we immediately obtain

$$\Theta(\tau;S \oplus T) = \Theta(\tau;S) \cdot \Theta(\tau;T). \tag{2.4}$$

If $S \in \operatorname{Pos}(n;\mathbb{Z}) := \operatorname{Pos}(n;\mathbb{R}) \cap \operatorname{Mat}(n;\mathbb{Z})$, then $S[g] \in \mathbb{Z}$ obviously holds for all $g \in \mathbb{Z}^n$. The definition then implies

$$\Theta(\tau+2;S) = \Theta(\tau;S), \quad \tau \in \mathbb{H}. \tag{2.5}$$

A rearrangement yields the FOURIER series expansion (cf. Corollary 1.15)

§ 2 Theta series as modular forms

$$\Theta(\tau; S) = \sum_{m=0}^{\infty} \sharp(S, m) \cdot e^{\pi i m \tau}, \quad \tau \in \mathbb{H}, \tag{2.6}$$

where, of course, $\sharp(S, 0) = 1$.

2. Relations with lattices. In this section we explore the possibility of introducing theta series using lattices equivalently to sect. 1. Let V be a real vector space of dimension $n < \infty$. A subset Λ of V is called a *lattice in V* if there exist linearly independent vectors $g_1, \ldots, g_n \in V$ satisfying

$$\Lambda = \mathbb{Z}g_1 + \ldots \mathbb{Z}g_n. \tag{2.7}$$

$g_1, \ldots, g_n$ is called a *basis of the lattice* Λ. Compare also with the characterization in I, §1. Now, if σ is a positive definite bilinear form on V, i.e. (V, σ) is a euclidean vector space, then we define the *theta series with respect to the lattice* Λ by

$$\Theta_\Lambda(\tau) := \sum_{\lambda \in \Lambda} e^{\pi i \tau \cdot \sigma(\lambda, \lambda)}, \quad \tau \in \mathbb{H}. \tag{2.8}$$

2.3 Lemma. *Let (V, σ) be a euclidean vector space and $\Lambda = \mathbb{Z}g_1 + \ldots + \mathbb{Z}g_n$ a lattice in V. Let $S = (\sigma(g_\nu, g_\mu))$ denote the GRAM matrix with respect to the basis $g_1, \ldots, g_n$. Then*

$$\Theta_\Lambda(\tau) = \Theta(\tau; S) \quad \textit{for all} \quad \tau \in \mathbb{H}.$$

Proof For $\lambda = \lambda_1 g_1 + \cdots + \gamma_n \lambda_n \in \Lambda$, we define $h = (\lambda_1, \ldots, \lambda_n)^{tr} \in \mathbb{Z}^n$. The bilinearity of σ leads to

$$\sigma(\lambda, \lambda) = S[h]. \qquad \square$$

Conversely, given an $S \in \text{Pos}(n; \mathbb{R})$, consider $\mathbb{R}^n$ as a euclidean vector space with the positive definite bilinear form given by S. Then we obtain

$$\Theta(\tau; S) = \Theta_{\mathbb{Z}^n}(\tau) \quad \text{for all} \quad \tau \in \mathbb{H}.$$

2.4 Remarks. Let Λ be a lattice with basis (2.7). In analogy with the Basis Lemma for Lattices I.1.12, we can show that $h_1, \ldots, h_n$ in V form a basis of Λ if and only if there exists a matrix $U = (u_{\nu\mu}) \in \mathcal{U}_n$ such that

$$h_\nu = \sum_{\mu=1}^{n} u_{\mu\nu} g_\mu, \quad \nu = 1, \ldots, n.$$

Proposition 2.2c) states the independence of the choice of the basis in Lemma 2.3.

3. The Theta Transformation Formula. As an essential tool for the proof of the transformation formula, we need the

2.5 Theorem on the FOURIER Series Expansion. *Let $\varphi : \mathbb{C}^n \to \mathbb{C}$ be holomorphic and periodic in each component with period 1, i.e.*

$$\varphi(w + g) = \varphi(w) \quad \textit{for all} \quad g \in \mathbb{Z}^n.$$

Then φ has a FOURIER *series expansion of the form*

$$\varphi(w) = \sum_{h \in \mathbb{Z}^n} c_h \cdot e^{2\pi i h^{tr} w}, \tag{2.9}$$

which is absolutely and, on each compact subset of $\mathbb{C}^n$, uniformly convergent. The FOURIER *coefficients are given by*

$$c_h := \int_{[0,1]^n} \varphi(z + \xi) \cdot e^{-2\pi i h^{tr}(z+\xi)} d\xi \tag{2.10}$$

and are independent of $z \in \mathbb{C}^n$.

Proof We use an induction on n, where the case $n = 1$ is known due to classical theory (cf. A.4). Thus, let $n > 1$ and

$$w = \begin{pmatrix} w' \\ w_n \end{pmatrix}, \quad z = \begin{pmatrix} z' \\ z_n \end{pmatrix}, \quad \xi = \begin{pmatrix} \xi' \\ \xi_n \end{pmatrix}, \quad h = \begin{pmatrix} h' \\ h_n \end{pmatrix}.$$

For fixed $w_n \in \mathbb{C}$, the function $w' \mapsto \varphi(w)$ is holomorphic. According to the induction hypothesis, there exists an absolutely convergent FOURIER series expansion

$$\varphi(w) = \sum_{h' \in \mathbb{Z}^{n-1}} c_{h'}(w_n) \cdot e^{2\pi i h'^{tr} w'},$$

$$c_{h'}(w_n) = \int_{[0,1]^{n-1}} \varphi\begin{pmatrix} z' + \xi' \\ w_n \end{pmatrix} \cdot e^{-2\pi i h'^{tr}(z'+\xi')} d\xi'.$$

As φ is holomorphic, $c_{h'}$ is also holomorphic and periodic with period 1 in w_n. From the classical theory we get a FOURIER series expansion

$$c_{h'}(w_n) = \sum_{h_n \in \mathbb{Z}} c_h \cdot e^{2\pi i h_n w_n},$$

$$c_h = \int_{[0,1]} c_{h'}(z_n + \xi_n) \cdot e^{-2\pi i h_n(z_n+\xi_n)} d\xi_n$$

$$= \int_{[0,1]} \left(\int_{[0,1]^{n-1}} \varphi(z+\xi) \cdot e^{-2\pi i h^{tr}(z+\xi)} d\xi' \right) d\xi_n.$$

Since the integrand is continuous, we obtain (2.10) by FUBINI's Theorem and

$$\varphi(w) = \sum_{h' \in \mathbb{Z}^{n-1}} \left(\sum_{h_n \in \mathbb{Z}} c_h \cdot e^{2\pi i h^{tr} w} \right).$$

Now let $R > 0$ and

§ 2 Theta series as modular forms

$$M := \max\{|\varphi(w)| \; ; \; |w_j| \leq 2R+1, j = 1,\ldots, n\}.$$

For $h \in \mathbb{Z}^n$, let $z = -i2R(\operatorname{sgn} h_1, \ldots, \operatorname{sgn} h_n)^{tr}$, thus

$$h^{tr} z = -i2R(|h_1| + \ldots + |h_n|).$$

With this z, (2.10) yields

$$|c_h \cdot e^{2\pi i h^{tr} w}| \leq |c_h| \cdot e^{2\pi (|h_1 w_1| + \ldots + |h_n w_n|)}$$

$$\leq \int_{[0,1]^n} M \cdot e^{-4\pi R(|h_1| + \ldots + |h_n|)} d\xi \cdot e^{2\pi R(|h_1| + \ldots + |h_n|)}$$

$$= M \cdot e^{-2\pi R(|h_1| + \ldots + |h_n|)}$$

for $|w_j| \leq R$, $j = 1, \ldots, n$. From this, the absolute and locally uniform convergence of the series (2.9) follows immediately, because we have

$$\sum_{h \in \mathbb{Z}^n} e^{-2\pi R(|h_1| + \ldots + |h_n|)} = \left(\frac{1 + e^{-2\pi R}}{1 - e^{-2\pi R}}\right)^n. \qquad \square$$

This leads to the announced

2.6 Theta Transformation Formula. *If $S \in \operatorname{Pos}(n; \mathbb{R})$, $p, q \in \mathbb{C}^n$ and $\tau \in \mathbb{H}$, then*

$$\Theta_{-q,p}(-1/\tau; S^{-1}) = (\tau/i)^{n/2} \cdot \sqrt{\det S} \cdot e^{-2\pi i p^{tr} q} \cdot \Theta_{p,q}(\tau; S) \qquad (2.11)$$

holds. If n is odd, one has to choose the branch of the square root, which is positive, whenever τ is purely imaginary.

Proof Because of Proposition 2.2 and Lemma 2.1, we can apply the Theorem on the FOURIER Series Expansion 2.5 to

$$\varphi(p) := \Theta_{p,q}(iy; S).$$

Because of the absolute convergence of the integral, we get for $h \in \mathbb{Z}^n$

$$c_h = \sum_{g \in \mathbb{Z}^n} \int_{[0,1]^n} e^{-\pi y S[g+p] + 2\pi i (g+p)^{tr} q} \cdot e^{-2\pi i h^{tr} p} dp$$

$$= \int_{\mathbb{R}^n} e^{-\pi y S[p] + 2\pi i p^{tr} (q-h)} dp$$

$$= e^{-\pi (yS)^{-1}[h-q]} \int_{\mathbb{R}^n} e^{-\pi y S[p + i(yS)^{-1}(h-q)]} dp.$$

We choose $W \in \mathrm{GL}(n; \mathbb{R})$ satisfying $yS[W] = I$ according to the Equivalence Theorem 1.10 and substitute $p = Wu$ such that

$$dp = |\det W| \, du = y^{-n/2}(\det S)^{-1/2} du.$$

Using $W^{tr}(h - q) = v$, we get

$$\int_{\mathbb{R}^n} e^{-\pi yS[p+i(yS)^{-1}(h-q)]} dp = |\det W| \cdot \int_{\mathbb{R}^n} e^{-\pi(u+iv)^{tr}(u+iv)} du$$

$$= y^{-n/2} \cdot (\det S)^{-1/2} \cdot \prod_{j=1}^{n} \int_{-\infty}^{\infty} e^{-\pi(u_j+iv_j)^2} du_j.$$

Due to Corollary A.14, we know that each single variable integral is 1. Thus we get

$$y^{n/2} \cdot \sqrt{\det S} \cdot \Theta_{p,q}(iy; S) = \sum_{h \in \mathbb{Z}^n} e^{-\pi(yS)^{-1}[h-q]} \cdot e^{2\pi i h^{tr} p}$$
$$= e^{2\pi i q^{tr} p} \cdot \Theta_{-q,p}(i/y; S^{-1}).$$

Thus, (2.11) holds for all $\tau = iy$, $y > 0$. Because both sides of (2.11) are holomorphic in $\tau \in \mathbb{H}$, the assertion follows by the Identity Theorem A.1. □

We write down the special case of theta zero values, i.e. $p = q = 0$, in the following

2.7 Corollary. *Given $S \in \mathrm{Pos}(n; \mathbb{R})$ and $\tau \in \mathbb{H}$,*

$$\Theta(-1/\tau; S^{-1}) = (\tau/i)^{n/2} \cdot \sqrt{\det S} \cdot \Theta(\tau; S)$$

holds.

We formulate another important special case in

2.8 Corollary. *If $S \in \mathrm{Pos}(n; \mathbb{Z})$ with $\det S = 1$, then*

$$\Theta(-1/\tau; S) = (\tau/i)^{n/2} \cdot \Theta(\tau; S) \quad \text{for all } \tau \in \mathbb{H}.$$

Proof $S^{-1}[S] = S$ holds as $S \in \mathrm{GL}(n; \mathbb{Z})$. Now we use Corollary 2.7 and Proposition 2.2c). □

2.9 Remarks. a) Let Λ be a lattice in the euclidean vector space (V, σ) of the form (2.7). After normalization of the measure, the volume of a fundamental parallelotope

$$\{\gamma_1 g_1 + \ldots + \gamma_n g_n \, ; \, 0 \leq \gamma_\nu \leq 1, \, 1 \leq \nu \leq n\}$$

of Λ is equal to $\sqrt{\det S}$, where S is the corresponding GRAM matrix. We use the abbreviation $\mathrm{vol}(\Lambda)$ for this. The set

$$\Lambda^\sigma := \{v \in V \, ; \, \sigma(v, \lambda) \in \mathbb{Z} \text{ for all } \lambda \in \Lambda\}$$

§ 2 Theta series as modular forms

is again a lattice, the *dual lattice* of Λ (with respect to σ). If we choose $h_1, \ldots, h_n \in V$ satisfying $\sigma(h_i, g_j) = \delta_{ij}$, then $h_1, \ldots, h_n$ is a basis of Λ^σ. Moreover

$$\sum_{\nu=1}^{n} \sigma(g_i, g_\nu) h_\nu = g_i \quad \text{for} \quad i = 1, \ldots, n$$

shows that

$$\sum_{\nu=1}^{n} \sigma(g_i, g_\nu) \sigma(h_\nu, h_j) = \delta_{ij} \quad \text{for} \quad i, j = 1, \ldots, n$$

and that S^{-1} is the associated GRAM matrix. Thus, we can also formulate Corollary 2.7 equivalently as

$$\Theta_{\Lambda^\sigma}(-1/\tau) = (\tau/i)^{n/2} \cdot \text{vol}(\Lambda) \cdot \Theta_\Lambda(\tau).$$

In the case $\Lambda = \Lambda^\sigma$ we call Λ *self-dual* and then we also have $\text{vol}(\Lambda) = 1$.

b) In the notation (0.2), $\Theta_{p,q}(\tau; S) = \vartheta(\tau S, p, q)$ holds. Thus, (2.11) becomes a special case of (0.3). On the other hand, we have proved (0.3) for all $Z = iS$, $S \in \text{Pos}(n; \mathbb{R})$, in (2.11), thus the statement follows with the Identity Theorem in several variables for Z.

4. Even matrices. A matrix $S \in \text{Sym}(n; \mathbb{R})$ is called *even* if

$$S[g] = g^{tr} S g \in 2\mathbb{Z} \quad \text{for all} \quad g \in \mathbb{Z}^n. \tag{2.12}$$

A characterization is given by the

2.10 Lemma. *For a matrix $S = (s_{\nu\mu}) \in \text{Sym}(n; \mathbb{R})$, the following assertions are equivalent*:

(i) *S is even.*

(ii) *$s_{\nu\nu} \in 2\mathbb{Z}$ for $\nu = 1, \ldots, n$ and $s_{\nu\mu} \in \mathbb{Z}$ for all $\nu \neq \mu$.*

(iii) *$\text{tr}(ST) \in 2\mathbb{Z}$ for all $T \in \text{Sym}(n; \mathbb{Z})$.*

Proof (i) $\Longrightarrow$ (ii): If we choose $g = e_\nu$, the ν-th unit vector in (2.12), we immediately get $s_{\nu\nu} \in 2\mathbb{Z}$. Then, for $\nu \neq \mu$

$$S[e_\nu + e_\mu] = s_{\nu\nu} + s_{\mu\mu} + 2s_{\nu\mu} \in 2\mathbb{Z}$$

holds and immediately yields $s_{\nu\mu} \in \mathbb{Z}$.

(ii) $\Longrightarrow$ (iii): Writing $T = (t_{\nu\mu})$, we calculate

$$\text{tr}(ST) = \sum_{1 \le \nu \le n} s_{\nu\nu} t_{\nu\nu} + 2 \sum_{1 \le \nu < \mu \le n} s_{\nu\mu} t_{\nu\mu} \in 2\mathbb{Z}.$$

(iii) $\Longrightarrow$ (i): If $g \in \mathbb{Z}^n$, then $T = gg^{tr} \in \text{Sym}(n; \mathbb{Z})$ holds and yields

$$\text{tr}(ST) = S[g].$$

□

In particular, we immediately conclude

$$S \text{ even}, \ G \in \text{Mat}(n, m; \mathbb{Z}) \implies S[G] \text{ even.} \tag{2.13}$$

Now we consider theta series with respect to even, positive definite matrices. As an immediate conclusion from (2.12) and (2.6), we note the

2.11 Proposition. *Let $S \in \text{Pos}(n; \mathbb{Z})$ be even. Then*

$$\Theta(\tau + 1; S) = \Theta(\tau; S) \quad \text{for all} \quad \tau \in \mathbb{H}$$

holds. The FOURIER *series expansion of $\Theta(\cdot; S)$ has the form*

$$\Theta(\tau; S) = \sum_{m=0}^{\infty} \#(S, 2m) \cdot e^{2\pi i m \tau}, \quad \tau \in \mathbb{H}.$$

5. Even, unimodular, positive definite matrices. In this section we describe necessary and sufficient conditions on the rank for the existence of such matrices. We obtain them with the help of the corresponding theta series.

2.12 Theorem. *If $S \in \text{Pos}(n; \mathbb{Z})$ is even and unimodular, then n is divisible by 8.*

Proof For $\tau \in \mathbb{H}$, we define

$$\tau_1 = -\frac{1}{\tau}, \quad \tau_2 = \tau_1 + 1 = \frac{\tau - 1}{\tau}, \quad \tau_3 = -\frac{1}{\tau_2} = \frac{\tau}{1 - \tau},$$

$$\tau_4 = \tau_3 + 1 = \frac{1}{1 - \tau}, \quad \tau_5 = -\frac{1}{\tau_4} = \tau - 1, \quad \tau_6 = \tau_5 + 1 = \tau.$$

Now we apply Proposition 2.11 and Corollary 2.8 and get

$$\Theta(\tau; S) = \Theta(\tau_6; S) = (\tau/i)^{n/2} \cdot (\tau_2/i)^{n/2} \cdot (\tau_4/i)^{n/2} \cdot \Theta(\tau; S).$$

If we now choose $\tau = i$ and use the fact that $\Theta(i; S)$ is a positive real number, it follows that

$$1 = (1 + i)^{n/2} \left(\frac{1 + i}{2}\right)^{n/2} = e^{\pi i n/4}$$

if we take into account that we have to choose the branch of the root which is positive for positive arguments. Thus, n is divisible by 8. □

We now give an example for $n = 8$.

2.13 Lemma. *The matrix*

$$S_8 := \begin{pmatrix} 2I & A \\ -A & 2I \end{pmatrix}, \quad I = I^{(4)}, \quad A = \begin{pmatrix} 0 & 1 & 1 & 1 \\ -1 & 0 & -1 & 1 \\ -1 & 1 & 0 & -1 \\ -1 & -1 & 1 & 0 \end{pmatrix}, \tag{2.14}$$

is an even, unimodular and positive definite 8×8 matrix.

§ 2 Theta series as modular forms

Proof Since A is skew symmetric, S_8 is symmetric and even by Lemma 2.10. We now easily calculate
$$-A^2 = A^{tr}A = AA^{tr} = 3I.$$
Hence, (1.8) yields
$$\begin{pmatrix} 2I & A \\ -A & 2I \end{pmatrix} \left[\begin{pmatrix} I & -\tfrac{1}{2}A \\ 0 & I \end{pmatrix} \right] = \begin{pmatrix} 2I & 0 \\ 0 & \tfrac{1}{2}I \end{pmatrix}.$$

Thus, obviously, det $S_8 = 1$ holds and the matrix is also positive definite according to the Equivalence Theorem 1.10. □

Forming direct sums in the sense of (2.3) with the matrix S_8, we see that for each n divisible by 8 there is an even, unimodular, positive definite $n \times n$ matrix. Whenever $n \equiv 0 \pmod{8}$, let h_n denote the number of equivalence classes of even, unimodular, positive definite $n \times n$ matrices. This number is finite by Corollary 1.19. It is known that
$$h_8 = 1, \quad h_{16} = 2, \quad h_{24} = 24.$$
However, according to J.-P. SERRE [73], chap. 5.2,
$$h_{32} > 80,000,000.$$

2.14 Remarks. a) A matrix equivalent to S_8 was constructed in 1873 by A. KORKIN and G. ZOLOTAREFF (*Math. Ann.* **6**, 366-389). The fact that there is only one class for $n = 8$ was proved in 1938 by L.J. MORDELL (*J. Math. Pures Appl.* **17**, 41–46).
b) The result $h_{16} = 2$ was given by E. WITT (*Abh. Math. Semin. Hans. Univ.* **14**, 323–337 (1941)). A classification of all even unimodular matrices in Pos$(24; \mathbb{Z})$ is due to H. NIEMEIER (*J. Number Theory* **5**, 142–178) in 1973. Confer also J.H. CONWAY and N.J.A. SLOANE [14] chap. 18.

6. Theta series as modular forms. If k is a multiple of 4, then $(\tau/i)^k = \tau^k$ holds for all $\tau \in \mathbb{H}$. Because of Theorem 2.12, we can formulate Proposition 2.11 and Corollary 2.8 as

2.15 Theorem. *If $S \in \mathrm{Pos}\,(n; \mathbb{Z})$ is even and unimodular, then the theta series $\Theta(\tau; S)$ is a modular form of weight $n/2$ with the* FOURIER *series expansion*
$$\Theta(\tau; S) = \sum_{m=0}^{\infty} \sharp(S, 2m) \cdot e^{2\pi i m \tau}, \quad \tau \in \mathbb{H}.$$

We consider the example $\Theta(\tau; S_8) \in \mathbb{M}_4 = \mathbb{C}E_4^*$. Because of $\sharp(S, 0) = 1$, we conclude from III(2.10) the

2.16 Corollary. *One has*
$$\Theta(\tau; S_8) = E_4^*(\tau) \quad \textit{for all} \quad \tau \in \mathbb{H}$$
and
$$\sharp(S_8, 2m) = 240 \cdot \sigma_3(m) \quad \textit{for all} \quad m \geq 1.$$

We can also formulate Corollary III.2.6 as a result about the growth behavior of representation numbers.

2.17 Corollary. *Let $S \in \text{Pos}(n; \mathbb{Z})$ be even and unimodular. Then*

$$\sharp(S, 2m) = -\frac{n}{B_{n/2}} \cdot \sigma_{\frac{n}{2}-1}(m) + O\left(m^{n/4}\right) \quad \text{for} \quad m \to \infty.$$

2.18 Remarks. a) We reformulate the result using the language of lattices. Let Λ be a lattice in a euclidean vector space (V, σ). We call Λ *even* if $\sigma(\lambda, \lambda) \in 2\mathbb{Z}$ for all $\lambda \in \Lambda$. This means that the corresponding GRAM matrix S (cf. Lemma 2.3) is even. Because of Remark 2.9a) and Theorem 2.12, even self-dual lattices exist if and only if the dimension n is a multiple of 8. If Λ is an even, self-dual lattice, then the theta series $\Theta_\Lambda(\tau)$ is a modular form of weight $n/2$.

b) The Corollaries 2.16 and 2.17 and Theorem 2.15 were first proved by B. SCHOENEBERG (*Math. Ann.* **116**, 511–523 (1939); Corrigendum 780) and E. HECKE [36], 867–868.

7. A construction of the LEECH lattice. In this section, we construct an even unimodular matrix in $\text{Pos}(24; \mathbb{Z})$ which does not represent 2. This matrix belongs to the so-called LEECH *lattice*. We give a construction which goes back to J. McKAY (cf. J.H. CONWAY and N.J.A. SLOANE [14], chap. 8.5). For a prime $p > 2$, $a \in \mathbb{Z}$ is called a *quadratic residue* (mod p) if $p \nmid a$ and the congruence $x^2 \equiv a \pmod{p}$ has a solution $x \in \mathbb{Z}$. If $x^2 \equiv a \pmod{p}$ has no solution $x \in \mathbb{Z}$, we call a a *quadratic non-residue* (mod p). Thus we define the LEGENDRE symbol for $a \in \mathbb{Z}$ by

$$\left(\frac{a}{p}\right) = \begin{cases} 0, & \text{if } p|a, \\ 1, & \text{if } a \text{ quadratic residue} \pmod{p}, \\ -1, & \text{if } a \text{ quadratic non-residue} \pmod{p}. \end{cases}$$

An analogon of the matrix A in (2.14) with $p = 11$ instead of $p = 3$ is now given by

$$A := \begin{pmatrix} 0 & e^{tr} \\ -e & B \end{pmatrix} \in \text{Mat}(12; \mathbb{Z}), \quad e = \begin{pmatrix} 1 \\ \vdots \\ 1 \end{pmatrix} \in \mathbb{Z}^{11}, \quad B = \left(\left(\frac{i-j}{11}\right)\right)_{1 \le i,j \le 11}.$$

Because of $\left(\frac{-a}{11}\right) = -\left(\frac{a}{11}\right)$, the matrix A is skew-symmetric. Moreover, we verify by a direct calculation or with simple properties of the LEGENDRE symbol (cf. T.M. APOSTOL [6], chap. 9) that

$$A^{tr}A = 11I. \tag{2.15}$$

2.19 Theorem. *The matrix*

$$L_{24} := \begin{pmatrix} 4I & A - 2I \\ A^{tr} - 2I & 4I \end{pmatrix}, \quad I = I^{(12)}, \tag{2.16}$$

is even, unimodular, positive definite and satisfies

§ 2 Theta series as modular forms   297

$$\mu(L_{24}) = 4. \tag{2.17}$$

Proof Because A is skew symmetric,

$$(A^{tr} - 2I)(A - 2I) = 15I$$

follows from (2.15). Therefore, completing the square in (1.8), we get

$$\begin{pmatrix} 4I & A-2I \\ A^{tr}-2I & 4I \end{pmatrix} \left[\begin{pmatrix} I & \frac{1}{4}(2I-A) \\ 0 & I \end{pmatrix} \right] = \begin{pmatrix} 4I & 0 \\ 0 & \frac{1}{4}I \end{pmatrix}. \tag{2.18}$$

Thus L_{24} is even, unimodular and positive definite. A calculation analogously to (2.18) yields

$$L_{24} > \frac{1}{8}I.$$

For $g \in \mathbb{Z}^{24}$ with $g^{tr}g \geq 16$,

$$L_{24}[g] > \frac{1}{8}I[g] \geq 2$$

holds due to (1.14). Thus, we are left to show that

$$L_{24}[g] > 2 \quad \text{for all} \quad g \in \mathbb{Z}^{24} \quad \text{with} \quad 0 < g^{tr}g \leq 15$$

in order to prove (2.17). This is a longer, more laborious calculation, which can easily be done by a computer program. □

By III, §4, there is a unique $f \in \mathbb{M}_{12}$ with $\alpha_f(0) = 1$ and $\alpha_f(1) = 0$. Therefore, Theorem 2.15, Corollary 2.16 and Theorem 2.19 imply

2.20 Corollary. *The theta series $\Theta(\tau; S_8)^3 = \Theta(\tau; S_8 \oplus S_8 \oplus S_8)$ and $\Theta(\tau; L_{24})$ are linearly independent. One has*

$$\Theta(\tau; L_{24}) = E_4^*(\tau)^3 - 720\Delta^*(\tau) = E_{12}^*(\tau) - \frac{65\,520}{691}\Delta^*(\tau)$$

$$= \frac{7}{12}E_4^*(\tau)^3 + \frac{5}{12}E_6^*(\tau)^2,$$

$$\Delta^*(\tau) = \frac{1}{720}(\Theta(\tau; S_8 \oplus S_8 \oplus S_8) - \Theta(\tau, L_{24}))$$

and

$$\sharp(L_{24}, 2m) = \frac{65\,520}{691}(\sigma_{11}(m) - \tau(m)) \quad \text{for all} \quad m \geq 1.$$

From Corollary 2.20, Corollary 2.16, Proposition III.4.11, and (2.4), we directly conclude

2.21 Corollary. *For positive $k \equiv 0 \pmod 4$, each $f \in \mathbb{M}_k$ is a linear combination of theta series associated with even, unimodular, positive definite $2k \times 2k$ matrices.*

2.22 Remark. The lattice belonging to the matrix L_{24} was discovered by J. LEECH in 1967 (*Canadian J. Math.* **19**, 251–265) in connection with the densest sphere packing in $\mathbb{R}^{24}$. From the classification of H. NIEMEIER mentioned in Remark 2.14, it follows that there is only one class $\langle S \rangle$ of even unimodular matrices in Pos $(24; \mathbb{Z})$ with $\mu(S) > 2$. So the class $\langle L_{24} \rangle$ is uniquely determined by the condition (2.17). Compare J.H. CONWAY and N.J.A. SLOANE [14] for the construction and classification of lattices. In particular, we find a construction of the LEECH lattice using codes and using the KNESER neighborhood method there.

8*. Extremal lattices. The minimum $\mu(S)$ (cf. (1.19)) of a matrix $S \in \text{Pos}\,(n; \mathbb{Z})$ with $\det S = 1$ cannot become arbitrarily large. From HERMITE's inequality in Corollary 1.18 it follows that

$$\mu(S) \leq \left(\tfrac{4}{3}\right)^{(n-1)/2}.$$

If S is even, we can improve this bound substantially.

2.23 Proposition. *Let $S \in \text{Pos}\,(n; \mathbb{Z})$ be even and unimodular. Then*

$$\mu(S) \leq 2 \left[\tfrac{n}{24}\right] + 2$$

holds.

Proof We give the proof indirectly and assume that $\mu(S) > 2 \left[\tfrac{n}{24}\right] + 2$ holds. Then we consider the theta series

$$\Theta(\cdot\,; S) \in \mathbb{M}_k\,, \ k = n/2,$$

by Theorem 2.15. The FOURIER coefficients satisfy

$$\sharp(S, 0) = 1\,, \ \sharp(S, 2m) = 0\,, \ 1 \leq m \leq \left[\tfrac{k}{12}\right] + 1.$$

$\dim \mathbb{M}_k = \left[\tfrac{k}{12}\right] + 1$ holds due to III, §4 for $k \equiv 0 \pmod 4$. This contradicts Corollary III.6.12. □

We call an even, unimodular matrix $S \in \text{Pos}\,(n; \mathbb{Z})$ with $\mu(S) = 2 \left[\tfrac{n}{24}\right] + 2$ *extremal*. If Λ is a self-dual even lattice in a euclidean vector space (V, σ), then Λ is called *extremal* if the associated GRAM-matrix is extremal, i.e.

$$\sigma(\lambda, \lambda) \geq 2 \left[\frac{n}{24}\right] + 2 \quad \text{for all} \quad \lambda \in \Lambda, \ \lambda \neq 0.$$

Two examples of extremal matrices are S_8, $S_8 \oplus S_8$. By Theorem 2.19, the LEECH lattice L_{24} is also extremal.

2.24 Exercises.
1) Let $k \equiv 0 \pmod 4$, $k \geq 12$. Then the $\mathbb{Z}$-module spanned by the theta series $\Theta(\tau; S)$ associated with even unimodular $S \in \text{Pos}\,(2k; \mathbb{Z})$ is a proper subset of $\mathbb{M}_k^{\mathbb{Z}}$.
2) Let $R \in \text{Pos}\,(24; \mathbb{Z})$ be even and unimodular. Then the action of the HECKE operator T_n, $n \geq 1$, on the corresponding theta series is given by:

§ 2 Theta series as modular forms

$$T_n\Theta(\tau; R) = \left(\sigma_{11}(n) - \frac{\#(R, 2n)}{720}\right)\Theta(\tau; L_{24}) + \frac{\#(R, 2n)}{720}\Theta(\tau; S_8)^3.$$

In particular, for $n > 1$ the theta series $\Theta(\tau; R)$ is not an eigenform with respect to T_n and $\#(R, 2n) > 0$ holds.

3) Calculate $E_k^*(\tau)$ for $k = 12, 16, 20$ as a linear combination of theta series.

4) Let $S \in \text{Pos}(2k; \mathbb{Q})$ and $q \in \mathbb{Z}$ such that qS and qS^{-1} are even. Then

$$\Theta(\cdot, S)|_k \begin{pmatrix} 1 & q \\ 0 & 1 \end{pmatrix} = \Theta(\cdot; S)|_k \begin{pmatrix} 1 & 0 \\ q & 1 \end{pmatrix} = \Theta(\cdot; S).$$

5) Let χ_ϑ be the character of Γ_ϑ defined in III, §7. Then

$$\vartheta^{4k} = \Theta(\cdot; I^{(4k)}) \in \mathbb{M}_{2k}(\Gamma_\vartheta, \chi_\vartheta^k).$$

6) The map $\chi: \Gamma_0[4] \to \mathbb{C}$, $\chi\begin{pmatrix} a & b \\ c & d \end{pmatrix} := i^{d-1}$, is an abelian character. For $k \in \mathbb{N}$ one has

$$\vartheta(2\tau)^{2k} = \Theta(\tau; 2I^{(2k)}) \in \mathbb{M}_k(\Gamma_0[4], \chi^k).$$

7) For all $m \in \mathbb{N}$ one has

$$\sum_{r+s=m} \#(S_8, 2r) \cdot \#(S_8, 2s) = 480 \cdot \sigma_7(m).$$

8) Let p be a prime, $p + 1 = k = 4r$, $I = I^{(k)}$, $e = (1, \ldots, 1)^{tr} \in \mathbb{Z}^p$ as well as $A = \begin{pmatrix} 0 & e^{tr} \\ -e & B \end{pmatrix}$, $B = \left(\left(\frac{i-j}{p}\right)\right)_{1 \leq i,j \leq p} \in \text{Mat}(p; \mathbb{Z})$. Then $S_{2k} = \begin{pmatrix} 2I & A \\ A^{tr} & 2rI \end{pmatrix}$ is even unimodular and positive definite. In particular, $\#(S_{16}, 2m) = 480 \cdot \delta_7(m)$ holds all $m \in \mathbb{N}$.

9) $S_8^{-1} = S_8$ and $J = \begin{pmatrix} 0 & -I \\ I & 0 \end{pmatrix} \in \text{Aut } S_8$, if $I = I^{(4)}$

10) Determine the characteristic polynomial of S_8, S_{16} and L_{24}.

11) Compute $S_8^{1/2}$, $S_{16}^{1/2}$ and $L_{24}^{1/2}$.

12) Let $K = \mathbb{Q}(\sqrt{-m})$, $m \in \mathbb{N}$ square-free, be an imaginary-quadratic number field of discriminant d_K and ring of integers O_K, i.e. $O_K = \mathbb{Z} + \mathbb{Z}(1 + \sqrt{-m})/2$ and $d_K = -m$, whenever $m \equiv 3 \pmod 4$, resp. $O_K = \mathbb{Z} + \mathbb{Z}\sqrt{-m}$, $d_K = -4m$ otherwise. Consider the $\mathbb{Z}$-lattice $\Lambda = O_K^k$ of rank $2k$ with the bilinear form

$$\sigma(\lambda, \mu) = \overline{\lambda}^{tr} H \mu, \quad \lambda, \mu \in \Lambda,$$

where $H = \overline{H}^{tr} \in \text{Mat}(k; \mathbb{C})$ is a positive definite Hermitian matrix, i.e. all eigenvalues of H are positive.

a) Show that

$$\Lambda^\sigma = H^{-1} \frac{2}{\sqrt{d_K}} O_K^k.$$

b) Show that Λ is an even lattice if and only if

$$H \in \text{Mat}(k; \frac{2}{\sqrt{d_K}} O_K).$$

c) If Λ is even and $\det H = \left(2/\sqrt{d_K}\right)^k$ then

$$\Theta_\Lambda \in \mathbb{M}_k \quad \text{cf. (2.8)}.$$

d) A matrix H such that Λ is even and $\det H = \left(2/\sqrt{d_K}\right)^k$ exists if and only if $k \equiv 0 \pmod 4$.

13) Consider Jacobi's theta series from sect. 2 in I, §8 and $\lambda(\tau)$ from Exercise III.7.20 26). Show that $f(\tau) = (\vartheta_2(0; \tau)/\vartheta_3(0; \tau))^4 \in \mathbb{K}(\Gamma[2])$ and represent f as a rational function in λ.

§ 3 A special case of Siegel's Main Theorem

In this paragraph, we shall prove a special case of the so-called Siegel's *Main Theorem*. For $k \equiv 0 \pmod 4$, the Eisenstein series is the weighted average of the theta series associated with the classes of the even, unimodular, positive definite quadratic forms of dimension $2k$. The proof is based on the Hecke theory.

1. Reduction to diagonal form. For $q \in \mathbb{N}$ and $A, B \in \text{Mat}(n, m; \mathbb{Z})$, we define the *congruence* (mod q) by

$$A \equiv B \pmod{q} \iff \frac{1}{q}(A - B) \in \text{Mat}(n, m; \mathbb{Z}), \tag{3.1}$$

thus if the congruence holds in each component. Under obvious conditions on the dimensions, we conclude

$$A \equiv A' \pmod{q}, \ B \equiv B' \pmod{q} \implies AB \equiv A'B' \pmod{q}. \tag{3.2}$$

The goal of this section is to find a simple representative in each class of integral symmetric matrices with respect to the congruence (mod q).

3.1 Proposition. *Let $q \in \mathbb{N}$ and $S = (s_{ij}) \in \text{Sym}(n; \mathbb{Z})$, $n \geq 2$, be such that q and s_{11} are coprime. Then there is a matrix $U \in \mathcal{U}_n$ and a matrix $R \in \text{Sym}(n-1; \mathbb{Z})$ satisfying*

$$S[U] \equiv \begin{pmatrix} s_{11} & 0 \\ 0 & R \end{pmatrix} \pmod{q}.$$

Proof Because $\gcd(s_{11}, q) = 1$ there are $u_j \in \mathbb{Z}$ with $s_{11} u_j + s_{1j} \equiv 0 \pmod{q}$ for $2 \leq j \leq n$. Then the assertion follows with

$$U = \begin{pmatrix} 1 & u^{tr} \\ 0 & I \end{pmatrix}, \quad u^{tr} = (u_2, \ldots, u_n). \qquad \square$$

§ 3 A special case of SIEGEL's Main Theorem

Now, if q is a power of an odd prime, we get a representative in diagonal form in each class.

3.2 Theorem. *Let p be an odd prime, $\ell \in \mathbb{N}$ and $S \in \text{Sym}(n; \mathbb{Z})$. Then there is a matrix $U \in \mathcal{U}_n$ and a diagonal matrix D such that*

$$S[U] \equiv D \pmod{p^\ell}.$$

Proof Let $n > 1$ and $S \neq 0$. Since S can be replaced by $\frac{1}{\delta(S)} S$, $\delta(S) = \gcd(s_{ij})$, we may assume $\delta(S) = 1$ without loss of generality.

If any diagonal element of S is not divisible by p, we choose a permutation matrix V, then apply Proposition 3.1 to $S[V]$. Otherwise, p is a divisor of s_{jj} for $j = 1, \ldots, n$. Because of $\delta(S) = 1$ there exist i, j, $i \neq j$, with $p \nmid s_{ij}$. Now we choose a unimodular $V \in \text{GL}(n; \mathbb{Z})$ with $e_i + e_j$ as its first column. Then, the first diagonal element of $S[V]$ is $S[e_i + e_j] = s_{ii} + s_{jj} + 2 s_{ij}$ and thus not divisible by p because of $p \neq 2$. Now we can again apply Proposition 3.1 to $S[V]$ and obtain the assertion by induction. □

3.3 Remarks. a) Theorem 3.2 was proved by H. MINKOWSKI ([58], vol. I, 22) in the context of his dissertation. For powers of 2 it is no longer possible to find a representative in diagonal form in general. Here we can only find a generalized diagonal form, i.e. having numbers or 2×2 blocks on the diagonal (cf. Exercise 3.20 1)).

b) Later on we need the statement of the Theorem 3.2 only for $\ell = 1$. In this case, a simple proof uses the normal form of quadratic forms over fields of characteristic $\neq 2$ and the fact that the mapping

$$\text{SL}(n; \mathbb{Z}) \longrightarrow \text{SL}(n; \mathbb{Z}/p\mathbb{Z}),$$

reducing in every component (mod p), is a surjective homorphism of groups. Compare W. SCHARLAU [70], 7–8.

2. Congruence mod p of classes. In this section, we show that, in all classes of unimodular symmetric matrices we can find a representative which has a different canonical form mod p.

First we answer the question under which conditions a matrix has an inverse mod q.

3.4 Lemma. *If $q \in \mathbb{N}$, the following assertions are equivalent for $A \in \text{Mat}(n; \mathbb{Z})$:*
(i) *There exists a matrix $B \in \text{Mat}(n; \mathbb{Z})$ with $AB \equiv BA \equiv I \pmod{q}$.*
(ii) $\det A$ *and q are coprime.*

Proof (i) $\Longrightarrow$ (ii): It follows that $(\det A) \cdot (\det B) \equiv 1 \pmod{q}$.
(ii) $\Longrightarrow$ (i): We choose $\alpha \in \mathbb{Z}$ with $\alpha \cdot \det A \equiv 1 \pmod{q}$. Then we get the assertion with $B = \alpha A^\sharp$, where $A^\sharp$ is the adjugate of A, i.e. the matrix with the $(n-1) \times (n-1)$ subdeterminants of A as entries satisfying $A A^\sharp = (\det A) I$. □

Now we show that certain quadratic congruences mod p always have a solution.

3.5 Lemma. *Let p be a prime number, $a, b, c \in \mathbb{Z}$ with $p \nmid a$ and $p \nmid b$. Then the congruence*

$$ax^2 + by^2 \equiv c \pmod{p}$$

has a solution $(x, y) \in \mathbb{Z} \times \mathbb{Z}$.

Proof The intersection of the sets

$$\{ax^2 + p\mathbb{Z} \, ; \, x \in \mathbb{Z}\} \cap \{c - by^2 + p\mathbb{Z} \, ; \, y \in \mathbb{Z}\}$$

is not empty because both sets consist of $\frac{p+1}{2}$ or 2 residue classes (mod p) for $p > 2$ and $p = 2$, respectively, because $p \nmid a$ and $p \nmid b$. □

We formulate the first reduction step in

3.6 Proposition. *Let* $q \in \mathbb{N}$ *be odd and* $S = (s_{ij}) \in \operatorname{Sym}(n;\mathbb{Z})$, $n \geq 2$. *Let* $s_{11} \equiv 0 \pmod{q}$ *and* $\alpha = \gcd(s_{12}, \ldots, s_{1n})$ *be coprime to* q. *Then there is a matrix* $U \in \mathcal{U}_n$ *with* $(1, 0, \ldots, 0)^{tr}$ *as its first column and a matrix* $R \in \operatorname{Sym}(n-2;\mathbb{Z})$ *with the property*

$$S[U] \equiv \begin{pmatrix} Q & 0 \\ 0 & R \end{pmatrix} \pmod{q}, \quad Q = \begin{pmatrix} 0 & \alpha \\ \alpha & 0 \end{pmatrix}.$$

Proof Let $s = (s_{12}, \ldots, s_{1n})^{tr}$. Then Lemma 1.3 leads to a matrix $V \in \mathcal{U}_{n-1}$ such that $V^{tr} s = (\alpha, 0, \ldots, 0)^{tr}$. Now we get

$$T = S\left[\begin{pmatrix} 1 & 0 \\ 0 & V \end{pmatrix}\right] = \begin{pmatrix} T_1 & T_2 \\ T_2^t & T_3 \end{pmatrix}, \quad T_1 \equiv \begin{pmatrix} 0 & \alpha \\ \alpha & \beta \end{pmatrix} \pmod{q}.$$

Due to Lemma 3.4, there exists some $B \in \operatorname{Mat}(2;\mathbb{Z})$ such that $T_1 B \equiv I \pmod{q}$. We choose $\gamma \in \mathbb{Z}$ satisfying $2\alpha\gamma + \beta \equiv 0 \pmod{q}$. Then we get

$$T[U] \equiv \begin{pmatrix} Q & 0 \\ 0 & R \end{pmatrix} \pmod{q} \quad \text{with} \quad U = \begin{pmatrix} A & -BT_2 \\ 0 & I \end{pmatrix}, \quad A = \begin{pmatrix} 1 & \gamma \\ 0 & 1 \end{pmatrix}. \quad \square$$

An obvious modification of the proof together with an induction on n yields an analogous result for the prime $q = 2$ (cf. Remark 3.3). We formulate the result as

3.7 Lemma. *Let* $S \in \operatorname{Sym}(n;\mathbb{Z})$ *be even. Then there exists a matrix* $U \in \mathcal{U}_n$ *such that*

$$S[U] = [Q_1, \ldots, Q_r, 0] \pmod{2}, \quad Q_j = \begin{pmatrix} 0 & 1 \\ 1 & 0 \end{pmatrix}, \quad j = 1, \ldots, r,$$

where $2r$ is the rank of S over $\mathbb{Z}/2\mathbb{Z}$.

As a generalization of Proposition 3.6, we now obtain the

3.8 Theorem. *Let* $p > 2$ *be a prime and* $S \in \operatorname{Sym}(n;\mathbb{Z})$ *with even* $n = 2k$. *If* $(-1)^k \det S$ *is a quadratic residue* (mod p), *then there exist* $\alpha_1, \ldots, \alpha_k \in \mathbb{Z}$ *and a matrix* $U \in \mathcal{U}_n$ *with the property*

$$S[U] \equiv \begin{pmatrix} 0 & D \\ D & 0 \end{pmatrix} \pmod{p}, \quad D = [\alpha_1, \ldots, \alpha_k]. \tag{3.3}$$

Proof We use an induction on k. By Theorem 3.2, we may assume that S is a diagonal matrix (mod p) with $d_1, \ldots, d_n$ as diagonal elements. We first show that a

$$\text{primitive } g \in \mathbb{Z}^n \text{ with } S[g] \equiv 0 \pmod{p} \tag{3.4}$$

exists. For this purpose, we distinguish two cases:
(i) There exists some $j > 1$ such that $-d_1 d_j$ is a quadratic residue (mod p). Then the congruence $\alpha^2 + d_1 d_j \equiv 0$ (mod p) has a solution. If necessary, we replace α by $\alpha + \beta p$ with a suitable $\beta \in \mathbb{Z}$. Thus, according to Lemma II.3.3, we may assume $\gcd(\alpha, d_1) = 1$. The vector $g = \alpha e_1 + d_1 e_j$ then satisfies (3.4).
(ii) $-d_1 d_j$ is a quadratic non-residue (mod p) for all $j = 2, \ldots, n$. Then $n \geq 4$ holds. By Lemma 3.5, there exist $\alpha, \beta \in \mathbb{Z}$ satisfying

$$d_1 \alpha^2 + d_2 \beta^2 + d_3 \equiv 0 \pmod{p}.$$

The vector $g = (\alpha, \beta, 1, 0, \ldots, 0)^{tr}$ then satisfies (3.4).

Now we choose g from (3.4) as the first column of a matrix $U \in \mathcal{U}_n$ and can then immediately assume

$$S[U] \equiv \begin{pmatrix} Q & 0 \\ 0 & R \end{pmatrix} \pmod{p}, \quad Q = \begin{pmatrix} 0 & \alpha_1 \\ \alpha_1 & 0 \end{pmatrix}, \quad R \in \text{Sym}(n-2; \mathbb{Z})$$

due to Proposition 3.6. This settles the case $k = 1$. Since $(-1)^{k-1} \det R$ is also a quadratic residue (mod p) in the case $k > 1$, we can already modify U by induction hypothesis such that R (mod p) has the desired form. If we multiply U from the right by a suitable permutation matrix P, we can obtain that $S[UP]$ has the form (3.3). □

3.9 Remark. In the language of quadratic forms, Theorem 3.8 can be formulated as follows (W. SCHARLAU [70], II, §3):

Let $\mathbb{F}_q$ be a finite field with char $\mathbb{F}_q \neq 2$ *and* $S \in \text{Sym}(4k; \mathbb{F}_q)$ *such that* $\det S \neq 0$ *is a square in* $\mathbb{F}_q$. *Then the quadratic space* $(\mathbb{F}_q^{4k}, S)$ *is hyperbolic.*

3. Representations mod p. In this section, we deal with representations and representation numbers (mod p), where p is always supposed to be an odd prime. We recall the definition of $\mathcal{D}(S, T)$ and Aut S, $S \in \text{Pos}(n; \mathbb{Z})$, from Corollary 1.15. In this section, we use the abbreviations
$$A_p(S) := \{H\mathcal{U}_n \ ; H \in \text{Mat}(n; \mathbb{Z}), \ |\det H| = p^{n/2}, \ S[H] \equiv 0 \pmod{p}\}. \tag{3.5}$$

3.10 Lemma. *Let $S_1, \ldots, S_h$ be representatives of the classes of even, unimodular, positive definite $n \times n$ matrices with $n = 2k$. Furthermore, let S be any matrix in one of the classes and $p > 2$ a prime.*
a) *If $H \in \text{Mat}(n; \mathbb{Z})$ satisfies $|\det H| = p^k$ and $S[H] \equiv 0 \pmod{p}$, then*

$$pH^{-1} \in \text{Mat}(n; \mathbb{Z}).$$

b) *The mapping*

$$\bigcup_{j=1}^{h} \{G\operatorname{Aut} S_j\,;\, G \in \mathcal{D}(S, pS_j)\} \longrightarrow A_p(S), \quad G\operatorname{Aut} S_j \longmapsto G\mathcal{U}_n,$$

is a bijection.

c) *The mapping*

$$\bigcup_{j=1}^{h} \{(\operatorname{Aut} S_j)G\,;\, G \in \mathcal{D}(S_j, pS)\} \longrightarrow A_p(S), \quad (\operatorname{Aut} S_j)G \longmapsto pG^{-1}\mathcal{U}_n,$$

is a bijection.

Proof a) Let $S[H] = pT$. From the calculation of the determinant, $T \in \mathcal{U}_n$ follows. Thus, we get

$$pH^{-1} = T^{-1}H^{tr}S \in \operatorname{Mat}(n; \mathbb{Z}).$$

b) From $S[G] = pS_j$ and $\det S = \det S_j = 1$, it follows that $|\det G| = p^k$. As $\operatorname{Aut} S_j \subseteq \mathcal{U}_n$, the mapping which we will call Ψ, is well-defined.

Let $G \in \mathcal{D}(S, pS_i)$ and $H \in \mathcal{D}(S, pS_j)$ be given such that $G\mathcal{U}_n = H\mathcal{U}_n$. Then there exists some $U \in \mathcal{U}_n$ satisfying $G = HU$. Thus, we obtain

$$S_i = \frac{1}{p}S[G] = \frac{1}{p}S[HU] = S_j[U].$$

Since $S_1, \ldots, S_h$ are representatives of the classes, we first get $i = j$ and then also $U \in \operatorname{Aut} S_i$. Hence, Ψ is injective.

Now let $H\mathcal{U}_n \in A_p(S)$ be given. Then, $T := \frac{1}{p}S[H]$ is even, unimodular and positive definite. Thus, there exists some $U \in \mathcal{U}_n$ and some $1 \leq j \leq h$ with $S_j = T[U]$. Therefore, $HU \in \mathcal{D}(S, pS_j)$ follows and

$$\Psi(HU\operatorname{Aut} S_j) = H\mathcal{U}_n.$$

Hence, ψ is also surjective.

c) The proof is carried out analogously to b). $\square$

If we calculate the order of the set $A_p(S)$, the above observations lead to

3.11 Corollary. *Under the assumptions of Lemma* 3.10 *one has*

$$\#A_p(S) = \sum_{j=1}^{h} \frac{\#(S, pS_j)}{\#(S_j, S_j)} = \sum_{j=1}^{h} \frac{\#(S_j, pS)}{\#(S_j, S_j)}.$$

The goal of this section is to explicitly determine the order of $A_p(S)$. For this, we need the

3.12 Proposition. *Let p be a prime and $H \in \operatorname{Mat}(n; \mathbb{Z})$ with the properties*

$$|\det H| = p^r, \quad r \leq n, \quad pH^{-1} \in \operatorname{Mat}(n; \mathbb{Z}).$$

§ 3 A special case of SIEGEL's Main Theorem   305

Then there are exactly p^{n-r} vectors $g \in \mathbb{Z}^n$ (mod p) such that $H^{-1}g \in \mathbb{Z}^n$.

Proof We use an induction on n, where $n = 1$ is trivial. Because of Theorem 1.6, we may assume that $H = (h_{ij})$ is in the upper triangular form (1.1). Then, $h_{11} = 1$ or $h_{11} = p$ holds.

(i) $H = \begin{pmatrix} 1 & 0 \\ 0 & H' \end{pmatrix}$, $H' \in \text{Mat}(n-1; \mathbb{Z})$, $\det H' = p^r$, $pH'^{-1} \in \text{Mat}(n-1; \mathbb{Z})$. Then we have exactly the possibilities

$$g = \begin{pmatrix} \gamma \\ g' \end{pmatrix}, \ \gamma \in \mathbb{Z}, \ g' \in \mathbb{Z}^{n-1}, \ H'^{-1}g' \in \mathbb{Z}^{n-1}$$

for g. But by the induction hypothesis, there are p^{n-1-r} possibilities for g' (mod p).

(ii) $H = \begin{pmatrix} p & h^{tr} \\ 0 & H' \end{pmatrix}$, $h \in \mathbb{Z}^{n-1}$, $\det H' = p^{r-1}$, $pH'^{-1} \in \text{Mat}(n-1; \mathbb{Z})$.

Since pH^{-1} is an integral matrix, it follows that $H'^{tr-1}h \in \mathbb{Z}^{n-1}$. So there are the possibilities

$$\begin{pmatrix} p\gamma + h^{tr}H'^{-1}g' \\ g' \end{pmatrix}, \ g' \in \mathbb{Z}^{n-1}, \ H'^{-1}g' \in \mathbb{Z}^{n-1}, \ \gamma \in \mathbb{Z}.$$

for g. However, according to the induction hypothesis, there are exactly p^{n-r} possibilities mod p for g. □

Furthermore, we need to know the representation numbers of 0 by S (mod p).

3.13 Proposition. *Let $p > 2$ be a prime number and $S \in \text{Sym}(n; \mathbb{Z})$, $n = 2k$ be even and such that $(-1)^k \det S$ is a quadratic residue (mod p). Then the number of vectors $g \in \mathbb{Z}^n$ (mod p) with $S[g] \equiv 0$ (mod p) is equal to*

$$p^{2k-1} + p^k - p^{k-1}.$$

Proof According to Theorem 3.8, we may assume without loss of generality that we have $S \equiv \begin{pmatrix} 0 & D \\ D & 0 \end{pmatrix}$ (mod p) with a diagonal matrix D and $p \nmid \det D$. Then, we write $g = \begin{pmatrix} a \\ b \end{pmatrix}$, $a, b \in \mathbb{Z}^k$, and get

$$S[g] \equiv 2a^{tr}Db \ (\text{mod } p).$$

If $a \equiv 0$ (mod p), we have p^k possibilities for b (mod p). But if $a \not\equiv 0$ (mod p) holds, then we have exactly p^{k-1} possibilities for b (mod p) because $p \nmid \det D$. In this case, there are $p^k - 1$ possibilities for a (mod p). □

Now we can explicitly calculate the number $\sharp A_p(S)$ we are looking for.

3.14 Theorem. *Let $p > 2$ be a prime, $n = 2k$ be even, and $S \in \text{Sym}(n; \mathbb{Z})$ such that $(-1)^k \det S$ is a quadratic residue (mod p).*
a) *One has*

$$\sharp A_p(S) = a(p,k) := \prod_{j=0}^{k-1}(p^j + 1).$$

b) *If $k > 1$ and $g \in \mathbb{Z}^n$, $g \not\equiv 0 \pmod{p}$, with $S[g] \equiv 0 \pmod{p}$, then there are exactly $a(p, k-1)$ left cosets $H\mathcal{U}_n \in A_p(S)$ with the property $H^{-1}g \in \mathbb{Z}^n$.*

Proof We use an induction on k. In the case $k = 1$, Theorem 3.8 allows us to assume $S \equiv \begin{pmatrix} 0 & s \\ s & 0 \end{pmatrix} \pmod{p}$ with $p \nmid s$. Now, using the normal form (1.1), we easily compute that only the two left cosets

$$\begin{pmatrix} 1 & 0 \\ 0 & p \end{pmatrix} \mathcal{U}_2 \quad \text{and} \quad \begin{pmatrix} p & 0 \\ 0 & 1 \end{pmatrix} \mathcal{U}_2$$

belong to $A_p(S)$.

Now we consider $k > 1$. We calculate the order m of the set $\mathcal{M}$ defined by

$$\{(H\mathcal{U}_n, g \pmod{p})) \, ; \, H\mathcal{U}_n \in A_p(S), \, g \in \mathbb{Z}^n, \, g \not\equiv 0 \pmod{p}, \, H^{-1}g \in \mathbb{Z}^n\}$$

in two different ways. First of all note that for the vectors g appearing in the set $\mathcal{M}$, $S[g] \equiv 0 \pmod{p}$ already holds because of $S[H] \equiv 0 \pmod{p}$.

(i) For $H\mathcal{U}_n \in A_p(S)$ it follows that $pH^{-1} \in \text{Mat}(n; \mathbb{Z})$ according to Lemma 3.10. By Proposition 3.12, there are exactly p^k integral vectors $g \pmod{p}$ with $H^{-1}g \in \mathbb{Z}^n$. Thus, we get

$$m = (p^k - 1) \cdot \sharp A_p(S).$$

(ii) Let $g \in \mathbb{Z}^n$, $g \not\equiv 0 \pmod{p}$, with $S[g] \equiv 0 \pmod{p}$. Following Lemma 1.3, we choose a matrix $U \in \mathcal{U}_n$ with $U^{-1}g = (\delta, 0, \ldots, 0)^{tr}$, thus $p \nmid \delta$. Then the $(1,1)$-coefficient of $S[U]$ is equal to $\delta^{-2}S[g]$ and therefore divisible by p. After a possible modification of U as in Proposition 3.6, we may immediately assume

$$g = \delta e_1 \quad \text{and} \quad S \equiv \begin{pmatrix} Q & 0 \\ 0 & R \end{pmatrix} \pmod{p}, \quad Q = \begin{pmatrix} 0 & \alpha \\ \alpha & 0 \end{pmatrix}, \quad p \nmid \alpha,$$

Here, $R \in \text{Sym}(n-2; \mathbb{Z})$ and $(-1)^{k-1} \det R$ is again a quadratic residue $\pmod{p}$. We now consider all the matrices H in the normal form (1.1) satisfying

$$\det H = p^k, \quad S[H] \equiv 0 \pmod{p} \quad \text{and} \quad H^{-1}g \in \mathbb{Z}^n.$$

The last condition and (1.1) state that H is of the form $\begin{pmatrix} 1 & 0 \\ 0 & * \end{pmatrix}$. If we now use $S[H] \equiv 0 \pmod{p}$ and the special form of $S \pmod{p}$, we can easily calculate that there are only the following possibilities for H:

$$H = \begin{pmatrix} A & 0 \\ 0 & G \end{pmatrix}, \quad A = \begin{pmatrix} 1 & 0 \\ 0 & p \end{pmatrix}; \quad \begin{array}{l} G \in \text{Mat}(n-2; \mathbb{Z}) \text{ of the form (1.1)}, \\ \det G = p^{k-1}, \quad R[G] \equiv 0 \pmod{p}. \end{array}$$

By the induction hypothesis, for fixed g, there are $a(p, k-1)$ possibilities for the left cosets $H\mathcal{U}_n$. In particular, b) has already been proved. By Proposition 3.13, we

§ 3 A special case of Siegel's Main Theorem

conclude that
$$m = (p^{2k-1} + p^k - p^{k-1} - 1) \cdot a(p, k-1).$$

The two equations for m lead to
$$\#A_p(S) = (p^{k-1} + 1) \cdot a(p, k-1) = a(p, k). \qquad \square$$

4. The action of the Hecke operators T_p on theta series. By Corollary 2.21, we already know that the image of a theta series under a Hecke operator can again be represented as a linear combination of theta series. Quite explicitly, we demonstrate

3.15 Theorem. *Let $S_1, \ldots, S_h$ be representatives of the classes of even, unimodular, positive definite $n \times n$ matrices with $n = 2k$. Let S be any matrix in one of the classes and $p > 2$ a prime. Then*

$$T_p \Theta(\cdot; S) = \left(\prod_{j=0}^{k-2} (p^j + 1) \right)^{-1} \cdot \sum_{j=1}^{h} \frac{\#(S, pS_j)}{\#(S_j, S_j)} \cdot \Theta(\cdot; S_j). \tag{3.6}$$

Proof We first define
$$f(\tau) := \sum_{j=1}^{h} \frac{\#(S, pS_j)}{\#(S_j, S_j)} \cdot \Theta(\tau; S_j) \in \mathbb{M}_k.$$

If G_j runs through a system of representatives of left cosets G Aut S_j with the property $G \in \mathcal{D}(S, pS_j)$, we conclude

$$f(\tau) = \sum_{j=1}^{h} \sum_{G_j} \sum_{g \in \mathbb{Z}^n} e^{\pi i \tau S[G_j][g]/p} = \sum_{H\mathcal{U}_n \in A_p(S)} \sum_{g \in \mathbb{Z}^n} e^{\pi i \tau S[Hg]/p},$$

where we use Lemma 3.10b) in the last step. We note that

$$S[Hg] \equiv 0 \pmod{2p} \quad \text{for all} \quad H\mathcal{U}_n \in A_p(S) \quad \text{and} \quad g \in \mathbb{Z}^n.$$

Let us denote the Fourier coefficients of f by $\alpha_f(m)$. Then

$$\alpha_f(m) = \#\{(H\mathcal{U}_n, g) \, ; \, g \in \mathcal{D}(S, 2mp), H\mathcal{U}_n \in A_p(S), H^{-1}g \in \mathbb{Z}^n\}.$$

There are exactly $\#(S, 2m/p)$ vectors $g \in \mathcal{D}(S, 2mp)$ with $g \equiv 0 \pmod{p}$. Any such g satisfies $H^{-1}g \in \mathbb{Z}^n$ for all $H\mathcal{U}_n \in A_p(S)$ due to Lemma 3.10a), so Theorem 3.14 shows that there are exactly $a(p, k)$ possibilities for $H\mathcal{U}_n$.

We have exactly $\#(S, 2mp) - \#(S, 2m/p)$ vectors $g \in \mathcal{D}(S, 2mp)$ with the property $g \not\equiv 0 \pmod{p}$. Given any such g, Theorem 3.14b) demonstrates that there are exactly $a(p, k-1)$ left cosets $H\mathcal{U}_n \in A_p(S)$ with $H^{-1}g \in \mathbb{Z}^n$.

Taking the value for $a(p, k)$ from Theorem 3.14, we conclude

$$\alpha_f(m) = \sharp(S, 2m/p) \cdot a(p, k) + (\sharp(S, 2mp) - \sharp(S, 2m/p)) \cdot a(p, k-1)$$
$$= a(p, k-1) \cdot \left(\sharp(S, 2mp) + p^{k-1}\sharp(S, 2m/p)\right).$$

Using Proposition 2.11, we conclude from Corollary IV.1.3 that the FOURIER coefficients of $\frac{1}{a(p,k-1)} \cdot f$ and $T_p\Theta(\cdot; S)$ coincide. From the uniqueness of the FOURIER expansion, we get (3.6). □

3.16 Remarks. a) The formula (3.6) is also valid for $p = 2$. However, the proof requires more effort. The proof is sketched in the Exercise 3.20.

b) The formula (3.6) is theoretically explicit. For practical calculations, however, it is challenging, due to the size of the class number in sect. 5 of §2. The linear combination in (3.6) is not unique in general, since the theta series are linearly dependent in general.

5. The EISENSTEIN series as a linear combination of theta series. In this section we prove a special case of the analytic version of SIEGEL's *Main Theorem*. The EISENSTEIN series is the weighted average of the theta series.

3.17 SIEGEL's Main Theorem. *Let $S_1, \ldots, S_h$ be representatives of the classes of even, unimodular, positive definite $n \times n$ matrices with $n = 2k$. Then*

$$\sum_{j=1}^{h} \frac{\frac{1}{\sharp(S_j, S_j)}}{\frac{1}{\sharp(S_1, S_1)} + \ldots + \frac{1}{\sharp(S_h, S_h)}} \cdot \Theta(\tau; S_j) = E_k^*(\tau), \quad \tau \in \mathbb{H}. \tag{3.7}$$

Proof We write $f(\tau) \in \mathbb{M}_k$ for the left-hand side of (3.7) and use N for the expression in the denominator as an abbreviation. If $p > 2$ is a prime, then we conclude

$$T_p f = \frac{1}{N} \cdot \sum_{j=1}^{h} \frac{1}{\sharp(S_j, S_j)} \cdot T_p \Theta(\cdot; S_j)$$

$$= \frac{1}{N \cdot a(p, k-1)} \cdot \sum_{i=1}^{h} \frac{1}{\sharp(S_i, S_i)} \cdot \left(\sum_{j=1}^{h} \frac{\sharp(S_j, pS_i)}{\sharp(S_j, S_j)}\right) \cdot \Theta(\cdot; S_i)$$

from Theorem 3.15. Corollary 3.11 and Theorem 3.14 imply

$$T_p f = \frac{a(p, k)}{N \cdot a(p, k-1)} \cdot \sum_{i=1}^{h} \frac{1}{\sharp(S_i, S_i)} \cdot \Theta(\cdot; S_i) = (p^{k-1} + 1) \cdot f.$$

Because of $\alpha_f(0) = 1$, the assertion follows from Theorem IV.2.12. □

If we compare the m-th FOURIER coefficients on both sides of (3.7), then III(2.9) leads to

3.18 Corollary. *For all $m \geq 1$, one has*

§ 3 A special case of Siegel's Main Theorem

$$\sum_{j=1}^{h} \frac{\sharp(S_j, 2m)}{\sharp(S_j, S_j)} = -\frac{2k}{B_k} \left(\frac{1}{\sharp(S_1, S_1)} + \cdots + \frac{1}{\sharp(S_h, S_h)} \right) \cdot \sigma_{k-1}(m).$$

3.19 Remarks. a) A much more general version of Theorem 3.17 was proved in 1935 by C.L. Siegel ([78], vol. I, 326–405). The special case of even, unimodular, positive definite matrices, which we consider here, was extracted from the general result in 1941 for so-called Siegel modular forms by E. Witt (*Abh. Math. Semin. Hans. Univ.* **14**, 323–337).

b) The quantity appearing in the denominator of (3.7),

$$M(n) := \frac{1}{\sharp(S_1, S_1)} + \cdots + \frac{1}{\sharp(S_h, S_h)},$$

is called *mass* of the genus of the even, unimodular, positive definite $n \times n$ matrices. The Minkowski-Siegel *mass formula* gives its size:

$$M(n) = \frac{|B_k|}{2k} \cdot \prod_{j=1}^{k-1} \frac{|B_{2j}|}{4j}, \quad n = 2k.$$

Numerically, we obtain the following values for the mass (cf. J.H. Conway and N.J.A. Sloane [14], chap. 16.2):

n	mass	
8	$\dfrac{1}{696729600}$	$\approx 1, 4 \cdot 10^{-9}$
16	$\dfrac{691}{277667181515243520000}$	$\approx 2, 5 \cdot 10^{-18}$
24	$\dfrac{1027637932586061520960267}{129477933340026851560636148613120000000}$	$\approx 7, 9 \cdot 10^{-15}$
32	$\dfrac{4890529010450384254108570593011950899382291953107314413193123}{121325280941552041649762780685623131486814208000000000}$	$\approx 4, 0 \cdot 10^{7}$

Because of $h(8) = 1$ the value of $M(8)$ leads to

$$\sharp(S_8, S_8) = \sharp \mathrm{Aut}\,(S_8) = 696,729,600.$$

c) In the case $n > 8$, the theta series appearing in Theorem 3.17 are linearly dependent, but we can consider (3.7) a 'canonical' representation of the Eisenstein series.

d) The Fourier coefficients of E_{12}^* are not integral. Thus, in the case $n = 24$, the existence of 2 linearly independent theta series (and thus also $h_{12} > 1$) follows from Theorem 3.17. Therefore, we obtain a new proof of Corollary 2.21 without constructing the lattice.

e) The proof of Theorem 3.17 given here essentially follows the lecture notes by H.–G. Quebbemann (*Geometrie der Zahlen*, Münster 1987). In a more general situation, R. Schulze–Pillot (*Invent. Math.* **75**, 282–299 (1984)) proceeds analogously. For lattices over orders in quaternion skew fields, a corresponding result is due to H.–G. Quebbemann (*Mathematika* **31**, 12–16 (1984)). A much more general result was derived in 1965 by A. WEIL (*Acta Math.* **113**, 1–87).

3.20 Exercises.
1) Let $S \in \mathrm{Sym}\,(n;\mathbb{Z})$ and $\ell \geq 1$. Then there are $U \in \mathcal{U}_n$, $\alpha_1,\ldots,\alpha_r \in \mathbb{Z}$ and matrices $A_1,\ldots,A_s \in \mathrm{Mat}\,(2;\mathbb{Z})$ satisfying

$$S[U] \equiv [\alpha_1,\ldots,\alpha_r, A_1,\ldots,A_s]\ (\mathrm{mod}\ 2^\ell), \quad r + 2s = n.$$

2) There exists an even matrix $S \in \mathrm{Sym}\,(n;\mathbb{Z})$ satisfying $\det S = 1$ if and only if $n \equiv 0\ (\mathrm{mod}\ 4)$.

3) Let $S \in \mathrm{Sym}\,(n;\mathbb{Z})$ be even with odd determinant. Then $n = 2k$ is even and there is some $U \in \mathcal{U}_n$ satisfying

$$S[U] \equiv [D_1,\ldots,D_k]\ (\mathrm{mod}\ 4), \quad D_j = \begin{pmatrix} 0 & 1 \\ 1 & 0 \end{pmatrix} \text{ or } \begin{pmatrix} 2 & 1 \\ 1 & 2 \end{pmatrix}, \quad j = 1,\ldots,k.$$

4) Let $S \in \mathrm{Sym}\,(n;\mathbb{Z})$, $n = 2k$, be even with odd determinant. Then there are $2^{2k-1} + 2^{k-1}$ vectors $g \in \mathbb{Z}^n\,(\mathrm{mod}\ 2)$ with $S[g] \equiv 0\,(\mathrm{mod}\ 4)$ and $2^{2k-1} - 2^{k-1}$ vectors $g \in \mathbb{Z}^n\,(\mathrm{mod}\ 2)$ with $S[g] \equiv 2\,(\mathrm{mod}\ 4)$.

5) Let $S \in \mathrm{Sym}\,(n;\mathbb{Z})$, $n = 2k$, be even with odd determinant.
a) One has

$$a_2(S) = a(2,k) = \prod_{j=0}^{k-1}(2^j + 1).$$

b) If $k > 1$ and $g \in \mathbb{Z}^n$, $g \not\equiv 0\,(\mathrm{mod}\ 2)$, with $S[g] \equiv 0\,(\mathrm{mod}\ 4)$, then there are exactly $a(2, k-1)$ left cosets $H\mathcal{U}_n$ with the properties

$$|\det H| = 2^k,\ H^{-1}g \in \mathbb{Z}^n, \text{ such that } \frac{1}{2}S[H] \text{ is even.}$$

6) Derive the analogon of Lemma 3.10 and Corollary 3.11 for $p = 2$.
7) (3.6) also holds for $p = 2$.
8) Calculate the number of solutions $S[g] \equiv \alpha(\mathrm{mod}\ p)$ for $\alpha \in \mathbb{Z}$, $p \nmid \alpha$ under the assumptions of Proposition 3.13.
9) For any prime p one has $\sharp(S_8, pS_8) = 2(p + 1)(p^2 + 1)(p^3 + 1) \cdot \sharp(S_8, S_8)$.

§ 4* Harmonic polynomials and quadratic forms of higher level

In this section, we first describe harmonic polynomials. They serve as a tool to obtain cusp forms from theta series. Moreover, it is shown that theta series for any rational positive definite quadratic form always yield modular forms with respect to congruence subgroups.

1. Harmonic polynomials. If $X_1,\ldots,X_n$ are indeterminates over $\mathbb{C}$, then for $\alpha \in \mathbb{N}_0^n$, we use the abbreviations

§ 4* Harmonic polynomials and quadratic forms of higher level

$$X = \begin{pmatrix} X_1 \\ \vdots \\ X_n \end{pmatrix}, \quad \alpha = \begin{pmatrix} \alpha_1 \\ \vdots \\ \alpha_n \end{pmatrix}, \quad X^\alpha := X_1^{\alpha_1} \cdot \ldots \cdot X_n^{\alpha_n}, \quad \alpha! := \alpha_1! \cdot \ldots \cdot \alpha_n!. \quad (4.1)$$

For $r \in \mathbb{N}_0$, we denote by $\mathcal{P}_r^{(n)}$ the $\mathbb{C}$–vector space of the *homogeneous polynomials of degree r* in $\mathbb{C}[X_1, \ldots, X_n]$, that is, with (4.1)

$$\mathcal{P}_r^{(n)} = \left\{ P(X) = \sum_{\alpha \in \mathbb{N}_0^n} p(\alpha) X^\alpha \, ; \, p(\alpha) \in \mathbb{C}, \, \alpha_1 + \ldots + \alpha_n = r \right\}. \quad (4.2)$$

As it is well-known,

$$\dim \mathcal{P}_r^{(n)} = \sharp\{\alpha \in \mathbb{N}_0^n \, ; \, \alpha_1 + \ldots + \alpha_n = r\} = \binom{n + r - 1}{r} \quad (4.3)$$

holds.

4.1 Lemma. *The set $\{(h^{tr} X)^r \, ; \, h \in \mathbb{R}^n\}$ spans $\mathcal{P}_r^{(n)}$.*

Proof In the standard notation (4.1) and (4.2), we define a scalar product on $\mathcal{P}_r^{(n)}$ by

$$\left\langle P(X), Q(X) \right\rangle = \sum_{\alpha \in \mathbb{N}_0^n} \alpha! \cdot p(\alpha) \cdot \overline{q(\alpha)} = P\left(\left(\tfrac{\partial}{\partial X_1}, \ldots, \tfrac{\partial}{\partial X_n} \right)^{tr} \right) \overline{Q(X)}.$$

In particular, then, all $P(X) \in \mathcal{P}_r^{(n)}$, $h \in \mathbb{R}^n$ satisfy

$$\left\langle P(X), (h^{tr} X)^r \right\rangle = r! \cdot P(h).$$

Let $\mathcal{U}$ be the $\mathbb{C}$–subspace spanned by $(h^{tr} X)^r$, $h \in \mathbb{R}^n$, of $\mathcal{P}_r^{(n)}$. Then for each $P(X) \in \mathcal{U}^\perp$,

$$P(h) = 0 \quad \text{for all } h \in \mathbb{R}^n$$

holds. An induction on n together with the Identity Theorem A.1 immediately shows that $P \equiv 0$, hence $\mathcal{U} = \mathcal{P}_r^{(n)}$. □

We call $P(X) \in \mathcal{P}_r^{(n)}$ *harmonic* if

$$\Delta P(X) \equiv 0, \quad \Delta = \frac{\partial^2}{\partial X_1^2} + \ldots + \frac{\partial^2}{\partial X_n^2}.$$

Thus, Δ is the usual LAPLACE *operator* and

$$\mathcal{H}_r^{(n)} := \{P(X) \in \mathcal{P}_r^{(n)}; \, \Delta P(X) \equiv 0\}$$

is a subspace of $\mathcal{P}_r^{(n)}$. Obviously,

$$\mathcal{H}_0^{(n)} = \mathbb{C}, \quad \mathcal{H}_1^{(n)} = \mathbb{C}X_1 + \ldots + \mathbb{C}X_n, \quad \mathcal{H}_r^{(1)} = \{0\} \quad \text{for } r \geq 2.$$

A description of $\mathcal{H}_r^{(n)}$ is now given by

4.2 Proposition. *Let $n \geq 2, r \geq 1$. Then the map*

$$\phi : \mathcal{H}_r^{(n)} \to \mathcal{P}_r^{(n-1)} \times \mathcal{P}_{r-1}^{(n-1)}, \quad P(X) \mapsto \left((P(X))\big|_{X_n=0}, \frac{\partial(P(X))}{\partial X_n}\bigg|_{X_n=0} \right),$$

is an isomorphism of vector spaces.

Proof Let $\widetilde{X} = (X_1, \ldots, X_{n-1})^{tr}$. We write $P(X) \in \mathcal{P}_r^{(n)}$ in the form

$$P(X) = \sum_{j=0}^{r} P_j(\widetilde{X}) X_n^j, \quad P_j(\widetilde{X}) \in \mathcal{P}_{r-j}^{(n-1)}.$$

Because of $\Delta P_r = \Delta P_{r-1} = 0$, it follows immediately that

$$\Delta P(X) = \sum_{j=0}^{r-2} \left(\Delta P_j(\widetilde{X}) \right) X_n^j + \sum_{j=2}^{r} j(j-1) P_j(\widetilde{X}) X_n^{j-2}$$

$$= \sum_{j=0}^{r-2} \left[\Delta P_j(\widetilde{X}) + (j+2)(j+1) P_{j+2}(\widetilde{X}) \right] X_n^j.$$

Thus, we obtain

$$P_{j+2}(\widetilde{X}) = \frac{-1}{(j+2)(j+1)} \Delta P_j(\widetilde{X}), \quad j = 0, \ldots, r-2,$$

and $P_0 \in \mathcal{P}_r^{(n-1)}, P_1 \in \mathcal{P}_{r-1}^{(n-1)}$ are arbitrary. The assertion now follows from

$$P_0(\widetilde{X}) = P(X)\big|_{X_n=0}, \quad P_1(\widetilde{X}) = \frac{\partial P(X)}{\partial X_n}\bigg|_{X_n=0}. \qquad \square$$

Because of (4.3), we directly obtain

4.3 Corollary. *For $n \geq 2, r \geq 1$, one has*

$$\dim \mathcal{H}_r^{(n)} = \binom{n+r-2}{r} + \binom{n+r-3}{r-1}.$$

Now we give to a useful description of the harmonic polynomials.

4.4 Theorem. *The following assertions are equivalent for $n \geq 2$ and $r \in \mathbb{N}_0$:*
  (i) $P(X) \in \mathcal{H}_r^{(n)}$.
  (ii) $P(X)$ *is a finite linear combination over $\mathbb{C}$ of polynomials of the form $(u^{tr} X)^r$, where $u \in \mathbb{C}^n$ with $u^{tr} u = 0$.*

Proof (ii) $\Longrightarrow$ (i): We calculate directly that

§ 4* Harmonic polynomials and quadratic forms of higher level

$$\Delta(u^{tr}X)^r = r(r-1) \cdot u^{tr}u \cdot (u^{tr}X)^{r-2} = 0, \quad \text{hence } (u^{tr}X)^r \in \mathcal{H}_r^{(n)}.$$

(i) $\Longrightarrow$ (ii): Let $\mathcal{U}$ be the subspace of $\mathcal{H}_r^{(n)}$ spanned by $(u^{tr}X)^r$, $u \in \mathbb{C}^n$, $u^{tr}u = 0$. For $h \in \mathbb{R}^{n-1}$, let $\gamma = \pm i\sqrt{h^{tr}h}$. Then

$$u = \binom{h}{\gamma} \in \mathbb{C}^n, \quad u^{tr}u = 0, \quad \phi((u^{tr}X)^r) = \left((h^{tr}\tilde{X})^r, \gamma r(h^{tr}\tilde{X})^{r-1}\right)$$

in the notation of Proposition 4.2. Thus, $\phi(\mathcal{U})$ contains the elements

$$((h^{tr}\tilde{X})^r, 0), \; (0, (g^{tr}\tilde{X})^{r-1}), \quad h, g \in \mathbb{R}^{n-1}.$$

By Lemma 4.1, these elements span $\mathcal{P}_r^{(n-1)} \times \mathcal{P}_{r-1}^{(n-1)}$. Therefore,

$$\phi(\mathcal{U}) = \mathcal{P}_r^{(n-1)} \times \mathcal{P}_{r-1}^{(n-1)}$$

implies $\mathcal{U} = \mathcal{H}_r^{(n)}$, due to Proposition 4.2. $\square$

2. Theta series with respect to harmonic polynomials. For $S \in \text{Pos}(n; \mathbb{R})$, let $S^{1/2}$ be the square root given by (1.13). If $P(X) \in \mathcal{H}_r^{(n)}$, then we define the *theta series in S with respect to the harmonic polynomial* $P(X)$ by

$$\Theta(\tau; S, P) := \sum_{g \in \mathbb{Z}^n} P(S^{1/2}g) \, e^{\pi i \tau S[g]}. \tag{4.4}$$

It is absolutely convergent due to the majorant

$$C \cdot \sum_{g \in \mathbb{Z}^n} (g^{tr}g)^{r/2} \cdot e^{-\varepsilon g^{tr}g} < \infty, \quad C > 0, \; \varepsilon > 0.$$

Hence, $\Theta(\cdot; S, P)$ is holomorphic in $\tau \in \mathbb{H}$. Because of

$$\Theta_{0,q}(\tau; S) = \sum_{g \in \mathbb{Z}^n} e^{\pi i \tau S[g] + 2\pi i g^{tr}q},$$

we also get

$$\Theta(\tau; S, P) = (2\pi i)^{-r} P\left(S^{1/2} \frac{\partial}{\partial q}\right) \Theta_{0,q}(\tau; S)\Big|_{q=0},$$

$$\frac{\partial}{\partial q} = \left(\frac{\partial}{\partial q_1}, \ldots, \frac{\partial}{\partial q_n}\right)^{tr}. \tag{4.5}$$

4.5 Theorem. *If $S \in \text{Pos}(n; \mathbb{Z})$ is even and unimodular and $P(X) \in \mathcal{H}_r^{(n)}$ with even $r > 0$, then*

$$\Theta(\cdot; S, P) \in \mathbb{S}_{r+n/2}$$

with the FOURIER *series expansion*

$$\Theta(\tau; S, P) = \sum_{m=1}^{\infty} \left(\sum_{g \in \mathcal{D}(S, 2m)} P(S^{1/2} g) \right) \cdot e^{2\pi i m \tau}. \tag{4.6}$$

Proof We obtain (4.6) by a rearrangement in (4.4) with $P(0) = 0$. Moreover, $\Theta(\tau + 1; S, P) = \Theta(\tau; S, P)$ is obvious. Using Theorem 2.12, (4.5) and the Theta Transformation Formula 2.6, we obtain

$$\Theta(-1/\tau; S, P) = (2\pi i)^{-r} \cdot P\left(S^{1/2} \tfrac{\partial}{\partial q}\right) \cdot \Theta_{0,q}(-1/\tau; S)\Big|_{q=0}$$
$$= (2\pi i)^{-r} \cdot P\left(S^{1/2} \tfrac{\partial}{\partial q}\right) \cdot \tau^{n/2} \cdot \Theta_{q,0}(\tau; S^{-1})\Big|_{q=0}.$$

For $u \in \mathbb{C}^n$ with $u^{tr} u = 0$, one has

$$\left(u^{tr} S^{1/2} \tfrac{\partial}{\partial q}\right)^2 e^{\pi i \tau S^{-1}[g+q]}$$
$$= (2\pi i \tau) \cdot \left(u^{tr} S^{1/2} \tfrac{\partial}{\partial q}\right) \left(u^{tr} S^{1/2} \cdot S^{-1}(g+q)\right) e^{\pi i \tau S^{-1}[g+q]}$$
$$= \left(2\pi i \tau \cdot u^{tr} S^{1/2} \cdot S^{-1} S^{1/2} u + (2\pi i \tau \cdot u^{tr} S^{1/2} S^{-1}(g+q))^2\right) \cdot e^{\pi i \tau S^{-1}[g+q]}$$
$$= (2\pi i \tau)^2 \cdot (u^{tr} S^{1/2} S^{-1}(g+q))^2 \cdot e^{\pi i \tau S^{-1}[g+q]}.$$

For $r \in \mathbb{N}$, an induction yields

$$\left(u^{tr} S^{1/2} \tfrac{\partial}{\partial q}\right)^r e^{\pi i \tau S^{-1}[g+q]} = (2\pi i)^r \cdot \left(u^{tr} S^{1/2} S^{-1}(g+q)\right)^r \cdot e^{\pi i \tau S^{-1}[g+q]}.$$

By Theorem 4.4, it follows that

$$P(S^{1/2} \tfrac{\partial}{\partial q}) e^{\pi i \tau S^{-1}[g+q]}\big|_{q=0}$$
$$= (2\pi i \tau)^r \cdot P(S^{1/2} S^{-1}(g+q)) \cdot e^{\pi i \tau S^{-1}[g+q]}\Big|_{q=0}$$
$$= (2\pi i \tau)^r \cdot P(S^{1/2} \cdot S^{-1} g) \cdot e^{\pi i \tau S[S^{-1} g]}.$$

Thus, the substitution $h = S^{-1} g$ yields

$$\Theta(-1/\tau; S, P) = \tau^{r+n/2} \cdot \sum_{g \in \mathbb{Z}^n} P(S^{1/2} \cdot S^{-1} g) \cdot e^{\pi i \tau S[S^{-1} g]}$$
$$= \tau^{r+n/2} \cdot \Theta(\tau; S, P). \qquad \square$$

Now consider the particular polynomial

$$P(X) = (u^{tr} X)^2 - \frac{1}{n} u^{tr} u \cdot X^{tr} X \in \mathcal{H}_2^{(n)}$$

for $u \in \mathbb{C}^n$. If we set $u = S^{1/2} v$, $v \in \mathbb{C}^n$, this P yields

4.6 Corollary. *For any even, unimodular* $S \in \mathrm{Pos}\,(n;\mathbb{Z})$ *and* $v \in \mathbb{C}^n$ *one has*

$$\Theta(\tau; S, P) = \sum_{g \in \mathbb{Z}^n} ((v^{tr} Sg)^2 - \frac{1}{n} S[v] \cdot S[g]) \cdot e^{\pi i \tau S[g]} \in \mathbb{S}_{2+n/2}.$$

$\Theta(\cdot; S, P) \equiv 0$ *holds for* $n = 8, 16, 24$ *and all* $m \in \mathbb{N}$ *satisfy*

$$\sum_{g \in \mathcal{D}(S,2m)} (v^{tr} Sg)^2 = \frac{1}{n} S[v] \cdot \sharp(S, 2m) .$$

Proof The result follows from Theorem 4.5 as well as $\mathbb{S}_6 = \mathbb{S}_{10} = \mathbb{S}_{14} = \{0\}$, according to III, §4. □

Now we consider especially $S = S_8$ from (2.14).

4.7 Corollary. *For any* $v \in \mathbb{C}^8$,

$$\sum_{g \in \mathbb{Z}^8} \left((v^{tr} S_8 g)^8 - \frac{1}{128} (S_8[v])^4 (S_8[g])^4 \right) \cdot e^{\pi i \tau S_8[g]} = c_v \cdot \Delta^*(\tau)$$

holds, where

$$c_v = \left(\sum_{g \in \mathcal{D}(S_8, 2)} (v^{tr} S_8 g)^8 \right) - 30 (S_8[v])^4.$$

Proof We easily verify $P(X) \in \mathcal{H}_8^{(8)}$ for $u \in \mathbb{C}^8$ and

$$P(X) = (u^{tr} X)^8 - \frac{7}{5}(u^{tr} u) \cdot (u^{tr} X)^6 \cdot (X^{tr} X) + \frac{7}{12}(u^{tr} u)^2 \cdot (u^{tr} X)^4 \cdot (X^{tr} X)^2$$

$$- \frac{7}{96}(u^{tr} u)^3 \cdot (u^{tr} X)^2 \cdot (X^{tr} X)^3 + \frac{1}{768}(u^{tr} u)^4 \cdot (X^{tr} X)^4 .$$

Because of $\mathbb{S}_{12} = \mathbb{C} \cdot \Delta^*$, Corollary 4.6 yields

$$\Theta(\tau; S_8, P) = c_u \Delta^*(\tau), \quad c_u = \sum_{g \in \mathcal{D}(S_8, 2)} P(S_8^{1/2} g).$$

Now consider

$$Q(X) = (u^{tr} X)^6 - \frac{15}{16}(u^{tr} u) \cdot (u^{tr} X)^4 \cdot (X^{tr} X) + \frac{45}{224}(u^{tr} u)^2 \cdot (u^{tr} X)^2 \cdot (X^{tr} X)^2$$

$$- \frac{5}{896}(u^{tr} u)^3 \cdot (X^{tr} X)^3 \in \mathcal{H}_6^{(8)},$$

$$Q(X) = (u^{tr} X)^4 - \frac{1}{2}(u^{tr} u) \cdot (u^{tr} X)^2 \cdot (X^{tr} X) + \frac{1}{40}(u^{tr} u)^2 \cdot (X^{tr} X)^2 \in \mathcal{H}_4^{(8)}$$

$$Q(X) = (u^{tr} X)^2 - \frac{1}{8}(u^{tr} u) \cdot (X^{tr} X) \in \mathcal{H}_2^{(8)}.$$

Because of $\mathbb{S}_6 = \mathbb{S}_8 = \mathbb{S}_{10} = \{0\}$, it follows that $\Theta(\cdot; S_8, Q) \equiv 0$. From the FOURIER series expansion in Corollary 4.6 we also get

$$\sum_{g \in \mathbb{Z}^8} (S_8[g])^r Q(S_8^{1/2} g) \cdot e^{\pi i \tau S_8[g]} \equiv 0 \quad \text{for } r \in \mathbb{N}_0.$$

Finally, by forming suitable linear combinations, it follows that

$$\Theta(\tau; S_8, P) = \sum_{g \in \mathbb{Z}^8} \left((u^{tr} S_8^{1/2} g)^8 - \frac{1}{128}(u^{tr} u)^4 (S_8[g])^4 \right) \cdot e^{\pi i \tau S_8[g]}.$$

Now we set $u = S_8^{1/2} v$ and use $\sharp(S_8, 2) = 240$ according to Corollary 2.16. □

If we especially choose $v \in \mathcal{D}(S_8, 2)$, we can show that

$$\sum_{g \in \mathcal{D}(S_8, 2)} (v^{tr} S_8 g)^8 = 624$$

holds, i.e. $c_v = 144$ in Corollary 4.7. Then a comparison of the FOURIER coefficients in Corollary 4.7 together with Corollary 2.12 yields

4.8 Corollary. *If $v \in \mathcal{D}(S_8, 2)$, then all $m \in \mathbb{N}$ satisfy*

$$\left(\sum_{g \in \mathcal{D}(S_8, 2m)} (v^{tr} S_8 g)^8 \right) - 480 \cdot m^4 \cdot \sigma_3(m) = 144 \cdot \tau(m).$$

The results of this paragraph are due to E. HECKE [36], 789–918. Corollary 4.6 is an essential tool in the classification of the 24–dimensional, even, unimodular lattices by B.B. VENKOV (cf. J.H. CONWAY and N.J.A. SLOANE [14], chap. 18).

3. The level of an even matrix. Let $S \in \text{Sym}(n; \mathbb{Z})$ be even with $\det S \neq 0$. Then

$$N := \min\{q \in \mathbb{N} ; \ qS^{-1} \text{ even}\} \tag{4.7}$$

is called the *level of S*. Because $S^{-1} \in \text{Sym}(n; \mathbb{Q})$, this N exists and NS^{-1} is even. We formulate the essential properties in

4.9 Proposition. *Let $S \in \text{Sym}(n; \mathbb{Z})$ be even, $\det S \neq 0$ with level N. Then the following holds:*
a) *Every matrix $S[U]$, $U \in \mathcal{U}_n$, also has level N.*
b) *If $q \in \mathbb{N}$ and qS^{-1} is even, then $N|q$.*
c) *$N|2 \det S$.*
d) *For a prime number p, $p|N$ holds if and only if $p| \det S$.*
e) *$N = 1$ holds if and only if S is unimodular.*

Thus, the level is a class invariant in the sense of (1.18) and it has the same prime divisors as $\det S$ (but possibly with different multiplicities).

Proof a) This is because $(S[V])^{-1} = S^{-1}[V^{tr-1}]$ and (2.13) qS^{-1} is even if and only

§ 4* Harmonic polynomials and quadratic forms of higher level

if $q(S[V])^{-1}$ is even.
b) Let qS^{-1} be even and $g = \gcd(N, q) \in \mathbb{N}$. Then there exist $\alpha, \beta \in \mathbb{Z}$ with $g = \alpha q + \beta N$. Thus
$$gS^{-1} = \alpha(q \cdot S^{-1}) + \beta(N \cdot S^{-1})$$
is also even. It follows that $g = N$, hence $N|q$ from (4.7).
c) The adjugate matrix $S^{\#} = \det S \cdot S^{-1}$ is integral, hence $2 \det S \cdot S^{-1}$ is even. Now we use b).
d) " $\Rightarrow$ " Let p be a prime with $p|N$. Then $p|\det S$ follows from c), except for the case $p = 2$, $\det S$ odd, which we want to consider now. Because of a) and Lemma 3.7, we may assume without loss of generality that
$$S \equiv [Q, \ldots, Q] \pmod{2}, \quad Q = \begin{pmatrix} 0 & 1 \\ 1 & 0 \end{pmatrix}.$$
Then $s^{\#}_{ii} \equiv 0 \pmod{2}$ follows. Therefore, $\det S \cdot S^{-1}$ is already even. Hence, this case cannot occur because of b).
" $\Leftarrow$ " Let p be a prime with $p \nmid N$. Then there exists some $\alpha \in \mathbb{Z}$ with the property $\alpha N \equiv 1 \pmod{p}$. Hence, $T := \alpha N S^{-1} \in \text{Sym}(n; \mathbb{Z})$ holds with
$$S \cdot T = \alpha N \cdot I \equiv I \pmod{p}.$$
Thus, $S \pmod p$ is invertible over $\mathbb{Z}/p\mathbb{Z}$, i.e. $p \nmid \det S$.
e) Use d). □

For further investigations of theta series with respect to even $S \in \text{Pos}(n; \mathbb{Z})$, we may restrict ourselves to $N > 1$ because of the results of §2.

4. A general theta transformation formula. We again use the general theta series with characteristic
$$\Theta_{p,q}(\tau; S) := \sum_{g \in \mathbb{Z}^n} e^{\pi i \tau S[g+p] + 2\pi i q^{tr}(g+p)} \tag{4.8}$$
and write $\Theta(\tau; S) = \Theta_{0,0}(\tau; S)$ for the theta zero value.

4.10 Lemma. *Let $S \in \text{Pos}(n; \mathbb{Z})$ be even and $N \in \mathbb{N}$ such that $N \cdot S^{-1}$ is even. Then all $M = \begin{pmatrix} a & b \\ c & d \end{pmatrix} \in \Gamma$ with $c \neq 0$ satisfy:*
$$\Theta(M\tau; S) = \sqrt{\frac{\tau + d/c}{i}}^n \cdot \frac{1}{\sqrt{\det S}} \cdot \sum_{q:\mathbb{Z}^n/S\mathbb{Z}^n} \varphi_S(M, q) \cdot \Theta_{S^{-1}q, 0}(\tau, S),$$

where
$$\varphi_S(M, q) := \sum_{p:\mathbb{Z}^n/c\mathbb{Z}^n} e^{\pi i(aS[p] - 2p^{tr}q + dS^{-1}[q])/c}. \tag{4.9}$$

Proof We use the decomposition

$$M\tau = \frac{a}{c} - \frac{1}{c^2} \cdot \frac{1}{\tau + d/c}.$$

Moreover, we choose $g \in \mathbb{Z}^n$ in the form $g = ch + p$, $h \in \mathbb{Z}^n$, $p : \mathbb{Z}^n/c\mathbb{Z}^n$ and use

$$S[g] = c^2 S[h] + 2ch^{tr} Sp + S[p] \equiv S[p] \pmod{2c}.$$

Then we get

$$\begin{aligned}
\Theta(M\tau; S) &= \sum_{g \in \mathbb{Z}^n} e^{\pi i S[g] \cdot M\tau} \\
&= \sum_{p:\mathbb{Z}^n/c\mathbb{Z}^n} \sum_{h \in \mathbb{Z}^n} e^{\pi i S[ch+p]((a/c)-(1/c)^2/(\tau+d/c))} \\
&= \sum_{p:\mathbb{Z}^n/c\mathbb{Z}^n} e^{\pi i a S[p]/c} \cdot \sum_{h \in \mathbb{Z}^n} e^{\pi i S[h+p/c] \cdot (-1/(\tau+d/c))} \\
&= \sum_{p:\mathbb{Z}^n/c\mathbb{Z}^n} e^{\pi i a S[p]/c} \cdot \Theta_{p/c,0}\left(\frac{-1}{\tau+d/c}; S\right) \\
&= \sqrt{\frac{\tau+d/c}{i}}^n \cdot \frac{1}{\sqrt{\det S}} \cdot \sum_{p:\mathbb{Z}^n/c\mathbb{Z}^n} e^{\pi i a S[p]/c} \cdot \Theta_{0,-p/c}(\tau+d/c; S^{-1}),
\end{aligned}$$

where we apply the Theta Transformation Formula 2.6 in the last step. Now, let $g \in \mathbb{Z}^n$ be given in the form $g = Sh + q$, $h \in \mathbb{Z}^n$, $q : \mathbb{Z}^n/S\mathbb{Z}^n$ and use

$$S^{-1}[g] = S[h] + 2h^{tr} q + S^{-1}[q].$$

This yields

$$\begin{aligned}
&\sum_{p:\mathbb{Z}^n/c\mathbb{Z}^n} e^{\pi i a S[p]/c} \cdot \Theta_{0,-p/c}(\tau+d/c; S^{-1}) \\
&= \sum_{p:\mathbb{Z}^n/c\mathbb{Z}^n} \sum_{q:\mathbb{Z}^n/S\mathbb{Z}^n} \sum_{h \in \mathbb{Z}^n} e^{\pi i a S[p]/c + \pi i S^{-1}[Sh+q](\tau+d/c) - 2\pi i p^{tr}(Sh+q)/c} \\
&= \sum_{q:\mathbb{Z}^n/S\mathbb{Z}^n} \sum_{h \in \mathbb{Z}^n} e^{\pi i \tau S[h+S^{-1}q]} \\
&\quad \cdot \sum_{p:\mathbb{Z}^n/c\mathbb{Z}^n} e^{\pi i (aS[p]+dS[h]+2dh^{tr}q+dS^{-1}[q]-2p^{tr}Sh-2p^{tr}q)/c}.
\end{aligned}$$

From $ad - bc = 1$, we obtain

$$aS[p - dh] \equiv aS[p] - 2p^{tr} Sh + dS[h] \pmod{2c}.$$

Thus, the last sum over p is calculated to be

$$\begin{aligned}
&\sum_{p:\mathbb{Z}^n/c\mathbb{Z}^n} e^{\pi i (aS[p-dh]-2(p-dh)^{tr}q+dS^{-1}[q])/c} \\
&= \sum_{r:\mathbb{Z}^n/c\mathbb{Z}^n} e^{\pi i (aS[r]-2r^{tr}q+dS^{-1}[r])/c} = \varphi_S(M, q).
\end{aligned}$$

§ 4* Harmonic polynomials and quadratic forms of higher level

Since the sum is independent of h, we obtain the theta series by summation over h and thus the assertion. □

We formulate an important special case as

4.11 Corollary. *If $d \equiv 0 \pmod{N}$ is additionally assumed in Lemma 4.10, then*

$$\Theta(M\tau; S) = \sqrt{\frac{\tau + d/c}{i}}^n \cdot \frac{1}{\sqrt{\det S}} \cdot \varphi_S(M, 0) \cdot \sum_{q:\mathbb{Z}^n/S\mathbb{Z}^n} \Theta_{S^{-1}q, 0}(\tau; S).$$

Proof Because of $ad - bc = 1$ and $dS^{-1}[q] \in 2\mathbb{Z}$, we have

$$aS[p - dS^{-1}q] \equiv aS[p] - 2p^{tr}q + dS^{-1}[q] \pmod{2c}.$$

From this, $dS^{-1}q \in \mathbb{Z}^n$ yields

$$\varphi_S(M, q) = \sum_{p:\mathbb{Z}^n/c\mathbb{Z}^n} e^{\pi i aS[p - dS^{-1}q]/c}$$

$$= \sum_{r:\mathbb{Z}^n/c\mathbb{Z}^n} e^{\pi i aS[r]/c} = \varphi_S(M, 0). \quad □$$

5. Theta series with respect to even matrices of level N. We recall

$$\Gamma_0[N] = \left\{ M = \begin{pmatrix} a & b \\ c & d \end{pmatrix} \in \Gamma ; \ c \equiv 0 \pmod{N} \right\}$$

from II (3.9). Note that $M \in \Gamma_0[N]$ and $N > 1$ imply $d \neq 0$.

4.12 Proposition. *Let $S \in \mathrm{Pos}\,(n;\mathbb{Z})$ be even of level $N > 1$. Then all $M \in \Gamma_0[N]$ satisfy*

$$\Theta(M\tau; S) = \chi_S(M) \cdot \sqrt{c\tau + d}^n \cdot \Theta(\tau; S),$$

where

$$\chi_S(M) = \left(\frac{\sqrt{(-1/\tau - c/d)/i} \cdot \sqrt{\tau/i}}{\sqrt{c\tau + d}} \right)^n \cdot \sum_{p:\mathbb{Z}^n/d\mathbb{Z}^n} e^{\pi i bS[p]/d}.$$

Of course, the choice of $\sqrt{c\tau + d}$ does not matter here.

Proof In the case $c = 0$, it follows that $M = \pm T^m$, $\chi_S(M) = \sqrt{d}^{-n}$ and

$$\Theta(M\tau; S) = \Theta(\tau + m; S) = \Theta(\tau; S),$$

thus the assertion is true. Therefore, let $c \neq 0$. Then

$$M\tau = (MJ)\langle -1/\tau \rangle, \quad MJ = \begin{pmatrix} b & -a \\ d & -c \end{pmatrix}.$$

Now we can apply Corollary 4.11 to MJ instead of M and $-1/\tau$ instead of τ. It follows that

$$\Theta(M\tau; S)$$
$$= \Theta(MJ\langle -1/\tau\rangle; S)$$
$$= \sqrt{\frac{-1/\tau - c/d}{i}}^n \cdot \frac{1}{\sqrt{\det S}} \cdot \varphi_S(MJ, 0) \cdot \sum_{q:\mathbb{Z}^n/S\mathbb{Z}^n} \Theta_{S^{-1}q, 0}(-1/\tau; S)$$
$$= \sqrt{\frac{-1/\tau - c/d}{i}}^n \cdot \sqrt{\frac{\tau}{i}}^n \cdot \varphi_S(MJ, 0) \cdot \frac{1}{\det S} \cdot \sum_{q:\mathbb{Z}^n/S\mathbb{Z}^n} \Theta_{0, -S^{-1}q}(\tau; S^{-1}),$$

where we use the Theta Transformation Formula 2.6 in the last step. Again we put $g \in \mathbb{Z}^n$ in the form $g = Sh + p$, $h \in \mathbb{Z}^n$, $p : \mathbb{Z}^n/S\mathbb{Z}^n$ and use

$$g^{tr}S^{-1}q \equiv p^{tr}S^{-1}q \mod \mathbb{Z}, \quad S^{-1}[g] = S[h + S^{-1}p].$$

From this we get

$$\sum_{q:\mathbb{Z}^n/S\mathbb{Z}^n} \Theta_{0, -S^{-1}q}(\tau; S^{-1}) = \sum_{q:\mathbb{Z}^n/S\mathbb{Z}^n} \sum_{g \in \mathbb{Z}^n} e^{\pi i \tau S^{-1}[g] - 2\pi i g^{tr} S^{-1} q}$$
$$= \sum_{p:\mathbb{Z}^n/S\mathbb{Z}^n} \sum_{h \in \mathbb{Z}^n} e^{\pi i \tau S[h + S^{-1}p]} \sum_{q:\mathbb{Z}^n/S\mathbb{Z}^n} e^{-2\pi i p^{tr} S^{-1} q}.$$

The last sum is independent of the choice of q. If we replace q by $q + r$, $r \in \mathbb{Z}^n$, then we obtain

$$0 = \left(\sum_{q:\mathbb{Z}^n/S\mathbb{Z}^n} e^{-2\pi i p^{tr} S^{-1} q} \right) \cdot \left(1 - e^{-2\pi i p^{tr} S^{-1} r} \right).$$

Thus, the sum is 0 except for the case $p^{tr}S^{-1}r \in \mathbb{Z}$ for all $r \in \mathbb{Z}^n$, i.e. for $S^{-1}p \in \mathbb{Z}^n$. In this case, each summand is 1, hence the sum is equal to

$$\sharp(\mathbb{Z}^n/S\mathbb{Z}^n) = \det S$$

according to Corollary 1.8. Since we can choose $p = 0$, for example, the sum over h exactly yields $\Theta(\tau; S)$. Now the assertion follows with (4.9). □

Our next goal is to determine $\chi_S(M)$ explicitly.

4.13 Proposition. *Let $S \in \mathrm{Pos}\,(n; \mathbb{Z})$ be even of level $N > 1$. Then the following holds:*
a) $\chi_S(M)^4 = 1$ *for all* $M \in \Gamma_0[N]$.
b) *If* $n \equiv 0 \pmod 2$, *then*

$$\chi_S : \Gamma_0[N] \to \{\pm 1\}$$

is an abelian character and satisfies

§ 4* Harmonic polynomials and quadratic forms of higher level

$$\chi_S(M) = \begin{cases} \left(\dfrac{d}{|d|}\right)^{n/2} & \text{for } c = 0, \\ \left(\dfrac{d}{|d|}\right)^{n/2} \cdot \left(\dfrac{(-1)^{n/2} \det S}{p}\right) & \text{for } c \ne 0, \end{cases}$$

if p is an odd prime satisfying $p \equiv d \pmod{c}$.

Here, the last expression $\left(\dfrac{\cdot}{p}\right)$ stands for the LEGENDRE symbol.

Proof a) Let

$$\phi_S(M) := |d|^{-n/2} \cdot \sum_{q:\mathbb{Z}^n/d\mathbb{Z}^n} e^{\pi i b S[q]/d}.$$

Then $\phi_S(M)^4 = \chi_S(M)^4$ holds by Proposition 4.12 as well as

$$\Theta^4(\cdot; S)\Big|_{2n} M = \phi_S(M)^4 \cdot \Theta^4(\cdot; S)$$

for all $M \in \Gamma_0[N]$. Since $\Theta(\cdot; S)$ does not vanish identically, III(1.2) yields

$$\phi_S(MM')^4 = \phi_S(M)^4 \cdot \phi_S(M')^4$$

for all $M, M' \in \Gamma_0[N]$. We have $\phi_S(T^m) = 1$ for all $m \in \mathbb{Z}$. Hence let $c \ne 0$. According to the DIRICHLET Prime Number Theorem (cf. T.M. APOSTOL [6], chap. 7), there exists an odd prime number p of the form $p = d + mc$, $m \in \mathbb{Z}$. It follows that $\phi_S(M)^4 = \phi_S(MT^m)^4$ as well as

$$\phi_S(MT^m) = p^{-n/2} \cdot \sum_{q:\mathbb{Z}^n/p\mathbb{Z}^n} e^{\pi i (b+ma) S[q]/p}.$$

By Theorem 3.2, we may assume without loss of generality that

$$S \equiv [2s_1, \ldots, 2s_n] \pmod{p}.$$

From the calculation of the classical GAUSS sums, (cf. T.M. APOSTOL [6], chap. 9) we conclude

$$\phi_S(MT^m) = p^{-n/2} \cdot \prod_{j=1}^{n} \left(\sum_{q_j=1}^{p} e^{2\pi i (b+ma) s_j q_j^2 / p} \right)$$

$$= \varepsilon_p^n \cdot \prod_{j=1}^{n} \left(\dfrac{s_j(b+ma)}{p} \right), \quad \varepsilon_p = \begin{cases} 1, & \text{if } p \equiv 1 \pmod{4} \\ i, & \text{if } p \equiv 3 \pmod{4}. \end{cases}$$

Hence, $\phi_S(M) \in \{\pm 1, \pm i\}$ follows and also $\chi_S(M)^4 = 1$ for all $M \in \Gamma_0[N]$.
b) If n is even, the properties of the LEGENDRE symbol lead to

$$\varepsilon_p^2 = \left(\frac{-1}{p}\right), \quad \phi_S(MT^m) = \left(\frac{-1}{p}\right)^{n/2} \cdot \prod_{j=1}^{n}\left(\frac{s_j}{p}\right) = \left(\frac{(-1)^{n/2}\det S}{p}\right).$$

In this case, all $M \in \Gamma_0[N]$ satisfy

$$\Theta(\cdot; S)\Big|_{n/2} M = \left(\frac{d}{|d|}\right)^{n/2} \phi_S(M) \cdot \Theta(\cdot; S).$$

Thus, $\phi_S : \Gamma_0[N] \to \{\pm 1\}$ and $\chi_S : \Gamma_0[N] \to \{\pm 1\}$ are abelian characters with the desired properties. □

As a direct consequence, III, §7 leads to

4.14 Corollary. *Let $S \in \text{Pos}(n; \mathbb{Z})$ be even of level $N > 1$ and $n \equiv 0 \pmod 2$. Then*

$$\Theta(\cdot; S) \in \mathbb{M}_{n/2}(\Gamma_0[N], \chi_S)$$

holds.

Proof The transformation behavior follows from Proposition 4.12, and χ_S is an abelian character according to Proposition 4.13. The boundedness requirement (MK.3) follows from Lemma 4.10. □

For example, let $S = \left(\begin{smallmatrix} 2 & 1 \\ 1 & 2 \end{smallmatrix}\right)$. Then $N = 3$ and

$$\Theta(\cdot; S) \in \mathbb{M}_1(\Gamma_0[3], \chi_S), \quad \chi_S(M) = \left(\frac{d}{3}\right).$$

4.15 Corollary. *Let $S \in \text{Pos}(n; \mathbb{Q})$, $n \equiv 0 \pmod 2$ and $N \in \mathbb{N}$ such that NS and NS^{-1} are even. Then*

$$\Theta(\cdot; S) \in \mathbb{M}_{n/2}(\Gamma[N], \tilde{\chi}_S).$$

Proof The level of NS divides N^2. If $M \in \Gamma[N]$, Corollary 4.14 therefore yields

$$\Theta(M\tau; S) = \Theta(\tilde{M} \langle \tau/N \rangle; NS) = \tilde{\chi}_S(M) \cdot (c\tau + d)^{n/2} \cdot \Theta(\tau; S),$$

where

$$\tilde{M} = \begin{pmatrix} a & b/N \\ cN & d \end{pmatrix} \in \Gamma_0[N^2], \quad \tilde{\chi}_S(M) = \chi_{NS}(\tilde{M}). \quad \square$$

The results of this paragraph are due to B. SCHOENEBERG ([71], chap. IX) and W. PFETZER (*Arch. Math.* **6**, 448–454 (1953)). Theta series with respect to harmonic polynomials are also considered there.

4.16 Exercises.
1) For $P(X) \in \mathcal{P}_r^{(n)}$ the following assertions are equivalent:
(i) $P(X) \in \mathcal{H}_r^{(n)}$.
(ii) $\int_K P(x)\overline{Q(x)}dx = 0$ for $K = \{x \in \mathbb{R}^n; x^{tr}x \leq 1\}$ and any $Q(X) \in \mathcal{P}_\nu^{(n)}$ with $\nu < r$.
2) Every $P(X) \in \mathcal{P}_r^{(n)}$ has a unique representation

$$P(X) = \sum_{0 \le j \le r/2} (X^{tr} X)^j \cdot P_j(X) \quad \text{with } P_j(X) \in \mathcal{H}^{(n)}_{r-2j}.$$

3) On $K = \{x \in \mathbb{R}^n; x^{tr}x = 1\}$, every polynomial in $\mathbb{C}[X_1, \ldots, X_n]$ is a sum of homogeneous polynomials.

4) $\Delta : \mathcal{P}^{(n)}_r \to \mathcal{P}^{(n)}_{r-2}$, $r \ge 2$, is an epimorphism and one has

$$\dim \mathcal{H}^{(n)}_r = \dim \mathcal{P}^{(n)}_r - \dim \mathcal{P}^{(n)}_{r-2}.$$

5) $\chi_S \equiv 1$ holds if and only if $n \equiv 0 \pmod 4$ and $\det S$ is a square in $\mathbb{N}$.

6) For an even $S \in \text{Sym}(n; \mathbb{Z})$, $n \equiv 0 \pmod 2$ one has $(-1)^{n/2} \det S \equiv 0, 1 \pmod 4$.

In the following exercises, the analytic continuation and functional equation for HECKE's L–series is derived. Let χ always be a primitive DIRICHLET character mod N, $N > 1$, and

$$L(\chi, s) := \sum_{m=1}^{\infty} \chi(m) \cdot m^{-s}, \quad \text{Re } s > 1.$$

7) a) The absolute convergence abscissa of $L(\chi, s)$ is $\sigma_0 = 1$.
b) Show that $L(\chi, s)$ is convergent for Re $s > 0$.
c) Calculate the EULER product of $L(\chi, s)$.

8) Let χ be even, i.e. $\chi(-1) = 1$, and $\mathbb{L}(\chi, s) := \left(\frac{\pi}{N}\right)^{-s/2} \Gamma\left(\frac{s}{2}\right) L(\chi, s)$. Then the integral representation

$$2\mathbb{L}(\chi, s) = N^{-s/2} \sum_{m=1}^{N} \chi(m) \cdot \int_1^{\infty} \left[y^{s/2} \Theta_{m/N, 0}(iy; 1) + y^{(1-s)/2} \Theta_{0, -m/N}(iy; 1) \right] \frac{dy}{y}$$

holds. $\mathbb{L}(\chi, s)$ has an analytic continuation to an entire function of s and satisfies the functional equation

$$\mathbb{L}(\chi, 1-s) = \varepsilon_\chi \cdot \mathbb{L}(\overline{\chi}, s) \quad \text{with } \varepsilon_\chi \in \mathbb{C}, \quad |\varepsilon_\chi| = 1.$$

$L(\chi, s)$ is an entire function satisfying $L(\chi, -2n) = 0$, $n \in \mathbb{N}_0$.

9) For $u, v \in \mathbb{C}$, $\tau \in \mathbb{H}$, let

$$\Theta^*_{u,v}(\tau) := \sum_{n \in \mathbb{Z}} (n+u) e^{\pi i (n+u)^2 \tau + 2\pi i (n+u)v} .$$

Then $\Theta^*_{-v,u}(-1/\tau) = \tau \cdot \sqrt{\tau/i} \cdot e^{-2\pi i u v} \cdot \Theta^*_{u,v}(\tau)$ holds.

10) Let χ be odd, i.e. $\chi(-1) = -1$, and $\mathbb{L}(\chi, s) := \left(\frac{\pi}{N}\right)^{-s/2} \Gamma\left(\frac{s+1}{2}\right) \cdot L(\chi, s)$. Then $2\mathbb{L}(\chi, s)$ satisfies the integral representation

$$N^{-s/2}\sqrt{\pi} \cdot \sum_{m=1}^{N} \chi(m) \cdot \int_{1}^{\infty} \left[y^{(s+1)/2} \Theta^*_{m/N,0}(iy) + iy^{1-s/2} \Theta^*_{0,-m,N}(iy) \right] \frac{dy}{y}.$$

$\mathbb{L}(\chi, s)$ has an analytic continuation to an entire function of s and satisfies the functional equation

$$\mathbb{L}(\chi, 1-s) = \varepsilon_\chi \cdot \mathbb{L}(\overline{\chi}, s) \quad \text{with } \varepsilon_\chi \in \mathbb{C}, \ |\varepsilon_\chi| = 1.$$

$L(\chi, s)$ is then an entire function satisfying $L(\chi, 1-2n) = 0, n \in \mathbb{N}$.

§ 5* The Epstein zeta function and applications

As a generalization of the Riemann zeta function, we introduce zeta functions associated to positive definite quadratic forms, the so-called Epstein zeta functions. As a further application of the Theta Transformation Formula 2.6, we derive the meromorphic continuation of the Epstein zeta function. From this as a special case, we obtain the real–analytic Eisenstein series. These in turn lead to the analytic continuation of the so-called Rankin convolution of the Dirichlet series of two modular forms.

1. The Epstein zeta function. In this section, we study a different type of class invariants. We consider a series of the form

$$\zeta(T; s) := \sum_{g \in \mathbb{Z}^n}{}' (T[g])^{-s}, \quad T \in \text{Pos}(n; \mathbb{R}), \ s \in \mathbb{C}, \ \text{Re}\, s > n/2, \quad (5.1)$$

which we call an Epstein *zeta function*. Here and later, the prime over the sum means that $g = 0$ is to be omitted from the summation. Of course, as a special case for $n = 1$, we get the Riemann zeta function

$$\zeta(1; s) = 2\zeta(2s).$$

To investigate the convergence we need the

5.1 Lemma. *The series*

$$\sum_{g \in \mathbb{Z}^n}{}' (g^{tr} g)^{-k} \quad (5.2)$$

converges for real $k > n/2$.

Proof We use an induction on n and abbreviate (5.2) by $\varphi(n; k)$. In the case $n = 1$, the assertion follows from

$$\varphi(1; k) = 2\zeta(2k)$$

and Theorem B.1. If $n > 1$, we distinguish whether or not 0 occurs as a coefficient of g. It follows that

§ 5* The Epstein zeta function and applications

$$\varphi(n;k) \leq n \cdot \varphi(n-1;k) + 2^n \cdot \sum_{g \in \mathbb{N}^n} (g^{tr}g)^{-k}.$$

From the inequality between the geometric and arithmetic mean, we conclude

$$g^{tr}g = \gamma_1^2 + \ldots + \gamma_n^2 \geq n(\gamma_1^2 \cdot \ldots \cdot \gamma_n^2)^{1/n}.$$

This yields

$$\varphi(n;k) \leq n \cdot \varphi(n-1;k) + 2^n n^{-k} \cdot \zeta(2k/n)^n.$$

Hence, we get convergence for $k > n/2$ due to Theorem B.1. □

The convergence behavior of the Epstein zeta function is described in

5.2 Proposition. *The series* (5.1) *is absolutely convergent for* $T \in \text{Pos}(n;\mathbb{R})$ *and* $s \in \mathbb{C}$ *with* $\text{Re } s > n/2$ *and satisfies*

$$\zeta(T[U];s) = \zeta(T;s) \tag{5.3}$$

for all $U \in \mathcal{U}_n$.

Proof By Proposition 1.13, there exists some $\beta > 0$ such that $T > \beta I$. Thus, absolute convergence follows from Lemma 5.1. The Equivalence Theorem for Unimodular Matrices 1.1 and a rearrangement yield (5.3). □

Thus, the Epstein zeta function is a class invariant in the sense of (1.18).

2. Analytic continuation of the Epstein zeta function. By means of the Theta Transformation Formula 2.6 we can derive a theorem for the Epstein zeta functions introduced in (5.1) that is similar to the theorem for the Riemann zeta function in IV, §4. We need the following

5.3 Proposition. *Let* $T \in \text{Pos}(n;\mathbb{R})$ *with* $T > \alpha I$, $\alpha > 0$. *Then there is a positive constant C such that*

$$|\Theta(iy;T) - 1| \leq C \cdot y^{-n/2} \cdot e^{-\alpha y} \quad \text{for all } y > 0.$$

Proof All $y > 0$ satisfy

$$|\Theta(iy;T) - 1| \leq \vartheta(i\alpha y)^n - 1. \tag{*}$$

(∗) immediately yields

$$|\Theta(iy;T) - 1| = O(e^{-2\alpha y}) \quad \text{for } y \geq 1.$$

Now the Theta Transformation Formula 2.6 yields

$$\vartheta(i\alpha y)^n = (\alpha y)^{-n/2} \cdot \vartheta(i/\alpha y)^n = O(y^{-n/2}) \quad \text{for } 0 < y \leq 1.$$

Due to (∗) the assertion follows. □

In analogy to IV(4.17), we set

$$\xi(T;s) := \pi^{-s}\Gamma(s) \cdot \zeta(T;s). \tag{5.4}$$

5.4 Theorem. a) *If $T \in \mathrm{Pos}\,(n;\mathbb{R})$, then the EPSTEIN zeta function $\zeta(T;s)$ possesses a continuation as a meromorphic function into the whole complex s-plane. It has a pole only at $s = n/2$, which is of 1st order with residue*

$$\mathrm{res}_{s=n/2}\zeta(T;s) = \frac{\pi^{n/2}}{\Gamma(n/2) \cdot \sqrt{\det T}}.$$

Moreover

$$\zeta(T;0) = -1 \quad \text{and} \quad \zeta(T;-m) = 0 \quad \text{for} \quad m = 1,2,3,\ldots. \tag{5.5}$$

b) *The function*

$$\xi(T;s) - \left(\frac{(\det T)^{-1/2}}{s - n/2} - \frac{1}{s}\right)$$

possesses a continuation to an entire function in the whole complex s- plane and satisfies the functional equation

$$(\det T)^{-1/2} \cdot \xi(T^{-1}; \frac{n}{2} - s) = \xi(T;s).$$

Proof In

$$\Gamma(s) := \int_0^\infty e^{-x} \cdot x^{s-1} dx, \quad \mathrm{Re}\,s > 0,$$

we substitute $x = \pi y T[g]$, $y > 0$, $0 \neq g \in \mathbb{Z}^n$. Because of the absolute convergence, we may interchange summation and integration. This leads to

$$\xi(T;s) = {\sum_{g \in \mathbb{Z}^n}}' \int_0^\infty e^{-\pi y T[g]} \cdot y^{s-1} dy = \int_0^\infty (\Theta(iy;T) - 1) \cdot y^{s-1} dy$$

for $\mathrm{Re}\,s > n/2$. The substitution $y \mapsto 1/y$ leads to

$$\xi(T;s) = \int_0^1 (\Theta(iy;T) - 1) \cdot y^{s-1} dy + \int_1^\infty (\Theta(iy;T) - 1) \cdot y^{s-1} dy$$

$$= \int_1^\infty (\Theta(i/y;T) - 1) \cdot y^{-s-1} dy + \int_1^\infty (\Theta(iy;T) - 1) \cdot y^{s-1} dy$$

Using the Theta Transformation Formula 2.6, it follows that

$$\int_1^\infty (\Theta(i/y;T) - 1) \cdot y^{-s-1} dy$$

$$= \int_1^\infty \left(\Theta(iy;T^{-1})y^{n/2}(\det T)^{-1/2} - 1\right) \cdot y^{-s-1} dy$$

$$= \int_1^\infty (\Theta(iy;T^{-1}) - 1)(\det T)^{-1/2} y^{\frac{n}{2}-s-1} dy + \int_1^\infty \left(y^{n/2}(\det T)^{-1/2} - 1\right) y^{-s-1} dy$$

$$= \int_1^\infty (\Theta(iy;T^{-1}) - 1) \cdot (\det T)^{-1/2} \cdot y^{\frac{n}{2}-s-1} dy + \left(\frac{(\det T)^{-1/2}}{s - n/2} - \frac{1}{s}\right).$$

This yields the representation

$$\xi(T;s) = \int_1^\infty \left[(\Theta(iy;T^{-1}) - 1) \cdot (\det T)^{-1/2} \cdot y^{\frac{n}{2}-s} + (\Theta(iy;T) - 1) \cdot y^s\right] \frac{dy}{y}$$

$$+ \left(\frac{(\det T)^{-1/2}}{s - n/2} - \frac{1}{s}\right) \quad (5.6)$$

for Re $s > n/2$. According to Proposition 5.3, the integral in (5.6) exists for all $s \in \mathbb{C}$ and thus is an entire function of s. The claim now follows directly from the integral representation (5.6). The trivial roots of $\zeta(T;s)$ in (5.5) follow from the poles of the gamma function according to Theorem A.13. Since $\xi(T;s)$ has the residue -1 at $s = 0$ and $\Gamma(s)$ has a simple pole with residue 1 at $s = 0$, the possible pole of $\zeta(T;s)$ at $s = 0$ is cancelled and leads to $\zeta(T;0) = -1$. □

5.5 Remarks. a) The series

$$\zeta(T;s) = {\sum_{m,n=-\infty}^\infty}' (am^2 + 2bmn + cn^2)^{-s}, \quad T = \begin{pmatrix} a & b \\ b & c \end{pmatrix} > 0, \quad (5.7)$$

was first studied by L. KRONECKER ([50], vol. IV, 495) in 1889 and considered as a class invariant in the sense of (1.18). He stated that $\zeta(T;s)$ has a pole of 1st order in the complex s-plane at $s = 1$ and calculated the constant term of the LAURENT series around $s = 1$ (the KRONECKER *limit formula*, cf. sect. 5). Later, $\zeta(T;s)$ was related to number theoretic problems, the so-called 'class number 1 problem' for imaginary-quadratic number fields. Compare the work of A. Selberg and S. Chowla (*J. Reine Angew. Math.* 227, 86–110 (1960)) from 1949.

b) The generalization of the series (5.7) to arbitrary positive definite matrices is due to P. EPSTEIN (*Math. Ann.* **56**, 614–644 (1903) and **63**, 205–216 (1907)).

3. The real–analytic EISENSTEIN series. For $\tau \in \mathbb{H}$ and $s \in \mathbb{C}$ with Re $s > 1$, we define the *real–analytic* EISENSTEIN *series* by

$$E(\tau; s) := \sum_{m,n \in \mathbb{Z}}{}' \left(\frac{y}{|m\tau + n|^2}\right)^2. \tag{5.8}$$

5.6 Convergence Lemma. *For $\varepsilon > 0$ and $s \in \mathbb{C}$ with $\operatorname{Re} s > 1$ the series (5.8) converges absolutely and uniformly in each vertical strip of height ε in $\mathbb{H}$*

$$\mathcal{V}_\varepsilon := \{\tau \in \mathbb{H} \; ; \; |x| \le 1/\varepsilon, \; y \ge \varepsilon\}.$$

There is some $C > 0$ with the property

$$|E(\tau; s)| \le C \cdot y^{\operatorname{Re} s} \quad \text{for all} \quad \tau \in \mathcal{V}_\varepsilon. \tag{5.9}$$

For fixed $\tau \in \mathbb{H}$, $E(\tau; s)$ is a holomorphic function in s.

Proof The proof follows from Exercise III.2.13 1) and the proof of the Convergence Lemma III.2.2. The convergence behavior also yields the holomorphy. □

It is again convenient to also consider the *normalized real–analytic* EISENSTEIN *series*, which is defined by

$$E^*(\tau; s) := \frac{1}{2\zeta(2s)} \cdot E(\tau; s) = \frac{1}{2} \sum_{\gcd(m,n)=1} \left(\frac{y}{|m\tau + n|^2}\right)^s. \tag{5.10}$$

If we again denote the modular group by Γ and define

$$\Gamma_\infty := \{\pm T^n \; ; \; n \in \mathbb{Z}\} = \{M \in \Gamma \; ; \; c = 0\}, \tag{5.11}$$

then II(1.14), in analogy to III(2.6), yields

$$E^*(\tau; s) = \sum_{M : \Gamma_\infty \backslash \Gamma} (\operatorname{Im} M\tau)^s = \sum_{M : \Gamma_\infty \backslash \Gamma} \varphi|_0 M(\tau), \quad \varphi(\tau) := y^s. \tag{5.12}$$

If $M \in \Gamma$, then also LM runs through a system of representatives for the right cosets of Γ modulo Γ_∞ as L does. Thus we get the

5.7 Invariance Lemma. *If $s \in \mathbb{C}$ with $\operatorname{Re} s > 1$ and $M \in \Gamma$, then one has*

$$E(M\tau; s) = E(\tau; s) \quad \text{and} \quad E^*(M\tau; s) = E^*(\tau; s) \quad \text{for all} \quad \tau \in \mathbb{H}.$$

As in II(1.36), we define the matrix

$$F_\tau := \frac{1}{y} \begin{pmatrix} 1 & -x \\ -x & x^2 + y^2 \end{pmatrix} \in \operatorname{Pos}(2; \mathbb{R}) \quad \text{for} \quad \tau \in \mathbb{H}. \tag{5.13}$$

A verification yields

$$\frac{|m\tau + n|^2}{y} = F_\tau[g], \quad g = \begin{pmatrix} -n \\ m \end{pmatrix}, \tag{5.14}$$

§ 5* The EPSTEIN zeta function and applications

and
$$F_\tau^{-1} = F_{-1/\tau} = F_\tau[J]. \tag{5.15}$$

Thus, it follows immediately by (5.14) that
$$E(\tau; s) = \zeta(F_\tau; s). \tag{5.16}$$

From (5.15) and (5.3), we get $\zeta(F_\tau^{-1}; s) = \zeta(F_\tau; s)$. Thus, we formulate a special case of Theorem 5.4 as

5.8 Theorem. *The real–analytic* EISENSTEIN *series $E(\tau; s)$ possesses a continuation to a meromorphic function into the whole complex s-plane. It has a pole only at $s = 1$, which is of 1st order with residue π. Moreover, one has*

$$E(\tau; 0) = -1 \quad \text{and} \quad E(\tau; -m) = 0 \quad \text{for} \quad m = 1, 2, 3, \ldots.$$

The function
$$\mathbb{E}(\tau; s) := \pi^{-s} \Gamma(s) \cdot E(\tau; s)$$

is meromorphic in $\mathbb{C}$, *satisfies the functional equation*
$$\mathbb{E}(\tau; 1-s) = \mathbb{E}(\tau; s)$$

and
$$\mathbb{E}(\tau; s) - \left(\frac{1}{1-s} - \frac{1}{s} \right)$$

possesses a continuation to an entire function into the whole complex s-plane.

For further applications, it is important to know the growth behavior of the continued function.

5.9 Proposition. *If C is a compact subset of $\mathbb{C} \setminus \{0, 1\}$, there exist positive constants C and χ with the property*

$$|\mathbb{E}(\tau; s)| \leq C \cdot y^\chi \quad \text{for all} \quad \tau \in \mathbb{F} \quad \text{and} \quad s \in C.$$

Proof For $\tau \in \mathbb{F}$ one has $|x| \leq \frac{1}{2}$ and $y \geq \frac{1}{2}\sqrt{3}$. Thus we easily verify

$$F_\tau > \frac{1}{4y} I \quad \text{and} \quad F_\tau^{-1} = F_\tau[J] > \frac{1}{4y} I.$$

We substitute these inequalities into Proposition 5.3 and we obtain

$$|\Theta(it; T) - 1| \leq C_1 \cdot e^{-t/4y} \quad \text{for all} \quad t \geq 1, \quad T = F_\tau \quad \text{and} \quad T = F_\tau^{-1}, \quad \tau \in \mathbb{F},$$

with a suitable constant C_1. Then we choose $\chi > 0$ such that $\operatorname{Re} s \leq \chi$ and also $1 - \operatorname{Re} s \leq \chi$ for all $s \in C$. With a constant C_2 depending only on C, (5.4) yields

$$|\mathbb{E}(\tau;s)| \leq C_2 + 2C_1 \cdot \int_1^\infty e^{-t/4y} \cdot t^{\chi-1} dt \leq C_2 + 2C_1 \cdot \Gamma(\chi) \cdot (4y)^\chi. \qquad \square$$

5.10 Remark. $E(\tau;s)$ is not holomorphic as a function of τ, but it is an eigenfunction of the *hyperbolic* LAPLACE *operator*

$$\tilde{\Delta} = y^2 \left(\frac{\partial^2}{\partial x^2} + \frac{\partial^2}{\partial y^2} \right).$$

Namely,

$$\tilde{\Delta} E(\tau;s) = s(s-1) \cdot E(\tau;s).$$

4. FOURIER series expansion. In analogy with the holomorphic EISENSTEIN series $E_k(\tau)$, the real–analytic EISENSTEIN series $E(\tau;s)$ also possess a FOURIER series expansion. Here, however, the so-called *K–BESSEL functions*

$$K_s(y) := \frac{1}{2} \int_0^\infty t^{s-1} \cdot e^{-(t+1/t)y/2} dt, \quad y > 0, \quad s \in \mathbb{C}, \tag{5.17}$$

occur. We formulate their essential properties in the

5.11 Lemma. *For every $\varepsilon > 0$, there exists some $C > 0$, such that*

$$|K_s(y)| \leq C \cdot e^{-y} \quad \text{for all} \quad y \geq \varepsilon, |\operatorname{Re} s| \leq 1/\varepsilon. \tag{5.18}$$

For any $y > 0$, $s \mapsto K_s(y)$ is an even entire function.

Proof The usual splitting trick yields

$$2K_s(y) = \int_1^\infty t^{s-1} \cdot e^{-(t+1/t)y/2} dt + \int_0^1 t^{s-1} \cdot e^{-(t+1/t)y/2} dt$$

$$= \int_1^\infty (t^s + t^{-s}) \cdot e^{-(t+1/t)y/2} \frac{dt}{t}.$$

Thus, $K_s(y) = K_{-s}(y)$ follows. For $y \geq \varepsilon$ and $\operatorname{Re} s \leq 1/\varepsilon$, using $t + 1/t \geq 2$, leads to

$$|K_s(y)| \cdot e^y \leq \frac{1}{2} \int_1^\infty \left(t^{\operatorname{Re} s} + t^{-\operatorname{Re} s} \right) \cdot e^{-(t+1/t-2)y/2} dt$$

$$\leq \int_1^\infty t^{1/\varepsilon} \cdot e^{-(t+1/t-2)\varepsilon/2} dt =: C.$$

§ 5* The EPSTEIN zeta function and applications

This yields (5.18) and the holomorphy of $s \mapsto K_s(y)$ using the WEIERSTRASS M-Test A.7. □

We now come to the announced

5.12 Theorem. *For $s \in \mathbb{C}$, $s \neq 0, 1$ and all $\tau \in \mathbb{H}$, we have the FOURIER series expansion*

$$\mathbb{E}(\tau; s) = 2\xi(2s) \cdot y^s + 2\xi(2 - 2s) \cdot y^{1-s} \tag{5.19}$$
$$+ 4 \sum_{n \in \mathbb{Z}}{}' |n|^{s-1/2} \sigma_{1-2s}(|n|) \cdot y^{1/2} \cdot K_{s-1/2}(2\pi|n|y) \cdot e^{2\pi i n x}.$$

Proof First suppose Re $s > 1$. Then, as in sect. 2, we get

$$\mathbb{E}(\tau; s) = \pi^{-s} \Gamma(s) \cdot \sum_{m,n \in \mathbb{Z}}{}' \left(\frac{y}{|m\tau + n|^2} \right)^s$$

$$= \sum_{n \in \mathbb{Z}}{}' \pi^{-s} \Gamma(s) (n^2)^{-s} y^s + \sum_{m \in \mathbb{Z}}{}' \sum_{n \in \mathbb{Z}} \pi^{-s} \Gamma(s) \left(\frac{y}{|m\tau + n|^2} \right)^s$$

$$= 2\xi(2s) \cdot y^s + \sum_{m \in \mathbb{Z}}{}' \int_0^\infty t^{s-1} \sum_{n \in \mathbb{Z}} e^{-\pi t ((mx+n)^2 + m^2 y^2)/y} \, dt$$

$$= 2\xi(2s) \cdot y^s + \sum_{m \in \mathbb{Z}}{}' \int_0^\infty t^{s-1} \Theta_{mx,0}(it/y; 1) e^{-\pi t m^2 y} \, dt$$

$$= 2\xi(2s) \cdot y^s + \sum_{m \in \mathbb{Z}}{}' y^{1/2} \cdot \int_0^\infty t^{s-3/2} \Theta_{0,-mx}(iy/t; 1) e^{-\pi t m^2 y} \, dt$$

$$= 2\xi(2s) \cdot y^s + y^{1/2} \sum_{m \in \mathbb{Z}}{}' \sum_{n \in \mathbb{Z}} e^{-2\pi i m n x} \cdot \int_0^\infty t^{s-3/2} e^{-\pi y (m^2 t + n^2/t)} \, dt,$$

where we have applied the Theta Transformation Formula 2.6. Now note that

$$\int_0^\infty t^{s-3/2} \cdot e^{-\pi y m^2 t} \, dt = \pi^{1/2-s} \Gamma\left(s - \frac{1}{2}\right) \cdot y^{1/2-s} \cdot |m|^{1-2s}.$$

We use the substitution $t = \left|\frac{n}{m}\right| r$ for $n \neq 0$. Then (5.17) yields

$$\int_0^\infty t^{s-3/2} \cdot e^{-\pi y (m^2 t + n^2/t)} \, dt = \left|\frac{n}{m}\right|^{s-1/2} \cdot \int_0^\infty r^{s-3/2} \cdot e^{-\pi |mn| y (r + 1/r)} \, dr$$

$$= 2 \left|\frac{n}{m}\right|^{s-1/2} \cdot K_{s-1/2}(2\pi |mn| y).$$

This leads to the following representation for $\mathbb{E}(\tau; s)$:

$$2\xi(2s)y^s + 2\xi(2s-1)y^{1-s} + 2\sum_{mn\neq 0} y^{1/2}\left|\frac{n}{m}\right|^{s-1/2} K_{s-1/2}(2\pi|mn|y)\cdot e^{-2\pi i mnx}.$$

Collecting the terms according to $-mn$ yields (5.19) if we moreover use the functional equation $\xi(2s-1) = \xi(2-2s)$. Now the assertion follows by analytic continuation, because the series in (5.19) converges for all $s \in \mathbb{C}$ by Lemma 5.11. □

5.13 Remarks. a) As $|n|^{s-1/2} \cdot \sigma_{1-2s}(|n|)$ and $K_{s-1/2}(2\pi|n|y)$ are invariant under $s \mapsto 1-s$, the functional equation thus also holds for each individual FOURIER coefficient.

b) One can also write the FOURIER series expansion in the form

$$\mathbb{E}(\tau; s) = 2\xi(2s)\cdot y^{-s} + 2\xi(2-2s)\cdot y^{1-s}$$

$$+ 8\sum_{n=1}^{\infty} n^{s-1/2}\sigma_{1-2s}(n)\cdot y^{1/2} K_{s-1/2}(2\pi ny)\cos(2\pi nx).$$

c) A more general FOURIER series expansion of EPSTEIN zeta functions in the variables of T can be found in A. TERRAS [82], vol. II, 4.5.1. One applies the Theorem on Completing the Square 1.9 in the form

$$T = \begin{pmatrix} T_1 & 0 \\ 0 & T_2 \end{pmatrix}\begin{bmatrix} I & T_{12} \\ 0 & I \end{bmatrix}$$

and uses the periodicity in T_{12}.

5. KRONECKER'S Limit Formula describes the constant coefficient of the LAURENT expansion of $E(\tau; s)$ around $s = 1$. Surprisingly, the DEDEKIND η-function (cf. III, §6) appears. In addition, we use

$$\lim_{s\to 1}\left(\zeta(s) - \frac{1}{s-1}\right) = C_E = \lim_{m\to\infty}\left(\sum_{k=1}^{m}\frac{1}{k} - \log m\right). \quad (5.20)$$

(cf. Theorem B.7), where C_E denotes the so-called EULER MASCHERONI *constant*.

5.14 KRONECKER'S Limit Formula. *For* $\tau \in \mathbb{H}$,

$$\lim_{s\to 1}\left(E(\tau; s) - \frac{\pi}{s-1}\right) = 2\pi\left(C_E - \log 2 - \log(\sqrt{y}\cdot|\eta(\tau)|^2)\right) \quad (5.21)$$

holds.

Proof From the identity

$$-\log(1-q) = \sum_{m=1}^{\infty}\frac{1}{m}q^m \quad \text{for } |q| < 1$$

§ 5* The Epstein zeta function and applications

and III(6.3), we get

$$-2\pi \log |\eta(\tau)|^2 = -2\pi \log(\eta(\tau) \cdot \overline{\eta(\tau)})$$

$$= -2\pi \log \left(e^{-\pi y/6} \prod_{n=1}^{\infty} (1 - e^{2\pi in\tau})(1 - e^{-2\pi in\overline{\tau}}) \right)$$

$$= \frac{\pi^2}{3} y + 2\pi \sum_{n=1}^{\infty} \sum_{m=1}^{\infty} \frac{1}{m} \cdot (e^{2\pi inm\tau} + e^{-2\pi inm\overline{\tau}})$$

$$= \frac{\pi^2}{3} y + 2\pi \sum_{r=1}^{\infty} \sigma_{-1}(r) \cdot e^{2\pi ir\tau} + 2\pi \sum_{r=-1}^{-\infty} \sigma_{-1}(|r|) \cdot e^{2\pi ir\overline{\tau}},$$

since we may rearrange the series arbitrarily because of the absolute convergence. Theorem 5.12 and Theorem 5.8 now lead to

$$E(\tau; s) = \frac{\pi^s}{\Gamma(s)} \mathbb{E}(\tau; s) = 2\zeta(2s) y^s + 2\sqrt{\pi} \frac{\Gamma(s - 1/2)}{\Gamma(s)} \zeta(2s - 1) y^{1-s}$$
$$+ \frac{4\pi^s}{\Gamma(s)} \sum_{n \in \mathbb{Z}}' |n|^{s-1/2} \sigma_{1-2s}(|n|) \sqrt{y} \, K_{s-1/2}(2\pi |n| y) \, e^{2\pi inx}.$$

Using (5.17), we calculate

$$K_{1/2}(y) = \frac{1}{2} \int_0^\infty t^{-1/2} \cdot e^{-(t+1/t)y/2} dt = \frac{1}{2} \int_1^\infty \left(t^{1/2} + t^{-1/2} \right) \cdot e^{-(t+1/t)y/2} \frac{dt}{t}$$

$$= e^{-y} \cdot \int_1^\infty \frac{1}{2} \left(t^{-1/2} + t^{-3/2} \right) \cdot e^{-(t^{1/2} - t^{-1/2})^2 y/2} dt$$

$$= e^{-y} \cdot \int_0^\infty e^{-r^2 y/2} dr = \sqrt{\frac{\pi}{2y}} \, e^{-y}$$

where we apply Corollary A.14 in the last step. This leads to

$$\lim_{s \to 1} \left(E(\tau; s) - \frac{\pi}{s-1} \right)$$
$$= R + 2\zeta(2) y + \frac{4\pi}{\Gamma(1)} \cdot \sum_{n \in \mathbb{Z}}' \sqrt{|n| y} \, \sigma_{-1}(|n|) \cdot K_{1/2}(2\pi |n| y) \cdot e^{2\pi inx}$$
$$= R + \frac{\pi^2}{3} y + 2\pi \sum_{n \in \mathbb{Z}}' \sigma_{-1}(|n|) \cdot e^{-2\pi |n| y + 2\pi inx}$$
$$= R - 2\pi \cdot \log |\eta(\tau)|^2$$

where

$$R = \lim_{s \to 1} \left(2\sqrt{\pi} \, \frac{\Gamma(s-1/2)}{\Gamma(s)} y^{1-s} \cdot \zeta(2s-1) - \frac{\pi}{s-1} \right)$$

$$= \lim_{s \to 1} \left(f(s)\zeta(2s-1) - \frac{f(1)/2}{s-1} \right)$$

$$= 2\pi \, C_E + \tfrac{1}{2} f'(1),$$

$$f(s) = 2\sqrt{\pi} \cdot \frac{\Gamma(s-1/2)}{\Gamma(s)} \cdot y^{1-s}, \quad f(1) = 2\pi$$

where we use (5.20). Now observe that

$$\tfrac{1}{2} f'(1) = \pi \left. \frac{f'(s)}{f(s)} \right|_{s=1} = \pi \cdot \left(\frac{\Gamma'(1/2)}{\Gamma(1/2)} - \frac{\Gamma'(1)}{\Gamma(1)} - \log y \right).$$

Logarithmic differentiation in the doubling formula IV(4.12) yields

$$\frac{\Gamma'(s)}{\Gamma(s)} = \log 2 + \frac{1}{2} \frac{\Gamma'(s/2)}{\Gamma(s/2)} + \frac{1}{2} \frac{\Gamma'((s+1)/2)}{\Gamma((s+1)/2)}.$$

When $s = 1$, we get

$$\frac{\Gamma'(1)}{\Gamma(1)} - \frac{\Gamma'(1/2)}{\Gamma(1/2)} = 2\log 2$$

and this yields the assertion (5.21). □

5.15 Remarks. a) Theorem 5.14 is due to L. KRONECKER ([50], vol. IV, 222, 347–495; vol. V, 1–132). An alternative proof was given by M. KOECHER (*Arch. Math.* **4**, 316–321 (1953)). There are applications of KRONECKER's Limit Formula, e.g., in the Theorem of COATES and WILES about the *L*-functions associated to elliptic curves with complex multiplication. Compare A. TERRAS [82], vol. I, §3.5.

b) More generally, we call Γ–invariant eigenfunctions of the hyperbolic LAPLACE operator $\tilde{\Delta}$ which still satisfy a certain growth condition MAASS *wave forms*. They can be regarded as a real-analytic analogon of modular forms. A detailed description of this theory can be found in H. MAASS [56] and A. TERRAS [82], and for the three-dimensional hyperbolic space in J. ELSTRODT, F. GRUNEWALD and J. MENNICKE [21].

6. The RANKIN convolution. In this section, we describe results of R.A. RANKIN (*Proc. Cambridge Phil. Soc.* **35**, 357–372) from 1939, with which we can obtain a sharper estimate for the growth behavior of the FOURIER coefficients of cusp forms. If there are two modular forms $f, g \in \mathbb{M}_k$ with FOURIER coefficients $\alpha_f(m)$ and $\alpha_g(m)$, we define the DIRICHLET series

$$D_{f,g}(s) := \sum_{m=1}^{\infty} \alpha_f(m) \cdot \overline{\alpha_g(m)} \cdot m^{-s}, \tag{5.22}$$

§ 5* The Epstein zeta function and applications

which converges absolutely for $s \in \mathbb{C}$ with $\operatorname{Re} s > 2k - 1$ because of III(2.13). $D_{f,g}(s)$ is called the Rankin *convolution* of $D_f(s)$ and $D_g(s)$ (cf. IV(4.16)). If $\langle \cdot, \cdot \rangle$ denotes again the Petersson scalar product IV(3.4), then we get

5.16 Theorem. *If $f, g \in \mathbb{M}_k$ and f or g is a cusp form, then $D_{f,g}(s)$ has a meromorphic continuation into the whole complex s-plane. For $s = k$, there may be a 1st order pole with the residue*

$$\operatorname{res}_{s=k} D_{f,g}(s) = 3 \cdot \frac{2^{2k} \pi^{k-1}}{(k-1)!} \cdot \langle f, g \rangle$$

If we define

$$\mathbb{D}_{f,g}(s) := (2\pi)^{-2s} \Gamma(s) \Gamma(s + 1 - k) \cdot \zeta(2s + 2 - 2k) \cdot D_{f,g}(s),$$

then

$$\mathbb{D}_{f,g}(s) - \frac{1}{2} \pi^{1-k} \langle f, g \rangle \cdot \left(\frac{1}{s-k} - \frac{1}{s+1-k} \right)$$

possesses a continuation as an entire function into the complex s-plane and satisfies the functional equation

$$\mathbb{D}_{f,g}(2k - 1 - s) = \mathbb{D}_{f,g}(s).$$

Proof For $\operatorname{Re} s > k$, we calculate the integral

$$\mathcal{I} = \left\langle f(\tau) \cdot E^*(\tau; s + 1 - k), g(\tau) \right\rangle.$$

Because of $f(\tau)\overline{g(\tau)} y^k = O(e^{-2\pi y})$ for $\tau \in \mathbb{F}$ by Lemma III.1.4, the absolute convergence of $\mathcal{I}$ follows from the Convergence Lemma 5.6. We fix a system of representatives $\mathcal{R}$ for the right cosets of Γ modulo Γ_∞. By (5.12) and Proposition IV.3.5, it follows that

$$\mathcal{I} = \int_{\mathbb{F}} y^k f(\tau)\overline{g(\tau)} \cdot \sum_{M \in \mathcal{R}} (\operatorname{Im} M\tau)^{s+1-k} dv$$

$$= \sum_{M \in \mathcal{R}} \int_{\mathbb{F}} f(M\tau)\overline{g(M\tau)} \cdot (\operatorname{Im} M\tau)^{s+1} dv = \int_{\mathbb{F}_\infty} f(\tau)\overline{g(\tau)} \cdot y^{s+1} dv,$$

where $\mathbb{F}_\infty$ is a fundamental domain with respect to Γ_∞. Since the integrand is invariant under Γ_∞, we may choose

$$\mathbb{F}_\infty = \{ \tau \in \mathbb{H} \,;\, 0 \leq x \leq 1 \}$$

by Lemma IV.3.10. Now we insert the Fourier series of f and g. Since at least one of the two functions f or g is a cusp form, we have absolute convergence due to the growth behavior of the Fourier coefficients (cf. Proposition III.1.7 and III(2.13)) for $\operatorname{Re} s > 3k/2$, such that we may interchange summation and integration for these values of s. Because $\alpha_f(0) \cdot \alpha_g(0) = 0$, we obtain

$$\mathcal{I} = \sum_{m\geq 0}\sum_{n\geq 0} \alpha_f(m)\overline{\alpha_g(n)} \cdot \int_0^\infty \left(\int_0^1 e^{2\pi i(m-n)x}\,dx \right) \cdot e^{-2\pi(m+n)y} \cdot y^{s-1}\,dy$$

$$= \sum_{m\geq 1} \alpha_f(m)\overline{\alpha_g(m)} \cdot \int_0^\infty y^{s-1} \cdot e^{-4\pi my}\,dy = (4\pi)^{-s}\Gamma(s) \cdot D_{f,g}(s) \,.$$

This immediately leads to

$$\mathbb{D}_{f,g}(s) = \frac{1}{2}\pi^{1-k} \cdot \left\langle f(\tau) \cdot \mathbb{E}(\tau; s+1-k),\, g(\tau) \right\rangle \tag{5.23}$$

for $\operatorname{Re} s > 3k/2$. By Theorem 5.8, the right-hand side of (5.23) possesses a meromorphic continuation into the whole complex s-plane with possible simple poles at $s = k$ and $s = k - 1$ and residues

$$\operatorname*{res}_{s=k} \mathbb{D}_{f,g}(s) = -\operatorname*{res}_{s=k-1} \mathbb{D}_{f,g}(s) = \frac{1}{2}\pi^{1-k}\langle f, g\rangle \,.$$

The functional equation of $\mathbb{D}_{f,g}(s)$ now follows from the functional equation of $\mathbb{E}(\tau; s)$ in Theorem 5.8. The assertions still to be proved are simple corollaries. □

We obtain a different result if neither f nor g is a cusp form. Since the definition of $D_{f,g}(s)$ is linear in f and g, it suffices to consider $f = g = E_k^*$ in this case.

5.17 Proposition. *Let $k \geq 4$ be even. Then, for $\operatorname{Re} s > 2k - 1$,*

$$D_{E_k^*,E_k^*}(s) = \left(\frac{2k}{B_k}\right)^2 \frac{\zeta(s) \cdot \zeta(s+1-k)^2 \cdot \zeta(s+2-2k)}{\zeta(2s+2-2k)}. \tag{5.24}$$

$D_{E_k^*,E_k^*}(s)$ *possesses a meromorphic continuation into the whole complex s-plane and has simple poles at $s = 2k - 1$ as well as $s = k$. The function*

$$\mathbb{D}_{E_k^*,E_k^*}(s) := (2\pi)^{-2s}\Gamma(s)\Gamma(s+1-k) \cdot \zeta(2s+2-2k) \cdot D_{E_k^*,E_k^*}(s)$$

has a meromorphic continuation into the whole s-plane. Its only singularities are simple poles at $s = 0, k-1, k, 2k-1$. It satisfies the functional equation

$$\mathbb{D}_{E_k^*,E_k^*}(2k-1-s) = \mathbb{D}_{E_k^*,E_k^*}(s).$$

Proof We use the FOURIER series expansion from III(2.9). Since the divisor sums are multiplicative, we obtain an EULER product which can easily be calculated with the summation formula for the geometric series:

$$D_{E_k^*,E_k^*}(s) = \left(\frac{2k}{B_k}\right)^2 \cdot \sum_{m\geq 1} \sigma_{k-1}(m)^2 \cdot m^{-s} = \left(\frac{2k}{B_k}\right)^2 \cdot \prod_p F_p$$

where

$$F_p = \sum_{n\geq 0} \sigma_{k-1}(p^n)^2 \cdot p^{-ns} = \sum_{n\geq 0} \left(\frac{p^{(k-1)(n+1)} - 1}{p^{k-1} - 1}\right)^2 \cdot p^{-ns}$$

$$= \frac{1 - p^{2k-2-2s}}{(1 - p^{2k-2-s})(1 - p^{k-1-s})^2(1 - p^{-s})}.$$

From the representation of $\zeta(s)$ as an EULER product according to Corollary IV.4.17, we obtain (5.24). The holomorphic continuation and the statement about the poles now follow from the Theorem IV.4.10, but it should be remarked that the possible pole at $s = k$ cancels, since $\zeta(s)$ has a simple root at $s = 2 - k$ due to the functional equation.

From the doubling IV(4.12) and the functional equation IV(4.10), we can easily obtain the representation

$$\left(\frac{k}{B_k}\right)^2 (2\pi)^{1-2k} \xi(s) \xi(s + 1 - k)^2 \xi(s + 2 - 2k) \prod_{j=1}^{k/2} (s + 1 - 2j)(s + 2 - k - 2j).$$

The missing assertions thus follow from Theorem IV.4.10. □

5.18 Remark. For $0 \neq f = g \in \mathbb{S}_k$, the DIRICHLET series $D_{f,f}(s)$ has a pole at $s = k$. From the calculation of the convergence abscissa of DIRICHLET series and LANDAU's Theorem (cf. e.g. D. ZAGIER [90], §1), we conclude that

$$\sum_{n=1}^{N} |\alpha_f(n)|^2 = O(N^{k+\varepsilon}) \quad \text{for} \quad N \to \infty \quad \text{and each} \quad \varepsilon > 0.$$

This statement improves the estimate of Proposition III.1.7. Compare R.A. RANKIN, *Proc. Cambridge Phil. Soc.* **35**, 357–372 (1939).

5.19 Exercises.
1) $\zeta(I^{(2)}; s) = 4\zeta(s) \cdot L(s; \chi) = E(i; s)$, where χ is the non–trivial DIRICHLET character mod 4.
2) $\zeta(I^{(4)}; s) = 8(1 - 2^{2-2s}) \cdot \zeta(s) \cdot \zeta(s - 1)$ (cf. sect. 7 in III, §7).
3) $\zeta(I^{(8)}; s) = 16(1 - 2^{1-s} + 2^{4-2s}) \cdot \zeta(s) \cdot \zeta(s - 3)$ (cf. sect. 7 in III, §7).
4) $\zeta(S_8; s) = 240 \cdot 2^{-s} \cdot \zeta(s) \cdot \zeta(s - 3)$.
5) $\zeta(L_{24}; s) = \frac{65\,520}{691} \cdot 2^{-s} \cdot (\zeta(s) \cdot \zeta(s - 11) - D_{\Delta^*}(s))$ (cf. IV.(4.16)).
6) For all $m \geq 2$ one has $\sharp(L_{24}, 2m) \equiv 0 \pmod{65\,520}$.
7) $K_{3/2}(y) = \sqrt{\frac{\pi}{2y}} \cdot e^{-y} \cdot (1 + \frac{1}{y})$ for $y > 0$.
8) $\frac{d}{dy} K_s(y) = -\frac{1}{2}(K_{s+1}(y) + K_{s-1}(y))$ for $y > 0$ and $s \in \mathbb{C}$.
9) $\zeta(3) = \frac{2}{\pi} E(i; 2) - \frac{\pi^3}{45} - 2 \sum_{n=1}^{\infty} \sigma_3(n) \cdot (2n\pi + 1) \cdot e^{-2\pi n}$.
10) Let $T \in \text{Pos}(2; \mathbb{R})$, $s \in \mathbb{C}$, $\text{Re}\, s > 1$. For $n \geq 1$, the expression

$$T'_n\zeta(T;s) := \sum_{A \in \Gamma:\Gamma_n} \zeta(T[A^{tr}];s)$$

(cf. Proposition IV.1.4) is well-defined. For a prime number p and $n \in \mathbb{N}$, one has

$$T'_p\zeta(T;s) = (1 + p^{1-2s}) \cdot \zeta(T;s), \quad T'_n\zeta(T;s) = \sigma_{1-2s}(n) \cdot \zeta(T;s).$$

11) For $n \geq 1$ and $s \in \mathbb{C}$ with $\mathrm{Re}\, s > 1$, one has

$$T'_n E(\tau;s) := \sum_{M \in \Gamma:\Gamma_n} E(M\tau;s) = \sigma_{1-2s}(n) \cdot E(\tau;s).$$

12) Let $T \in \mathrm{Pos}(2;\mathbb{R})$ and $\xi^*(T;s) := (\det T)^{s/2} \pi^{-s} \Gamma(s) \cdot \zeta(T;s)$. Then $\xi^*(T;s)$ satisfies the functional equation $\xi^*(T; 1-s) = \xi^*(T;s)$.

13) Demonstrate the convergence behavior of the EPSTEIN zeta function by an integral test.

14) Given $p, q \in \mathbb{R}^n$ and $T \in \mathrm{Pos}(n;\mathbb{R})$, define the *generalized* EPSTEIN *zeta function*

$$\zeta_{p,q}(T;s) := \sum_{g \in \mathbb{Z}^n,\, g+p \neq 0} e^{2\pi i (g+p)^{tr} q} (T[g+p])^{-s}, \quad s \in \mathbb{C}.$$

Investigate the convergence behavior and holomorphic continuation, as well as the functional equation in analogy with Theorem 5.4.

15) Let $K = \mathbb{Q}(\sqrt{d})$, $d < 0$ be square-free, an imaginary-quadratic number field of discriminant D and $\mathfrak{o}$ the corresponding ring of integers, i.e. $\mathfrak{o} = \mathbb{Z} + \mathbb{Z}\frac{1}{2}(1 + \sqrt{d})$ and $D = d$ if $d \equiv 1 \pmod 4$, resp. $\mathfrak{o} = \mathbb{Z} + \mathbb{Z}\sqrt{d}$ and $D = 4d$ otherwise. For $a \in \mathbb{C}$, let $N(a) = a\bar{a} = |a|^2$ and $\zeta_\mathfrak{o}(s) := \sum_{0 \neq a \in \mathfrak{o}} N(a)^{-s}$. This series converges absolutely for $\mathrm{Re}\, s > 1$ and has a holomorphic continuation into the s-plane except for a simple pole at $s = 1$ with the residue $\frac{2\pi}{\sqrt{-D}}$. The function $\xi_\mathfrak{o}(s) := |D|^{s/2} (2\pi)^{-s} \Gamma(s) \cdot \zeta_\mathfrak{o}(s)$ satisfies the functional equation

$$\xi_\mathfrak{o}(1-s) = \xi_\mathfrak{o}(s).$$

Appendix A.
Complex analytic foundations

In this appendix A, we collect the most important results from a complex analysis course which are cited in the text.

A.1 Identity Theorem. *Let $D \subseteq \mathbb{C}$ be a connected domain, $f : D \to \mathbb{C}$ holomorphic and $f \not\equiv 0$. The set of roots $\{z \in D;\ f(z) = 0\}$ is closed and discrete in D, i.e. does not possess a limit point in D.*

Proof J.B. CONWAY [13], IV.3.7. □

A.2 RIEMANN's Theorem on Removable Singularities. *Let $D \subseteq \mathbb{C}$ be open, $a \in \mathbb{C}$ and $f : D\setminus\{a\} \to \mathbb{C}$ holomorphic. Then the following assertions are equivalent:*
(i) *f possesses a holomorphic continuation to a.*
(ii) *f is bounded on some punctured neighborhood of a.*
(iii) *f possesses a continuous continuation to a.*
(iv) $\lim_{z \to a}(z - a)f(z) = 0$.

Proof J.B. CONWAY [13], V §1. □

A.3 Theorem on the LAURENT Series Development. *Let f be holomorphic on an annulus*
$$D = \{z \in \mathbb{C};\ r < |z - a| < R\},\ 0 \leq r < R \leq \infty,\ a \in \mathbb{C}.$$

Then f possesses an absolutely convergent LAURENT series development at a,

$$f(z) = \sum_{n=-\infty}^{\infty} a_n(z - a)^n, \quad z \in D,$$

which converges uniformly on compact subsets of D.

Proof J.B. CONWAY [13], V.1.11. □

In particular, A.3 contains the *power series development* of holomorphic functions as a special case if we apply A.2.

A.4 Theorem on the FOURIER Series Development. *Let f be holomorphic on stripe*
$$D = \{z \in \mathbb{C};\ a < \operatorname{Im} z < b\},\ -\infty \leq a < b \leq \infty$$
and periodic with period 1, i.e. $f(z+1) = f(z)$. Then f possesses an absolutely convergent FOURIER series development
$$f(z) = \sum_{n=-\infty}^{\infty} a_n e^{2\pi i n z},\ z \in D,$$
where the FOURIER coefficients are uniquely determined and given by
$$a_n = \int_0^1 f(z) e^{-2\pi i n z} dx,\ z = x + iy \in D,\ n \in \mathbb{Z}.$$

Proof E. FREITAG and R. BUSAM [27], III.5.4. Just consider the LAURENT series development of
$$F: \{q \in \mathbb{C};\ e^{-2\pi b} < |q| < e^{-2\pi a}\} \to \mathbb{C},\ q \mapsto f\left(\frac{\log q}{2\pi i}\right),\ q = e^{2\pi i z}. \qquad \square$$

A.5 Residue Theorem. *Let $D \subseteq \mathbb{C}$ be an open domain and f holomorphic on D except for isolated singularities at the points $a_1, \ldots, a_N$ in D. If γ is a closed path in D which is piecewise continously differentiable and homologous to zero in D, then*
$$\frac{1}{2\pi i} \int_\gamma f(z) dz = \sum_{k=1}^N n_{a_k}(\gamma) \cdot \operatorname{res}_{a_k}(f).$$
Here, $n_a(\gamma)$ denotes the *winding number* of a and $\operatorname{res}_a(f)$ the *residue* of f at a.

Proof J.B. CONWAY [13], V.2.2. $\qquad \square$

The particular case of a holomorphic f on D is formulated as

A.6 CAUCHY's Integral Theorem. *Let $D \subseteq \mathbb{C}$ be an open domain and $f : D \to \mathbb{C}$ holomorphic. If γ is a closed path in D satisfying $n_a(\gamma) = 0$ for all $a \in \mathbb{C} \backslash D$, then*
$$\int_\gamma f(z) dz = 0.$$

Proof J.B. CONWAY [13], IV. 5.7. $\qquad \square$

A.7 WEIERSTRASS M-Test. *Let $D \subseteq \mathbb{C}$ be an open domain, f_n, $f: D \to \mathbb{C}$, $n \in \mathbb{N}$.*
a) *If f_n, $n \in \mathbb{N}$, are holomorphic and the sequence $(f_n)_n$ converges uniformly to f on each compact subset of D, then f is holomorphic and*
$$\lim_{n \to \infty} f'_n = f'$$

Appendix A

holds.

b) *If f_n, $n \in \mathbb{N}$ are holomorphic, $|f_n(z)| \leq M_n$ for $z \in D$ and $\sum_{n=1}^{\infty} M_n < \infty$, then $\sum_{n=1}^{\infty} f_n$ is holomorphic and*

$$\left(\sum_{n=1}^{\infty} f_n\right)' = \sum_{n=1}^{\infty} f_n'.$$

Proof J.B. CONWAY [13], VII.2.1 and VII.2.4. □

We describe an immediate application to improper integrals.

A.8 Corollary. *Let $D \subseteq \mathbb{C}$ be an open domain, and let $f : D \times \mathbb{R} \to \mathbb{C}$ be continuous and holomorphic in the first variable. If there is a continuous function $g : \mathbb{R} \to \mathbb{R}$ such that*

$$|f(z,t)| \leq g(t) \text{ for all } (z,t) \in D \times \mathbb{R} \text{ and } \int_{-\infty}^{\infty} g(t)dt < \infty,$$

then the function

$$F : D \to \mathbb{C}, \quad F(z) := \int_{-\infty}^{\infty} f(z,t)dt,$$

is holomorphic and satisfies

$$F'(z) = \int_{-\infty}^{\infty} \frac{\partial}{\partial z} f(z,t)dt.$$

Proof Apply A.7 to the uniformly convergent series $\left(\int_{-n}^{n} f(z,t)dt\right)_{n \geq 1}$. □

An infinite product $\prod_{n=1}^{\infty}(1 + a_n)$, $a_n \in \mathbb{C}$, is said to *converge absolutely* if $\sum_{n=1}^{\infty} |a_n| < \infty$. In this case,

$$P_N := \prod_{n=1}^{N}(1 + a_n) \xrightarrow[N \to \infty]{} \prod_{n=1}^{\infty}(1 + a_n).$$

A.9 WEIERSTRASS M-Test for Products. *Let $D \subseteq \mathbb{C}$ be an open domain and $f_n : D \to \mathbb{C}$ holomorphic such that $\sum_{n=1}^{\infty}(f_n(z) - 1)$ converges absolutely and uniformly on all compact subsets of D. Then*

$$f(z) = \prod_{n=1}^{\infty} f_n(z)$$

is holomorphic and satisfies

$$\frac{f'(z)}{f(z)} = \sum_{n=1}^{\infty} \frac{f_n'(z)}{f_n(z)} \quad \text{for all } z \in D \text{ with } f_n(z) \neq 0 \text{ for all } n \in \mathbb{N}.$$

Proof J.B. CONWAY [13], VII.2.1 and VII.5.9. □

Moreover, let $E_0(z) = 1 - z$ and
$$E_k(z) = (1-z)\exp\left(z + \frac{z^2}{2} + \ldots + \frac{z^k}{k}\right), \quad k \in \mathbb{N}.$$

A.10 WEIERSTRASS Product Theorem. *Let $(a_n)_{n \geq 1}$ be a sequence in $\mathbb{C}\backslash\{0\}$ satisfying $\lim_{n \to \infty} a_n = \infty$. Then there exist $p_n \in \mathbb{N}_0$ such that*
$$\sum_{n=1}^{\infty} \left(\frac{r}{|a_n|}\right)^{p_n+1} < \infty \quad \text{for all} \quad r > 0,$$
e.g. $p_n = n - 1$. In this case,
$$f(z) = \prod_{n=1}^{\infty} E_{p_n}(z/a_n)$$
is an entire function with zeros exactly at the points a_n, $n \in \mathbb{N}$. The multiplicity is given by the number of repetitions in the sequence $(a_n)_{n \geq 1}$.

Proof J.B. CONWAY [13], VII. 2.1 and VII. 5.12. □

A.11 Corollary. *Each meromorphic function on $\mathbb{C}$ can be represented as a quotient of two entire functions.*

Proof Construct a product by A.10 which has zeros at the poles of f with the appropriate multiplicities. Then gf is an entire function. □

A.12 LIOUVILLE's Theorem. *If f is an entire bounded function, then f is constant.*

Proof J.B. CONWAY [13], IV.3. 4. □

A.13 Theorem on the Gamma Function. *The gamma function*
$$\Gamma(z) = \int_0^\infty t^{z-1} e^{-t} dt, \quad z \in \mathbb{C}, \ \operatorname{Re} z > 0$$
is holomorphic on $\{z \in \mathbb{C}; \ \operatorname{Re} z > 0\}$. It possesses a meromorphic continuation to the whole z-plane and is holomorphic except for simple poles at $z = -n$ with residue $(-1)^n/n!$, $n \in \mathbb{N}_0$. It satisfies the functional equations
$$\Gamma(z+1) = z \cdot \Gamma(z) \quad \Gamma(1) = 1,$$
$$\Gamma(z) = \frac{1}{\sqrt{\pi}} \cdot 2^{z-1} \cdot \Gamma\left(\frac{z}{2}\right) \cdot \Gamma\left(\frac{z+1}{2}\right).$$

Furthermore,
$$\Gamma(n) = (n-1)!, \ n \in \mathbb{N}, \quad \Gamma(1/2) = \sqrt{\pi}$$
holds. $1/\Gamma(z)$ is an entire function. Moreover, it satisfies the complex STIRLING formula, i.e. for $\varepsilon > 0$ there exists some $C > 0$ such that

Appendix A

$$|\log \Gamma(z) - (z - \frac{1}{2})(\log z) + z| \leq C \quad \text{for } \operatorname{Re} z \geq 0, \ |z| \geq \varepsilon.$$

Proof J.B. CONWAY [13], VII. §7. For the complex STIRLING formula, compare E. FREITAG and R. BUSAM [27], IV.1.14. □

A.14 Corollary. *For $z \in \mathbb{C}$*

$$\int_{-\infty}^{\infty} e^{-\pi(z+t)^2} dt = 1$$

holds.

Proof $F(z) := \int_{-\infty}^{\infty} e^{-\pi(z+t)^2} dt$ is an entire function due to A.8, satisfying

$$F'(z) = \int_{-\infty}^{\infty} -2\pi(z+t) e^{-\pi(z+t)^2} dt = \lim_{N \to \infty} e^{-\pi(z+t)^2} \bigg|_{t=-N}^{t=N} = 0.$$

Thus F is constant and the substitution $s = \pi t^2$ yields

$$F(0) = 2 \int_0^{\infty} e^{-\pi t^2} dt = \frac{1}{\sqrt{\pi}} \int_0^{\infty} s^{-1/2} e^{-s} ds = \frac{\Gamma(1/2)}{\sqrt{\pi}} = 1,$$

if we use A.13. □

A.15 Maximum Principle. *Let $D \subseteq \mathbb{C}$ be open and connected, $C \subseteq D$ compact and $f : D \to \mathbb{C}$ holomorphic. Then there exists some $z_0 \in \partial C$ such that*

$$|f(z)| \leq |f(z_0)| \quad \text{for all} \quad z \in C. \tag{$*$}$$

If there exists some z_0 in the interior of C satisfying $(*)$, then f is constant.

Proof J.B. CONWAY [13], VI §1. □

We need a generalization which is not standard in all complex analysis courses.

A.16 PHRAGMEN-LINDELÖF Theorem. *For $\alpha < \beta$, consider the vertical strip*

$$\mathcal{V}_{\alpha,\beta} := \{s \in \mathbb{C}; \ \alpha \leq \operatorname{Re} s \leq \beta\}.$$

Let ϕ be a holomorphic function on a domain containing $\mathcal{V}_{\alpha,\beta}$ satisfying

$$|\phi(s)| \leq c e^{|t|^{\gamma}} \quad \text{for all} \quad s = \sigma + it \in \mathcal{V}_{\alpha,\beta}$$

with some positive constants c, γ. If there are $\delta \in \mathbb{R}$ and $m > 0$ such that

$$|\phi(s)| \leq m(1 + |t|)^{\delta} \quad \text{for all} \quad \operatorname{Re} s = \alpha, \beta,$$

then there is some $M > 0$ such that

$$|\phi(s)| \leq M(1 + |t|)^{\delta} \quad \text{for all} \quad s \in \mathcal{V}_{\alpha,\beta}.$$

Proof (Cf. J.B. Conway [13], VI §4 or T. Miyake [59], 4.3.4)
Case 1. Let $\delta = 0$, i.e. $|\phi(s)| \leq m$ for $\operatorname{Re} s = \alpha, \beta$. First of all we choose some $n \in \mathbb{N}$, $n > \gamma$, $n \equiv 2 \pmod 4$. Then we obtain some $T > 0$ such that

$$-2t^n \leq \operatorname{Re}(s^n) \leq -t^n/2 \quad \text{for all} \quad s \in \mathcal{V}_{\alpha,\beta}, |t| \geq T.$$

Thus, we get an $N \in \mathbb{R}$ such that

$$\operatorname{Re}(s^n) \leq N \quad \text{for all} \quad s \in \mathcal{V}_{\alpha,\beta}.$$

Hence we have

$$\left|\phi(s)e^{\varepsilon s^n}\right| \leq me^{\varepsilon N} \quad \text{for } \operatorname{Re} s = \alpha, \beta$$

for all $\varepsilon > 0$, as well as

$$\left|\phi(s)e^{\varepsilon s^n}\right| \leq c \exp(|t|^\gamma - \varepsilon t^n/2) \xrightarrow[|t|\to\infty]{} 0$$

for all $s \in \mathcal{V}_{\alpha,\beta}$, $|t| \geq T$. Thus the Maximum Principle A.15, applied to the rectangle $\{s \in \mathcal{V}_{\alpha,\beta}; |t| \leq T\}$, yields

$$\left|\phi(s)e^{\varepsilon s^n}\right| \leq \max\{me^{\varepsilon N}, c\exp(T^\gamma - \varepsilon T^n/2)\} \quad \text{for all} \quad s \in \mathcal{V}_{\alpha,\beta}.$$

Letting $\varepsilon \to 0$, we obtain

$$|\phi(s)| \leq \max\{m, c\exp(T^\gamma)\} =: M \quad \text{for all} \quad s \in \mathcal{V}_{\alpha,\beta}.$$

Case 2. Let δ be arbitrary. Then consider

$$\psi(s) = (1 - \alpha + s)^\delta.$$

As we have

$$\frac{1}{2}(1 + |t|) \leq |1 - \alpha + s| \leq c_1(1 + |t|), \ c_1 = 1 + \beta - \alpha,$$

for all $s \in \mathcal{V}_{\alpha,\beta}$, we conclude that $\tilde{\phi}(s) = \phi(s)/\psi(s)$ fulfills the assumptions of case 1. Thus $\tilde{\phi}(s)$ is bounded on $\mathcal{V}_{\alpha,\beta}$, which yields the claim. □

B. The RIEMANN zeta function

In this appendix, we collect results on the RIEMANN zeta function.

B.1 Theorem. *The* RIEMANN *zeta function*

$$\zeta(s) = \sum_{n=1}^{\infty} n^{-s}$$

is absolutely convergent and holomorphic for Re $s > 1$ *and satisfies*

$$\zeta(s) = \prod_{p \text{ prime}} (1 - p^{-s})^{-1} \neq 0, \quad \text{Re } s > 1.$$

Proof From calculus, we know that the series $\zeta(\sigma)$ is absolutely convergent due to

$$\zeta(\sigma) \leq 1 + \int_1^{\infty} t^{-\sigma} dt = \frac{\sigma}{\sigma - 1} < \infty$$

for all $\sigma \in \mathbb{R}$, $\sigma > 1$. Hence, it is holomorphic by the WEIERSTRASS M-Test A.7. The infinite product is absolutely convergent due to

$$\sum_p \left| 1 - (1 - p^{-s})^{-1} \right| \leq \sum_p \sum_{k=1}^{\infty} p^{-k\sigma} \leq \zeta(\sigma) < \infty \quad \text{for } \sigma = \text{Re } s > 1,$$

using the geometric series. Hence, the EULER product is absolutely convergent and $\neq 0$ for Re $s > 1$. The Fundamental Theorem of Arithmetic therefore yields

$$\prod_p (1 - p^{-s})^{-1} = \prod_p \sum_{k=0}^{\infty} p^{-ks} = \sum_{n=1}^{\infty} n^{-s} = \zeta(s). \qquad \square$$

For further applications, we need

B.2 Theorem on the Partial Fractions Development of the Cotangent.

$$\pi\cot(\pi z) = \frac{1}{z} + \sum_{n=1}^{\infty}\left(\frac{1}{z+n} + \frac{1}{z-n}\right), \quad z \in \mathbb{C}\setminus\mathbb{Z},$$

holds, where the series converges absolutely and uniformly on every compact subset of $\mathbb{C}\setminus\mathbb{Z}$.

Proof (Cf. J.B. CONWAY [1973], VII. 6.1) If $c > 0$, $|z| \leq c, n \geq 2c$, we have

$$\left|\frac{1}{z+n} + \frac{1}{z-n}\right| \leq \frac{2|z|}{n^2 - |z|^2} \leq \frac{8c}{3n^2},$$

such that $\zeta(2)$ is a majorant and

$$f(z) = \frac{1}{z} + \sum_{n=1}^{\infty}\left(\frac{1}{z+n} + \frac{1}{z-n}\right)$$

is holomorphic in $\mathbb{C}\setminus\mathbb{Z}$ with simple poles with residue 1 at $n \in \mathbb{Z}$. We obtain

$$f(z+1) - f(z)$$
$$= \frac{1}{z+1} - \frac{1}{z} + \lim_{N\to\infty}\sum_{n=1}^{N}\left(\frac{1}{z+n+1} + \frac{1}{z+1-n} - \frac{1}{z+n} - \frac{1}{z-n}\right)$$
$$= \lim_{N\to\infty}\left(\frac{1}{z+N+1} - \frac{1}{z-N}\right) = 0.$$

Moreover, we have for $|y| = |\text{Im } z| \geq 1$, $|x| = |\text{Re } z| \leq 1$,

$$|z - n^2/z| = |z - \bar{z}n^2/|z|^2| \geq |y|(1 + n^2/|z|^2),$$

thus

$$|f(z)| \leq \frac{1}{|z|} + \sum_{n=1}^{\infty}\left|\frac{2z}{z^2 - n^2}\right|$$
$$\leq 1 + \frac{2}{|y|}\sum_{n=1}^{\infty}\frac{1}{1 + n^2/|z|^2}$$
$$\leq 1 + \frac{2}{|y|}\int_0^{\infty}\frac{1}{1 + t^2/|z|^2}dt$$
$$= 1 + \frac{2|z|}{|y|}\lim_{N\to\infty}\arctan(t/|z|)\,|_0^N$$
$$\leq 1 + 2\pi.$$

Appendix B

As $\cot(\pi(z+1)) = \cot(\pi z)$ and $|\pi \cot(\pi z)| = \pi\left|\left(e^{\pi iz} + e^{-\pi iz}\right)/\left(e^{\pi iz} - e^{-\pi iz}\right)\right| \leq 2\pi$ for $|y| \geq 1$ and $\pi \cot(\pi z)$ has the same poles and residues as $f(z)$ we conclude that the difference $F(z) = \pi \cot(\pi z) - f(z)$ is an entire, bounded function. LIOUVILLE's classical Theorem A.12 shows that $F(z)$ is constant. This constant is 0 because F is odd. □

Differentiation of B.2 yields

B.3 Lemma. *For $z \in \mathbb{C}\backslash\mathbb{Z}$, one has*

$$\left(\frac{\pi}{\sin(\pi z)}\right)^2 = \sum_{n \in \mathbb{Z}} \frac{1}{(z-n)^2}.$$

B.4 Lemma. *There exist $B_k \in \mathbb{Q}$ such that*

$$\frac{z}{e^z - 1} = \sum_{k=0}^{\infty} \frac{B_k}{k!} z^k \quad \text{for} \quad 0 < |z| < 2\pi.$$

$B_0 = 1$, $B_1 = -1/2$, $B_{2k+1} = 0$ *hold for $k \geq 1$ as well as*

$$\sum_{k=0}^{n} \binom{n+1}{k} B_k = 0 \quad \text{for all} \quad n \in \mathbb{N}.$$

Proof $z/(e^z - 1)$ has a removable singularity at $z = 0$ and poles of first order at $z = 2\pi in$, $n \in \mathbb{Z}$, $n \neq 0$. Thus, the power series expansion at $z = 0$ has convergence radius 2π. We have

$$B_0 = \lim_{z \to 0} \frac{z}{e^z - 1} = \frac{1}{\frac{d}{dz} e^z |_{z=0}} = 1$$

as well as

$$z = (e^z - 1) \left(\sum_{k=0}^{\infty} \frac{B_k}{k!} z^k\right)$$

$$= z + \sum_{n=1}^{\infty} \left(\sum_{k=0}^{n} \frac{1}{k!(n+1-k)!} B_k\right) z^{n+1}.$$

This yields $B_1 = -1/2$ and the recursion formula, which also shows that $B_k \in \mathbb{Q}$ for all $k \in \mathbb{N}_0$. As

$$\frac{z}{e^z - 1} + \frac{z}{2}$$

is even, we conclude that $B_{2k+1} = 0$ for all $k \in \mathbb{N}$. □

The B_k are called BERNOULLI *numbers*. The recursion formula yields

$$B_2 = \frac{1}{6}, \quad B_4 = \frac{-1}{30}, \quad B_6 = \frac{1}{42}, \quad B_8 = \frac{-1}{30},$$

$$B_{10} = \frac{5}{66}, \quad B_{12} = \frac{-691}{2730}, \quad B_{14} = \frac{7}{6}.$$

We obtain EULER's *formula* for the values of the RIEMANN zeta function at all positive even integer arguments.

B.5 Corollary. *If* $k \in \mathbb{N}$, *then*

$$\zeta(2k) = (-1)^{k+1} \frac{1}{2(2k)!} B_{2k} (2\pi)^{2k} \in \mathbb{Q}\pi^{2k}.$$

In particular, $(-1)^{k+1} B_{2k} > 0$ *holds for all* $k \in \mathbb{N}$, *as well as*

$$\zeta(2) = \frac{\pi^2}{6}, \quad \zeta(4) = \frac{\pi^4}{90}, \quad \zeta(6) = \frac{\pi^6}{945}.$$

Proof The definition of the cotangent yields

$$\pi z \cdot \cot(\pi z) = \pi i z \frac{e^{\pi i z} + e^{-\pi i z}}{e^{\pi i z} - e^{-\pi i z}}$$

$$= \pi i z \frac{e^{2\pi i z} + 1}{e^{2\pi i z} - 1}$$

$$= \pi i z + \frac{2\pi i z}{e^{2\pi i z} - 1}$$

$$= \pi i z + \sum_{k=0}^{\infty} \frac{B_k}{k!} (2\pi i z)^k$$

$$= 1 + \sum_{k=1}^{\infty} (-1)^k \frac{B_{2k}}{(2k)!} (2\pi)^{2k} z^{2k},$$

if we apply B.4. On the other hand, B.2 yields

$$\pi z \cdot \cot(\pi z) = 1 + \sum_{n=1}^{\infty} \frac{2z^2}{z^2 - n^2}$$

$$= 1 - 2 \sum_{n=1}^{\infty} \sum_{k=1}^{\infty} \frac{1}{n^{2k}} z^{2k}$$

$$= 1 - 2 \sum_{k=1}^{\infty} \zeta(2k) z^{2k}.$$

A comparison of the coefficients yields the formula. The recursion formula in B.4 leads to the special values. □

Next we introduce a particular sequence.

Appendix B

B.6 Proposition. *The limit*
$$C_E := \lim_{n \to \infty} \left(\sum_{k=1}^{\infty} \frac{1}{k} - \log n \right) = \int_1^{\infty} \left(\frac{1}{[t]} - \frac{1}{t} \right) dt$$

exists.
$[t] = \max\{n \in \mathbb{Z}; n \leq t\}$ stands for the *greatest integer function*. $C_E = 0.577215\ldots$ is called the EULER MASCHERONI *constant*.

Proof The sequence $c_n := \sum_{k=1}^{n} \frac{1}{k} - \log n$, $n \in \mathbb{N}$, satisfies

$$c_n - c_{n+1} = -\frac{1}{n+1} - \log n + \log(n+1) = \int_n^{n+1} \left(\frac{1}{t} - \frac{1}{n+1} \right) dt \geq 0.$$

$$c_n = \frac{1}{n} + \sum_{k=1}^{n-1} \left(\frac{1}{k} + \log k - \log(k+1) \right)$$

$$= \frac{1}{n} + \sum_{k=1}^{n-1} \int_k^{k+1} \left(\frac{1}{k} - \frac{1}{t} \right) dt$$

$$= \frac{1}{n} + \int_1^n \left(\frac{1}{[t]} - \frac{1}{t} \right) dt \geq 0.$$

Hence, $(c_n)_{n \geq 1}$ is monotonically decreasing and bounded, thus convergent. □

We demonstrate the meromorphic continuation of $\zeta(s)$ and describe the constant term of the LAURENT series development of $\zeta(s)$ at $s = 1$.

B.7 Theorem. $\zeta(s)$ *possesses a meromorphic continuation on the right half-plane* $\{s \in \mathbb{C}; \operatorname{Re} s > 0\}$. *It is holomorphic except for a simple pole at* $s = 1$ *with residue* 1 *and satisfies*

$$\lim_{s \to 1} \left(\zeta(s) - \frac{1}{s-1} \right) = C_E.$$

Proof For $n \in \mathbb{N}$ and $\sigma = \operatorname{Re} s > 1$, we obtain

$$\sum_{k=1}^{n} \frac{1}{k^s} + \frac{n^{1-s}}{s-1} - s \int_n^{\infty} \frac{t - [t]}{t^{s+1}} dt$$

$$= \sum_{k=1}^{n} \frac{1}{k^s} + \frac{n^{1-s}}{s-1} - \sum_{k=n}^{\infty} s \int_k^{k+1} \left(t^{-s} - kt^{-s-1} \right) dt$$

$$= \sum_{k=1}^{n} \frac{1}{k^s} + \frac{n^{1-s}}{s-1} + \sum_{k=n}^{\infty} \frac{s}{s-1} \left((k+1)^{1-s} - k^{1-s} \right) - k \left((k+1)^{-s} - k^{-s} \right)$$

$$= \sum_{k=1}^{\infty} \frac{1}{k^s} = \zeta(s).$$

A simple rearrangement yields

$$\zeta(s) - \frac{1}{s-1}$$
$$= \left(\sum_{k=1}^{n} \frac{1}{k} - \log n\right) + \sum_{k=1}^{n}\left(\frac{1}{k^s} - \frac{1}{k}\right) - \int_{1}^{n}\left(\frac{1}{t^s} - \frac{1}{t}\right)dt - s\int_{n}^{\infty}\frac{t-[t]}{t^{s+1}}dt. \quad (*)$$

Given $\varepsilon > 0$, Proposition B.6 leads to some $N \in \mathbb{N}$ such that

$$\left|\sum_{k=1}^{n}\frac{1}{k} - \log n - C_E\right| < \varepsilon/2$$

as well as

$$\left|s\int_{n}^{\infty}\frac{t-[t]}{t^{s+1}}dt\right| \leq |s|\int_{n}^{\infty}t^{-\sigma-1}dt = \frac{|s|}{\sigma}n^{-\sigma} < \varepsilon/2$$

for all $\sigma > 1$, $n \geq N$. Thus the right-hand side of $(*)$ is holomorphic for $\operatorname{Re} s > 0$. We get

$$\left|\zeta(s) - \frac{1}{s-1} - C_E\right| \leq \varepsilon + \sum_{k=1}^{N}\left|\frac{1}{k^s} - \frac{1}{k}\right| + \int_{1}^{N}\left|\frac{1}{t^s} - \frac{1}{t}\right|dt$$

and therefore

$$\lim_{s\to 1}\left|\zeta(s) - \frac{1}{s-1} - C_E\right| < \varepsilon.$$

This yields the claim. $\square$

Bibliography

[1] N.H. ABEL: *Œuvres complètes I, II*. Gröndahl, Christiania 1839; reprint, Cambridge University Press, Cambridge 2012.
[2] N.H. ABEL: *Abel on analysis*. Ed. by P. HOROWITZ. Kendrick Press, Heber City, UT 2007.
[3] C. ALFES-NEUMANN: *Modular forms*. Springer-Verlag, Cham 2021.
[4] G. ANDREWS, R. ASKEY, R. ROY: *Special functions*. Cambridge University Press, Cambridge 1999.
[5] T.M. APOSTOL: *Modular functions and Dirichlet series in number theory*. 2nd ed., Springer–Verlag, Berlin–Heidelberg–New York 1990.
[6] T.M. APOSTOL: *Introduction to analytic number theory*. 5th printing, Springer–Verlag, Berlin–Heidelberg–New York 1998.
[7] B. BERNDT, M. KNOPP: *Hecke's theory of modular forms and Dirichlet series*. World Scientific, Hackensack, N.J. 2008.
[8] H. BURKHARDT: *Elliptische Funktionen*. 3rd ed., edited by G. FABER, Ver. wiss. Verlag, Berlin 1920.
[9] A. CAYLEY: *Collected mathematical papers I–XIV*. Cambridge University Press, Cambridge 1889–1898; reprint, 2009.
[10] K. CHANDRASEKHARAN: *Elliptic functions*. Springer–Verlag, Berlin–Heidelberg–New York 1985.
[11] H. COHEN: *A course in computational algebraic number theory*. Springer–Verlag, Berlin–Heidelberg–New York 1993.
[12] H. COHEN, F. STRÖMBERG: *Modular forms: a classical approach*. Amer. Math. Soc., Providence, R.I. 2017.
[13] J.B. CONWAY: *Functions of one complex variable*. Springer-Verlag, New York 1973.
[14] J.H. CONWAY, N.J.A. SLOANE: *Sphere packings, lattices and groups*. 3rd ed., Springer–Verlag, Berlin–Heidelberg–New York 1999.
[15] A. DEITMAR: *Automorphic forms*. Springer-Verlag, London 2013.
[16] R. DEDEKIND: *Gesammelte mathematische Werke I–III*. Vieweg, Braunschweig 1930–1932; reprint, Chelsea, New York 1969.

[17] F. DIAMOND, J. SHURMAN: *A first course in modular forms.* Springer-Verlag, Berlin 2005.
[18] H.M. EDWARDS: *Fermat's last theorem.* Springer-Verlag, New York 1977.
[19] M. EICHLER, D. ZAGIER: *The theory of Jacobi forms.* Birkäuser, Boston-Basel-Stuttgart 1985.
[20] G. EISENSTEIN: *Mathematische Werke I, II.* Chelsea, New York 1975.
[21] J. ELSTRODT, F. GRUNEWALD, J. MENNICKE: *Groups acting on hyperbolic space. Harmonic analysis and number theory.* Springer–Verlag, Berlin–Heidelberg–New York 1998.
[22] L. EULER: *Opera Omnia.* Ser. I, 1–29, Teubner, Leipzig–Berlin 1911–1956.
[23] G.C. FAGNANO: *Opere matematiche I-III.* Mailand-Rom-Neapel 1911-1912.
[24] W. FISCHER, I. LIEB: *A course in complex analysis.* Vieweg + Teubner, Wiesbaden 2012.
[25] O. FORSTER: *Lectures on Riemann surfaces.* Springer–Verlag, Berlin–Heidelberg–New York 1981.
[26] J. FOURIER: *Œuvres I, II.* Gauthier–Villars, Paris 1888, 1890; reprint, Cambridge University Press, Cambridge 2013.
[27] E. FREITAG, R. BUSAM: *Complex analysis.* 3rd ed., Springer–Verlag, Berlin–Heidelberg–New York 2009.
[28] R. FRICKE: *Elliptische Funktionen; Automorphe Funktionen unter Einschluss der elliptischen Modulfunktionen.* In: *Enzyklopädie der Mathematischen Wissenschaften,* vol. II, Teubner, Leipzig 1901–1921.
[29] R. FRICKE: *Die elliptischen Funktionen und ihre Anwendungen I-III.* Edited by C. ADELMANN, J. ELSTRODT, E. KLIMENKO, Springer-Verlag, Berlin 2011.
[30] R. FUETER: *Vorlesungen über die singulären Moduln und die komplexe Multiplikation der elliptischen Funktionen I, II.* Teubner, Leipzig 1924, 1927.
[31] C.F. GAUSS: *Werke I–XII.* Ges. d. Wiss. Göttingen, Teubner, Leipzig 1863–1933.
[32] E. GRAESER: *Einführung in die Theorie der elliptischen Funktionen und deren Anwendungen.* Oldenbourg, München 1950.
[33] A.G. GREENHILL: *The applications of elliptic functions.* Macmillan, London 1892; reprint, Dover, New York 1959.
[34] G.H. HALPHÉN: *Traité des fonctions elliptiques et de leurs applications I–VIII.* Gauthier–Villars, Paris 1886–1891.
[35] H. HANCOCK: *Lectures on the theory of elliptic functions.* Wiley, New York 1910; reprint, Dover, New York 1958.
[36] E. HECKE: *Mathematische Werke.* Vandenhoeck & Ruprecht, Göttingen 1959.
[37] D. HILBERT: *Gesammelte Abhandlungen I–III.* 2nd ed., Springer–Verlag, Berlin–Heidelberg–New York 2015.
[38] E. HILLE: *Ordinary differential equations in the complex domain.* Republication, Dover, Mineola, NY 1997.
[39] A. HURWITZ: *Mathematische Werke I, II.* Birkhäuser, Basel 1932, 1933.
[40] A. HURWITZ, R. COURANT: *Vorlesungen über allgemeine Funktionentheorie und elliptische Funktionen.* 5th ed., Springer–Verlag, Berlin–Heidelberg–New York 2000.

[41] D.H. HUSEMÖLLER: *Elliptic curves.* 2nd ed., Springer–Verlag, Berlin–Heidelberg–New York 2004.
[42] C.G.J. JACOBI: *Gesammelte Werke I–VII.* Reimer, Berlin 1881–1891.
[43] F. KLEIN: *Gesammelte mathematische Abhandlungen I–III.* Springer–Verlag, Berlin 1921–1923.
[44] F. KLEIN, M. BRENDEL: *Materialien für eine wissenschaftliche Biographie von Gauss II, III.* Teubner, Leipzig 1911.
[45] F. KLEIN, R. FRICKE: *Vorlesungen über die Theorie der elliptischen Modulfunktionen I, II.* Teubner, Leipzig 1890, 1892.
[46] L.J.P. KILFORD: *Modular forms.* 2nd ed., World Scientific, Hackensack, N. J. 2015.
[47] N. KOBLITZ: *Introduction to elliptic curves and modular forms.* 2nd ed., Springer–Verlag, Berlin–Heidelberg–New York 1993.
[48] A. KRAZER: *Lehrbuch der Thetafunktionen.* Teubner, Leipzig 1903; reprint, Chelsea, New York 1970.
[49] A. KRIEG: *Hecke algebras.* Mem. Amer. Math. Soc. **435**, Providence, R.I. 1990.
[50] L. KRONECKER: *Werke I–V.* Teubner, Leipzig 1895–1930; reprint, Chelsea, New York 1968.
[51] J.-L. LAGRANGE: *Œuvres I–XIV.* Olms, Hildesheim 1973.
[52] S. LANG: *Elliptic functions.* 2nd ed., Springer–Verlag, Berlin–Heidelberg–New York 1987.
[53] S. LANG: *Algebra.* 3rd ed., Springer–Verlag, New York–Heidelberg–Berlin 2002.
[54] A.-M. LEGENDRE: *Traité des fonctions elliptiques I–III.* Paris 1825–1828.
[55] J. LEHNER: *Discontinous groups and automorphic functions.* Math. Surv. Monogr. VIII, Amer. Math. Soc., Providence, R.I. 1964.
[56] H. MAASS: *Lectures on modular functions of one complex variable.* Tata Institute, Bombay 1964; revised ed., Springer–Verlag, Berlin–Heidelberg–New York 1983.
[57] J.S. MILNE: *Fields and Galois theory.* Kea Books, Ann Arbor, MI 2022.
[58] H. MINKOWSKI: *Gesammelte Abhandlungen I, II.* Teubner, Leizig–Berlin 1911; reprint, Chelsea, New York 1967.
[59] T. MIYAKE: *Modular forms.* 2nd printing, Springer–Verlag, Berlin–Heidelberg–New York 2006.
[60] M. NEWMAN: *Integral matrices.* Academic Press, New York–London 1972.
[61] W.F. OSGOOD: *Lehrbuch der Funktionentheorie I.* 5th ed., Teubner, Leipzig 1929; reprint, Chelsa Publishing 1965.
[62] H. PETERSSON: *Modulfunktionen und quadratische Formen.* Springer–Verlag, Berlin–Heidelberg–New York 1982.
[63] E. PICARD: *Selecta.* Gauthier–Villars, Paris 1928.
[64] H. POINCARÉ: *Œuvres 1–11.* Gauthier–Villars, Paris 1916–1996.
[65] B. V. QUERENBURG: *Mengentheoretische Topologie.* 3rd ed., Springer–Verlag, Berlin–Heidelberg–New York 2001.
[66] R.A. RANKIN: *Modular forms and functions.* Cambridge University Press, Cambridge 1977; reprint 2008.

[67] R. REMMERT, G. SCHUMACHER: *Funktionentheorie I.* 5th ed., Springer–Verlag, Berlin–Heidelberg–New York 2002.
[68] B. RIEMANN: *Collected papers.* Kendrick Press, Heber City, UT 2004.
[69] R. ROY: *Elliptic and modular functions from Gauss to Dedekind to Hecke.* Cambridge University Press, Cambridge 2017.
[70] W. SCHARLAU: *Quadratic and hermitian forms.* Springer–Verlag, Berlin–Heidelberg–New York 1985.
[71] B. SCHOENEBERG: *Elliptic modular functions.* Springer–Verlag, Berlin–Heidelberg–New York 1974.
[72] H.A. SCHWARZ: *Formeln und Lehrsätze zum Gebrauche der elliptischen Funktionen.* Springer–Verlag, Berlin 1893.
[73] J.-P. SERRE: *A course in arithmetic.* 2nd printing, Springer–Verlag, Berlin–Heidelberg–New York 1978.
[74] G. SHIMURA: *Introduction to the arithmetic theory of automorphic functions.* Iwanami Publishers and Princeton University Press, Tokyo–Princeton 1971; reprint 1994.
[75] G. SHIMURA: *Modular forms: basics and beyond.* Springer-Verlag, Berlin 2012.
[76] C.L. SIEGEL: *Lectures on advanced analytic number theory.* Tata Institute of Fundamental Research, Bombay 1965.
[77] C.L. SIEGEL: *Topics in complex function theory I–III.* Reprint, J. Wiley, New York 1988, 1989.
[78] C.L. SIEGEL: *Gesammelte Abhandlungen I–IV.* Springer–Verlag, Berlin–Heidelberg–New York 2015–2016.
[79] J.H. SILVERMAN: *The arithmetic of elliptic curves.* 2nd ed., Springer–Verlag, Berlin–Heidelberg–New York 2009.
[80] P. STÄCKEL, W. AHRENS: *Der Briefwechsel zwischen C.G.J. Jacobi und P.H. von Fuss über die Herausgabe der Werke Leonhard Eulers.* Teubner, Leipzig 1908.
[81] W. STEIN: *Modular forms, a computational approach.* Amer. Math. Soc., Providence, R.I. 2007.
[82] A. TERRAS: *Harmonic analysis on symmetric spaces and applications I, II.* Springer–Verlag, Berlin–Heidelberg–New York 1985, 1988.
[83] E.C. TITCHMARSH: *The theory of functions.* 2nd ed., Oxford University Press, London 1975.
[84] F. TRICOMI, M. KRAFFT: *Elliptische Funktionen.* Akad. Verl. Ges., Leipzig 1948.
[85] W. WALTER: *Ordinary differential equations.* Springer–Verlag, Berlin–Heidelberg–New York 1998.
[86] L.C. WASHINGTON: *Introduction to cyclotomic fields.* 2nd ed., Springer–Verlag, Berlin–Heidelberg–New York 1997.
[87] H. WEBER: *Algebra III.* 2nd ed., Vieweg, Braunschweig 1908.
[88] K.T.W. WEIERSTRASS: *Mathematische Werke I–VI.* Mayer & Müller, Berlin 1894–1915.
[89] A. WEIL: *Elliptic functions according to Eisenstein and Kronecker.* Springer–Verlag, Berlin–Heidelberg–New York 1976; reprint 1999.
[90] D. ZAGIER: *Zetafunktionen und quadratische Zahlkörper.* Springer–Verlag, Berlin–Heidelberg–New York 1981.

List of mathematicians

ABEL, Niels Hendrik (1802 – 1829)
APOSTOL, Tom Mike (1923 – 2016)
ATKIN, Arthur Oliver Lonsdale (1925 – 2008)
BERNOULLI, Jakob (1655 – 1705)
BURKHARDT, Heinrich (1861 – 1914)
CAUCHY, Augustin-Louis (1798 – 1857)
CAYLEY, Arthur (1821 – 1895)
CHANDRASEKHARAN, Komaravolu (1920 – 2017)
COURANT, Richard (1888 – 1972)
CRELLE, August Leopold (1780 – 1855)
DEDEKIND, Richard (1831 – 1916)
DIRICHLET, Peter Gustav Lejeune (1805 – 1859)
EISENSTEIN, Gotthold (1823 – 1852)
EPSTEIN, Paul (1871 – 1939)
EULER, Leonhard (1707 – 1783)
FAGNANO, Guilio Carlo (1682 – 1766)
FERMAT, Pierre de (1607 – 1665)
FOURIER, Joseph (1768 – 1830)
FRICKE, Robert (1861 – 1930)
FUETER, Rudolf (1880 – 1950)
GAUSS, Carl Friedrich (1777 – 1855)
GLAISHER, James Whitbread Lee (1848 – 1928)
GUDERMANN, Christoph (1798 – 1851)
HARDY, Godfrey Harold (1877 – 1947)
HECKE, Erich (1887 – 1947)
HERMITE, Charles (1822 – 1901)
HILBERT, David (1862 – 1943)
HURWITZ, Adolf (1859 – 1919)
JACOBI, Carl Gustav Jacob (1804 – 1851)
KLEIN, Felix (1849 – 1925)
KOECHER, Max (1924 – 1990)

Korkin, Alexander Nikolajewitsch (1837 – 1908)
Krazer, Adolf (1858 – 1926)
Kronecker, Leopold (1823 – 1891)
Lagrange, Joseph-Louis (1736 – 1813)
Landau, Edmund (1877 – 1938)
Lang, Serge (1927 – 2005)
Leech, John (1926 – 1992)
Legendre, Adrien-Marie (1752 – 1833)
Liouville, Joseph (1809 – 1882)
Lipschitz, Rudolf (1832 – 1903)
Maass, Hans (1911 – 1992)
Minkowski, Hermann (1864 – 1909)
Möbius, August Ferdinand (1790 – 1868)
Mordell, Louis Joel (1888 – 1972)
Newman, Morris (1924 – 2007)
Petersson, Hans (1902 – 1984)
Poincaré, Henri (1854 – 1912)
Radermacher, Hans (1892 – 1969)
Ramanujan, Srinivasa (1887 – 1920)
Rankin, Robert Alexander (1915 – 2001)
Riemann, Bernhard (1826 – 1866)
Schoeneberg, Bruno (1906 – 1995)
Schwarz, Hermann Amandus (1843 – 1921)
Shimura, Goro (1930 – 2019)
Siegel, Carl Ludwig (1896 – 1981)
Wallis, John (1616 – 1703)
Weber, Heinrich (1842 – 1913)
Weierstrass, Karl Theodor Wilhelm (1815 – 1897)
Weil, André (1906 – 1998)
Witt, Ernst (1911 – 1991)
Zolotareff, Jegor Iwanowitsch (1847 – 1878)

Table of symbols

$\mathcal{A}_\chi$	space of functions in MELLIN transforms, 253
Aut D	automorphism group of a domain D, 113
Aut S	automorphism group of matrix S, 283
$\mathcal{B}_\chi$	space of functions in MELLIN transforms, 254
B_k	BERNOULLI number, 165, 347
$\mathbb{C}$	field of complex numbers
C_E	EULER MASCHERONI constant, 349
χ_Γ	abelian character of Γ, 210
χ_S	character, 319
$\mathbb{C}/\Omega$	factor group, 20
$dv, dv(\tau)$	invariant volume element, 242
$\mathbb{D}$	unit disk, 113, 157
D_f	domain of definition, 11
$D_f(s), \mathbb{D}_f(s)$	DIRICHLET series associated with f, 257
$dv = y^{-2}dxdy$	hyperbolic volume element, 242
$[D_1, \ldots, D_r]$	square matrix with blocks $D_1, \ldots, D_r$, 276
$\Delta, \Delta(\Omega), \Delta(\tau)$	discriminant, 41, 53, 166
Δ^*	normalized discriminant, 180
$\Delta, \widetilde{\Delta}$	(hyperbolic) LAPLACE operator, 311, 330
$\delta(\omega_1, \omega_2)$	diameter, 21
deg φ	degree of a divisor φ, 33
div($\mathbb{C}/\Omega$)	divisor group of $\mathbb{C}/\Omega$, 33
$\Diamond(u;\omega_1,\omega_2), \Diamond(\omega_1,\omega_2)$	period parallelogram, 19
$D_{f,g}(s)$	DIRICHLET series, 334
$D(z,w)$	cross-ratio of z and w, 121
det M	determinant, 111
$\mathcal{D}_{i,r}$	hyperbolic disk, 121, 149
$\mathcal{D}(S,T)$	representation set of T by S, 283
$\sharp\mathcal{D}(S,T) =: \sharp(S,T)$	representation number of T by S, 283
$E_2(\tau)$	conditionally convergent EISENSTEIN series, 194

Table of symbols

e_1, e_2, e_3	half-period values of the $\wp$-function, 29
$E_k, E_k(\Omega), E_k(\tau)$	EISENSTEIN series, 22, 49, 163
E_k^*	normalized EISENSTEIN series, 164
$E(\tau; s)$	real-analytic EISENSTEIN series, 327
$\mathcal{E}(\Omega)$	field of elliptic functions with respect to Ω, 25
$\mathbb{E}(\Omega)$	affine elliptic curve, 64
$\eta(\omega), \eta(\omega; \Omega)$	LEGENDRE quantity, 76
$\eta(\tau)$	DEDEKIND eta function, 84, 196
$\mathbb{F}$	exact fundamental domain, 127
$\mathcal{F}(\Lambda)$	fundamental domain for Λ, 136
$\mathcal{F}_q$	multiplicative invariant functions, 90
$\tilde{f}(\tau)$	Γ-invariant function, 160
$(f|M)(\tau)$	slash operator, 156
$\langle f, g \rangle$	PETERSSON scalar product of f and g, 243
$\Gamma(s)$	gamma function, 256
$\Gamma = \mathrm{SL}(2; \mathbb{Z})$	modular group, 109, 126
$\Gamma \backslash \mathbb{H}$	orbit space, 110
$\Gamma : \Gamma_n$	standard set of representatives of Γ_n modulo Γ, 222
$\Gamma_0[n], \Gamma^0[n], \Gamma_1[n]$	congruence subgroups, 140
Γ_n	transformations of order n, 221
$\Gamma_N[2]$	congruence subgroup of index 2, 143
$\Gamma[n]$	principal congruence group of level n, 137, 203
$\Gamma \tau$	Γ-orbit of τ, 132
Γ_∞	stabilizer subgroup of ∞, 164
$\Gamma(\varphi)$	invariance group of the function φ, 244
Γ_τ	stabilizer subgroup of τ, 131
Γ_ϑ	theta group, 142
$\Gamma \backslash \Gamma_n$	right cosets of Γ_n modulo Γ, 223
$\delta_k(m)$	number of representations of m as a sum of k squares, 208
g_2, g_3	WEIERSTRASS invariants, 38, 53
$G(\lambda)$	HECKE subgroup of $\mathrm{SL}(2; \mathbb{R})$, 151
$\mathrm{GL}(n; \mathbb{Z}) = \mathcal{U}_n$	unimodular group of degree n, 17, 276
$\mathrm{GL}(2; \mathbb{C})$	group of invertible complex 2×2 matrices, 111
$\mathbb{H}$	upper half-plane in $\mathbb{C}$, 49, 109, 113
$\mathcal{H}_k$	algebra of HECKE operators in End $\mathbb{M}_k$, 215, 235
$\mathcal{H}_r^{(n)}$	space of harmonic polynomials, 311
h_n	number of equivalence classes, 295
I	identity matrix, 16, 111, 276
$\mathrm{Im}\, z$	imaginary part of $z \in \mathbb{C}$
j_m	FOURIER coefficient of j, 55, 168
J	$\begin{pmatrix} 0 & -1 \\ 1 & 0 \end{pmatrix} \in \Gamma$, 126
$j, j(\Omega), j(\tau)$	absolute invariant, 41, 54, 168
$\mathbb{K} = \mathbb{V}_0$	field of modular functions, 159
$K_s(y)$	K–BESSEL function, 330
$\mathbb{L}$	left half of $\overline{\mathbb{F}}$, 133

Table of symbols

$L(\gamma)$	$\mathbb{H}$-length of γ, 119
$L \equiv M(\mathrm{mod}\ n)$	congruence for matrices, 137
Λ^σ	dual lattice of Λ, 293
L_{24}	Gram matrix of Leech lattice, 296
$\mathcal{M}$	field of meromorphic functions on $\mathbb{C}$, 11
$M_g(s)$	Mellin transform, 253
$\mathbb{M}_k$	space of modular forms of weight k, 159
$M\infty$	action of M at ∞, 112
$M\tau$	fractional linear transformation, 111
$M(n)$	Minkowski-Siegel mass formula, 309
M^{tr}	transposed matrix, 111
$M^\#$	adjugate matrix, 111
$\mathbb{M}_k^!$	space of weakly holomorphic modular forms of weight k, 192
$\mathbb{M}_k^{\mathbb{Z}}$	module of modular forms with integer Fourier coefficients, 182
$\mathbb{M}$	graded ring of modular forms, 182
$\mathbb{M}_k(\Lambda, \chi)$	space of modular forms for Λ, χ, 203
$|M|$	norm, 147
$\mathcal{M}(\tau)$	subring of $\mathrm{Mat}\,(2;\mathbb{Z})$, 100
$\mathrm{Mat}\,(n;\mathbb{Z})$	ring of $n \times n$ matrices over $\mathbb{Z}$, 276
$\mathrm{Mat}\,(2;\mathbb{Z})$	ring of 2×2 matrices over $\mathbb{Z}$, 16
$\mu(\Omega)$	minimum of the lattice Ω, 37
$\mu(S)$	minimum of the matrix S, 283
$\mathbb{N}, \mathbb{N}_0$	set of natural numbers, 20
$O(z^k)$	Landau symbol, 38
$\mathrm{ord}_c f$	order of f at c, 11, 172
ord_w	order of the stabilizer subgroup of w, 173
$\mathbb{P}(\mathbb{C})$	$\mathbb{C} \cup \{\infty\}$, 11
$\wp, \wp(z;\omega_1,\omega_2), \wp_\Omega(z)$	Weierstrass $\wp$-function, 28, 35, 36
$P_k(\tau, w)$	Poincaré series in τ and w, 171
$\mathcal{P}_\Lambda(z)$	normal polygon, 150
$\mathcal{P}_r^{(n)}$	$\mathbb{C}$-vector space of the homogneeous polynomials of degree r, 311
$p(n)$	partition function, 198
Per f	set of periods of f, 12
Φ_M	Möbius transformation, 111
$\mathrm{PSL}\,(2;\mathbb{C}), \mathrm{PSL}\,(2;\mathbb{R})$	group of fractional linear transformations, 112, 115
$\mathrm{Pos}\,(n;\mathbb{R})$	set of positive definite matrices in $\mathrm{Sym}\,(n;\mathbb{R})$, 123, 280
$\mathbb{Q}$	field of rational numbers
$\mathbb{R}$	field of real numbers
$\mathcal{R}(\tau)$	multiplier ring of τ, 99
$\mathrm{Re}\,z$	real part of $z \in \mathbb{C}$
$\mathrm{res}_c f$	residue of f at c, 25
ρ	$(1 + i\sqrt{3})/2$, 128
$Q_{k,m}(\tau)$	Poincaré series 170
$S^{1/2}$	square root of $S \in \mathrm{Pos}\,(n;\mathbb{R})$, 281
$\langle S \rangle$	class of S, 283

S_3	permutation group, 141
S_8	even, unimodular, positive definite 8×8 matrix, 294
$\mathbb{S}$	ideal of cusp forms in $\mathbb{M}$, 187
$\mathbb{S}_k$	space of cusp forms of weight k, 160
$\mathbb{S}_k^{\mathbb{Z}}$	$\mathbb{Z}$-module of cusp forms with integer FOURIER coefficients, 184
$S > T$	partial ordering, 281
$\sigma(z), \sigma(z; \Omega)$	WEIERSTRASS σ-function, 75
$SL(n; \mathbb{Z})$	special linear group of degree n over $\mathbb{Z}$, 277
$SL(2; \mathbb{R})$	special linear group of degree 2, over $\mathbb{R}$, 115
$SL(2; \mathbb{Z}) = \Gamma$	modular group, 17, 109, 126
$SO(2)$	special orthogonal group of degree 2, 117
$\mathrm{Sym}(n; \mathbb{R})$	symmetric 2×2 matrices over $\mathbb{R}$, 280
$[t]$	greatest integer function, 349
T	$\begin{pmatrix} 1 & 1 \\ 0 & 1 \end{pmatrix} \in \Gamma$, 126
$T_n f, T_n^{(k)}$	HECKE operator, 219, 224
$\tau = x + iy$	element of the upper half-plane $\mathbb{H}$
$\tau(m)$	RAMANUJAN tau function, 54, 166
$\vartheta(\tau)$	theta series, 184, 275
$\vartheta(z; \tau)$	JACOBI theta series, 85
$\Theta_\Lambda(\tau)$	theta series with respect to the lattice Λ, 289
$\Theta_{p,q}(\tau; S)$	theta series associated with S and the characteristic (p, q), 287
$\Theta(\tau; S, P)$	theta series in S with respect to the harmonic polynomial P, 313
$\mathrm{tr}\,(f)$	trace of f, 205
$\mathrm{tr}\,M$	trace of a matrix, 111
U	$\begin{pmatrix} -1 & 1 \\ -1 & 0 \end{pmatrix} \in \Gamma$, 127
$\mathcal{U}_n$	$= GL(n; \mathbb{Z})$ unimodular group, 276
$V(\mathbb{H})$	vector space of periodic meromorphic functions on $\mathbb{H}$ and at ∞, 2
$v(\Omega)$	$\mathbb{H}$–area, 242
$\mathcal{V}_{\alpha,\beta}$	vertical strip, 258
$\mathcal{V}_\varepsilon$	vertical strip of height ε in $\mathbb{H}$, 132
$\mathbb{V}_k$	vector space of meromorphic modular forms of weight k, 159
$\mathrm{vol}\,\Omega$	volume of a fundamental parallelogram of Ω, 19
$\mathbb{Z}$	ring of integers
z^s	principal value, 256
$\|z, w\|$	hyperbolic distance between z and w, 120
$\zeta(s)$	RIEMANN zeta function, 345
$\zeta(T; s)$	EPSTEIN zeta function, 324
$\zeta(z), \zeta(z; \Omega)$	WEIERSTRASS zeta function, 75
$\Diamond(u; \omega_1, \omega_2), \Diamond(\omega_1, \omega_2)$	period parallelogram, 19
$\sharp(S, T)$	representation number of T by S, 283

Index

ABEL, 10
ABEL's Relation, 27
abelian character, 203
absolute invariant, 41, 54, 153, 168, 226
absolutely convergent, 21
absolutely convergent product, 341
Addition Theorem of $\wp$, 61
adjugate, 111
algebra of HECKE operators, 215, 235
algebraic addition theorem, 63
amplitude, 9
arc length, 6
arithmetic-geometric mean, 104
ATKIN-LEHNER involution, 274
automorphism group, 283

base point, 19
basis, 14
Basis Lemma for Lattices, 18
basis transformation, 47
BERNOULLI number, 165, 347
biholomorphic automorphism, 109, 113
binomial differential equation, 40
birthday, 5

CAUCHY's Integral Theorem, 340
CAYLEY transformation, 113
center, 112
CHEBYSHEV polynomial, 236
circle, 112
class, 283
class invariant, 283
class number, 285
Cocycle Condition, 153
commutator subgroup, 210
complete system of inequivalent matrices, 221
Completion Lemma, 17, 126

complex multiplication, 99
conditionally convergent EISENSTEIN series, 194
conformal mapping, 190
congruence, 137, 300
congruence subgroup, 140, 203
congruent number, 71
conjugation stable lattice, 42
Construction Theorem for the $\wp$-Function, 36
contour integral, 174
Convergence Lemma, 21, 163, 328
Convergence Theorem for the $\wp$-Function, 35
cross-ratio, 3, 121
cusp, 137
cusp class, 137
cusp form, 160, 206

DEDEKIND eta function, 84, 196
DEDEKIND sum, 198
DEDEKIND's Theorem, 198
degree, 33
Δ–function, 41, 53, 166
Delta Product Formula, 83
determinant, 111
diameter, 21
differential calculus, 185
differential equation, 29, 38
Dimension Formula, 179
DIRICHLET series, 252
discontinuous, 147
discrete, 11, 147
discriminant, 41, 53, 166
divisor, 38
doubling formula, 256
doubly periodic, 5, 25
dual lattice, 293

eigenform of HECKE operator, 224
EISENSTEIN, 45
EISENSTEIN series, 22, 50, 163, 194, 300, 328
elliptic curve, 64
elliptic function, 9, 25
elliptic integral, 2
elliptic matrix, 118
EPSTEIN zeta function, 324
Equivalence Theorem for GL (2, $\mathbb{Z}$), 17
Equivalence Theorem for Positive Definite Matrices, 281
Equivalence Theorem for Sublattices, 94
Equivalence Theorem for Unimodular Matrices, 276
equivalent, 221, 241, 278
eta function, 84, 196
Eta Transformation Formula, 197, 267
EULER MASCHERONI constant, 349
EULER product, 265
EULER's Addition Theorem, 5
EULER's formula, 165, 348
EULER's Pentagonal Number Theorem, 87
even matrix, 293
exact fundamental domain, 127, 129
Existence and Representation Theorem, 91
Existence Theorem, 28, 78
extremal, 298
extremal modular form, 202

factor group, 20
FAGNANO integral, 3, 7, 58
FAGNANO's Theorem, 4
FERMAT curve, 66, 70
finite subgroups of Γ, 145
fixed point, 118, 131, 137, 149
Four-Squares-Theorem, 209
FOURIER series expansion, 50, 159, 160
fractional linear transformation, 111
fundamental domain, 19, 136
Fundamental Lemma, 12
fundamental parallelogram, 19

gamma function, 57, 255
GAUSS's Formula, 106
general linear group over $\mathbb{Z}$, 16
generating function, 155
generators of the modular group, 127
graded ring of modular forms, 182
GRAM matrix, 289
greatest integer function, 349

$\mathbb{H}$–area, 242
$\mathbb{H}$–center, 122
$\mathbb{H}$–circle, 122

$\mathbb{H}$–distance, 120
$\mathbb{H}$–length, 119
$\mathbb{H}$–metric, 120
$\mathbb{H}$–radius, 122
$\mathbb{H}$–volume element, 242
half-period values, 29
HAMBURGER's Theorem, 263
harmonic polynomial, 311
heat equation, 87
HECKE algebra, 217, 218, 235
HECKE eigenform, 224
HECKE group, 151
HECKE operator, 215, 219, 223, 235
HECKE's Converse Theorem, 261
HERMITE normal form, 279
HERMITE's inequality, 285
HERON number, 71
hexagonal lattice, 42
holomorphic at ∞, 158
homogeneity, 47
homogeneous polynomial, 311
homogeneous space, 117
HUA Identity, 117
HURWITZ Identity, 51
hyperbolic area, 242
hyperbolic disk, 120, 144
hyperbolic distance, 120
hyperbolic length, 119
hyperbolic matrix, 119
hyperbolic metric, 120
hyperbolic volume element, 242

Identity Theorem, 339
integral, 3
integral equivalence, 283
intersection formula, 66, 67
invariance group, 244
Invariance Lemma, 328
invariant metric, 120
Inversion Theorem, 57

j–function, 41, 54, 110, 153, 168, 226
JACOBI, 10
JACOBI decomposition, 286
JACOBI form, 106
JACOBI theta series, 85, 93
JACOBI's Lemma, 15
JACOBI's Triple Product Identity, 86

K–BESSEL function, 330
KRONECKER Approximation, 15
KRONECKER's Jugendtraum, 230
KRONECKER's Limit Formula, 332

lacunary polynomial, 89
LAGRANGE's Theorem, 209
LANDAU symbol, 38
LAPLACE operator, 311, 330
lattice, 14
LAURENT series development, 25, 30, 37
LEECH lattice, 296
left coset, 278
LEGENDRE normal form, 3
LEGENDRE Relation, 76
LEHMER conjecture, 167
LEHNER's Theorem, 169
lemniscate, 6
lemniscate function, 5
level of an even matrix, 316
LIE algebra, 187
linear combination over $\mathbb{Z}$, 14
LIOUVILLE's First Theorem, 26
LIOUVILLE's Fourth Theorem, 27
LIOUVILLE's Second Theorem, 26
LIOUVILLE's Theorem, 342
LIOUVILLE's Third Theorem, 26

MAASS wave form, 334
mapping behavior of j, 190
mapping behavior of $\wp$, 45
maximal discrete, 270
Maximum Principle, 343
MELLIN inversion formula, 254
MELLIN transformation, 253
meromorphic, 11
meromorphic modular form, 158
minimum, 284
MINKOWSKI-SIEGEL mass formula, 309
modular equation, 227, 228
modular form, 154, 156, 159, 203
modular function, 153, 159, 209
modular group, 109
modular of weight k, 156
modular substitution, 109, 153
modular tesselation, 130
modular triangle, 130
moduli space, 110
modulus, 3
MÖBIUS transformation, 111
monster, 170
Monstrous Moonshine, 169
multiplication of the $\wp$-function, 89
multiplicativity, 11
multiplicativity of HECKE operators, 232
multiplier ring, 99

n–division equation, 98
n–division polynomial, 98

natural numbers, 20
neighbor, 131
NIELSEN-SCHREIER Theorem, 143
norm, 147
normal polygon, 149
normalized, 183
normalized discriminant, 166
normalized EISENSTEIN series, 164
normalizer, 271

orbit, 110, 132
order, 11, 25, 158, 172
order of an elliptic function, 28
orthogonal circle, 116

$\wp$–function, 28, 36
$\wp$–partial values, 98
parabolic matrix, 119
partition function, 198
partition number, 87
path, 119
pentagonal number, 87
period, 12
period lattice, 14
period parallelogram, 19
period set, 12, 14
periodic function, 157
permutation matrix, 277
PETERSSON scalar product, 243
PHRAGMEN-LINDELÖF Theorem, 343
PICARD's Little Theorem, 192
POINCARÉ series, 170, 171
pole, 11
pole at ∞, 157
positive definite matrix, 123, 280
power series development, 339
primitive, 277
primitive minimum, 284
principal congruence group, 137, 203
principal divisor, 33
principal minor, 280
principal value, 256
projective elliptic curve, 65
projective space, 11, 65

quadratic non-residue, 296
quadratic residue, 296

radius, 112
RAMANUJAN, 167
RAMANUJAN Congruence, 181
RAMANUJAN Conjecture, 168
RAMANUJAN tau function, 53, 168
RANKIN convolution, 335

rational point, 70
real–analytic EISENSTEIN series, 327
rectangular lattice, 42
recursion formula, 233
reduction theory, 133, 134, 283
representation number, 283
Representation Theorem, 78
residue, 25, 340
Residue Theorem, 340
resultant, 42
RIEMANN zeta function, 23, 252, 260, 345
RIEMANN's Theorem on Removable Singularities, 339
root at ∞, 158
Roots Lemma, 166

SCHWARZ's Lemma, 114
SCHWARZian derivative, 113
self–adjoint, 246
self-dual, 293
SIEGEL's Main Theorem, 308
Sigma Relation, 80
simultaneous eigenform, 224
singular value, 229, 230
slash operator, 156
SMITH normal form, 228
special linear group of degree 2 over $\mathbb{Z}$, 17
special linear group over $\mathbb{R}$, 115
special orthogonal group, 117
special unimodular group, 277
Spitzenform, 160
Square Root Theorem, 281
stabilizer subgroup, 131
STIRLING formula, 256, 342
sublattice, 93
system of representatives for the right cosets, 221
system of representatives of the coset classes, 93

tetrahedral number, 75
Theorem of KRONECKER-WEBER, 230
Theorem of MALMQUIST and YOSIDA, 40
Theorem on Completing the Square, 280
Theorem on the FOURIER Series Development, 340
Theorem on the FOURIER Series Expansion, 289

Theorem on the Gamma Function, 342
Theorem on the LAURENT Series Development, 339
Theorem on the Partial Fractions Development of the Cotangent, 346
Theta Nullwert, 143, 154, 287
theta series, 85, 103, 154, 184, 287, 289, 313
theta series for a lattice, 289
theta series with characteristic, 275, 287
theta series with harmonic polynomial, 313
Theta Transformation Formula, 154, 198, 291, 317
theta zero value, 143
torus, 20
trace, 111, 205, 216
transcendence, 107
Transformation Lemma, 164
transformations of order n, 221
transposed, 111
transversal, 146
triangular number, 75
trivial character, 203

unimodular group, 276
unimodular matrix, 276
unit disk, 113
upper half-plane, 49, 109, 110

vertical strip in $\mathbb{C}$, 258
vertical strip of height ε, 132, 170

weakly holomorphic modular form, 192
WEIERSTRASS invariants, 38, 153
WEIERSTRASS M-Test, 340
WEIERSTRASS M-Test for Products, 341
WEIERSTRASS normal form, 2
WEIERSTRASS $\wp$–function, 28
WEIERSTRASS Product Theorem, 342
WEIERSTRASS sigma function, 75
WEIERSTRASS zeta function, 76
weight formula, 174, 189
winding number, 340

zero at infinity, 158
zero value, 154
zeros of the $\wp$-function, 102

MIX
Papier aus verantwortungsvollen Quellen
Paper from responsible sources
FSC® C105338

If you have any concerns about our products,
you can contact us on
ProductSafety@springernature.com

In case Publisher is established outside the EU,
the EU authorized representative is:
**Springer Nature Customer Service Center GmbH
Europaplatz 3, 69115 Heidelberg, Germany**

Printed by Libri Plureos GmbH
in Hamburg, Germany